AF410810

Crosstalk between MicroRNA and Oxidative Stress in Physiology and Pathology

Crosstalk between MicroRNA and Oxidative Stress in Physiology and Pathology

Special Issue Editors

Antonella Fioravanti
Francesco Dotta
Antonio Giordano
Luigi Pirtoli

MDPI • Basel • Beijing • Wuhan • Barcelona • Belgrade • Manchester • Tokyo • Cluj • Tianjin

Special Issue Editors

Antonella Fioravanti
Azienda Ospedaliera
Universitaria Senese
Italy

Francesco Dotta
University of Siena
Italy

Antonio Giordano
University of Siena
Italy

Luigi Pirtoli
University of Siena
Italy

Editorial Office
MDPI
St. Alban-Anlage 66
4052 Basel, Switzerland

This is a reprint of articles from the Special Issue published online in the open access journal *International Journal of Molecular Sciences* (ISSN 1422-0067) (available at: https://www.mdpi.com/journal/ijms/special_issues/microRNA_Oxidative_Stress).

For citation purposes, cite each article independently as indicated on the article page online and as indicated below:

LastName, A.A.; LastName, B.B.; LastName, C.C. Article Title. *Journal Name* **Year**, *Article Number*, Page Range.

ISBN 978-3-03936-330-8 (Pbk)
ISBN 978-3-03936-331-5 (PDF)

Contents

About the Special Issue Editors

Antonella Fioravanti MD, Sbarro Institute for Cancer Research and Molecular Medicine—Center for Biotechnology; Adjunct Professor, Department of Biology, College of Science and Technology, Temple University, Philadelphia PA 19122 (USA). Former Full Professor and Head, Radiation Oncology, University Hospital, Siena (Italy). Fioravanti graduated from Medical School at the University of Siena (Italy) with amajors in Endocrinology and Rheumatology in 1980, where she was also awarded her PhD in Experimental Rheumatology. She has served as Assistant Professor at the Rheumatology Unit of the General Hospital "Le Scotte" of Siena since 1989. Here, she has held the office for High Specialization in diagnosis and therapy of degenerative rheumatic diseases since 2004. Since 1982, Fioravanti has been a partner of Italian Society of Rheumatology (SIR), serving as Vice President of Tuscany Section of Rheumatology Italian Society. She was President of the Tuscany Section of Italian League Against Rheumatic Diseases (LIMAR) during 2000–2004 and, since 2006, is member of the Scientific Board of ISMH (International Society of Medical Hydrology), and was promoted to Vice-President (Scientific Leader) in 2014. In 2020, she was elected President of OMTH. Fioravanti teaches at Rheumatology Postgraduate School, Medical Hydrology, Genetics and Endocrinology and Internal Medicine. It has been some years since she was involved in clinical and laboratory research, and her work has led to authorship of 214 papers published in national and international journals. She is also co-author of 12 books dedicated to rheumatologic disease and medical hydrology. Her major research areas concern osteoarthritis, chondrocyte and synoviocyte culture, microRNA, intravenous immunoglobulin therapy in rheumatic diseases, and balneotherapy.

Francesco Dotta is currently Head of the Department of Medicine, Surgery and Neuroscience, Università degli Studi di Siena, Professor of Endocrinology and Director of the Diabetes and Clinical Nutrition Units at Policlinico University Hospital in Siena.

Antonio Giordano is Professor of Pathology at Siena University, Italy. He is currently also serving as Professor of Molecular Biology at Temple University in Philadelphia, where he is also Director and founder of the Sbarro Institute for Cancer Research and Molecular Medicine. Giordano was born in Naples, Italy, where he obtained his medical degree and PhD. He then moved to the US and worked at Cold Spring Harbor, under Nobel Laureate James Watson. Giordano's research activity is mainly devoted to the study of the mechanisms responsible for cell cycle deregulation in cancer. Prof. Giordano discovered numerous key regulators of the cell cycle and these pioneering studies contributed to understanding of some of the central mechanisms of cancer development. In 1993, he discovered the tumor suppressor gene RBL2/p130, whose expression is altered in a wide variety of human cancers. He also identified CDK9 and CDK10, two other key players in cell cycle regulation and cell differentiation, and made important contributions to the study of the cyclin-dependent kinase inhibitor p27, which is an important prognostic factor in cancer patients. His current research activity is focused on dissection of the mechanisms underlying cell cycle deregulation in a variety of solid tumors, including lung cancer and mesothelioma and how to exploit these findings to identify new therapeutic strategies. As a researcher, he has authored more than 600 articles, received over 30 awards for his contributions to cancer research and is the holder of 17 patents. Giordano has trained numerous students and written various academic texts, educational books, and book chapters. Giordano is a member of the editorial boards of numerous international journals, has been an Invited

Lecturer at hundreds of scientific meetings held within the US and abroad, and has been awarded numerous national and international research grants. In Italy, he received the honor of Knight of the Republic and Commander of the Order of Merit of the Italian Republic. Finally, included among his many research activities are those aimed at activism in the reporting of environmental factors causing an increase in tumor pathologies.

Luigi Pirtoli Born September 1, 1948. Researcher in Radiological Sciences (Radiation Therapy and Radiation Biology) and Oncology. MD (1973) and specialization degrees (Radiology, Radiotherapy, and Nuclear Medicine, 1977) earned at the Florence University, Italy. Present position: Adjunct Professor, College of Science of and Technology, Sbarro Health Research Organization (SHRO), and Department of Biology, Temple University (TU), Philadelphia, USA (since 2014). Affiliated programs: Temple Summer in Italy Siena Research Program. Lab/Translational research programs on lung cancer and mesothelioma (framework agreement Temple University (Philadelphia)/Siena University, at the SHRO/TU Prof. Antonio Giordano's Lab in Siena (Italy). Previous positions and commitments: Served numerous positions at Siena University until this retirement on in October, 2018, including Full Professor and Head, Radiation Oncology, Department of Medicine, Surgery, and Neurosciences; Director, Postgraduate School, Radiation Oncology; Deputy Director of the above Department; Former Director of the Department of Human Pathology and Oncology. Former Director of Department of Imaging, Siena Hospital. Former President of the Italian Society of Radiation Biology. His recent scientific activity has mainly been dedicated to autophagy as an actionable biologic mechanism in brain tumors and to immunotherapy in cancer. His indexed papers (Pirtoli L) are listed in Scopus, Web of Science, PubMed, Google Scholar, and ResearchGate.

International Journal of
Molecular Sciences

Editorial

Crosstalk between MicroRNA and Oxidative Stress in Physiology and Pathology

Antonella Fioravanti [1,*]**, Luigi Pirtoli** [2]**, Antonio Giordano** [2,3] **and Francesco Dotta** [4]

1   Department of Medicine, Surgery and Neuroscience, Rheumatology Unit, Azienda Ospedaliera Universitaria senese, Policlinico Le Scotte, 53100 Siena, Italy

2   Sbarro Institute for Cancer Research and Molecular Medicine, Department of Biology, College of Science and Technology, Temple University, Philadelphia, PA 19122, USA; luigipirtoli@gmail.com (L.P.); giordano@temple.edu (F.D.)

3   Department of Medical Biotechnologies, University of Siena, 53100 Siena, Italy

4   Diabetes Unit, Department of Medicine, Surgery and Neurosciences, University of Siena, Policlinico Le Scotte, 53100 Siena, Italy; francesco.dotta@unisi.it

*   Correspondence: fioravanti7@virgilio.it; Tel.: +39 0577233345

Received: 4 February 2020; Accepted: 11 February 2020; Published: 13 February 2020

MicroRNAs (miRNA), are short regulatory RNA molecules that regulate gene expression by binding specific sequences within target messenger RNA (mRNA). Increasing evidence revealed their involvement in important physiological cellular processes as well as in the pathophysiology of different disorders, including cancer, cardiovascular diseases, diabetes mellitus, and rheumatic and neurological disorders.

miRNA in different body fluids are considered new candidate biomarkers for diagnosis, classification, prognosis, and responsiveness to treatment, although none have been proposed for daily clinical use. Furthermore, the development of therapeutic strategies either restoring or repressing miRNA expression and activity has attracted much attention. Notwithstanding miRNA have been extensively studied, their detailed mechanisms of action have not yet been fully understood.

Increasing evidence has shown a crosstalk between miRNA and components of redox signaling. miRNA may regulate the expression of redox sensors and other reactive oxygen species (ROS) modulators, such as the key components of cellular antioxidant machinery, while ROS can induce or suppress miRNA expression and contribute to downstream biological function through the regulation of target genes.

The Special Issue entitled "Crosstalk between MicroRNA and Oxidative Stress in Physiology and Pathology" of the International Journal of Molecular Sciences includes three Original Articles and eleven Reviews providing new insights on the interaction between miRNA and oxidative stress under normal and diseased conditions.

A Review by Tsai et al. [1] discusses the crosstalk between excessive oxidative stress induced by mitochondrial dysfunction in tissues/cells and noncoding RNAs, highlighting the role of the epigenetic modulation and of the antioxidant therapy as possible new therapeutic strategies for patients with systemic lupus erythematosus.

Cheleschi et al. [2], in an in vitro study on human osteoarthritic synovial fibroblasts, confirm the presence of a complex relationship between the adipokines, visfatin, resistin, and some miRNA (miR-34a, miR-146a, and miR-181a) in the regulation of oxidative stress balance.

Furthermore, a study on rat cardiomyoblast cells performed by Zhang et al. [3] identifies miR-27a-5p as a cardioprotective agent on hypoxia-induced H9c2 cell injury, suggesting it may be a novel target for the treatment of hypoxia-related heart diseases.

Klieser et al. [4] provide a comprehensive overview of the interactions of oxidative stress and miRNA in pathological processes of the liver. Both, miRNA and oxidative stress are involved in the multifactorial development and progression of acute and chronic liver diseases, and carcinogenesis, by influencing numerous signaling and metabolic pathways.

Quadir et al. [5] extensively review the recent progress in the field of oxidative stress in diabetes mellitus, specifically focusing on the relationship between miRNA and oxidative stress during disease progression as well as on the role of miRNA as candidate biomarkers for the prediction and staging of diabetic chronic complications.

The role of individual miRNA in oxidative stress and related pathways has been further reviewed and confirmed in different neurodegenerative conditions by Konovalova et al. [6], who also raise some criticisms associated with the use of oversimplified cellular models and highlight the ways of studying miRNA regulation and oxidative stress in human stem cell-derived neurons.

A large contribution has been provided on cancer research. Cosentino et al. [7], dealing with Breast Cancer; Huang et al. [8] with Human Hepatocellular Aarcinoma; Zhang [9] for therapeutic tolerance and resistance as a general subject; and Lin. [10] and Babu and Tay [11] on the overall ROS–miRNA relationship domain. All these Authors thoroughly address and analyze from different perspectives the genomic, epigenetic, transcriptional, signaling, and metabolic levels at which the interplay occurs, on the grounds of a systematic and updated check of the evidence emerging from the related literature. Furthermore, Yamakawa et al. [12] address the subject of the possible development of Clinical Trials of Nucleic Acid Medicine, and their delivery systems for Pancreatic Cancer. Of note, among the above quoted contributors, Zhang [9] points out the opportunity of Large-Scale Screenings and Artificial Intelligence-based technology to optimize the therapeutic approach, that is, in accordance with our considerations about complexity in the conclusive remarks. Additionally, a very interesting overview comes from the paper by Marí-Alexandre et al. [13], who analyze the role of oxidative stress and miRNA in the pathophysiology of endometriosis and its possible evolution towards Ovarian Cancers: with their paper, they also provide a valuable educational contribution to this subject. The only oncological original research paper in this Special Issue [14] is dedicated to the overexpressed miR526b/miR655 upregulation of Thioredoxin Reductase 1 (TXNRD1) in Breast Cancer cells, identifying, through a bioinformatic analysis on external datasets, some negative regulators of TXNRD1 as direct targets. Their experiments show that oxidative stress induces miR526b/miR655 overexpression, thus establishing the dynamic function of these miRNA in oxidative stress induction in breast cancer. The adopted in silico procedure has allowed to deepen the knowledge of the involved transcription factors.

It is noteworthy that an exceeding majority of the articles included in this Special Issue—after an extensive call for papers and a rigorous peer-review process—are Reviews of the literature, which are available thanks to the previous, intensive work carried out over many years on miRNA, oxidative stress, and their reciprocal crosstalk. A possible interpretation of this remark is that, at the present time, the involved researchers and scholars are still pondering the overall and ultimate contribution of this scientific domain to the medical sciences. It seems that this field of biology and pathophysiology still preserves an apparent "opacity" regarding its possible practical development, that is, reliable markers and actionable targets for developing a cure. The hallmarks of complexity are as follows: the emergence of unsatisfactorily explained phenomena; the incomplete adequacy of the reductionistic experimental approach; the non-linearity of relationships; dynamic interactive variations. Indeed, complex systems are not completely reducible to direct cause–effect deterministic approaches, and new investigation toolsets are necessary.

The Editors hope that these articles will help readers to update their knowledge about the role of miRNA and oxidative stress in physiology and pathology. Finally, the Editors deeply appreciate all the Authors who contributed excellent Articles to this Special Issue.

References

1. Tsai, C.Y.; Hsieh, S.C.; Lu, C.S.; Wu, T.H.; Liao, H.T.; Wu, C.H.; Li, K.-J.; Kuo, Y.-M.; Lee, H.-T.; Shen, C.-Y.; et al. Cross-Talk between Mitochondrial Dysfunction-Provoked Oxidative Stress and Aberrant Noncoding RNA Expression in the Pathogenesis and Pathophysiology of SLE. *Int. J. Mol. Sci.* **2019**, *20*, 5183. [CrossRef] [PubMed]
2. Cheleschi, S.; Gallo, I.; Barbarino, M.; Giannotti, S.; Mondanelli, N.; Giordano, A.; Tenti, S.; Fioravanti, A. MicroRNA Mediate Visfatin and Resistin Induction of Oxidative Stress in Human Osteoarthritic Synovial Fibroblasts Via NF-κB Pathway. *Int. J. Mol. Sci.* **2019**, *20*, 5200. [CrossRef] [PubMed]
3. Zhang, J.; Qiu, W.; Ma, J.; Wang, Y.; Hu, Z.; Long, K.; Wang, X.; Jin, L.; Tang, Q.; Tang, G.; et al. miR-27a-5p Attenuates Hypoxia-induced Rat Cardiomyocyte Injury by Inhibiting Atg7. *Int. J. Mol. Sci.* **2019**, *20*, 2418. [CrossRef] [PubMed]
4. Klieser, E.; Mayr, C.; Kiesslich, T.; Wissniowski, T.; Fazio, P.D.; Neureiter, D.; Ocker, M. The Crosstalk of miRNA and Oxidative Stress in the Liver: From Physiology to Pathology and Clinical Implications. *Int. J. Mol. Sci.* **2019**, *20*, 5266. [CrossRef] [PubMed]
5. Qadir, M.M.F.; Klein, D.; Álvarez-Cubela, S.; Domínguez-Bendala, J.; Pastori, R.L. The Role of MicroRNAs in Diabetes-Related Oxidative Stress. *Int. J. Mol. Sci.* **2019**, *20*, 5423. [CrossRef] [PubMed]
6. Konovalova, J.; Gerasymchuk, D.; Parkkinen, I.; Chmielarz, P.; Domanskyi, A. Interplay between MicroRNAs and Oxidative Stress in Neurodegenerative Diseases. *Int. J. Mol. Sci.* **2019**, *20*, 6055. [CrossRef] [PubMed]
7. Cosentino, G.; Plantamura, I.; Cataldo, A.; Iorio, M.V. MicroRNA and Oxidative Stress Interplay in the Context of Breast Cancer Pathogenesis. *Int. J. Mol. Sci.* **2019**, *20*, 5143. [CrossRef] [PubMed]
8. Huang, P.S.; Wang, C.S.; Yeh, C.T.; Lin, K.H. Roles of Thyroid Hormone-Associated microRNAs Affecting Oxidative Stress in Human Hepatocellular Carcinoma. *Int. J. Mol. Sci.* **2019**, *20*, 5220. [CrossRef] [PubMed]
9. Zhang, W.C. microRNAs Tune Oxidative Stress in Cancer Therapeutic Tolerance and Resistance. *Int. J. Mol. Sci.* **2019**, *20*, 6094. [CrossRef] [PubMed]
10. Lin, Y.H. MicroRNA Networks Modulate Oxidative Stress in Cancer. *Int. J. Mol. Sci.* **2019**, *20*, 4497. [CrossRef] [PubMed]
11. Babu, K.R.; Tay, Y. The Yin-Yang Regulation of Reactive Oxygen Species and MicroRNAs in Cancer. *Int. J. Mol. Sci.* **2019**, *20*, 5335. [CrossRef] [PubMed]
12. Yamakawa, K.; Nakano-Narusawa, Y.; Hashimoto, N.; Yokohira, M.; Matsuda, Y. Development and Clinical Trials of Nucleic Acid Medicines for Pancreatic Cancer Treatment. *Int. J. Mol. Sci.* **2019**, *20*, 4224. [CrossRef] [PubMed]
13. Marí-Alexandre, J.; Carcelén, A.P.; Agababyan, C.; Moreno-Manuel, A.; García-Oms, J.; Calabuig-Fariñas, S.; Gilabert-Estellés, J. Interplay Between MicroRNAs and Oxidative Stress in Ovarian Conditions with a Focus on Ovarian Cancer and Endometriosis. *Int. J. Mol. Sci.* **2019**, *20*, 5322. [CrossRef] [PubMed]
14. Shin, B.; Feser, R.; Nault, B.; Hunter, S.; Maiti, S.; Ugwuagbo, K.C.; Majumder, M. miR526b and miR655 Induce Oxidative Stress in Breast Cancer. *Int. J. Mol. Sci.* **2019**, *20*, 4039. [CrossRef] [PubMed]

International Journal of
Molecular Sciences

Article

MicroRNA Mediate Visfatin and Resistin Induction of Oxidative Stress in Human Osteoarthritic Synovial Fibroblasts Via NF-κB Pathway

Sara Cheleschi [1,*,†], Ines Gallo [1], Marcella Barbarino [2], Stefano Giannotti [3], Nicola Mondanelli [3], Antonio Giordano [2], Sara Tenti [1,†] and Antonella Fioravanti [1,†]

[1]   Department of Medicine, Surgery and Neuroscience, Rheumatology Unit, Azienda Ospedaliera Universitaria senese, Policlinico Le Scotte, 53100 Siena, Italy; ins.gll3@gmail.com (I.G.); sara_tenti@hotmail.it (S.T.); fioravanti7@virgilio.it (A.F.)

[2]   Sbarro Institute for Cancer Research and Molecular Medicine, Department of Biology, College of Science and Technology, Temple University, Philadelphia, PA 19122, USA; marcella.barbarino@unisi.it (M.B.); giordano@temple.edu (A.G.)

[3]   Department of Medicine, Surgery and Neurosciences, Section of Orthopedics and Traumatology, University of Siena, Policlinico Le Scotte, 53100 Siena, Italy; stefano.giannotti@unisi.it (S.G.); nicola@nicolamondanelli.it (N.M.)

*   Correspondence: saracheleschi@hotmail.com or sara.cheleschi@unisi.it; Tel.: +39-0577-233471

†   These authors contributed equally to this work.

Received: 21 September 2019; Accepted: 17 October 2019; Published: 20 October 2019

Abstract: Synovial membrane inflammation actively participate to structural damage during osteoarthritis (OA). Adipokines, miRNA, and oxidative stress contribute to synovitis and cartilage destruction in OA. We investigated the relationship between visfatin, resistin and miRNA in oxidative stress regulation, in human OA synovial fibroblasts. Cultured cells were treated with visfatin and resistin. After 24 h, we evaluated various pro-inflammatory cytokines, metalloproteinases (*MMPs*), type II collagen (*Col2a1*), *miR-34a, miR-146a, miR-181a*, antioxidant enzymes, and B-cell lymphoma (*BCL)2* by qRT-PCR, apoptosis and mitochondrial superoxide production by cytometry, p50 nuclear factor (NF)-κB by immunofluorescence. Synoviocytes were transfected with miRNA inhibitors and oxidative stress evaluation after adipokines stimulus was performed. The implication of NF-κB pathway was assessed by the use of a NF-κB inhibitor (BAY-11-7082). Visfatin and resistin significantly up-regulated gene expression of interleukin *(IL)-1β, IL-6, IL-17*, tumor necrosis factor *(TNF)-α, MMP-1, MMP-13* and reduced *Col2a1*. Furthermore, adipokines induced apoptosis and superoxide production, the transcriptional levels of *BCL2*, superoxide dismutase (*SOD)-2*, catalase (*CAT*), nuclear factor erythroid 2 like 2 (*NRF2*), *miR-34a, miR-146a,* and *miR-181a*. MiRNA inhibitors counteracted adipokines modulation of oxidative stress. Visfatin and resistin effects were suppressed by BAY-11-7082. Our data suggest that miRNA may represent possible mediators of oxidative stress induced by visfatin and resistin via NF-κB pathway in human OA synoviocytes.

Keywords: microRNA; visfatin; resistin; osteoarthritis; oxidative stress; apoptosis; synovial fibroblasts; synovitis; NF-κB

---

1. Introduction

Osteoarthritis (OA) is the most prevalent musculoskeletal disease characterized by a progressive degradation of articular cartilage, osteophyte formation, subchondral sclerosis and synovitis [1,2]. Increasing evidence suggests that synovial membrane inflammation is implicated in the pathophysiology of the disease; prostaglandins, leukotrienes, reactive oxygen species (ROS), cytokines, chemokines and adipokines, produced by inflamed synovium, induced cartilage degradation and further bolster inflammation [3–5].

Adipokines, including adiponectin, chemerin, leptin, resistin, and visfatin, are secreted by white adipose tissue and are known to be involved in multiple biological processes, as immunity, inflammation, cartilage and bone metabolism. Much attention has been paid regarding their implication in the pathogenesis of many rheumatic diseases, even OA [6–10].

Visfatin has originally identified as an insulin-mimetic factor, with pro-inflammatory and immunomodulating functions [11], while resistin is implicated in obesity-associated insulin resistance and involved in inflammatory response [12].

Visfatin and resistin serum levels and synovial fluid were found to be increased in patients with knee and hand OA [9,13–15]; moreover, it has been highlighted the pro-inflammatory effect of these adipokines on the expression of different cytokines and chemokines, as well as their role in mediating the production of matrix degrades enzymes in human OA chondrocytes and synovial fibroblasts [16–19].

Recent studies demonstrated a complex interaction between adipokines and microRNAs (miRNA) [17,18,20,21]. miRNA are an abundant class of conserved double stranded non-coding RNA molecules of 22–25 nucleotides that are classified as important post-transcriptional regulators of gene expression of target gene messenger RNA [22]. They are implicated in important physiological cellular processes as well as in the pathophysiology of different disorders, including OA [23–26]. Some miRNA, also known as oxidative stress-responsive factors, can be induced or suppressed by ROS, and their biological function, through regulation of target genes, should be influenced [27]; besides, a specific modulation of oxidative stress balance by specific miRNA has been postulated [28].

In the present study, we investigated the complex cross-talk between visfatin, resistin and some miRNA (*miR-34a*, *miR-146a*, and *miR-181a*) in the regulation of oxidative stress, in human OA synovial fibroblasts.

In particular, we analyzed the effect of visfatin and resistin in gene expression of interleukin (*IL*)-*1β*, *IL-6*, *IL-17A*, tumor necrosis factor (*TNF*)-*α*, metalloproteinases (*MMP*)-*1*, *MMP-13*, collagen type II (*Col2a1*). Furthermore, the apoptotic cells and the transcriptional levels of the anti-apoptotic marker B-cell lymphoma (*BCL*) 2, as well as the production of mitochondrial superoxide anion and the gene levels of antioxidant enzymes [superoxide dismutase (*SOD*)-2, catalase (*CAT*)] and nuclear factor erythroid 2 like 2 (*NRF2*) were also investigated.

To examine the potential role of *miR-34a*, *miR-146a*, and *miR-181a* as mediators of the visfatin and resistin effects on oxidative stress, we transfected synovial fibroblasts with miRNA specific inhibitors.

Finally, the possible implication of nuclear factor (NF)-κB pathway in adipokines-mediated effects was assessed.

2. Results

2.1. Cell viability Evaluation in Visfatin and Resistin Treated Cells

Cell viability assay was analyzed by 3-(4,4-dimethylthiazol-2-yl)-2,5-diphenyl-tetrazoliumbromide (MTT) test and the results are represented in Figure S1. A significant reduction of the percentage of survival cells was observed in human OA synovial fibroblasts incubated with visfatin 5 μg/mL and 10 μg/mL ($p < 0.05$) and resistin 50 ng/mL and 100 ng/mL ($p < 0.05$), in comparison to basal condition.

2.2. Visfatin and Resistin Promote Inflammation and Regulate Cartilage Turnover

The effect of adipokines on gene expression of the main pro-inflammatory mediators IL-1β, IL-6, Il-17A and TNF-α in human OA synovial fibroblasts is reported in Figure 1.

Visfatin, tested at both concentrations, 5 μg/mL and 10 μg/mL, significantly increased the mRNA expression of *IL-1β*, *IL-6*, *IL-17A*, and *TNF-α* ($p < 0.01$, $p < 0.001$) (Figure 1A), in a dose dependent manner. Similarly, resistin 50 and 100 ng/mL induced a significant up-regulation ($p < 0.001$) of gene levels of the studied cytokines compared with the un-stimulated cells (Figure 1B).

In Figure 1C,D we summarized the regulation of the main extracellular matrix (ECM) degrading enzyme, MMP-1, MMP-13, and of the main component of articular ECM, Col2a1.

In human OA synovial fibroblasts stimulated with visfatin 5 and 10 μg/mL (Figure 1C) and resistin 50 ng/mL and 100 ng/mL (Figure 1D) we showed a significant increase of *MMP-1, MMP-13* ($p < 0.01$, $p < 0.001$) and a reduction of *Col2a1* ($p < 0.01$, $p < 0.001$) expression levels, in comparison to basal time.

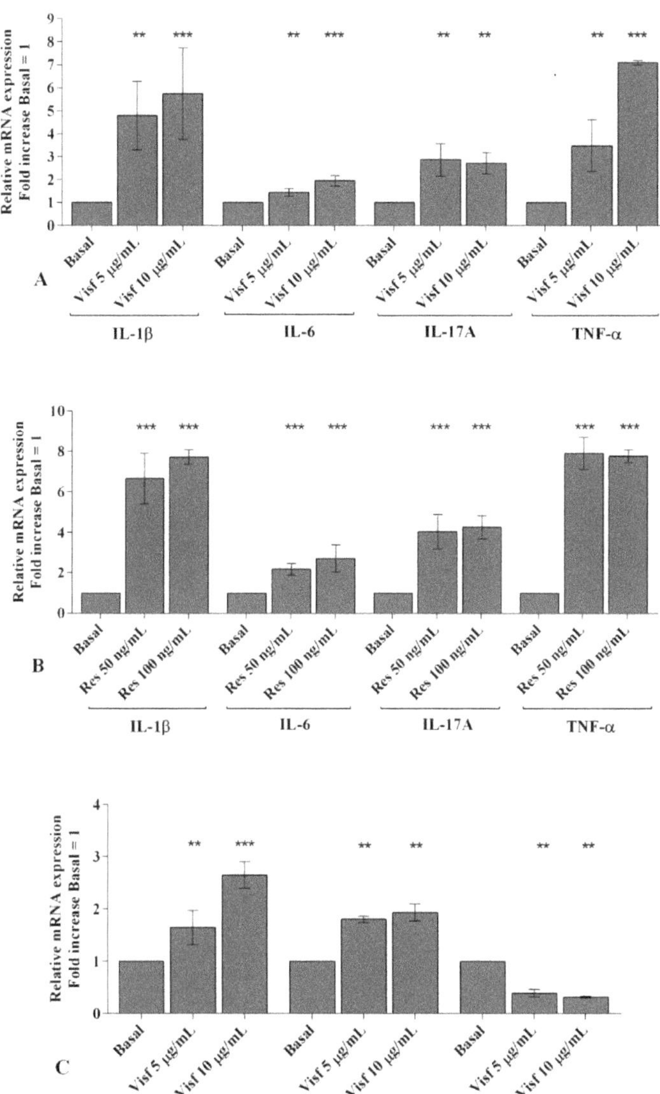

Figure 1. *Cont.*

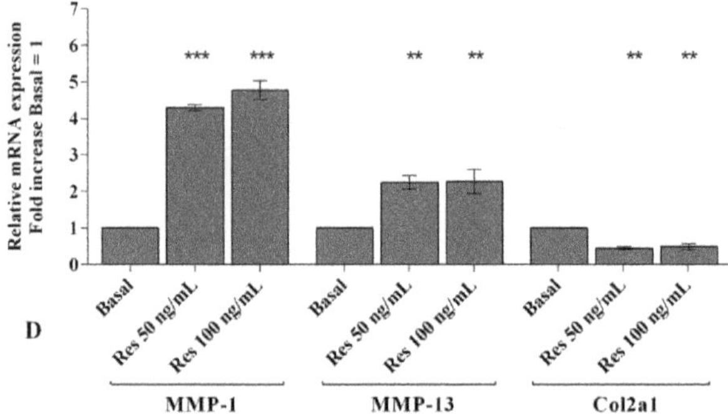

Figure 1. (**A–D**) Expression levels of interleukin (*IL*)-*1β*, *IL-6*, *IL-17A*, tumor necrosis factor (*TNF*)-*α*, metalloproteinases (*MMP*)-*1*, *MMP-13*, and collagen type II (*Col2a1*) by real-time PCR. Human osteoarthritic (OA) synovial fibroblasts were evaluated at basal condition and after incubation with visfatin (5 and 10 μg/mL) and resistin (50 and 100 ng/mL) for 24 h. The gene expression was referenced to the ratio of the value of interest and the value of basal condition (basal, cells without treatment), reported equal to 1. Data were expressed as mean ± SD of triplicate values. ** $p < 0.01$, *** $p < 0.001$ versus basal condition. Visf = visfatin, Res = resistin.

2.3. Adipokines Induce Apoptosis and Regulate BCL2 Expression

Visfatin (5 and 10 μg/mL) and resistin (50 and 100 ng/mL) stimulation induced a significant and dose-dependent increase ($p < 0.01$, $p < 0.001$) of apoptotic OA synovial fibroblasts in comparison to baseline (Figure S2 and Figure 2A).

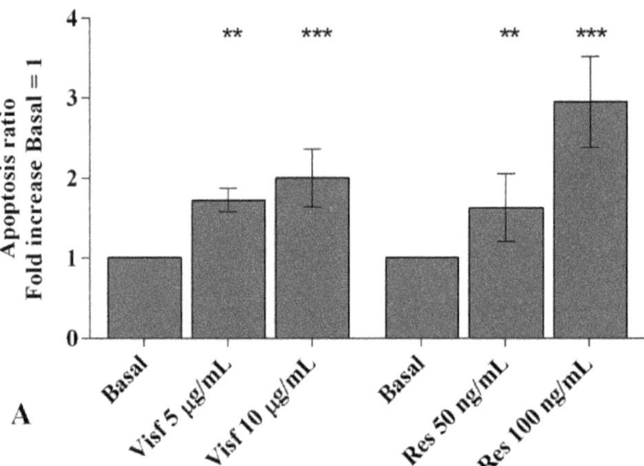

Figure 2. *Cont.*

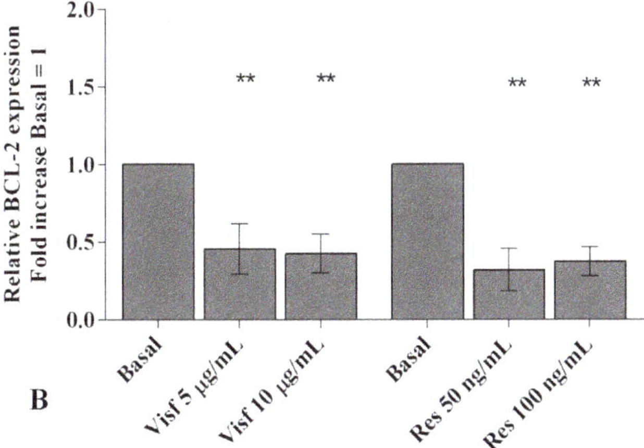

Figure 2. (**A**) Apoptosis detection performed by the analysis at flow cytometry and measured with Annexin Alexa fluor 488 assay. Data were expressed as the percentage of positive cells for Annexin-V and propidium iodide (PI) staining. (**B**) Expression levels of gene B-cell lymphoma (*BCL*)2 by real-time PCR. Human osteoarthritic (OA) synovial fibroblasts were evaluated at basal condition and after incubation with visfatin (5 and 10 µg/mL) and resistin (50 and 100 ng/mL) for 24 h. The apoptosis ratio and the gene expression were referenced to the ratio of the value of interest and the value of basal condition (basal, cells without treatment), reported equal to 1. Data were expressed as mean ± SD of triplicate values. ** $p < 0.01$, *** $p < 0.001$ versus basal condition. Visf = visfatin, Res = resistin.

Real-time PCR analysis underlines a significant reduction of the expression levels of the anti-apoptotic marker *BCL2* ($p < 0.01$) in cells incubated with visfatin and resistin, at both tested concentrations, when compared to un-treated cells (Figure 2B).

2.4. Visfatin and Resistin Regulate Oxidant/Antioxidant Balance

To investigate the potential role of the studied adipokines in the regulation of oxidant/antioxidant balance, we assessed the production of superoxide anion and the analysis of the gene expression of the main antioxidant enzymes implicated in ROS scavenge (Figure S3 and Figure 3).

The stimulus of the cells with the higher concentration of visfatin (10 µg/mL) caused a significant increase of mitochondrial superoxide anion production ($p < 0.05$, Figure 3A); resistin 50 and 100 ng/mL significantly induced a dose-dependent activation of oxidative stress condition ($p < 0.05$, $p < 0.01$, respectively) in comparison to basal time (Figure 3A).

Both concentrations of the tested adipokines significantly up-regulated the expression levels of the antioxidant enzymes *SOD-2* ($p < 0.01$, $p < 0.001$), *CAT* ($p < 0.01$, $p < 0.001$), and *NRF2* ($p < 0.001$) (Figure 3B,C).

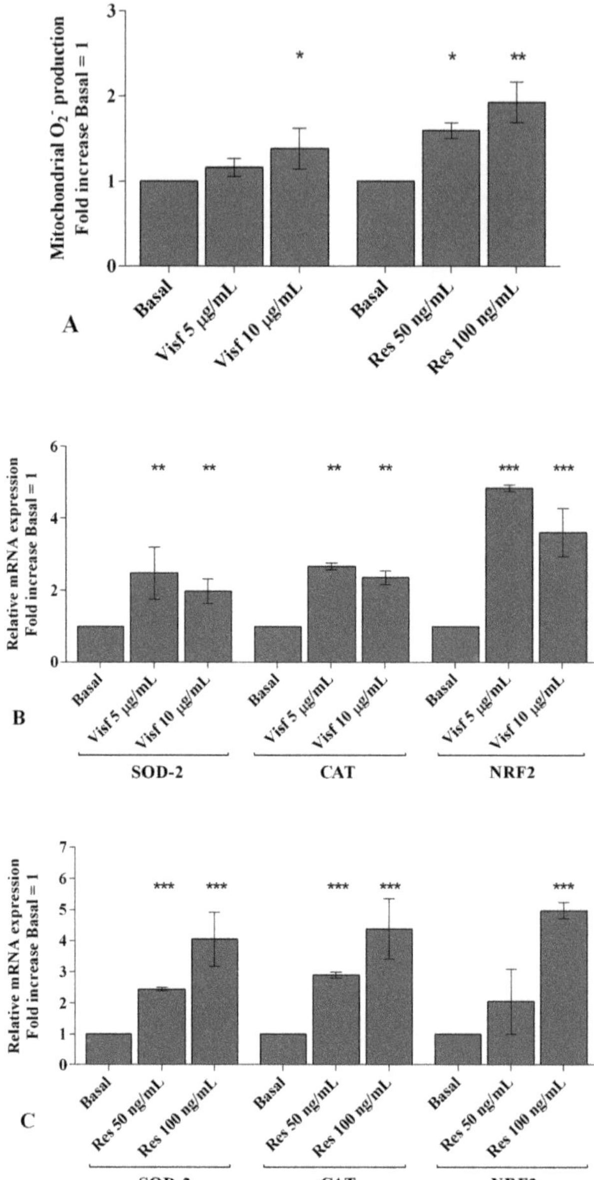

Figure 3. (**A**) Mitochondrial superoxide anion production was assessed by the analysis at flow cytometry using MitoSox Red staining. (**B**,**C**) Expression levels of superoxide dismutase (*SOD-2*), catalase (*CAT*), nuclear factor erythroid 2 like 2 (*NRF2*) by real-time PCR. Human osteoarthritic (OA) synovial fibroblasts were evaluated at basal condition and after incubation with visfatin (5 and 10 µg/mL) and resistin (50 and 100 ng/mL) for 24 h. The superoxide anion production and the gene expression were referenced to the ratio of the value of interest and the value of basal condition (basal, cells without treatment), reported equal to 1. Data were expressed as mean ± SD of triplicate values. * $p < 0.05$, ** $p < 0.01$, *** $p < 0.001$ versus basal condition. Visf = visfatin, Res = resistin.

2.5. Visfatin and Resistin Modulate miRNA Gene Expression

A real-time PCR analysis has been performed in order to evaluate the modulation of *miR-34a*, *miR-146a*, and *miR-181a* gene expression induced by adipokines. Visfatin at a concentration of 5 and 10 µg/mL ($p < 0.01$, $p < 0.001$) up-regulated *miR-34a* and *miR-146a* transcriptional levels in comparison to basal condition, while it did not influence *miR-181a* levels (Figure 4A). Resistin 50 and 100 ng/mL significantly increased the gene expression of *miR-34a* ($p < 0.001$), *miR-146a* ($p < 0.01$), and *miR-181a* ($p < 0.05$, $p < 0.01$) (Figure 4B).

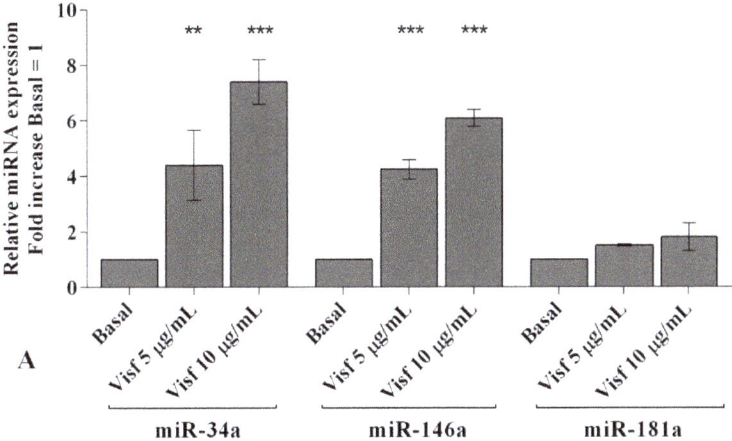

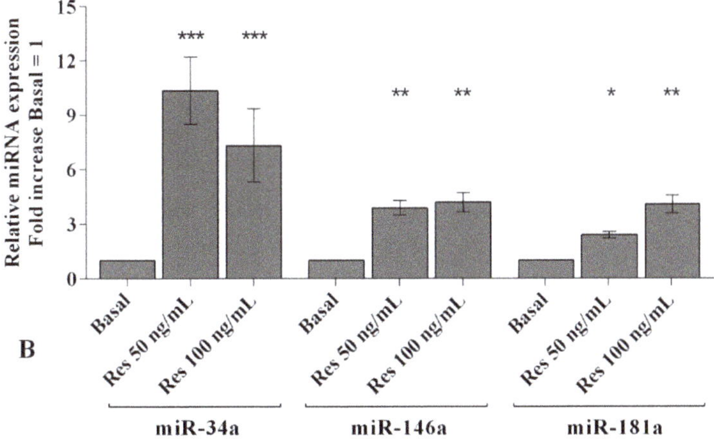

Figure 4. (**A,B**) Expression levels of *miR-34a*, *miR-146a*, and *miR-181a* by real-time PCR. Human osteoarthritic (OA) synovial fibroblasts were evaluated at basal condition and after incubation with visfatin (5 and 10 µg/mL) and resistin (50 and 100 ng/mL) for 24 h. The gene expression was referenced to the ratio of the value of interest and the value of basal condition (basal, cells without treatment), reported equal to 1. Data were expressed as mean ± SD of triplicate values. * $p < 0.05$, ** $p < 0.01$, *** $p < 0.001$ versus basal condition. Visf = visfatin, Res = resistin.

2.6. MiRNA Regulate Oxidative Stress Induced by Visfatin and Resistin

To confirm the involvement of miRNA in modulating oxidative stress induced by visfatin and resistin, we transfected OA synoviocytes with *miR-34a, miR-146a,* and *miR-181a* specific inhibitors (Figure 5).

Real-time PCR showed a significant reduction of gene expression levels of the studied miRNA ($p < 0.01$) in transfected OA cells with respect to basal condition and NC (Figure 5A).

Visfatin (5 and 10 µg/mL) and resistin (50 and 100 ng/mL) significantly up-regulated transcriptional levels of *miR-34a, miR-146a,* and *miR-181a* ($p < 0.01$, Figure 5B–G) in OA synoviocytes incubated with NC. After the transfection with miRNA inhibitors, the treatment with visfatin or resistin did not show any significant modification in *miR-34a, miR-146a,* and *miR-181a* expression in comparison to what is observed in synoviocytes transfected with the inhibitors alone (Figure 5B–G). In addition, the inhibition of *miR-34a, miR-146a,* and *miR-181a* significantly reduced the increase of miRNA transcriptional levels induced by visfatin and resistin incubation ($p < 0.01$, Figure 5B–G).

In Figures 6–8 we reported the modulation of redox balance induced by visfatin and resistin after the transfection of OA synoviocytes with *miR-34a, miR-146a,* and *miR-181a* inhibitors.

MiRNA silencing determined a significant reduction of mitochondrial superoxide anion production ($p < 0.05$, $p < 0.01$, Figures 6A, 7A and 8A) as well as a down-regulation of *SOD-2, CAT,* and *NRF2* expression levels ($p < 0.05$, $p < 0.01$, Figures 6B, 7B and 8B) in comparison to basal condition and NC.

The production of superoxide anion and the expression of *SOD-2, CAT,* and *NRF2* were increased, in a significant manner, in OA cells transfected with NC after stimulus with visfatin ($p < 0.01$, $p < 0.001$, Figure 6C,E, Figure 7C,E and Figure 8C,E) and resistin ($p < 0.01$, $p < 0.001$, Figure 6D,F, Figure 7D,F and Figure 8D,F), while their effect was significantly inhibited by *miR-34a, miR-146a,* and *miR-181a* specific inhibitors ($p < 0.01$, Figure 6C–F, Figure 7C–F and Figure 8C–F).

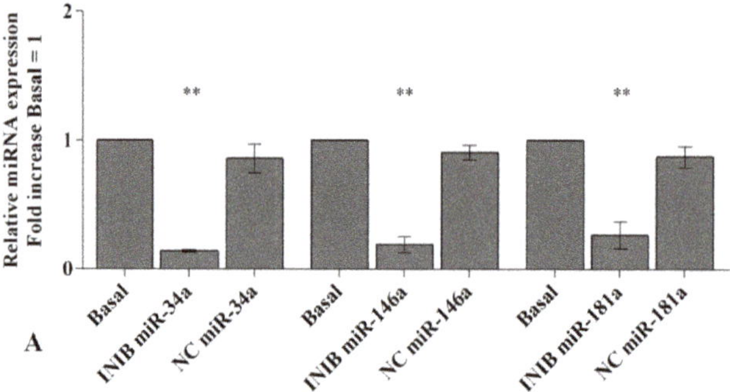

Figure 5. *Cont.*

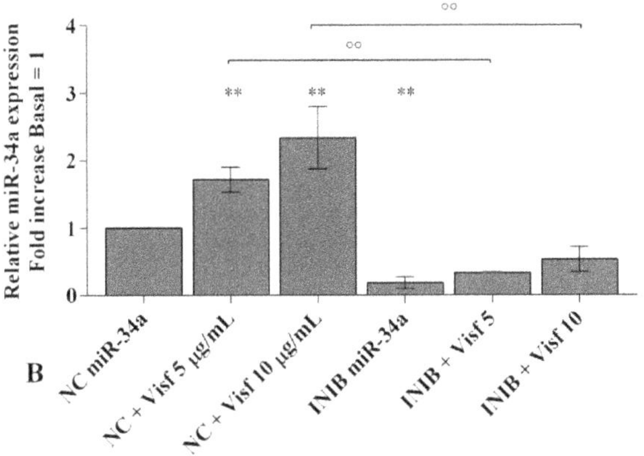

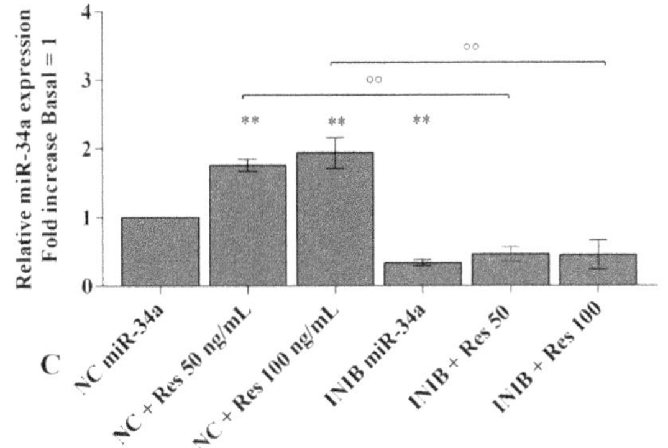

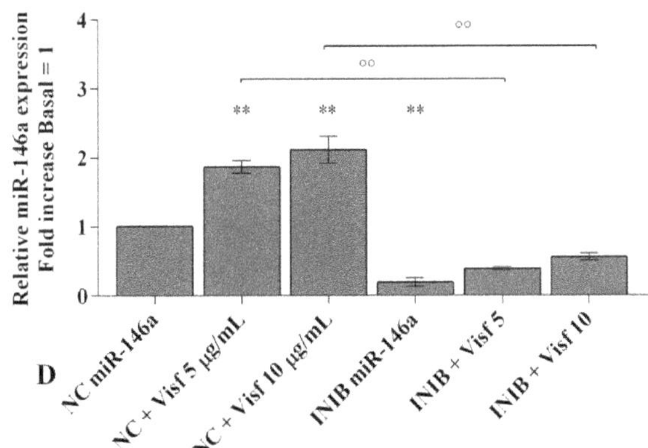

Figure 5. *Cont.*

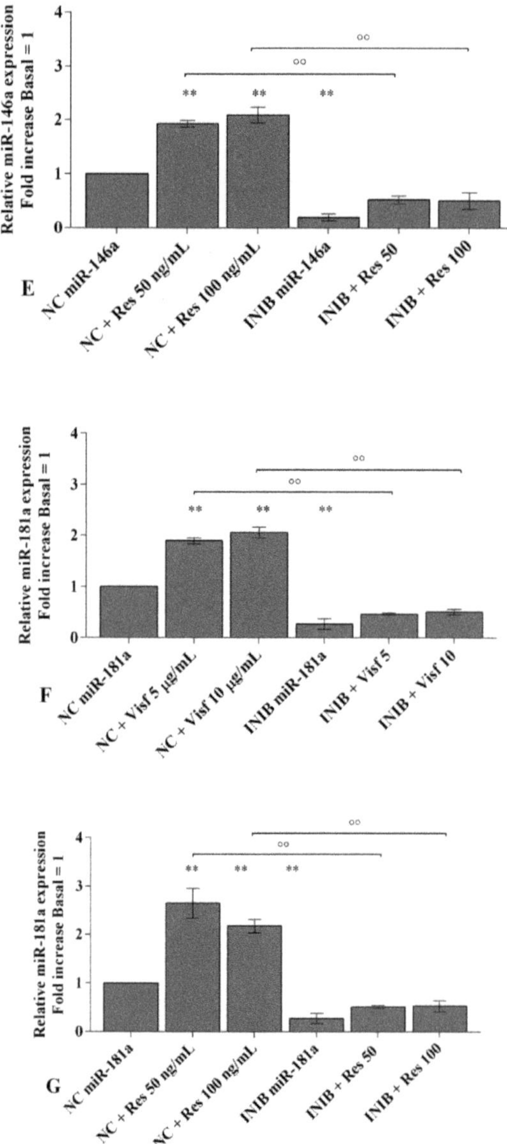

Figure 5. (A–G) Expression levels of *miR-34a*, *miR-146a*, and *miR-181a* by real-time PCR. Human osteoarthritic (OA) synovial fibroblasts were evaluated at basal condition, after 24 h of transfection with *miR-34a*, *miR-146a*, and *miR-181a* inhibitors or NC, and after incubation with visfatin (5 and 10 µg/mL) and resistin (50 and 100 ng/mL). The gene expression was referenced to the ratio of the value of interest and the value of basal condition (basal, cells without treatment) or NC, reported equal to 1. Data were expressed as mean ± SD of triplicate values. ** $p < 0.01$ versus basal condition or NC. °° $p < 0.01$ versus inhibitor. INIB= inhibitor, NC= negative control siRNA, Visf= visfatin, Res = resistin.

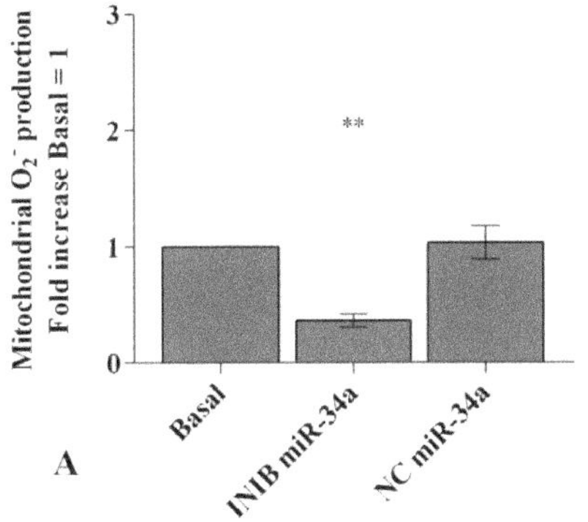

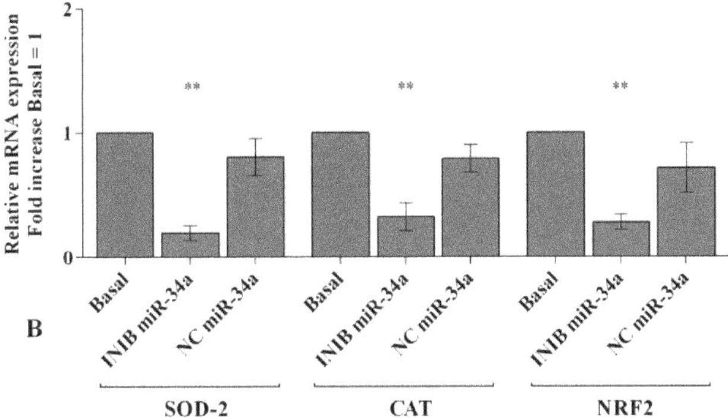

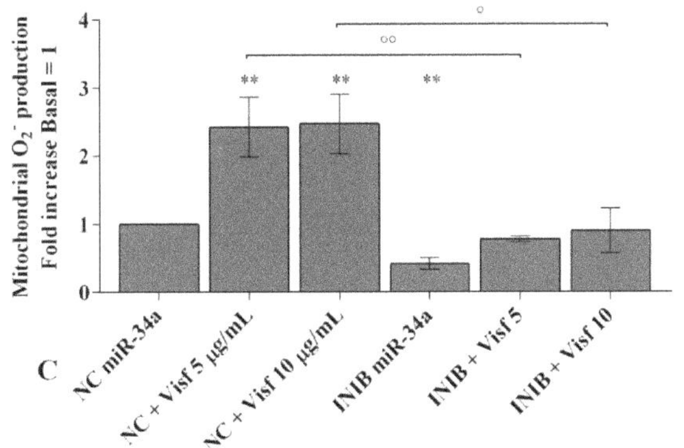

Figure 6. *Cont.*

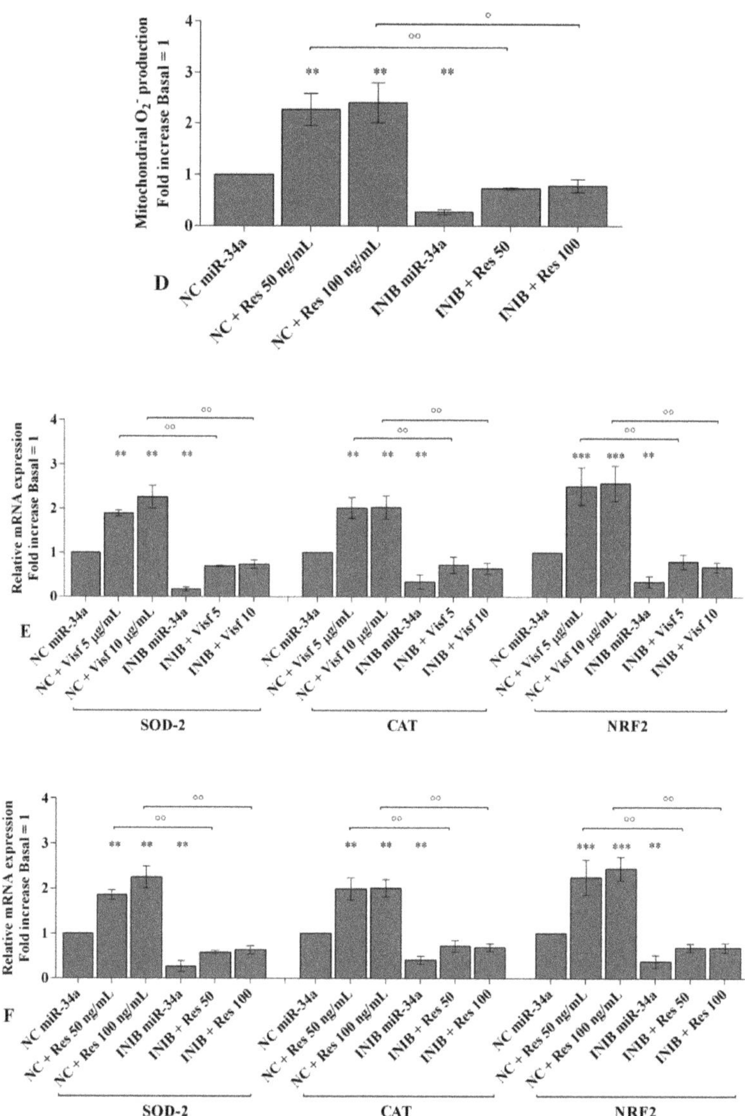

Figure 6. (**A,C,D**) Mitochondrial superoxide anion production was assessed by the analysis at flow cytometry using MitoSox Red staining. (**B,E,F**) Expression levels of superoxide dismutase (*SOD-2*), catalase (*CAT*), nuclear factor erythroid 2 like 2 (*NRF2*) by real-time PCR. Human osteoarthritic (OA) synovial fibroblasts were evaluated at basal condition, after 24 h of transfection with *miR-34a* inhibitor or NC, and after incubation with visfatin (5 and 10 μg/mL) and resistin (50 and 100 ng/mL). The superoxide anion production and the gene expression were referenced to the ratio of the value of interest and the value of basal condition (basal, cells without treatment) or NC, reported equal to 1. Data were expressed as mean ± SD of triplicate values. ** $p < 0.01$, *** $p < 0.001$ versus basal condition or NC. ° $p < 0.05$, °° $p < 0.01$ versus inhibitor. INIB= inhibitor, NC= negative control siRNA, Visf= visfatin, Res = resistin.

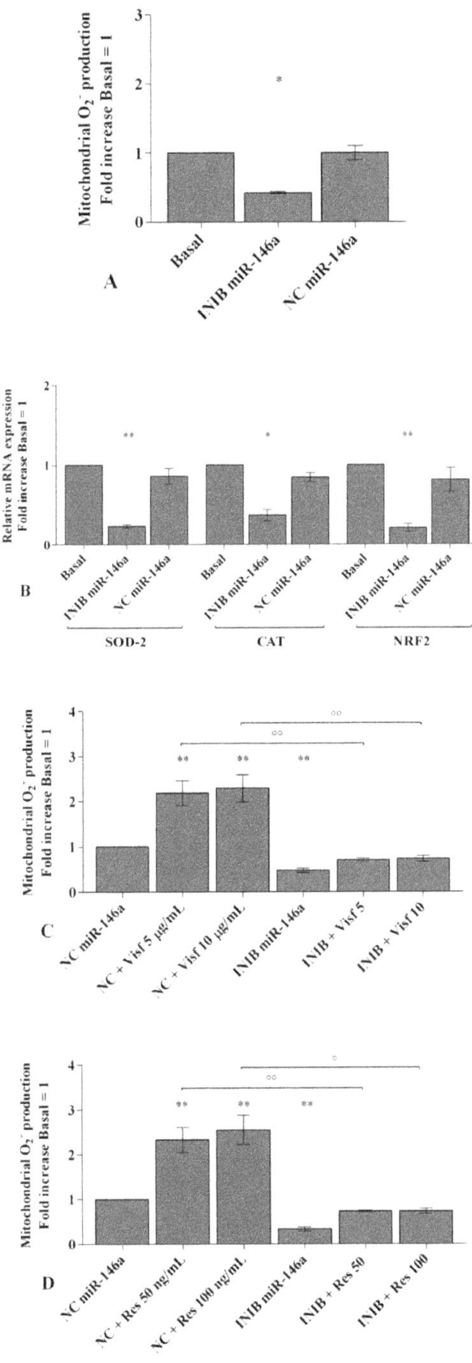

Figure 7. *Cont.*

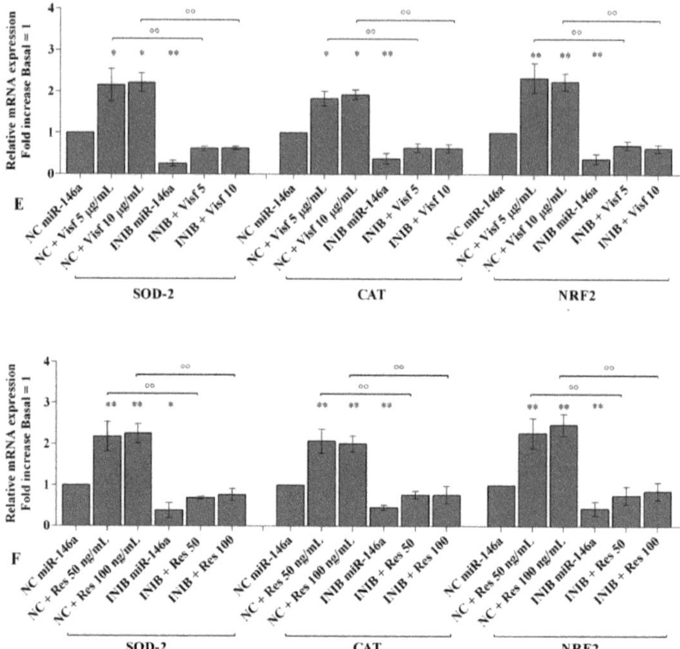

Figure 7. (**A,C,D**) Mitochondrial superoxide anion production was assessed by flow cytometry using MitoSox Red staining. (**B,E,F**) Expression levels of superoxide dismutase (*SOD-2*), catalase (*CAT*), nuclear factor erythroid 2 like 2 (*NRF2*) by real-time PCR. Human osteoarthritic (OA) synovial fibroblasts were evaluated at basal condition, after 24 h of transfection with *miR-146a* inhibitor or NC, and after incubation with visfatin (5 and 10 μg/mL) and resistin (50 and 100 ng/mL). The superoxide anion production and the gene expression were referenced to the ratio of the value of interest and the value of basal condition (basal, cells without treatment) or NC, reported equal to 1. Data were expressed as mean ± SD of triplicate values. * $p < 0.05$, ** $p < 0.01$ versus basal condition or NC. ° $p < 0.05$, °° $p < 0.01$ versus inhibitor. INIB= inhibitor, NC= negative control siRNA, Visf= visfatin, Res = resistin.

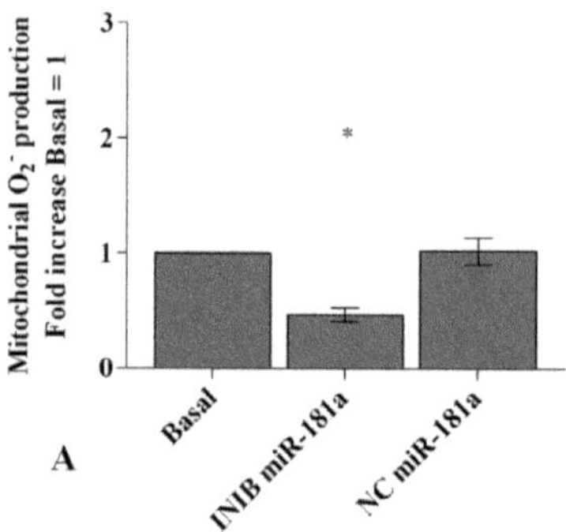

Figure 8. *Cont.*

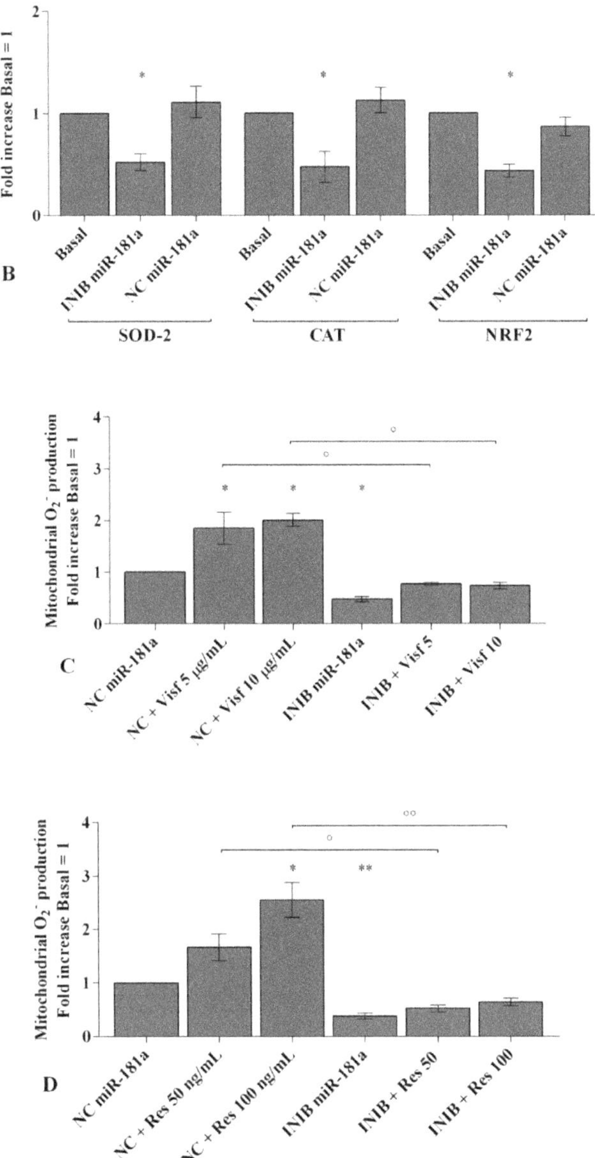

Figure 8. *Cont.*

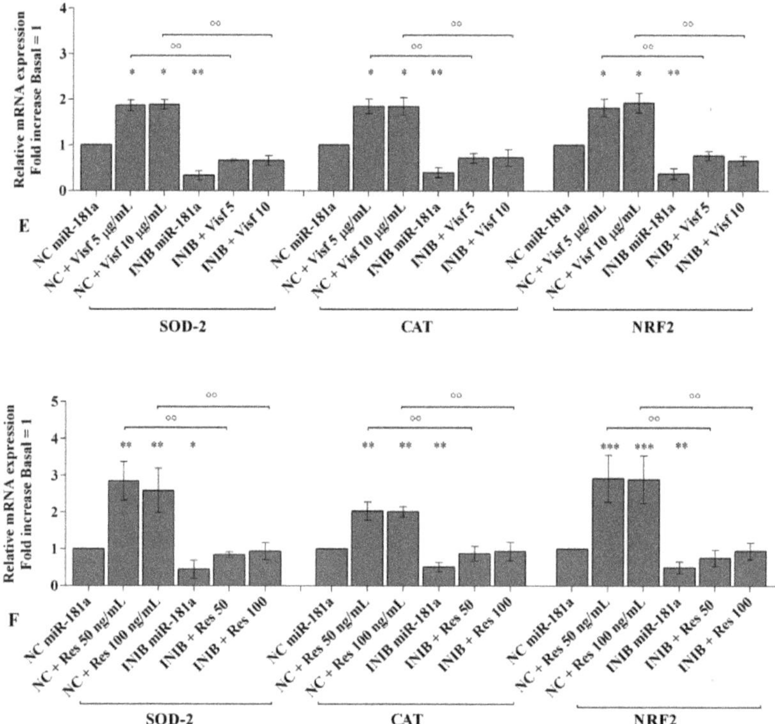

Figure 8. (**A,C,D**) Mitochondrial superoxide anion production was assessed by flow cytometry using MitoSox Red staining. (**B,E,F**) Expression levels of superoxide dismutase (*SOD-2*), catalase (*CAT*), nuclear factor erythroid 2 like 2 (*NRF2*) by real-time PCR. Human osteoarthritic (OA) synovial fibroblasts were evaluated at basal condition, after 24 h of transfection with *miR-181a* inhibitor or NC, and after incubation with visfatin (5 and 10 µg/mL) and resistin (50 and 100 ng/mL). The superoxide anion production and the gene expression were referenced to the ratio of the value of interest and the value of basal condition (basal, cells without treatment) or NC reported equal to 1. Data were expressed as mean ± SD of triplicate values. * $p < 0.05$, ** $p < 0.01$, *** $p < 0.001$ versus basal condition or NC. ° $p < 0.05$, °° $p < 0.01$ versus inhibitor. INIB= inhibitor, NC= negative control siRNA, Visf= visfatin, Res = resistin.

2.7. Visfatin and Resistin Activate NF-κB Signaling Pathway

Figure 9A,B shows the cytoplasmic and nuclear signal intensity of p50 NF-κB subunit in synovial fibroblasts stimulated with visfatin and resistin for 30 min and 4 h. The signal of p50 NF-κB was low mainly detected in the cytoplasm of the cells, with a minimum translocation into the nucleus, at basal condition. After 30 min of incubation with visfatin and resistin we observed a significant increase of p50 subunit cytoplasmic synthesis and nuclear translocation ($p < 0.05$, $p < 0.01$, respectively), in comparison to baseline, while no significant modifications of p50 subunit signal were found after 4 h of adipokines incubation.

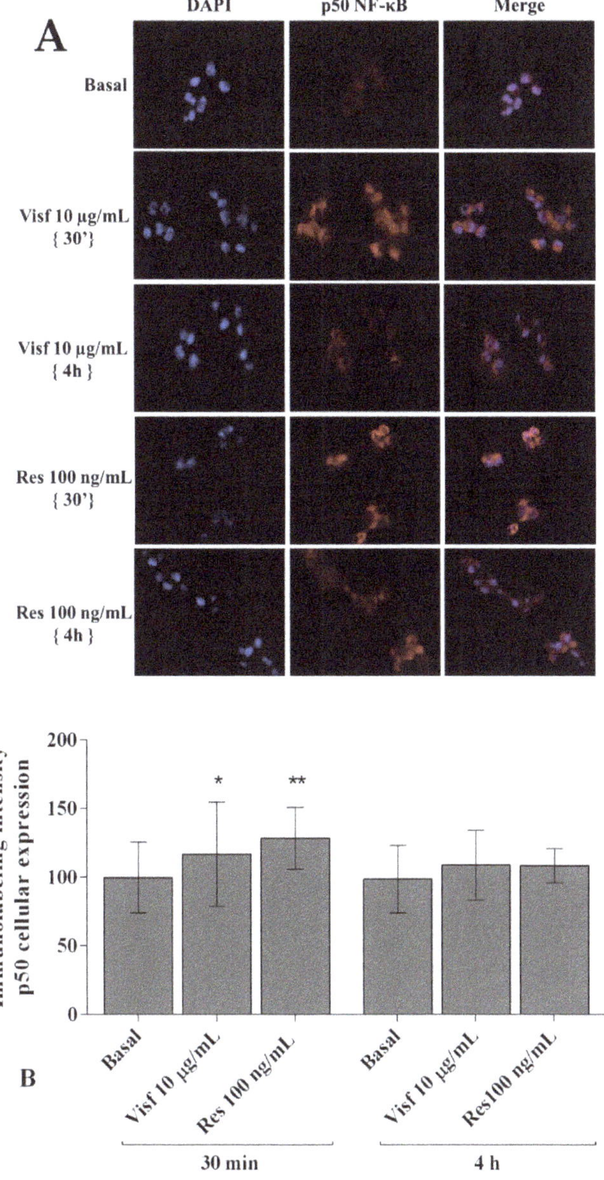

Figure 9. Immunofluorescence labeling of p50 NF-κB subunit localization. Human osteoarthritic (OA) synovial fibroblasts were evaluated at basal condition and after 30 min or 4 h of incubation with visfatin (10 µg/mL) and resistin (100 ng/mL). (**A**) Representative immunocytochemical images of the cells showing localization of p50 NF-κB (red); nuclei were stained with DAPI (blue). Original Magnification 400×. Scale bar: 20 µm. (**B**) The histogram of immunolabeling intensity was plotted for the nuclear and cytoplasmic expression for p50 subunit. Data were expressed as mean ± SD of triplicate values. * $p < 0.05$, ** $p < 0.01$ versus basal condition. Visf = visfatin, Res = resistin.

2.8. NF-κB Signaling Pathway Inhibits Visfatin and Resistin Effects

The involvement of NF-κB pathway in mediating the adipokines-induced effects on inflammatory, apoptotic and oxidative stress mediators is summarized in Figure 10.

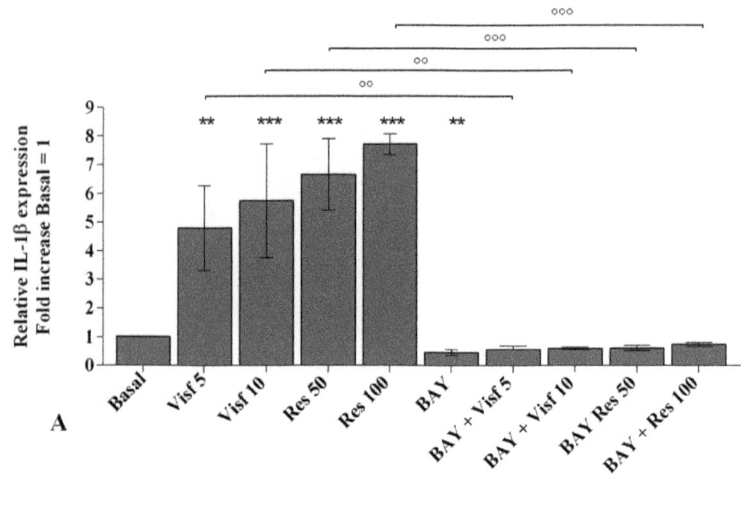

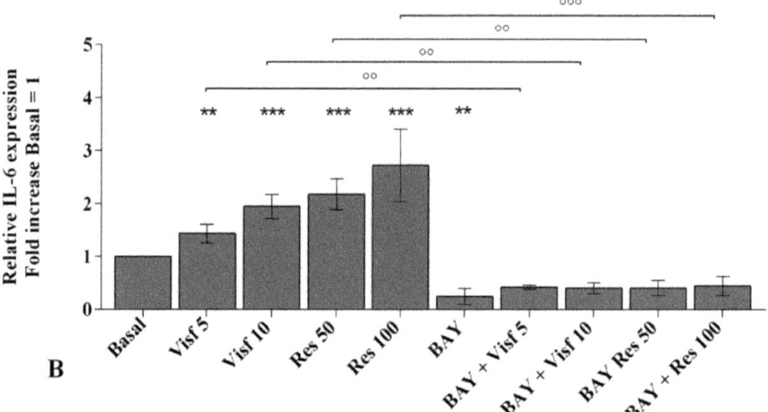

Figure 10. *Cont.*

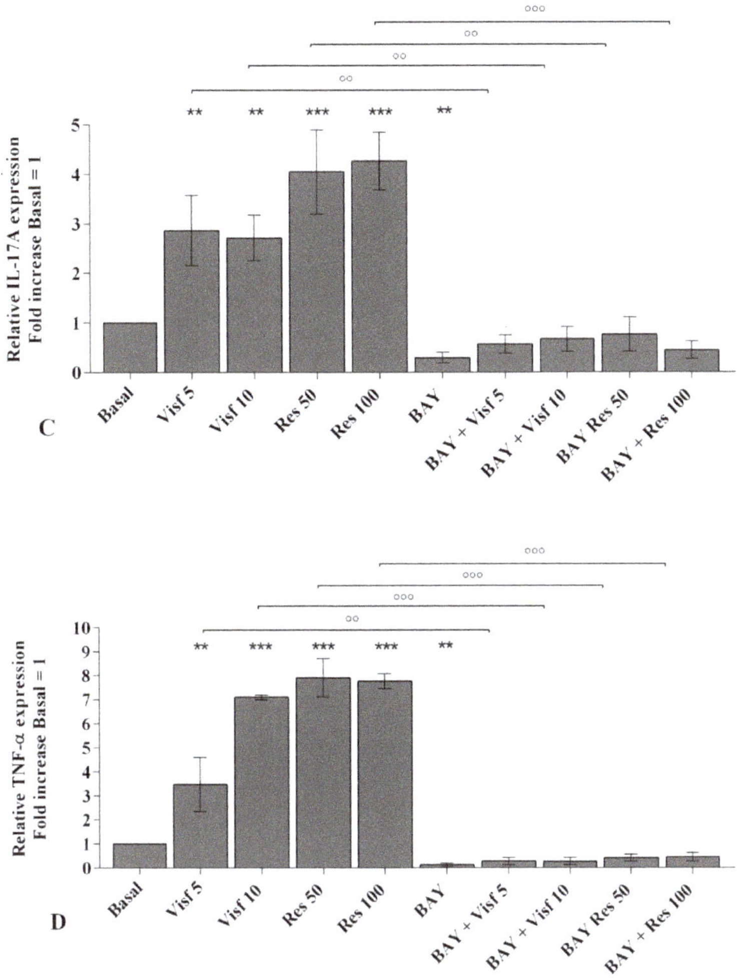

Figure 10. Expression levels of interleukin (*IL*)-*1β* (**A**), *IL-6* (**B**), *IL-17A* (**C**), tumor necrosis factor (*TNF*)-*α* (**D**) by real-time PCR. Human osteoarthritic (OA) synovial fibroblasts were evaluated at basal condition, after 2 h pre-incubation with a specific nuclear factor (NF)-κB inhibitor (BAY 11-7082, IKKα/β, 1 μM) and after 24 h of stimulus with visfatin (5 and 10 μg/mL) and resistin (50 and 100 ng/mL). The gene expression was referenced to the ratio of the value of interest and the value of basal condition (basal, cells without treatment) reported equal to 1. Data were expressed as mean ± SD of triplicate values, ** $p < 0.01$, *** $p < 0.001$ versus basal condition. °° $p < 0.01$, °°° $p < 0.001$ versus BAY. BAY = BAY 11-7082, Visf = visfatin, Res= resistin.

A specific NF-κB inhibitor (IKKα/β, BAY 11-7082) was used to analyze the modulation of the signaling pathway in the gene expression of selected target genes (Figures 10–12) and the studied miRNA (Figure 13).

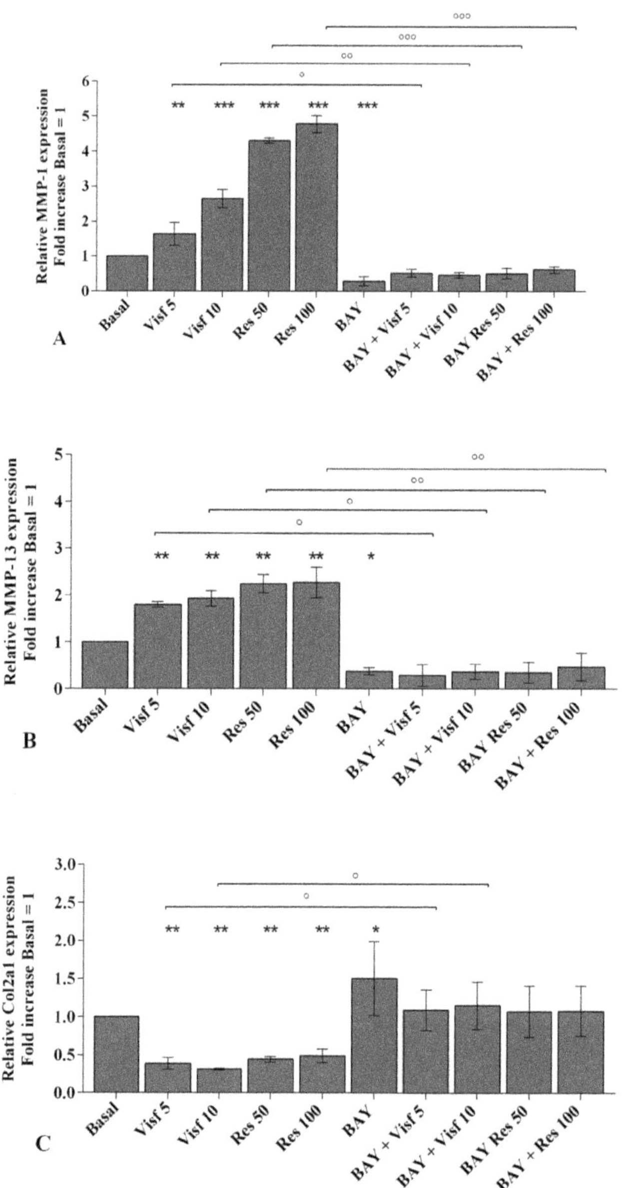

Figure 11. Expression levels metalloproteinases (*MMP*)-1 (**A**), *MMP-13* (**B**), collagen type II (*Col2a1*) (**C**) by real-time PCR. Human osteoarthritic (OA) synovial fibroblasts were evaluated at basal condition, after 2 h pre-incubation with a specific nuclear factor (NF)-κB inhibitor (BAY 11-7082, IKKα/β, 1 μM) and after 24 h of stimulus with visfatin (5 and 10 μg/mL) and resistin (50 and 100 ng/mL). The gene expression was referenced to the ratio of the value of interest and the value of basal condition (basal, cells without treatment) reported equal to 1. Data were expressed as mean ± SD of triplicate values. * $p < 0.05$, ** $p < 0.01$, *** $p < 0.001$ versus basal condition. ° $p < 0.05$, °° $p < 0.01$, °°° $p < 0.001$ versus BAY. BAY = BAY 11-7082, Visf = visfatin, Res= resistin.

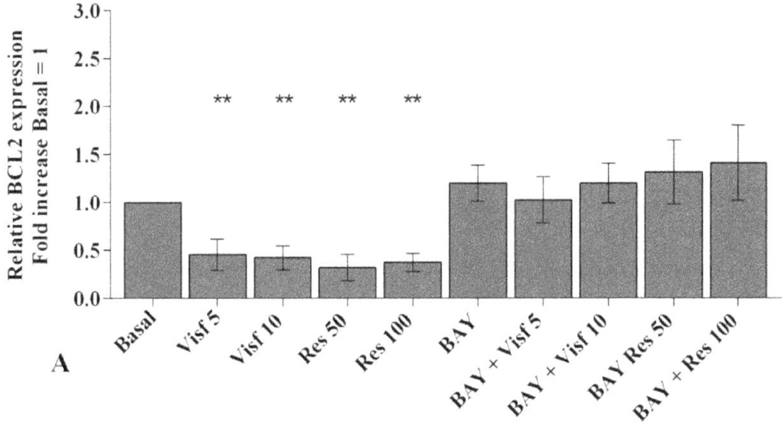

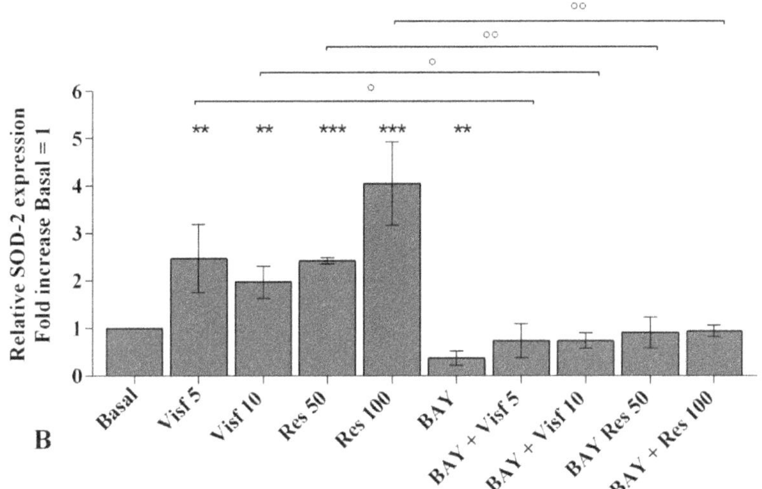

Figure 12. *Cont.*

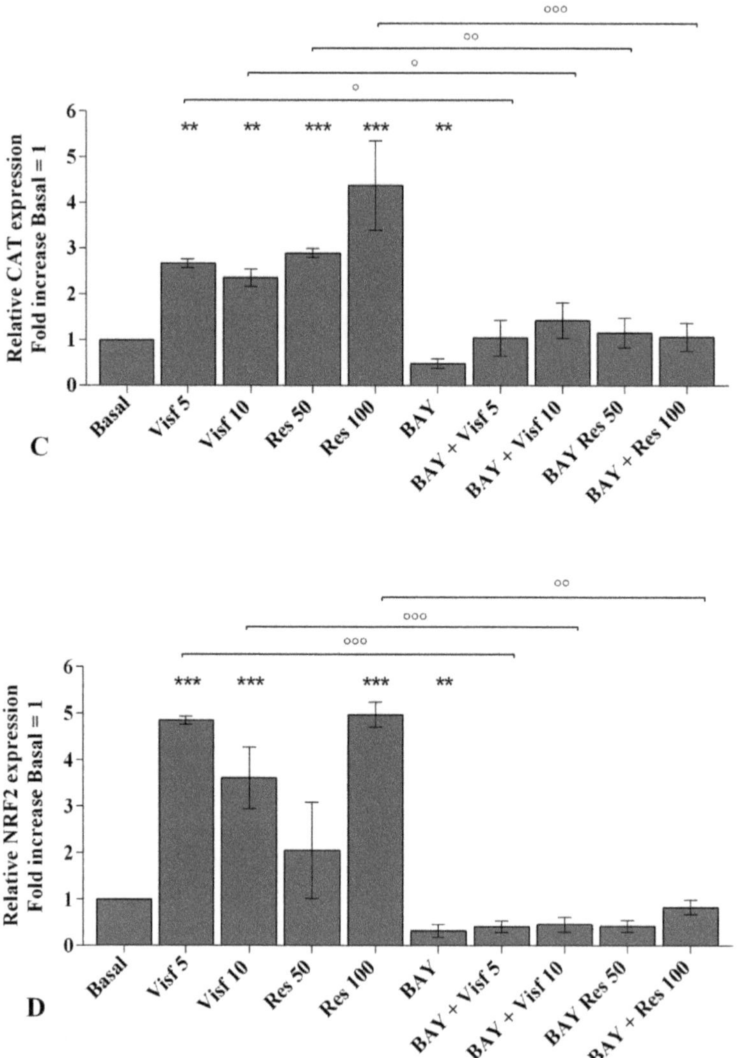

Figure 12. Expression levels of B-cell lymphoma (*BCL*)2 (**A**), superoxide dismutase (*SOD-2*) (**B**), catalase (*CAT*) (**C**), nuclear factor erythroid 2 like 2 (*NRF2*) (**D**) by real-time PCR. Human osteoarthritic (OA) synovial fibroblasts were evaluated at basal condition, after 2 h pre-incubation with a specific nuclear factor (NF)-κB inhibitor (BAY 11-7082, IKKα/β, 1 μM) and after 24 h of stimulus with visfatin (5 and 10 μg/mL) and resistin (50 and 100 ng/mL). The gene expression was referenced to the ratio of the value of interest and the value of basal condition (basal, cells without treatment) reported equal to 1. Data were expressed as mean ± SD of triplicate values. ** $p < 0.01$, *** $p < 0.001$ versus basal condition. ° $p < 0.05$, °° $p < 0.01$, °°° $p < 0.001$ versus BAY. BAY = BAY 11-7082, Visf = visfatin, Res= resistin.

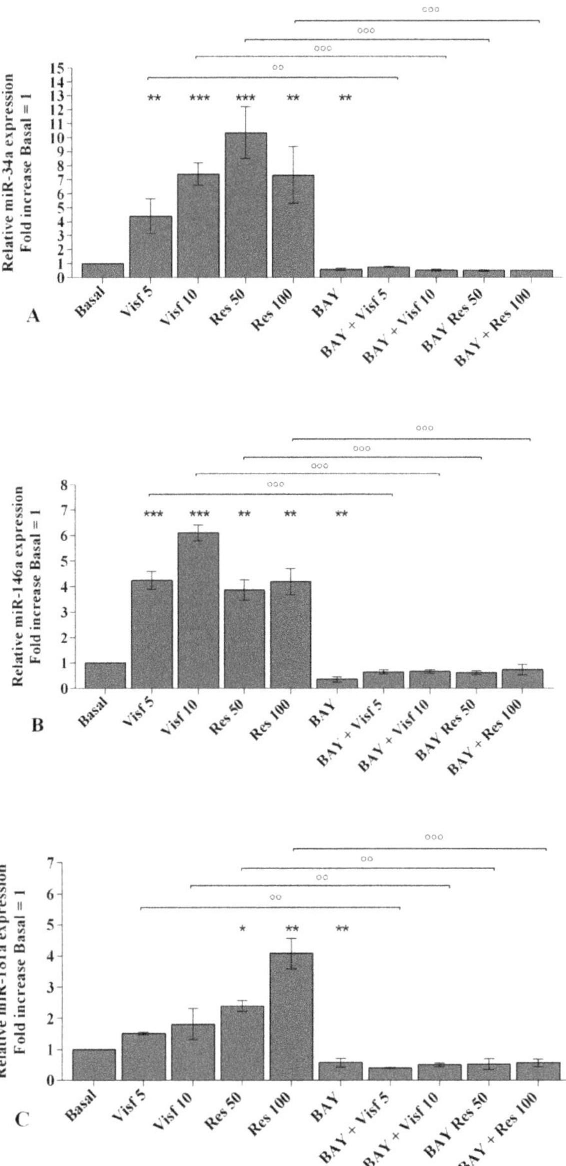

Figure 13. Expression levels of *miR-34a* (**A**), *miR-146a* (**B**), and *miR-181a* (**C**) by real-time PCR. Human osteoarthritic (OA) synovial fibroblasts were evaluated at basal condition, after 2 h pre-incubation with a specific nuclear factor (NF)-κB inhibitor (BAY 11-7082, IKKα/β, 1 μM) and after 24 h of stimulus with visfatin (5 and 10 μg/mL) and resistin (50 and 100 ng/mL). The gene expression was referenced to the ratio of the value of interest and the value of basal condition (basal, cells without treatment) reported equal to 1. Data were expressed as mean ± SD of triplicate values. * $p < 0.05$, ** $p < 0.01$, *** $p < 0.001$ versus basal condition. °° $p < 0.01$, °°° $p < 0.001$ versus BAY. BAY = BAY 11-7082, Visf = visfatin, Res= resistin.

The transcriptional levels of *IL-1β, IL-6, IL-17A, TNF-α* (Figure 10A–D), *MMP-1, MMP-13* (Figure 11A,B), *SOD-2, CAT, NRF2* (Figure 12B–D), *miR-34a, miR-146a,* and *miR-181a* (Figure 13A–C) were significantly decreased ($p < 0.01$, $p < 0.001$) in OA synovial fibroblasts incubated with BAY 11-7082, while an up-regulation of *Col2a1* mRNA levels was observed ($p < 0.05$, Figure 11C), in comparison to basal condition.

The co-treatment of the cells with BAY 11-7082 and visfatin or resistin did not exhibit any difference in miRNA and target genes expression with respect to what is observed in OA synoviocytes incubated with BAY 11-7082 alone (Figures 10–13).

Furthermore, the pre-treatment of the cells with the NF-κB inhibitor significantly limited the effect of visfatin and resistin on the expression levels of the analyzed target genes (Figures 10–13).

No modifications in mRNA levels of *BCL2*, after the treatment, were observed (Figure 12A).

3. Discussion

OA is a musculoskeletal condition mainly characterized by articular cartilage degeneration, however, in recent years, the role of synovial inflammation in the development and in the progression of the disease has been gradually recognized [2,4].

Fibroblast-like synoviocytes actively participate in the synovitis-structural damage cycle of OA through the production of inflammatory cytokines, including IL-6, IL-1β, and TNF-α, and cartilage-degrading enzymes and proteases, such as MMPs [2,29].

Growing evidence demonstrated that adipokines, mainly produced by adipose tissue and by other adipose tissue depots as infrapatellar fat pad, are potentially involved in OA pathophysiology [30]. Indeed, the adipokines may participate in synovium-bone and synovium-cartilage interactions [7,31], however, their exact effect in OA synovial cells have not been completely elucidated [16,32,33] and the results on *in vitro* studies are sparse [19,34].

In the present study, performed in human OA synovial fibroblast cultures, we confirmed previous evidence about the role of visfatin and resistin in inflammation. Furthermore, we demonstrated their impact on apoptosis and oxidative stress processes, as well as in the modulation of some miRNA and target genes, implicated in OA pathogenesis, through the activation of NF-κB pathway. Finally, we hypothesized the direct cross-talk between miRNA and adipokines in mediating oxidative stress induction, via NF-κB signaling.

It is well established that IL-1β, IL-6, IL-17, and TNF-α are the main important cytokines involved in the pathogenesis of OA [35]; they have been found elevated in serum and synovial fluid of patients with knee OA [36,37] and play synergistic effects in OA chondrocytes and synovial fibroblasts stimulating the synthesis and secretion of other cytokines and proteases [16,29].

Our data showed a significant increase of *IL-1β, IL-6,* and *TNF-α* gene expression levels in human OA synovial fibroblast cultures stimulated with visfatin and resistin, according to what is observed by other authors [16,18,38]. On the other hand, we demonstrated, for the first time, the up-regulation of *IL-17* expression levels induced by the studied adipokines in our cultures.

MMPs are the main proteases implicated in cartilage turnover, playing a significant role in the degradation of cartilage ECM that occur during OA damage [39]. MMP-1 and MMP-13 are expressed in chondrocytes and in synoviocytes and contribute to promoting cartilage breakdown inducing the destruction of proteoglycans and Col2a1, the major structural protein of articular ECM [40]. The exposure of OA chondrocytes and fibroblast-like synoviocytes to pro-inflammatory cytokines, such as IL-1β, and adipokines, as visfatin and resistin, determined a markedly increase of matrix-degrading enzymes and a down-regulation of *Col2a1* gene levels [16,17,29,41,42]. In agreement with the current literature we reported the up-regulation of *MMP-1, MMP-13* and a reduction of *Col2a1* expression levels in visfatin and resistin-stimulated OA synovial cells.

These results highlight the role of the studied adipokines in mediating the pro-inflammatory cascade in synovial cells and their consequent implication in articular cartilage destruction that occur in course of OA. Previous evidence reporting that chondrocytes and synovial cells express membrane toll-like receptors (TLRs) which are identified as putative receptors for visfatin and resistin mechanism of action. Adipokines bind to TLRs and stimulate phosphorylation of ERK/p38/mitogen-activated protein kinase (MAPK) signaling, inducing the expression of cytokines, chemokines and degrading proteases [10,14,38,43,44].

The regulation of chondrocytes and fibroblast-like synoviocytes survival is important for the maintenance of a proper cartilage and synovium structure and function [17,45]. Indeed, apoptosis is a complex multi-step process playing a critical role in maintaining the homeostasis of various tissues and cells, and an increasing number of genes have been identified as controller and inductors of this mechanism. Among them, BCL-2 family, anti-apoptotic proteins, are responsible for many biochemical processes driving apoptosis [45].

Dysregulation of apoptosis, thus, is related to a variety of diseases including autoimmune and degenerative disorders as rheumatoid arthritis (RA) and OA [45,46]. The over-expression of BCL-2 family proteins protects OA chondrocytes and human synovial fibroblasts from the programmed cell death [47,48].

The results of our research revealed an increased percentage of apoptosis and a down-regulation of *BCL-2* gene expression in human OA synovial fibroblasts stimulated with visfatin and resistin. Similar data were previously obtained by other authors in endothelial cell lines and in human OA chondrocyte cultures [17,49]. However, we first observed the effect of resistin in the regulation of BCL-2 protein in this cell type.

Oxidative stress and inflammation have been increasingly recognized as being closely integrated with OA pathology. Under physiological conditions, the production of endogenous ROS is balanced by the antioxidant defense system, mainly controlled by NRF2 [50]. The latter is translocated to the nucleus, when released from its repressive cytosolic protein Kelch-like ECH associated protein 1 (KEAP1), and activates the expression of cytoprotective genes, including enzymes involved in the biosynthesis, activity, and detoxification of different ROS species, such as SOD-2 and CAT [50,51]. Various inflammatory mediators, such as cytokines, chemokines, prostaglandins, and growth factors participate to increase oxidative stress in the joint with accumulation of ROS, and nitric oxide (NO), and concomitant failure in the expression of antioxidant scavenging systems [50]. At the cellular level, oxidative stress causes mitochondrial and nuclear DNA damage, lipid peroxidation, alterations in cell signaling and transcription, and epigenetic changes in gene expression contributing to exacerbate synovitis, destruction of matrix components and cell apoptosis [50,52].

In this paper, the analysis of endogenous production of ROS reported an increase of mitochondrial superoxide anion content in OA synoviocytes cultures after visfatin and resistin stimulation, with a concomitant up-regulation of *SOD-2*, *CAT*, and *NRF2* gene expression. There is no evidence about the effects of the studied adipokines on oxidative stress induction in synovial fibroblasts; however, a number of studies, performed in different cell lines incubated with visfatin, resistin and leptin, are in agreement with our data [53–55].

The observed rapid increase of the studied detoxificant factors and NRF2 in adipokines-stimulated human synoviocytes confirm what is observed in a previous study on OA chondrocyte cultures [56]. In our opinion, this result could be explained as an acute adaptive response to protect mitochondria from the deleterious effects of the raised oxidant agents after adipokines stimulus [27,52,56].

Taken together, these findings underline the involvement of visfatin and resistin in the regulation of apoptosis and oxidative stress balance. This conclusion could be supported by the effects of adipokines in stimulating p38 phosphorylation to further activate PI3K/Art signaling and NADPH oxidase (NOX), a major source of ROS generation. Indeed, NOX activation cause the ROS-forming cascade signaling, induces NF-κB translocation into the nucleus, leading to likewise inflammation, cell proliferation, survival and apoptosis [53,55].

MiRNA has been widely investigated for their role in gene regulation; by binding to mRNA 3′-UTRs, miRNA can affect many protein-encoding genes at the post-transcriptional levels [22,24,57].

It is proved that some miRNA are differentially expressed in OA cartilage samples with respect to normal ones, demonstrating their role in the development and progression of OA [23,24,26].

MiR-34a is largely known to be an anti-proliferative factor regulating cell cycle arrest or senescence [58]. Some authors reported the involvement of *miR-34a* in activating apoptosis signaling and limiting cell proliferation in human OA chondrocytes and RA synovial fibroblasts [59,60], as well as its role in modulation of oxidative stress balance in HUVEC lines [61].

MiR-181a was found highly expressed in circulating PBMC of OA patients and in human OA chondrocytes [62,63], and its results implicated the regulation of apoptosis and oxidative stress signaling by targeting multiple anti-apoptotic BCL2 members and modulating mitochondria metabolism in different cell types [63–65].

Data from the current literature concerning the involvement of miR-146a in OA pathogenesis are controversial [27,66,67]. Yamasaki et al. [66] demonstrated that this miRNA is up-regulated in OA cartilage with a low grade on the Mankin scale, or after the stimulus of OA chondrocytes with IL-1β [67]. On the contrary, its reduced expression in hydrogen peroxide-stimulated OA cells was observed [27]. Additionally, this miRNA resulted implicated in oxidative stress regulation by its direct effect on NRF2 transcriptional factor [68].

In this study we showed a significant increase of *miR-34a*, *miR-146a*, and *miR-181a* gene expression after the incubation of OA synoviocytes with visfatin and resistin, consistently with the results of other in vitro studies [17,18,69–71]. On the basis of the results obtained by Wu et al. [18] we can hypothesize the modulation of miRNA gene expression through the phosphorylation of ERK/p38/MAPK signaling induced by visfatin and resistin.

Accumulating evidence has shown a cross-talk between miRNA and components of redox signaling [27,28,57,72]. The transcription, biogenesis, translocation, and function of miRNA are highly correlated with ROS, and, meanwhile, miRNA can regulate the expression of redox factors and other ROS modulators, such as the key components of cellular antioxidant machinery [27,28,57].

Recently, some miRNA were identified as oxidative stress-responsive factors after the treatment of OA chondrocytes with H_2O_2 [27,73], on the other hand, cellular mechanisms regulating oxidative stress were fine-tuned by particular miRNA [28,56].

A number of studies demonstrated the regulation of *miR-34a*, *miR-146a*, and *miR-181a* expression by oxidative stress in PC12, cardiac and carcinoma cell lines and in OA chondrocytes [27,74–76]; furthermore, the inhibition of these miRNA decreased the expression of the main antioxidant enzymes and reduced the mitochondrial intracellular ROS levels [56,61,64,74,76]. According to this evidence, in the present study, the transient transfection of OA synovial fibroblasts with *miR-34a*, *miR-146a*, and *miR-181a* specific inhibitors significantly reduced the production of mitochondrial superoxide anion as well as the expression of *SOD-2*, *CAT*, and *NRF2*, limiting the negative effects of visfatin and resistin. In a similar manner, other authors revealed the involvement of miRNA in mediating visfatin and resistin effects in HepG2 cells and in human synovial fibroblasts [18,71]. The ability of these miRNA in regulating oxidative stress has been reported in different in vitro studies and seems to be related to the regulation of NRF2 activity [57]. Huang et al. [77] showed the implication of miR-34a in modulating *NRF2* expression and NRF2-dependent antioxidant pathway through the direct targeting of *miR-34a* with the 3′UTR of *NRF2* mRNA. Furthermore, *miR-146a* resultingly involved in the regulation of *NRF2* activation by targeting the 3′-UTR of IL-1R-associated kinase (*IRAK*)1 and TNFR-associated factor (*TRAF*)6 mRNA, the downstream adaptors of TLRs [68]. These data suggest the presence of a regulatory network between miRNA and NRF2 in regulating oxidative stress.

However, in the present study we observed a reduction in the gene expression of antioxidant enzymes when the miRNA were inhibited. This finding could be due to the fact that *miR-34a* and *miR-181a* also directly bind the 3′UTR of silent mating type information regulation 2 homolog (*SIRT*)1 mRNA, inducing a decrease in the protein and/or mRNA expression of this gene.

SIRT1 and SIRT6 are putative anti-ageing molecules that regulate the expression of several antioxidant genes and are classified as regulator of oxidative stress balance. Elevated oxidative stress decreased both the protein and mRNA levels of *SIRT1*, whilst up-regulating the expression of miR-34a and miR-181a.

In view of these reports, we can postulate that the obtained results concerning the gene expression of antioxidant enzymes could be related to the up-regulation of *SIRT1* after of *miR-34* and *miR-181a* inhibition [77].

We finally supposed that the complex crosstalk found between adipokines and miRNA, in OA synovial fibroblasts, could be regulated by NF-κB signaling pathway.

NF-κB proteins constitute a family of ubiquitously expressed transcription factors playing essential roles in phlogistic events, immune and stress responses, and in cartilage degradation [78,79]. Accumulation data indicate NF-κB signaling as the most prominent mechanism in the pathogenesis of OA [78,79]. Furthermore, the importance of NF-κB signaling pathway for visfatin and resistin-induced inflammation, as well as for miRNA-related post-transcriptional regulation has been reported [16–19, 56,70].

Our results showed an increase of NF-κB activation and of p50 subunit nuclear translocation in OA synoviocytes stimulated with visfatin and resistin, in agreement with other researches performed in various cell cultures [16,17,33,49,55,80,81]. Besides, these studies also affirmed that NF-κB is involved in regulation of visfatin and resistin-mediated effects in human OA chondrocytes and endothelial progenitor cells incubated with a specific NF-κB inhibitor [16,33,49,55]. Our data support these findings demonstrating that the inhibition of NF-κB signaling limits inflammation and oxidative stress induced by visfatin and resistin, in human OA synovial fibroblasts. The current literature establishes the activation of NF-κB signaling after phosphorylation of ERK/p38/ MAPK pathway induced by visfatin and resistin, triggering the downstream up-regulation of pro-inflammatory and pro-catabolic-related genes, which contribute to inflammatory and degrading processes of OA. Hence, the inhibition of NF-κB transcriptional factor could represent one of the molecular mechanisms to limit adipokines effects on joint injury.

In addition, we also observed that the modulation of *miR-34a*, *miR-146a*, and *miR-181a* expression induced by the studied adipokines was strongly limited by NF-κB inhibition. Similar results were found by other authors, showing an increased gene expression of *miR-34a* and *miR-146a* after IL-1β stimulus through activation of NF-kB; in turn, *miR-34a* and *miR-146a* were found to be able to inhibit the activation of NF-kB via suppressing their target genes expression such as *NRF2*, *IRAK1* and *TRAF6* [68,77,82].

These data suggest that the cross-talk between visfatin, resistin and miRNA could be mediated by NF-κB signaling pathway, highlighting the mutual interaction between miRNA and NF-kB.

However, the present study presents some limitations that need to take into consideration.

First of all, additional experiments on healthy primary cells are recommended; further transfection experiments with specific miRNA mimic could be useful to confirm the regulation induced by the studied miRNA. In addition, the protein levels of the antioxidant enzymes and of the transcriptional factor NRF2 should be detected as well to elucidate if transcription modifications reflect a translational regulation.

Finally, a simultaneous miRNA and NF-κB inhibition could help to deeper investigate their direct interaction in mediating adipokines effects.

4. Materials and Methods

4.1. Sample Collection and Cell Culture

Synovial tissue samples were obtained from three non-obese (BMI from 20 to 25 Kg/m^2) and non-diabetic patients (two men and three women, age from 67 to 75) with primary knee OA defined by the clinical and radiological ACR criteria [83], during their total knee arthroplasty. The tissues were supplied by the Orthopaedic Surgery, University of Siena, Italy. The human articular samples protocols used in this work were evaluated and approved by the Ethic Committee of Azienda Ospedaliera

Universitaria Senese/Siena University Hospital (Prot n 13931_2018, 15 October 2018), and all patients signed a free and informed consent form.

Synovial tissue was separated from adjacent cartilaginous and adipose structures, and isolated immediately after surgery. Briefly, samples were aseptically dissected from each donor, cut into small thick pieces and processed by an enzymatic digestion by using trypsin-EDTA Solution 10× (Sigma–Aldrich, Milan, Italy) for 15 min at 37 °C and then, washed and incubated with type IV collagenase (Sigma–Aldrich, Milan, Italy) in Dulbecco's Modified Eagle Medium (DMEM) (Euroclone, Milan, Italy) medium with shaking for 12–16 h at 37 °C.

The obtained cell suspension was filtered using 70-μm nylon meshes, washed, and centrifuged for 5 min at 700× g. The viability was assessed by Trypan Blue (Sigma–Aldrich, Milan, Italy) test and a percentage of 90% to 95% of cell survival was assessed. Cells were collected, seeded into 10-cm diameter tissue culture plates, and expanded for a minimum of two weeks in a monolayer in incubator with 5% CO_2 and 90% humidified atmosphere at 37 °C, until a confluence of 80% to 85% was reached.

Human OA synovial fibroblasts were grown in DMEM containing 10% fetal bovine serum (FBS) (Euroclone, Milan, Italy), with 200 U/mL penicillin and 200 μg/mL streptomycin (P/S) (Sigma–Aldrich, Milan, Italy). The culture medium was changed two times for week. The morphology was examined daily with an inverted microscope (Olympus IMT-2, Tokyo, Japan), and the cells from passages 3 to 6 were employed for the experimental procedures. A cell culture derived from a unique donor was used for each single experiment, for a total of three independent experiments.

4.2. Stimulus of Synovial Cell Cultures

Human OA synovial fibroblasts were transferred and plated in 6-well dishes at a starting density of 1×10^5 cells/well until they became confluent. Human recombinant visfatin (Sigma–Aldrich, Milan, Italy) and human recombinant resistin (BioVendor, Rome, Italy) were dissolved in phosphate buffered saline (PBS) (Euroclone, Milan, Italy), according to the manufacturer's instructions, and then directly diluted in the culture medium for the treatment in order to obtain the final concentration required.

The cells were immersed in DMEM medium enriched with 0.5% FBS and 2% P/S and stimulated for 24 h with visfatin at concentration of 5 and 10 μg/mL or resistin 50 and 100 ng/mL. The concentrations of the adipokines used in our in vitro study were selected according to those used by other authors and in our previous report [16,17,84]; the final concentrations were chosen based on the best results obtained in terms of viability (Figure S1).

After the treatment, the cells were recovered and immediately processed to carry out flow cytometry analysis and quantitative real-time PCR.

In addition, the synovial cells were pre-incubated for 2 h with 1 μM BAY 11-7082 (NF-κB inhibitor, IKKα/β, Sigma–Aldrich, Milan, Italy) and then stimulated 24 h with the selected concentrations of visfatin (5 and 10 μg/mL) and resistin (50 and 100 ng/mL). Then, the gene expression of the target genes (*IL-1β, IL-6, IL-17A, TNF-α, MMP-1, MMP-13, Col2a1, BCL2, SOD-2, CAT* and *NRF2*) and miRNA (*miR-34a, miR-146a,* and *miR-181a*) was evaluated.

4.3. MTT Assay

The viability of the cells was evaluated, by MTT test, after the treatment of the cells with visfatin and resistin at the tested concentrations.

Chondrocytes were incubated for 3 h at 37 °C in a culture medium containing 10% of 5 mg/mL of MTT (Sigma–Aldrich, Milan, Italy). At the end of this period, the medium was removed and 0.2 mL of dimethyl sulfoxide (DMSO) (Rottapharm Biotech, Monza, Italy) was added to the wells to solubilize the formazan crystals. The absorbance was measured at 570 nm in a microplate reader (BioTek Instruments, Inc., Winooski, VT, USA). A control well without cells was employed for blank measurement.

The percentage of survival cells was evaluated as (absorbance of considered sample) / (absorbance of control) × 100.

The experiments were performed on cell cultures at 80% to 85% of confluence in order to prevent contact inhibition which can alter the results. Data were reported as OD units per 10^4 adherent cells.

4.4. Transfection of Synovial Cells

The cells were grown in 6-well dishes at a starting density of 1×10^5 cells/well until a confluence of 85% in DMEM supplemented with 10% FBS; then, the media were replaced with DMEM 0.5% FBS for 6 h before transfection. Afterwards, synoviocytes were transfected with specific inhibitors of *miR-34a, miR-146a*, and *miR-181a* (Qiagen, Hilden, Germany), at the concentration of 50 nM, or with their relative negative controls siRNA (NC) (Qiagen, Hilden, Germany), at the concentration of 5 nM, in serum-free medium for a period of 24 h. Supernatants were removed and synoviocytes immediately harvested or incubated with visfatin (5 and 10 μg/mL) or resistin (50 and 100 ng/mL) for additional 24 h.

4.5. Quantitative Real-Time PCR of mRNA and miRNA

Synovial fibroblasts were grown in 6-well dishes at a starting density of 1×10^5 cells/well in DMEM supplemented with 10% FBS. Then, the supernatant was removed, and the cells were cultured in DMEM with 0.5% FBS used for the treatment procedure.

Total RNA, including miRNA, was extracted using TriPure Isolation Reagent (Euroclone, Milan, Italy) according to the manufacturer's instructions, and was stored at −80 °C. The concentration, purity, and integrity of RNA were evaluated by measuring the OD at 260 nm and the 260/280 and 260/230 ratios by Nanodrop-1000 (Celbio, Milan, Italy). The quality of RNA was verified by electrophoresis on agarose gel (FlashGel System, Lonza, Rockland, ME, USA). Reverse transcription for miRNA was carried out by the cDNA miScript PCR Reverse Transcription kit (Qiagen, Hilden, Germany), while for target genes the QuantiTect Reverse Transcription kit (Qiagen, Hilden, Germany) was used, according to the manufacturer's instructions.

MiRNA and target genes were examined by real-time PCR using, miScript SYBR Green (Qiagen, Hilden, Germany) and QuantiFast SYBR Green PCR (Qiagen, Hilden, Germany) kits, respectively. A list of the used primers is reported in Table 1.

Table 1. Primers used for RT-qPCR.

miRNA Genes	Cat. No. (Qiagen)
miR-34a	MS00003318
miR-146a	MS00003535
miR-181a	MS00006692
SNORD-25	MS00014007
Target Genes	**Cat. No. (Qiagen)**
IL-1β	QT00021385
IL-6	QT00083720
IL-17A	QT00009233
TNF-α	QT00029162
MMP-1	QT00014581
MMP-13	QT00001764
Col2a1	QT00049518
BCL2	QT00000721
SOD-2	QT01008693
CAT	QT00079674
NRF2	QT00027384
ACTB	QT00095431

Abbreviations: miRNA = microRNA; SNORD-25 = Small Nucleolar RNA, C/D Box 25; IL-1β = interleukin 1β; IL-6 = interleukin 6; IL-17A = interleukin 17A; TNF-α = tumor necrosis factor-α; MMP-1 = matrix metalloproteinase 1; MMP-13 = matrix metalloproteinase 13; Col2a1 = type II collagen alpha 1 chain; BCL2 = B-cell lymphoma; SOD-2 = superoxide dismutase 2; CAT = catalase; NRF2 = nuclear factor erythroid 2 like 2; ACTB = actin beta.

All qPCR reactions were achieved in glass capillaries by a LightCycler 1.0 (Roche Molecular Biochemicals, Mannheim, Germany) with LightCycler Software Version 3.5. The reaction procedure for miRNA consisted of 95 °C for 15 min for HotStart polymerase activation, followed by 40 cycles of 15 s at 95 °C for denaturation, 30 s at 55 °C for annealing, and 30 s at 70 °C for elongation, according to the protocol. Target genes amplification was performed at 5 in at 95 °C, 40 cycles of 15 s at 95 °C, and 30 s at 60 °C. In the final step of both protocols, the temperature was raised from 60 °C to 95 °C at 0.1 °C/step to plot the melting curve.

The analysis of the dissociation curves was performed by visualizing the amplicons lengths in agarose gel to confirm the correct amplification of the resulting PCR products.

For the data analysis, the C_t values of each sample and the efficiency of the primer set were calculated through LinReg Software [85] and then converted into relative quantities and normalized using the Pfaffl model [86].

The normalization was performed considering Small Nucleolar RNA, C/D Box 25 (SNORD-25) for miRNA and Actin Beta (ACTB) for target genes, as the housekeeping genes. The choice of the genes was carried out by using geNorm software version 3.5 [87].

4.6. Apoptosis Detection

Apoptotic cells were evaluated by using Annexin V-FITC and propidium iodide (PI) (ThermoFisher Scientific, Milan, Italy). Human OA synovial fibroblasts were seeded in 12-well plates (8×10^4 cells/well) for 24 h in DMEM with 10% FBS. Then, the medium was discarded, and the cells were cultured in DMEM with 0.5% FBS used for the treatment procedure. Afterwards, the synovial cells were washed and harvested by using trypsin, collected into cytometry tubes, and centrifuged at 1500 rpm for 10 min. The supernatant was replaced, and the pellet was resuspended in 100 µL of 1× Annexin-binding buffer, 5 µL of Alexa Fluor 488 annexin-V conjugated to fluorescein (green fluorescence) and 1 µL of 100 µg/mL PI working solution. Markers were added to 100 µL of cell suspension. Cells were incubated at room temperature for 15 min in the dark. Then, 600 µL of 1× Annexin-binding buffer were added before the analysis at flow cytometer. A total of 10,000 events (1×10^4 cells per assay) were measured by the instrument. The obtained results were analyzed with Cell Quest software (Version 4.0, Becton Dickinson, San Jose, CA, USA). The evaluation of apoptosis was carried out considering staining cells simultaneously with Alexa Fluor 488 annexin-V and PI; a discrimination of intact cells (annexin-V and PI-negative), early apoptosis (annexin-V-positive and PI-negative), and late apoptosis (annexin-V and PI-positive) is allowed [88].

The results were expressed as percentage of positive cells to each dye (total apoptosis), and the data were represented as the mean of three independent experiments (mean ± SD).

4.7. Mitochondrial Superoxide Anion ($\bullet O_2$-) Production

Human OA synovial fibroblasts were seeded in a density of 8×10^4 cells per well in 12 multi-plates for 24 h in DMEM with 10% FCS. Then, the medium was eliminated, and the cells were cultured in DMEM with 0.5% FBS used for the treatment procedure. Then, the cells were incubated in Hanks' Balanced Salt Solution (HBSS) and MitoSOX Red for 15 min at 37°C in dark, to assess mitochondrial superoxide anion ($\bullet O_2$-) production. MitoSOX was dissolved in DMSO, at a final concentration of 5 µM. Cells were then harvested by trypsin and collected into cytometry tubes and centrifuged at 1500 rpm for 10 min. Besides, cells were suspended in saline solution before being analyzed by flow cytometry. A density of 1×10^4 cells per assay (a total of 10,000 events) were measured by flow cytometry and data were analyzed with CellQuest software (Version 4.0, Becton Dickinson, San Jose, CA, USA). Results were collected as median of fluorescence (AU) and represented the mean of three independent experiments (mean ± SD).

4.8. Immunofluorescence Analysis

Human OA synovial fibroblasts were plated in coverslips in Petri dishes (35×10 mm) at a starting low density of 4×10^4 cells/chamber, to prevent possible cell overlapping, and re-suspended in 2 mL of culture medium until 80% of confluence. The cells were processed after 2 h of stimulus with adipokines to evaluate the potential activation of the NF-κB pathway. The synovial cells were washed in PBS and then fixed in 4% paraformaldehyde (ThermoFisher Scientific, Milan, Italy) (pH 7.4) for 10 min at room temperature. Afterwards, the cells were permeabilized with a blocking solution (PBS, 1% bovine serum albumin (BSA) (Sigma–Aldrich, Milan, Italy) and 0.2% Triton X-100 (ThermoFisher Scientific, Milan, Italy) for 20 min at room temperature, and then incubated overnight at 4 °C with mouse monoclonal anti-p50 subunit primary antibody (Santa Cruz Biotechnology, Italy) diluted at 1:100 in PBS, 1% BSA and 0.05% Triton X-100. Three washes in PBS of the coverslips were followed by 1 h incubation with goat anti-mouse IgG-Texas Red conjugated antibody (Southern Biotechnology, Italy) diluted at 1:100 in PBS, 1% BSA and 0.05% Triton X-100. Finally, the coverslips were washed three times in PBS and submitted to nuclear counterstain by 4,6-diamidino-2-phenylindole (DAPI), and then mounted with Vecta shield (Vector Labs). Fluorescence was examined under an AxioPlan (Zeiss, Oberkochen, Germany) light microscope equipped with epifluorescence at 200× and 400× magnification. The negative controls were obtained by omitting the primary antibody. Immunoreactivity of p50 was semi-quantified as the mean densitometric area of p50 signal into the nucleus and into the cytoplasm, by AxioVision 4.6 software measure program [89]. At least 100 synovial cells from each group were evaluated.

4.9. Statistical Analysis

Three independent experiments were carried out and the results were expressed as the mean ± SD of triplicate values for each experiment. Data normal distribution was evaluated by Shapiro–Wilk, D'Agostino and Pearson, and Kolmogorov–Smirnov tests.

Data from real-time PCR were evaluated by one-way ANOVA with a Tukey's post-hoc test using $2^{-\Delta\Delta CT}$ values for each sample. Flow cytometry results were analyzed by ANOVA with Bonferroni post-hoc test.

All analyses were performed through the SAS System (SAS Institute Inc., Cary, NC, USA) and GraphPad Prism 6.1. A significant value was defined with a p-value < 0.05.

5. Conclusions

Growing evidence supports the relevance of synovitis in OA pathophysiology. Among the various factor involved in synovial membrane inflammation and in cartilage degradation during the development and the progression of OA, adipokines, miRNA, and oxidative stress play a crucial role. These findings induced us to deeper investigate the possible link between adipokines and some miRNA in oxidative stress regulation in human OA synovial cultures.

We firstly demonstrated the ability of visfatin and resistin to induce the gene expression of a pattern of pro-inflammatory cytokines (*IL-1β, IL-6, IL-17A* and *TNF-α*), MMPs (*MMP-1, MMP-13*), anti-oxidant enzymes (*SOD-2, CAT* and *NRF2*), as well as *miR-34a, miR-146a,* and *miR-181a*. Furthermore, they caused apoptosis and superoxide anion production, down-regulated the transcriptional levels of Col2a1 and the anti-apoptotic marker BCL2 and increased the p50 NF-κB activation.

Furthermore, we investigated the implication of *miR-34a, miR-146a,* and *miR-181a* as possible regulators of adipokines effects on the modulation of oxidative stress.

Finally, the use of NF-κB specific inhibitor points out the involvement of the pathway in adipokines-mediated effects.

In conclusion, altogether, these results confirm the role of visfatin and resistin in the induction of inflammation and cartilage degradation, and contribute to elucidate the existing crosstalk among adipokines, miRNA and oxidative stress.

However, further studies are required to deeper investigate this complex network and how this evidence can be useful to identify new possible therapeutic targets to reduce synovitis and cartilage degradation in OA.

Supplementary Materials: Supplementary materials can be found at http://www.mdpi.com/1422-0067/20/20/5200/s1.

Author Contributions: The authors declare to have participated in the drafting of this paper as specified below: Conceptualization, S.T. and A.F.; Data curation, S.C., I.G. and M.B.; Funding acquisition, S.G.; Investigation, S.C.; Methodology, S.C., I.G. and M.B.; Project administration, A.F.; Resources, A.F.; Supervision, A.F.; Visualization, S.T.; Writing – original draft, S.C. and A.F.; Writing – review & editing, S.C., S.G., N.M., A.G., S.T. and A.F. SC, ST and AF contributed equally to this work.

Funding: This research was partially funded by a grant (funds for the research) of University of Siena.

Acknowledgments: We thank the University of Siena for economical contribution.

Conflicts of Interest: The authors declare no conflict of interest.

References

1. Vos, T.; Allen, C.; Arora, M.; Barber, R.M.; Bhutta, Z.A.; Brown, A.; Cater, A.; Casey, C.C.; Charlson, J.F.; Chen, Z.A.; et al. Global, regional, and national incidence, prevalence, and years lived with disability for 310 diseases and injuries, 1990-2015: A systematic analysis for the Global Burden of Disease Study 2015. *Lancet* **2016**, *388*, 1545–1602. [CrossRef]

2. Poulet, B.; Staines, K.A. New developments in osteoarthritis and cartilage biology. *Curr. Opin. Pharmacol.* **2016**, *28*, 8–13. [CrossRef] [PubMed]

3. Sellam, J.; Berenbaum, F. The role of synovitis in pathophysiology and clinical symptoms of osteoarthritis. *Nat. Rev. Rheumatol.* **2010**, *6*, 625–635. [CrossRef] [PubMed]

4. Scanzello, C.R.; Goldring, S.R. The role of synovitis in osteoarthritis pathogenesis. *Bone* **2012**, *51*, 249–257. [CrossRef]

5. Sachdeva, M.; Aggarwal, A.; Sharma, R.; Randhawa, A.; Sahni, D.; Jacob, J.; Sharma, V.; Aggarwal, A. Chronic inflammation during osteoarthritis is associated with an increased expression of CD161 during advanced stage. *Scand J. Immunol* **2019**, e12770. [CrossRef]

6. Neumann, E.; Junker, S.; Schett, G.; Frommer, K.; Müller-Ladner, U. Adipokines in bone disease. *Nat. Rev. Rheumatol.* **2016**, *12*, 296–302. [CrossRef]

7. Azamar-Llamas, D.; Hernández-Molina, G.; Ramos-Ávalos, B.; Furuzawa-Carballeda, J. Adipokine Contribution to the Pathogenesis of Osteoarthritis. *Mediators Inflamm.* **2017**, *2017*, 5468023. [CrossRef]

8. Tenti, S.; Palmitesta, P.; Giordano, N.; Galeazzi, M.; Fioravanti, A. Increased serum leptin and visfatin levels in patients with diffuse idiopathic skeletal hyperostosis: A comparative study. *Scand J. Rheumatol* **2017**, *46*, 156–158. [CrossRef]

9. Fioravanti, A.; Cheleschi, S.; De Palma, A.; Addimanda, O.; Mancarella, L.; Pignotti, E.; Pulsatelli, L.; Galezzi, M.; Meliconi, R. Can adipokines serum levels be used as biomarkers of hand osteoarthritis? *Biomarkers* **2018**, *23*, 265–270. [CrossRef]

10. Carrión, M.; Frommer, K.W.; Pérez-García, S.; Müller-Ladner, U.; Gomariz, R.P.; Neumann, E. The Adipokine Network in Rheumatic Joint Diseases. *Int. J. Mol. Sci.* **2019**, *20*, 4091. [CrossRef]

11. Sun, Z.; Lei, H.; Zhang, Z. Pre-B cell colony enhancing factor (PBEF), a cytokine with multiple physiological functions. *Cytokine Growth F. R.* **2013**, *24*, 433–442. [CrossRef] [PubMed]

12. Senolt, L.; Housa, D.; Vernerová, Z.; Jirásek, T.; Svobodová, R.; Veigl, D.; Anderlova, K.; Muller-Ladner, U.; Pavelka, K.; Haluzik, M. Resistin in rheumatoid arthritis synovial tissue, synovial fluid and serum. *Ann. Rheum. Dis.* **2007**, *66*, 458–463. [CrossRef] [PubMed]

13. Fioravanti, A.; Giannitti, C.; Cheleschi, S.; Simpatico, A.; Pascarelli, N.A.; Galeazzi, M. Circulating levels of adiponectin, resistin, and visfatin after mud-bath therapy in patients with bilateral knee osteoarthritis. *Int. J. Biometeorol.* **2015**, *59*, 1691–1700. [CrossRef] [PubMed]

14. Liao, L.; Chen, Y.; Wang, W. The current progress in understanding the molecular functions and mechanisms of visfatin in osteoarthritis. *J. Bone Miner. Metab.* **2016**, *34*, 485–490. [CrossRef]

15. Calvet, J.; Orellana, C.; Gratacós, J.; Berenguer-Llergo, A.; Caixàs, A.; Chillarón, J.J.; Pedro-Botet, J.; Garcia-Manrique, M.; Navarro, N.; Larrosa, M. Synovial fluid adipokines are associated with clinical severity in knee osteoarthritis: A cross-sectional study in female patients with joint effusion. *Arthritis Res. Ther.* **2016**, *18*, 207. [CrossRef]

16. Zhang, Z.; Xing, X.; Hensley, G.; Chang, L.W.; Liao, W.; Abu-Amer, Y.; Sandell, L.J. Resistin induces expression of proinflammatory cytokines and chemokines in human articular chondrocytes via transcription and messenger RNA stabilization. *Arthritis Rheum.* **2010**, *62*, 1993–2003. [CrossRef]

17. Cheleschi, S.; Giordano, N.; Volpi, N.; Tenti, S.; Gallo, I.; Di Meglio, M.; Giannotti, S.; Fioravanta, A. A Complex Relationship between Visfatin and Resistin and microRNA: An In Vitro Study on Human Chondrocyte Cultures. *Int. J. Mol. Sci.* **2018**, *19*, 3909. [CrossRef]

18. Wu, M.H.; Tsai, C.H.; Huang, Y.L.; Fong, Y.C.; Tang, C.H. Visfatin Promotes IL-6 and TNF-α Production in Human Synovial Fibroblasts by Repressing miR-199a-5p through ERK, p38 and JNK Signaling Pathways. *Int. J. Mol. Sci.* **2018**, *19*, 190. [CrossRef]

19. Chen, W.C.; Wang, S.W.; Lin, C.Y.; Tsai, C.H.; Fong, Y.C.; Lin, T.Y.; Wang, S.L.; Huang, H.D.; Liao, K.W.; Tang, C.H. Resistin Enhances Monocyte Chemoattractant Protein-1 Production in Human Synovial Fibroblasts and Facilitates Monocyte Migration. *Cell Physiol. Biochem.* **2019**, *52*, 408–420. [CrossRef]

20. Gerin, I.; Bommer, G.T.; McCoin, C.S.; Sousa, K.M.; Krishnan, V.; MacDougald, O.A. (2010) Roles for miRNA-378/378* in adipocyte gene expression and lipogenesis. *Am. J. Physiol. Endocrinol. Metab.* **2010**, *299*, E198–E206. [CrossRef]

21. Maurizi, G.; Babini, L.; Della Guardia, L. Potential role of microRNAs in the regulation of adipocytes liposecretion and adipose tissue physiology. *J. Cell Physiol.* **2018**, *233*, 9077–9086. [CrossRef] [PubMed]

22. Malemud, C.J. MicroRNAs and Osteoarthritis. *Cells* **2018**, *7*, 92. [CrossRef] [PubMed]

23. Díaz-Prado, S.; Cicione, C.; Muiños-López, E.; Hermida-Gómez, T.; Oreiro, N.; Fernández-López, C.; Blanco, F.J. Characterization of microRNA expression profiles in normal and osteoarthritic human chondrocytes. *BMC Musculoskelet. Disord.* **2012**, *13*, 144. [CrossRef] [PubMed]

24. De Palma, A.; Cheleschi, S.; Pascarelli, N.A.; Tenti, S.; Galeazzi, M.; Fioravanti, A. Do MicroRNAs have a key epigenetic role in osteoarthritis and in mechanotransduction? *Clin. Exp. Rheumatol.* **2017**, *35*, 518–526.

25. Cheleschi, S.; De Palma, A.; Pecorelli, A.; Pascarelli, N.A.; Valacchi, G.; Belmonte, G.; Carta, S.; Galeazzi, M.; Fioravanti, A. Hydrostatic Pressure Regulates MicroRNA Expression Levels in Osteoarthritic Chondrocyte Cultures via the Wnt/β-Catenin Pathway. *Int. J. Mol. Sci.* **2017**, *18*, 133. [CrossRef]

26. Fathollahi, A.; Aslani, S.; Jamshidi, A.; Mahmoudi, M. Epigenetics in osteoarthritis: Novel spotlight. *J. Cell Physiol.* **2019**, *234*, 12309–12324. [CrossRef]

27. Cheleschi, S.; De Palma, A.; Pascarelli, N.A.; Giordano, N.; Galeazzi, M.; Tenti, S.; Fioravanti, A. Could Oxidative Stress Regulate the Expression of MicroRNA-146a and MicroRNA-34a in Human Osteoarthritic Chondrocyte Cultures? *Int. J. Mol. Sci.* **2017**, *18*, 2660. [CrossRef]

28. Bu, H.; Wedel, S.; Cavinato, M.; Jansen-Dürr, P. MicroRNA Regulation of Oxidative Stress-Induced Cellular Senescence. *Oxid. Med. Cell Longev.* **2017**, *2017*, 2398696. [CrossRef]

29. Machado, C.R.L.; Resende, G.G.; Macedo, R.B.V.; do Nascimento, V.C.; Branco, A.S.; Kakehasi, A.M.; Andrade, M.V. Fibroblast-like synoviocytes from fluid and synovial membrane from primary osteoarthritis demonstrate similar production of interleukin 6, and metalloproteinases 1 and 3. *Clin. Exp. Rheumatol.* **2019**, *37*, 306–309.

30. Ioan-Facsinay, A.; Kloppenburg, M. An emerging player in knee osteoarthritis: The infrapatellar fat pad. *Arthritis Res. Ther.* **2013**, *15*, 225. [CrossRef]

31. Tilg, H.; Moschen, A.R. Adipocytokines: Mediators linking adipose tissue, inflammation and immunity. *Nat. Rev. Immunol.* **2006**, *6*, 772–783. [CrossRef] [PubMed]

32. Francin, P.J.; Abot, A.; Guillaume, C.; Moulin, D.; Bianchi, A.; Gegout-Pottie, P.; Jouzeau, J.-Y.; Mainard, D.; Presle, D. Association between adiponectin and cartilage degradation in human osteoarthritis. *Osteoarthr. Cartilage* **2014**, *22*, 519–526. [CrossRef] [PubMed]

33. Su, Y.P.; Chen, C.N.; Chang, H.I.; Huang, K.C.; Cheng, C.C.; Chiu, F.Y.; Lee, K.C.; Lo, C.M.; Chang, S.F. Low Shear Stress Attenuates COX-2 Expression Induced by Resistin in Human Osteoarthritic Chondrocytes. *J. Cell Physiol.* **2017**, *232*, 1448–1457. [CrossRef] [PubMed]

34. Laiguillon, M.C.; Houard, X.; Bougault, C.; Gosset, M.; Nourissat, G.; Sautet, A.; Jacques, C.; Berenbaum, F.; Sellam, J. Expression and function of visfatin (Nampt), an adipokine-enzyme involved in inflammatory pathways of osteoarthritis. *Arthritis Res. Ther.* **2014**, *16*, R38. [CrossRef]

35. Wang, T.; He, C. Pro-inflammatory cytokines: The link between obesity and osteoarthritis. *Cytokine Growth F. R.* **2018**, *44*, 38–50. [CrossRef]

36. Pelletier, J.P.; McCollum, R.; Cloutier, J.M.; Martel-Pelletier, J. Synthesis of metalloproteases and interleukin 6 (IL-6) in human osteoarthritic synovial membrane is an IL-1 mediated process. *J. Rheumatol. Suppl.* **1995**, *43*, 109–114.

37. Altobelli, E.; Angeletti, P.M.; Piccolo, D.; De Angelis, R. Synovial Fluid and Serum Concentrations of Inflammatory Markers in Rheumatoid Arthritis, Psoriatic Arthritis and Osteoarthitis: A Systematic Review. *Curr. Rheumatol. Rev.* **2017**, *13*, 170–179. [CrossRef]

38. Sato, H.; Muraoka, S.; Kusunoki, N.; Masuoka, S.; Yamada, S.; Ogasawara, H.; Imai, T.; Akasaka, Y.; Tochigi, N.; Takahashi, H.; et al. Resistin upregulates chemokine production by fibroblast-like synoviocytes from patients with rheumatoid arthritis. *Arthritis Res. Ther.* **2017**, *19*, 263. [CrossRef]

39. Cawston, T.E.; Young, D.A. Proteinases involved in matrix turnover during cartilage and bone breakdown. *Cell Tissue Res.* **2010**, *339*, 221–235. [CrossRef]

40. Troeberg, L.; Nagase, H. Proteases involved in cartilage matrix degradation in osteoarthritis. *Biochim Biophys Acta* **2012**, *1824*, 133–145. [CrossRef]

41. Pérez-García, S.; Carrión, M.; Jimeno, R.; Ortiz, A.M.; González-Álvaro, I.; Fernández, J.; Gomariz, R.P.; Juarranz, Y. Urokinase plasminogen activator system in synovial fibroblasts from osteoarthritis patients: Modulation by inflammatory mediators and neuropeptides. *J. Mol. Neurosci.* **2014**, *52*, 18–27. [CrossRef] [PubMed]

42. Pérez-García, S.; Gutiérrez-Cañas, I.; Seoane, I.V.; Fernández, J.; Mellado, M.; Leceta, J.; Tío, L.; Villanueva-Romero, R.; Juarranz, Y.; Gomariz, R.P. Healthy and Osteoarthritic Synovial Fibroblasts Produce a Disintegrin and Metalloproteinase with Thrombospondin Motifs 4, 5, 7, and 12: Induction by IL-1β and Fibronectin and Contribution to Cartilage Damage. *Am. J. Pathol.* **2016**, *186*, 2449–2461. [CrossRef] [PubMed]

43. Goldring, M.B.; Otero, M. Inflammation in osteoarthritis. *Curr. Opin. Rheumatol.* **2011**, *23*, 471–478. [CrossRef] [PubMed]

44. Meier, F.M.; Frommer, K.W.; Peters, M.A.; Brentano, F.; Lefèvre, S.; Schröder, D.; Kyburz, D.; Steinmeyer, J.; Rehart, S.; Gay, S.; et al. Visfatin/pre-B-cell colony-enhancing factor (PBEF), a proinflammatory and cell motility-changing factor in rheumatoid arthritis. *J. Biol. Chem.* **2012**, *287*, 28378–28385. [CrossRef] [PubMed]

45. Hwang, H.S.; Kim, H.A. Chondrocyte Apoptosis in the Pathogenesis of Osteoarthritis. *Int. J. Mol. Sci.* **2015**, *16*, 26035–26054. [CrossRef] [PubMed]

46. Wang, N.; Lu, H.S.; Guan, Z.P.; Sun, T.Z.; Chen, Y.Y.; Ruan, G.R.; Chen, Z.K.; Jiang, J.; Bai, C.J. Involvement of PDCD5 in the regulation of apoptosis in fibroblast-like synoviocytes of rheumatoid arthritis. *Apoptosis* **2007**, *12*, 1433–1441. [CrossRef]

47. Feng, L.; Precht, P.; Balakir, R.; Horton, W.E., Jr. Evidence of a direct role for Bcl-2 in the regulation of articular chondrocyte apoptosis under the conditions of serum withdrawal and retinoic acid treatment. *J. Cell Biochem.* **1998**, *71*, 302–309. [CrossRef]

48. Jiao, Y.; Ding, H.; Huang, S.; Liu, Y.; Sun, X.; Wei, W.; Ma, J.; Zheng, F. Bcl-XL and Mcl-1 upregulation by calreticulin promotes apoptosis resistance of fibroblast-like synoviocytes via activation of PI3K/Akt and STAT3 pathways in rheumatoid arthritis. *Clin. Exp. Rheumatol.* **2018**, *36*, 841–849.

49. Sun, L.; Chen, S.; Gao, H.; Ren, L.; Song, G. Visfatin induces the apoptosis of endothelial progenitor cells via the induction of pro-inflammatory mediators through the NF-κB pathway. *Int. J. Mol. Med.* **2017**, *40*, 637–646. [CrossRef]

50. Marchev, A.S.; Dimitrova, P.A.; Burns, A.J.; Kostov, R.V.; Dinkova-Kostova, A.T.; Georgiev, M.I. Oxidative stress and chronic inflammation in osteoarthritis: Can NRF2 counteract these partners in crime? *Ann. N Y Acad Sci* **2017**, *1401*, 114–135. [CrossRef]

51. Marampon, F.; Codenotti, S.; Megiorni, F.; Del Fattore, A.; Camero, S.; Gravina, G.L.; Festuccia, C.; Musio, D.; Felice, F.D.; Nardone, V.; et al. NRF2 orchestrates the redox regulation induced by radiation therapy, sustaining embryonal and alveolar rhabdomyosarcoma cells radioresistance. *J. Cancer Res. Clin. Oncol.* **2019**, *145*, 881–893. [CrossRef]

52. Huang, M.L.; Chiang, S.; Kalinowski, D.S.; Bae, D.H.; Sahni, S.; Richardson, D.R. The Role of the Antioxidant Response in Mitochondrial Dysfunction in Degenerative Diseases: Cross-Talk between Antioxidant Defense, Autophagy, and Apoptosis. *Oxid. Med. Cell Longev.* **2019**, *2019*, 6392763. [CrossRef]

53. Raghuraman, G.; Zuniga, M.C.; Yuan, H.; Zhou, W. PKCε mediates resistin-induced NADPH oxidase activation and inflammation leading to smooth muscle cell dysfunction and intimal hyperplasia. *Atherosclerosis* **2016**, *253*, 29–37. [CrossRef]

54. Teixeira, T.M.; da Costa, D.C.; Resende, A.C.; Soulage, C.O.; Bezerra, F.F.; Daleprane, J.B. Activation of Nrf2-Antioxidant Signaling by 1,25-Dihydroxycholecalciferol Prevents Leptin-Induced Oxidative Stress and Inflammation in Human Endothelial Cells. *J. Nutr.* **2017**, *147*, 506–513. [CrossRef]

55. Lin, Y.T.; Chen, L.K.; Jian, D.Y.; Hsu, T.C.; Huang, W.C.; Kuan, T.T.; Wu, S.Y.; Kwok, C.F.; Ho, L.T.; Juan, C.C. Visfatin Promotes Monocyte Adhesion by Upregulating ICAM-1 and VCAM-1 Expression in Endothelial Cells via Activation of p38-PI3K-Akt Signaling and Subsequent ROS Production and IKK/NF-κB Activation. *Cell Physiol. Biochem.* **2019**, *52*, 1398–1411. [CrossRef]

56. Cheleschi, S.; Tenti, S.; Mondanelli, N.; Corallo, C.; Barbarino, M.; Giannotti, S.; Gallo, I.; Giordano, A.; Fioravanti, A. MicroRNA-34a and MicroRNA-181a Mediate Visfatin-Induced Apoptosis and Oxidative Stress via NF-κB Pathway in Human Osteoarthritic Chondrocytes. *Cells* **2019**, *8*, 874. [CrossRef]

57. Lin, Y.H. MicroRNA Networks Modulate Oxidative Stress in Cancer. *Int. J. Mol. Sci.* **2019**, *20*, 4497. [CrossRef]

58. Yan, S.; Wang, M.; Zhao, J.; Zhang, H.; Zhou, C.; Jin, L.; Zhang, Y.; Qiu, X.; Ma, B.; Fan, Q. MicroRNA-34a affects chondrocyte apoptosis and proliferation by targeting the SIRT1/p53 signaling pathway during the pathogenesis of osteoarthritis. *Int. J. Mol. Med.* **2016**, *38*, 201–209. [CrossRef]

59. Abouheif, M.M.; Nakasa, T.; Shibuya, H.; Niimoto, T.; Kongcharoensombat, W.; Ochi, M. Silencing microRNA-34a inhibits chondrocyte apoptosis in a rat osteoarthritis model in vitro. *Rheumatology* **2010**, *49*, 2054–2060. [CrossRef]

60. Niederer, F.; Trenkmann, M.; Ospelt, C.; Karouzakis, E.; Neidhart, M.; Stanczyk, J.; Kolling, C.; Gay, R.E.; Detmar, M.; Gay, S.; et al. Down-regulation of microRNA-34a* in rheumatoid arthritis synovial fibroblasts promotes apoptosis resistance. *Arthritis Rheum.* **2012**, *64*, 1771–1779. [CrossRef]

61. Zhong, X.; Li, P.; Li, J.; He, R.; Cheng, G.; Li, Y. Downregulation of microRNA-34a inhibits oxidized low-density lipoprotein-induced apoptosis and oxidative stress in human umbilical vein endothelial cells. *Int. J. Mol. Med.* **2018**, *42*, 1134–1144. [CrossRef]

62. Okuhara, A.; Nakasa, T.; Shibuya, H.; Niimoto, T.; Adachi, N.; Deie, M.; Ochi, M. Changes in microRNA expression in peripheral mononuclear cells according to the progression of osteoarthritis. *Mod. Rheumatol.* **2012**, *22*, 446–457. [CrossRef]

63. Zheng, H.; Liu, J.; Tycksen, E.; Nunley, R.; McAlinden, A. MicroRNA-181a/b-1 over-expression enhances osteogenesis by modulating PTEN/PI3K/AKT signaling and mitochondrial metabolism. *Bone* **2019**, *123*, 92–102. [CrossRef]

64. Chen, K.L.; Fu, Y.Y.; Shi, M.Y.; Li, H.X. Down-regulation of miR-181a can reduce heat stress damage in PBMCs of Holstein cows. *In Vitro Cell Dev. Biol. Anim.* **2016**, *52*, 864–871. [CrossRef]

65. Feng, X.; Zhang, C.; Yang, Y.; Hou, D.; Zhu, A. Role of miR-181a in the process of apoptosis of multiple malignant tumors: A literature review. *Adv. Clin. Exp. Med.* **2018**, *27*, 263–270. [CrossRef]

66. Yamasaki, K.; Nakasa, T.; Miyaki, S.; Ishikawa, M.; Deie, M.; Adachi, N.; Yasunaga, Y.; Asahara, H.; Ochi, M. Expression of microRNA-146a in osteoarthritis cartilage. *Arthritis Rheumatol.* **2009**, *60*, 1035–1041. [CrossRef]

67. Li, L.; Chen, X.P.; Li, Y.J. MicroRNA-146a and human disease. *Scand. J. Immunol.* **2010**, *71*, 227–231. [CrossRef]

68. Xie, Y.; Chu, A.; Feng, Y.; Chen, L.; Shao, Y.; Luo, Q.; Deng, X.; Wu, M.; Shi, X.; Chen, Y. MicroRNA-146a: A Comprehensive Indicator of Inflammation and Oxidative Stress Status Induced in the Brain of Chronic T2DM Rats. *Front. Pharmacol.* **2018**, *9*, 478. [CrossRef]

69. Li, J.; Huang, J.; Dai, L.; Yu, D.; Chen, Q.; Zhang, X.; Dai, K. miR-146a, an IL-1β responsive miRNA, induces vascular endothelial growth factor and chondrocyte apoptosis by targeting Smad4. *Arthritis Res. Ther.* **2012**, *14*, R75. [CrossRef]

70. Wang, J.H.; Shih, K.S.; Wu, Y.W.; Wang, A.W.; Yang, C.R. Histone deacetylase inhibitors increase microRNA-146a expression and enhance negative regulation of interleukin-1β signaling in osteoarthritis fibroblast-like synoviocytes. *Osteoarthr. Cartilage* **2013**, *21*, 1987–1996. [CrossRef]

71. Wen, F.; Li, B.; Huang, C.; Wei, Z.; Zhou, Y.; Liu, J.; Zhang, H. MiR-34a is Involved in the Decrease of ATP Contents Induced by Resistin Through Target on ATP5S in HepG2 Cells. *Biochem. Genet.* **2015**, *53*, 301–309. [CrossRef]

72. Gong, Y.Y.; Luo, J.Y.; Wang, L.; Huang, Y. MicroRNAs regulating reactive oxygen species in cardiovascular diseases. *Antioxid. Redox Signal.* **2018**, *29*, 1092–1107. [CrossRef]

73. D'Adamo, S.; Cetrullo, S.; Guidotti, S.; Borzì, R.M.; Flamigni, F. Hydroxytyrosol modulates the levels of microRNA-9 and its target sirtuin-1 thereby counteracting oxidative stress-induced chondrocyte death. *Osteoarthr. Cartilage* **2017**, *25*, 600–610. [CrossRef]

74. Ji, G.; Lv, K.; Chen, H.; Wang, T.; Wang, Y.; Zhao, D.; Qu, L.; Li, Y. MiR-146a regulates SOD2 expression in H2O2 stimulated PC12 cells. *PLoS ONE* **2013**, *8*, e69351. [CrossRef]

75. Wang, L.; Huang, H.; Fan, Y.; Kong, B.; Hu, H.; Hu, K.; Guo, J.; Mei, Y.; Liu, W.L. Effects of downregulation of microRNA-181a on H2O2-induced H9c2 cell apoptosis via the mitochondrial apoptotic pathway. *Oxid. Med. Cell Longev.* **2014**, *2014*, 960362. [CrossRef]

76. Baker, J.R.; Vuppusetty, C.; Colley, T.; Papaioannou, A.I.; Fenwick, P.; Donnelly, L.; Ito, K.; Barnes, P.J. Oxidative stress dependent microRNA-34a activation via PI3Kα reduces the expression of sirtuin-1 and sirtuin-6 in epithelial cells. *Sci. Rep.* **2016**, *6*, 35871. [CrossRef]

77. Huang, X.; Gao, Y.; Qin, J.; Lu, S. The role of miR-34a in the hepatoprotective effect of hydrogen sulfide on ischemia/reperfusion injury in young and old rats. *PLoS ONE* **2014**, *9*, e113305. [CrossRef]

78. Rigoglou, S.; Papavassiliou, A.G. The NF-κB signaling pathway in osteoarthritis. *Int. J. Biochem. Cell Biol.* **2013**, *45*, 2580–2584. [CrossRef]

79. Zhang, Q.; Lenardo, M.J.; Baltimore, D. 30 Years of NF-κB: A blossoming of relevance to human pathobiology. *Cell* **2017**, *168*, 37–57. [CrossRef]

80. Zhang, Z.; Zhang, Z.; Kang, Y.; Hou, C.; Duan, X.; Sheng, P.; Sandell, L.J.; Liao, W. Resistin stimulates expression of chemokine genes in chondrocytes via combinatorial regulation of C/EBPβ and NF-κB. *Int. J. Mol. Sci.* **2014**, *15*, 17242–17255. [CrossRef]

81. Aslani, M.R.; Keyhanmanesh, R.; Alipour, M.R. Increased Visfatin Expression Is Associated with Nuclear Factor-κB in Obese Ovalbumin-Sensitized Male Wistar Rat Tracheae. *Med. Princ. Pract.* **2017**, *26*, 351–358. [CrossRef]

82. Xu, B.; Li, Y.Y.; Ma, J.; Pei, F.X. Roles of microRNA and signaling pathway in osteoarthritis pathogenesis. *J. Zhejiang Univ. Sci. B* **2016**, *17*, 200–208. [CrossRef]

83. Altman, R.; Asch, E.; Bloch, D.; Bole, G.; Borenstein, D.; Brandt, K.; Christy, W.; Cooke, T.D.; Greenwald, R.; Hochberg, M.; et al. Development of criteria for the classification and reporting of osteoarthritis. Classification of osteoarthritis of the knee. Diagnostic and Therapeutic Criteria Committee of the American Rheumatism Association. *Arthritis Rheum.* **1986**, *29*, 1039–1049. [CrossRef]

84. Gosset, M.; Berenbaum, F.; Salvat, C.; Sautet, A.; Pigenet, A.; Tahiri, K.; Jacques, C. Crucial role of visfatin/pre-B cell colony-enhancing factor in matrix degradation and prostaglandin E2 synthesis in chondrocytes: Possible influence on osteoarthritis. *Arthritis Rheum.* **2008**, *58*, 1399–1409. [CrossRef]

85. Ramakers, C.; Ruijter, J.M.; Deprez, R.H.; Moorman, A.F. Assumption-free analysis of quantitative real-time polymerase chain reaction (PCR) data. *Neurosci. Lett* **2003**, *339*, 62–66. [CrossRef]

86. Pfaffl, M.W. A new mathematical model for relative quantification in real RT-PCR. *Nucleic Acid Res.* **2001**, *29*, e45. [CrossRef]

87. Vandesompele, J.; de Preter, K.; Pattyn, F.; Poppe, B.; van Roy, N.; de Paepe, A.; Speleman, F. Accurate normalization of real-time quantitative RT-PCR data by geometric averaging of multiple internal control genes. *Genome Biol.* **2002**, *3*, research0034.1. [CrossRef]

88. Cheleschi, S.; Calamia, V.; Fernandez-Moreno, M.; Biava, M.; Giordani, A.; Fioravanti, A.; Anzini, M.; Blanco, F. In vitro comprehensive analysis of VA692 a new chemical entity for the treatment of osteoarthritis. *Int. Immunopharmacol.* **2018**, *64*, 86–100. [CrossRef]

89. Cheleschi, S.; Fioravanti, A.; De Palma, A.; Corallo, C.; Franci, D.; Volpi, N.; Bedogni, G.; Giannotti, S.; Giordano, N. Methylsulfonylmethane and mobilee prevent negative effect of IL-1β in human chondrocyte cultures via NF-κB signaling pathway. *Int. Immunopharmacol.* **2018**, *65*, 129–139. [CrossRef]

International Journal of
Molecular Sciences

Article

miR526b and miR655 Induce Oxidative Stress in Breast Cancer

Bonita Shin [†], Riley Feser [†], Braydon Nault [‡], Stephanie Hunter [‡], Sujit Maiti,
Kingsley Chukwunonso Ugwuagbo and Mousumi Majumder *

Department of Biology, Brandon University, 3rd Floor, John R. Brodie Science Centre, 270—18th Street, Brandon,
MB R7A6A9, Canada
* Correspondence: majumderm@brandonu.ca
† These authors contributed equally to this work.
‡ These authors contributed equally to this work.

Received: 23 July 2019; Accepted: 15 August 2019; Published: 19 August 2019

Abstract: In eukaryotes, overproduction of reactive oxygen species (ROS) causes oxidative stress, which contributes to chronic inflammation and cancer. MicroRNAs (miRNAs) are small, endogenously produced RNAs that play a major role in cancer progression. We established that overexpression of miR526b/miR655 promotes aggressive breast cancer phenotypes. Here, we investigated the roles of miR526b/miR655 in oxidative stress in breast cancer using in vitro and in silico assays. miRNA-overexpression in MCF7 cells directly enhances ROS and superoxide (SO) production, detected with fluorescence assays. We found that cell-free conditioned media contain extracellular miR526b/miR655 and treatment with these miRNA-conditioned media causes overproduction of ROS/SO in MCF7 and primary cells (HUVECs). Thioredoxin Reductase 1 (TXNRD1) is an oxidoreductase that maintains ROS/SO concentration. Overexpression of *TXNRD1* is associated with breast cancer progression. We observed that miR526b/miR655 overexpression upregulates *TXNRD1* expression in MCF7 cells, and treatment with miRNA-conditioned media upregulates *TXNRD1* in both MCF7 and HUVECs. Bioinformatic analysis identifies two negative regulators of TXNRD1, TCF21 and PBRM1, as direct targets of miR526b/miR655. We validated that *TCF21* and *PBRM1* were significantly downregulated with miRNA upregulation, establishing a link between miR526b/miR655 and TXNRD1. Finally, treatments with oxidative stress inducers such as H_2O_2 or miRNA-conditioned media showed an upregulation of miR526b/miR655 expression in MCF7 cells, indicating that oxidative stress also induces miRNA overexpression. This study establishes the dynamic functions of miR526b/miR655 in oxidative stress induction in breast cancer.

Keywords: MicroRNA (miRNA); miR526b; miR655; oxidative stress; reactive oxygen species (ROS); superoxide (SO); Thioredoxin Reductase 1 (TXNRD1); breast cancer

1. Introduction

Breast cancer is the most common cancer affecting women and is responsible for the highest number of cancer-related deaths among women worldwide [1]. Breast cancer progression follows a complex multistep process, which depends on multiple exogenous and endogenous factors. The production of reactive oxygen species (ROS) such as superoxide (SO) leads to the induction of oxidative stress, which has been largely associated with breast cancer [2]. Oxidative stress is the result of cellular inability to neutralize and eliminate excess ROS, which is frequently associated with cancer development and progression. Under normal physiological conditions, cells endogenously produce ROS such as H_2O_2, $ONOO^-$, OH^-, $HClO^-$, NO^-, ROO^-, and SO, during metabolism, respiration, and biosynthesis of macromolecules. Thus, cell metabolites are great resources for understanding oxidative stress. Excessive ROS production can induce inflammation, regulate the cell cycle, and stimulate intracellular transduction pathways, which leads to the promotion of cancer [3]. Specifically, SO production is the consequence of oxygen (O_2) acting as the final electron acceptor in the electron transport chain, and has been shown to regulate signaling cascades that lead to cell survival and proliferation [4]. Within the cell, there is a homeostatic balance of various protective molecules and ROS. However, in cancer, tumor cells demonstrate deviations in oxidative metabolism and signaling pathways as a result of the constitutive activation of growth signaling pathways, leading to increased levels of ROS and induction of oxidative stress [5].

A high concentration of ROS is a signature feature of the tumor microenvironment. Cells have a natural defense mechanism to reduce damage caused by oxidative stress. Antioxidants, which are stable molecules that donate electrons to neutralize free radicals, belong to this natural defense mechanism of the cell [6]. Cellular detoxification pathways are regulated by enzymes that eliminate ROS, which include SO dismutase, catalase, glutathione peroxidase, cysteine, and thioredoxin (TXN). Specifically, TXN is a ubiquitous antioxidant protein that is responsible for the regulation of dithiol/disulfide balance [7,8]. TXN is active when it is in its reduced form. When active, it will participate in a reaction catalyzed by peroxiredoxin to neutralize H_2O_2 and peroxynitrate, both of which are products of oxidative stress activity [9]. TXNRD1 is responsible for the conversion of TXN into its active state (Figure 1). Malfunctions in antioxidant pathways can lead to increased oxidative stress and consequential damage to the cells. High expression of *TXNRD1* is associated with increased oxidative stress and correlates with poor prognosis in breast cancer [10]. In cancers, excessive production of ROS can cause mutations in the DNA, overexpression of tumor-promoting microRNAs (miRNAs, miRs), release of inflammatory molecules, and inactivation of oxidoreductive enzymes; making antioxidant pathways dysfunctional. Overexpression of oncogenic miRNAs leads to the regulation and promotion of tumor growth; however, the regulation of oxidative stress in cancer by miRNAs remains unclear.

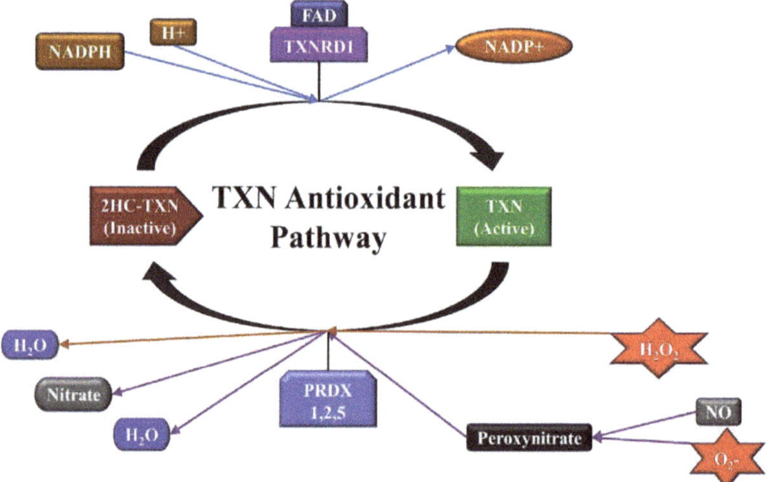

Figure 1. Thioredoxin (TXN) is a main constituent in an antioxidant pathway that neutralizes Hydrogen Peroxide (H_2O_2) and superoxide (O_2-), to prevent oxidative damage. TXN exists in active (reduced) and inactive (oxidized) states. Thioredoxin Reductase 1 (TXNRD1) is responsible for reducing 2HC-TXN (TXN with attached double hydrocarbon) into its active form. Therefore, in the presence of more ROS, an increased expression of *TXNRD1* occurs to protect the cells from oxidative damage.

miRNAs are small, endogenously produced RNAs which regulate gene expression at the post-transcriptional level [11]. Release of circulating miRNAs in the tumor microenvironment can regulate tumor growth and metastasis. Previously, miR526b and miR655 have been established as oncogenic and tumor-promoting miRNAs in human breast cancer [12–14]. The roles of miR526b and miR655 have been implicated in many hallmarks of cancer, including: Driving primary tumor growth, induction of stem-like cell (SLC) phenotypes, epithelial-to-mesenchymal transition (EMT), invasion and migration, distant metastasis. We have shown that cell metabolites and cell-free conditioned media of these two miRNA-high cells induce tumor-associated angiogenesis and lymphangiogenesis in breast cancer [15]. It has also been shown that cellular stress and ROS production can also induce oncogenic miRNA expression in tumors, and it is well-established that both ROS and miRNA expression signatures are associated with tumor development, progression, metastasis, and therapeutic response [16]. Thus, we wanted to investigate the relationship between ROS and miR526b/miR655 in breast cancer.

In this study, we investigate the roles of oncogenic miR526b and miR655 in oxidative stress in breast cancer. First, we show that both miR526b/miR655 directly and indirectly regulate oxidative stress. Next, we use the expression of *TXNRD1* as a molecular marker of oxidative stress to further validate the link between miRNA and ROS production. Moreover, we identify a positive feedback loop between oxidative stress and miRNA expression in breast cancer, showing that while the upregulation of miR526b and miR655 led to the induction of ROS production, the induction of oxidative stress also further upregulated miR526b and miR655 expression in breast tumor cells. Hence, we establish the dynamic roles of miR526b and miR655 in oxidative stress in breast cancer.

2. Results

To test the effects of miR526b and miR655 in oxidative stress in breast cancer, we used an estrogen receptor (ER)-positive, poorly metastatic breast cancer cell line, MCF7, and highly aggressive, miR526b/miR655-overexpressing MCF7-miR526b and MCF7-miR655 cell lines. We also used a primary endothelial cell line, human umbilical vein endothelial cells (HUVEC), to test the indirect or paracrine effects of miR526b and miR655 on oxidative stress induction. Finally, we used a breast epithelial cell line MCF10A and breast cancer cell lines T47D, MCF7, SKBR3, MCF7-COX2, Hs578T, and MDA-MB-231 to measure *TXNRD1* expression.

2.1. miR526b and miR655 Directly Induce Oxidative Stress by Overproduction of ROS and SO

2.1.1. Fluorescence Microplate Assay

Previously, studies have used a total ROS detection kit for the measurement of ROS and SO in triple negative breast cancer cell lines, colon cancer cells, colorectal cancer cell lines, and in hepatocellular carcinoma cells [17–21]. We used the same ROS-ID Total ROS/SO detection kit (Enzo Life Sciences, Farmingdale, NY, USA) to measure fluorescence due to ROS/SO production following manufacturer's protocol. Microplate readings were carried out at 1 and 21 h following Pyocyanin (ROS inducer) treatment and addition of non-fluorescent, cell-permeable ROS detection dyes. We monitored cellular morphology at various time points from 1–24 h after the addition of the ROS inducer in MCF7 cells (data not shown). With minimum dosage of ROS inducer, we observed oxidative stress in the cell within an hour, and after 21 h a decrease in cell viability was recorded due to the toxicity of the ROS inducer. Therefore, fluorescence was measured at two different timepoints; at 1 and 21 h. Fluorescence emissions were captured using two different filters to detect green (Fluorescein) and red (Rhodamine) emissions. ROS/SO production was calculated by subtracting the negative control emissions (basal emissions) from the test group emissions (with treatment) (Figure S1A). Overall, we found that ROS and SO production was greater in miRNA-high cells compared to MCF7 cells. Specifically, ROS production was found to be statistically significant at 21 h for MCF7-miR655 (Figure 2A). Similarly, SO production was found to be significantly greater in the MCF7-miR655 cell line compared to MCF7 at both 1 h and 21 h. ROS and SO production was not statistically significant for MCF7-miR526b compared to MCF7 (Figure 2A,B).

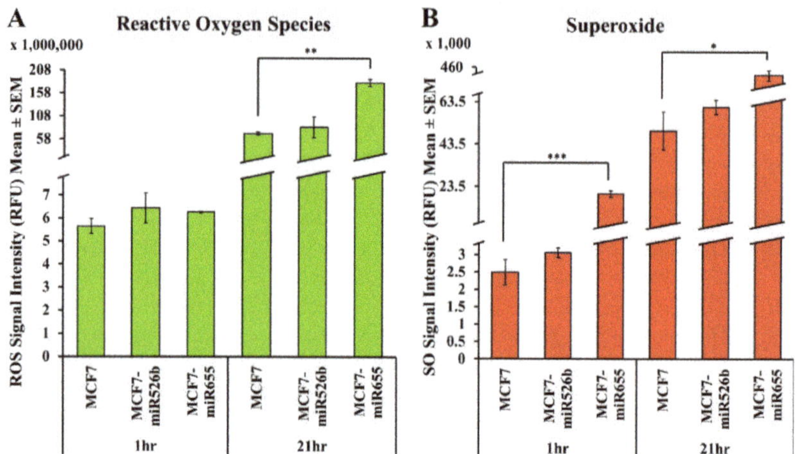

Figure 2. Fluorescence microplate assays to quantify ROS (Green) and SO (Red) production by MCF7, MCF7-miR526b, and MCF7-miR655 cell lines. (**A**) Quantitative data represents the ROS signal intensity in MCF7, MCF7-miR526b, and MCF7-miR655 cell lines at 1 and 21 h. (**B**) Quantitative data represents SO signal intensity in MCF7, MCF7-miR526b, and MCF7-miR655 cell lines at the 1 and 21 h. Data presented as the mean ± SEM of triplicate replicates; * $p < 0.05$, ** $p < 0.01$, *** $p < 0.001$.

2.1.2. Fluorescence Microscopy Assay

Fluorescence microscopy assays were conducted to measure the difference in cellular fluorescence expression with individual fluorescent cell quantification, determining the fraction of cells producing ROS and SO. Using the green (Fluorescein) and red (Rhodamine) fluorescence filter sets, photos of the fluorescent cells were captured with an inverted fluorescence microscope 1 h after the detection dyes were added. We also captured bright field images of cells without using fluorescence filters to quantify total number of viable cells (Figure S2M,R,W). Results show that wells containing MCF7-miR526b (Figure 3D,E) or MCF7-miR655 cell lines (Figure 3G,H) had more fluorescing cells than MCF7 (Figure 3A,B) under both red and green filters. Similarly, quantifications show significantly higher green (Figure 3J) and red (Figure 3K) cells in both MCF7-miR526b and MCF7-MCF7-miR655-high cells compared to MCF7 cells. Furthermore, we measured the ratio of cells positive for both ROS and SO production using merged channels and found that miRNA-high cell lines (Figure 3F,I) had a significantly higher ratio of fluorescing cells under both filters compared to MCF7 cells (Figure 3C,L).

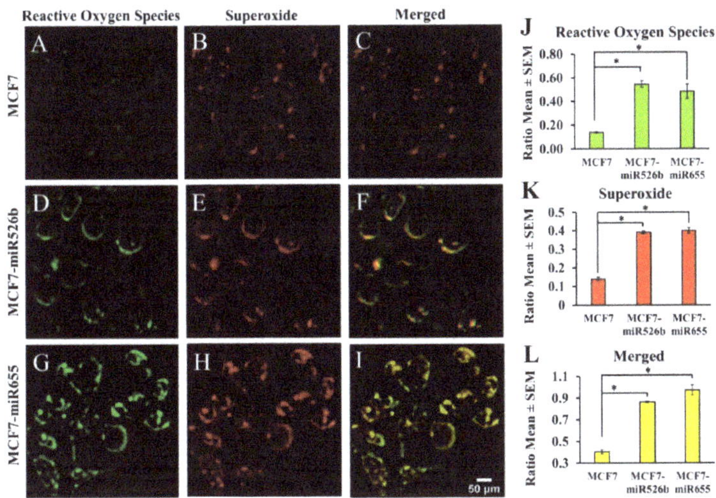

Figure 3. Fluorescence microscopy images and quantification of ROS/SO production in MCF7, MCF7-miR526b, and MCF7-miR655 cell lines. (**A–C**) Images of fluorescent MCF7 cells with green, red, or merged filters. (**D–F**) Images of fluorescent MCF7-miR526b cells with green, red, or merged filters. (**G–I**) Images of fluorescent MCF7-miR655 cells with green, red, or merged filters. Scale bar: 50 µm. (**J**) Quantification of ratios of cells positive for ROS detection. (**K**) Quantification ratios of cells positive for SO detection. (**L**) Quantification ratios of cells showing both ROS and SO production. Quantitative data presented as the mean ± SEM of triplicate replicates. Quantifications presented in ratios of fluorescence-positive cells to the total number of cells; * $p < 0.05$.

2.2. Cell-Free Conditioned Media from miR526b/miR655-High Cells Indirectly Induce Production of ROS and SO

The tumor microenvironment is very heterogeneous, containing tumor cells, endothelial cells, macrophages, miRNAs, cell metabolites, inflammatory molecules, growth factors, and also ROS. In the following assays, we first tested the paracrine effect of miRNA in oxidative stress. To test the paracrine effect of miRNA, we used the cell-free conditioned media from miR526b/miR655-high cells as an ROS inducer using MCF7 (tumor model) and HUVEC (primary endothelial model) cell lines. Next, we quantified pri-miR526b and pri-miR655 in the conditioned media to investigate if the indirect induction of oxidative stress in breast cancer is due to the presence of miR526b and miR655 in the cell secretions, and to justify our use of conditioned media as an ROS inducer.

2.2.1. Fluorescence Microplate Assay with MCF7 Cells

MCF7 cells were grown and then treated with basal media or cell-free conditioned media (containing cell metabolites and secretory proteins) collected from MCF7-miR526b and MCF7-miR655 cells for 24 h. Then we added the ROS inducer as described earlier and fluorescence data were collected at 1 and 21 h. These two time points were selected to remain consistent with our previous experiments that used the ROS inducer. MCF7 cells treated with miRNA-conditioned media showed significantly higher ROS production than the basal media treated MCF7 control group at both 1 and 21 h. Specifically, the change is extremely significant for MCF7 cells treated with MCF7-miR526b conditioned media at 1 h, and with MCF7-miR655 conditioned media at 21 h (Figure 4A). Similarly, MCF7 cells treated with MCF7-miR655 cell-free conditioned media had significantly higher SO production than the basal media treated cells at 1 h (Figure 4B). At 21 h, MCF7 cells treated with MCF7-miR526b cell-free conditioned media had a significantly higher SO production than MCF7 cells treated with basal media. While MCF7 cells treated with MCF7-miR655 cell-free conditioned media did show slightly higher SO production than MCF7 treated with basal media, this was not statistically significant (Figure 4B).

MCF7 Cells in Various Conditioned Media

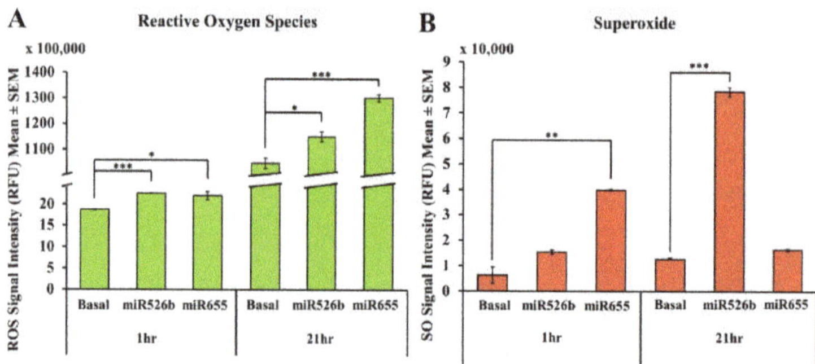

HUVECs in Various Conditioned Media

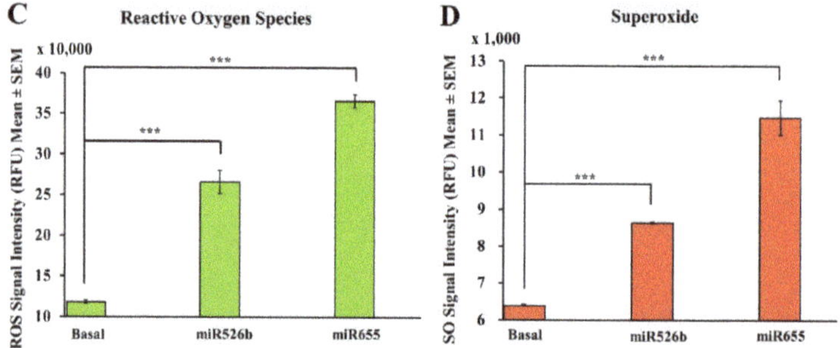

Figure 4. Fluorescence microplate assay with MCF7 and HUVEC cells cultured in miRNA conditioned media. (**A**) MCF7 cells treated with MCF7-miR526b or MCF7-miR655 conditioned media show an overproduction of ROS as compared to basal media treated cells at both 1 and 21 h. (**B**) MCF7 cells treated with MCF7-miR655 conditioned media show a significant overproduction of SO at 1 h, and MCF7 cells treated with MCF7-miR526b conditioned media show a significant overproduction of SO at 21 h compared to MCF7 cells treated with basal media. (**C**) HUVECs treated with MCF7-miR526b or MCF7-miR655 conditioned media show overproduction of ROS compared to HUVECs treated with basal media after 30 min. (**D**) HUVECs treated with MCF7-miR526b or MCF7-miR655 conditioned media show a significant overproduction of SO as compared to non-treated MCF7 cells after 30 min. Data presented as the mean ± SEM of triplicate replicates; * $p < 0.05$, ** $p < 0.01$, *** $p < 0.001$.

2.2.2. Fluorescence Microplate Assay with HUVECs

Previously, we have shown that cell-free conditioned media from miRNA-high cells induce angiogenic potential in HUVECs [15]. Here, we tested if cell-free conditioned media containing all secretory proteins and metabolites from miR526b/miR655-high cells can induce oxidative stress in HUVECs. HUVECs treated with MCF7-miR526b or MCF7-miR655 conditioned media for 12–18 h had significantly higher ROS/SO production compared to HUVECs treated with basal media (Figure 4C,D). It should be noted that HUVECs are very sensitive to changes in growth conditions and treatments, as they can only survive for 12–18 h without native growth condition. Thus, HUVECs were treated with conditioned media from miRNA-high cells for 12 h. We found that HUVECs were extremely stressed, observing cell death after an hour following the addition of the ROS inducer (Figure S4). Therefore, the microplate assay was done only 30 min after ROS inducer was added.

2.2.3. Fluorescence Microscopy Assay with MCF7 Cells in miRNA- Conditioned Media

In this experiment, cell-free conditioned media was used as an inducer of oxidative stress. MCF7 cells were grown and treated with basal media or cell-free conditioned media from MCF7-miR526b or MCF7-miR655 cells for 12–18 h. No other ROS inducer was added, only cell-permeable dyes from the ROS detection kit were added to detect cell-free conditioned media-induced oxidative stress. Images were captured after 1 h, and the number of fluorescent cells were measured with ImageJ as mentioned above. Results show that MCF7 cells treated with MCF7-miR526b (Figure 5D–F) or MCF7-miR655 conditioned media (Figure 5G–I) had more fluorescing cells than basal media treated MCF7 cells (Figure 5A–C) for both Fluorescein and Rhodamine filters. Quantification of MCF7 cells treated with MCF7-miR526b or MCF7-miR655 conditioned media show a significant increase in ROS production (Figure 5J) and SO production (Figure 5K). The ratio of cells positive for both ROS and SO production was also significantly higher in cells treated with MCF7-miR526b or MCF7-miR655 conditioned media than those treated with basal media (Figure 5L).

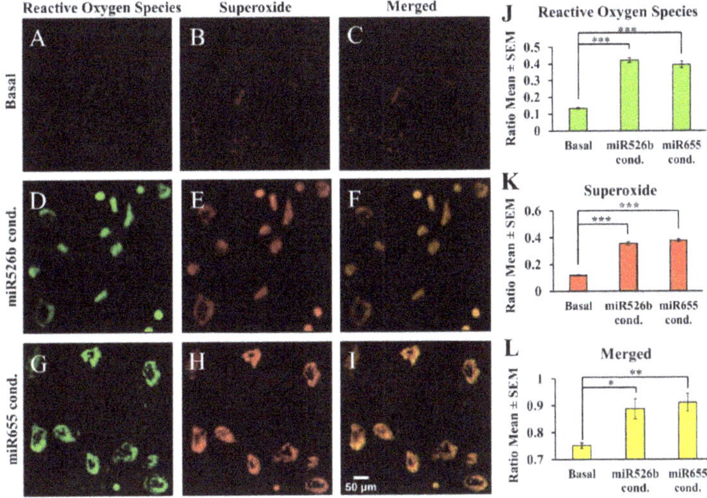

Figure 5. Fluorescence microscopy with MCF7 cell line treated with basal, MCF7-miR526b, or MCF7-miR655 cell-free conditioned media to quantify ROS/SO producing cells. MCF7 treated with basal media under the Rhodamine filter was used as a threshold to quantify ROS positive cells. (**A–C**) Images of MCF7 cells treated with basal media in green, red, or merged filters. (**D–F**) Images of MCF7 cells treated with cell-free conditioned media from MCF7-miR526b cells in green, red, or merged filters. (**G–I**) Images of MCF7 cells treated with cell-free conditioned media from MCF7-miR655 cells in green, red, or merged filters. Scale bar: 50 µm. (**J**) Quantification of cells positive for ROS detection presented as ratios. (**K**) Quantification of cells positive for SO detection presented as ratios. (**L**) Ratio of cells showing both ROS and SO production. Quantitative data presented as the mean ± SEM of quadruplicate replicates; * $p < 0.05$, ** $p < 0.01$, *** $p < 0.001$.

2.2.4. miRNA-High Cells Release miR526b and miR655 in Cell-Free Conditioned Media

To test if the indirect induction of oxidative stress with conditioned media is due to the presence of miRNA itself, we measured pri-miR526b and pri-miR655 expression in MCF7, MCF7-miR526, and MCF7-miR655 cell-free conditioned media. We found that both pri-miRNAs' expressions were significantly higher in MCF7-miR526b conditioned media compared to MCF7 conditioned media (Figure 6). The expression of pri-miR526b was significantly higher and the expression of pri-miR655 was marginally higher in MCF7-miR655 conditioned media. It should be noted that in the MCF7-miR526b conditioned media, the overall expression of pri-miR526b was higher than pri-miR655, while in

MCF7-miR655 conditioned media, the overall expression of pri-miR655 was higher than pri-miR526b (Figure 6). This result confirms that due to the release of miRNA in the conditioned media of serum starved cells, extracellular miR526b and miR655 act as an ROS inducer, therefore indirectly inducing oxidative stress in nearby cells.

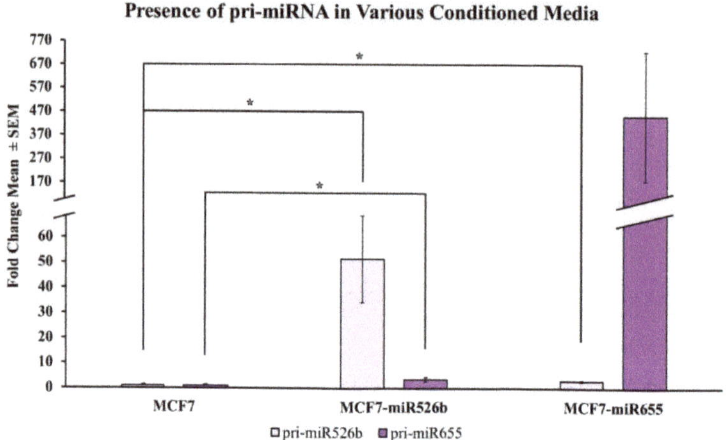

Figure 6. Expression of pri-miR526b and pri-miR655 in various conditioned media measured using qRT-PCR. MCF7-miR526b conditioned media show a significantly higher expression of both pri-miRNAs with prominent change in pri-miR526b expression compared to MCF7 conditioned media. MCF7-miR655 conditioned media show a significantly higher expression of miR526b, and very high expression of pri-miR655, which was not significant. Data is presented as the mean ± SEM of duplicate replicates; * $p < 0.05$.

2.3. TXNRD1 is a Marker for Oxidative Stress

TXN is an antioxidant protein that is responsible for neutralizing ROS within the cell [8]. TXNRD1 is the enzyme responsible for reducing TXN into its active form. Previous analyses of *TXNRD1* expression have shown that *TXNRD1* is upregulated in pancreatic, colon, lung, prostate, and breast cancers, and is associated with poor cancer prognosis [10]. To further investigate the direct and indirect roles of miR526b and miR655 in the induction of oxidative stress, *TXNRD1* was validated as a marker of oxidative stress using various breast cancer cell lines and its expression was measured in miR526b/miR655-high cell lines. Furthermore, bioinformatic analysis was done to investigate the regulation of *TXNRD1* by miR526b and miR655, which showed that miR526b and miR655 target two transcription factors that regulate *TXNRD1* expression. The expression of these transcription factors was then measured in miR526b/miR655-high cell lines. Moreover, with the success of using miRNA-conditioned media as an ROS inducer in our previous assays, we tested to see if cell-free conditioned media from miR526b/miR655-high cells regulate *TXNRD1* expression in both tumor and endothelial cells.

2.3.1. Highly Metastatic Breast Cancer Cell Lines Show Upregulation of TXNRD1

MCF10A, T47D, MCF7, SKBR3, MCF7-COX2, Hs578T, and MDA-MB-231 cell lines were used to quantify the expression of *TXNRD1* using qRT-PCR. Since MCF10A is a breast epithelial cell line, gene expression changes for all breast cancer cell lines were measured and compared to MCF10A. Results show that *TXNRD1* was significantly downregulated in the poorly metastatic MCF7 and SKBR3 cell lines, while the T47D cell line showed no change in expression (Figure 7A). *TXNRD1* was significantly upregulated in all highly metastatic cell lines, MCF7-COX2, Hs578T, and MDA-MB-231; with maximum upregulation seen in MDA-MB-231 (Figure 7A). We have previously found that these aggressive breast cancer cell lines (MCF7-COX2, Hs578T, and MDA-MB-231) show overexpression of both miR526b and miR655; while poorly metastatic cells (MCF7, T47D) show low expression of both miRNAs [12,13]. These observations validate the use of *TXNRD1* as a marker of oxidative stress in breast cancer, and show a link between *TXNRD1*, miR526b, and miR655 expression.

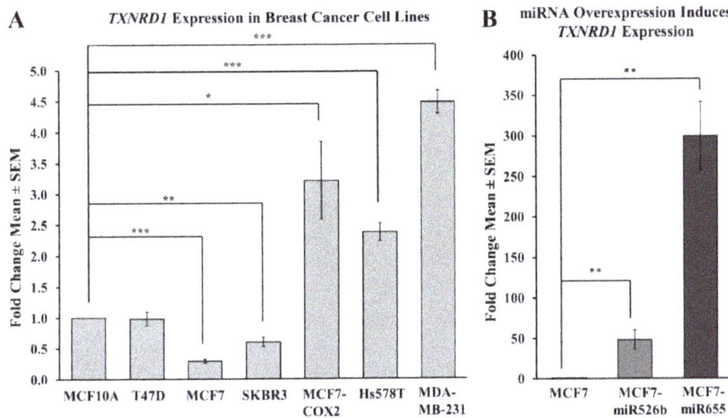

Figure 7. Expression of oxidative stress marker *TXNRD1* in various cell lines measured using qRT-PCR. (**A**) Breast cancer cell lines with various degrees of metastatic potential show a difference in *TXNRD1* expression. The more metastatic cell lines including MCF7-COX2, Hs578T, and MDA-MB-231 show the greatest fold change of *TXNRD1* expression and MCF7 cells showing lowest *TXNRD1* expression compared to the breast epithelial MCF10A cell line. (**B**) Expression of *TXNRD1* is quantified in MCF7 cells, MCF7-miR526b, and MCF7-miR655 cell lines, showing how these oncogenic miRNAs impact the expression of this oxidative stress marker. Large fold change increases are seen in both miRNA cell lines. Data is presented as the mean ± SEM of triplicate replicates; * $p < 0.05$, ** $p < 0.01$, *** $p < 0.001$.

2.3.2. miRNA Overexpression Directly Upregulates TXNRD1 Expression

To establish the direct role of miRNA in oxidative stress, total RNA extraction followed by qRT-PCR was carried out with MCF7, MCF7-miR526b, and MCF7-miR655 cell lines to quantify the expression of *TXNRD1*. Results show that *TXNRD1* was significantly upregulated in both MCF7-miR526b and MCF7-miR655 cell lines compared to MCF7, with greater fold change in *TXNRD1* expression measured in the MCF7-miR655 cell line (Figure 7B).

2.3.3. Bioinformatic Analysis to Identify a Link between miRNAs and TXNRD1

Since we observed that miRNA overexpression results in the upregulation of *TXNRD1* in breast cancer, we further wanted to investigate this mechanism in silico. Thus, we conducted bioinformatic analysis to investigate how miR526b and miR655 regulate *TXNRD1* expression. Both miRNA target gene lists were extracted from the miRBase database, using TargetScan analysis tool which can predict miRNA target genes in mammalian mRNA pool [22–26]. By virtue, miRNAs bind to target genes, degrading the corresponding mRNA at the post-transcriptional level, and thus block the protein expression of the target. We found that *TXNRD1* is not a direct target of miR526b and miR655, so we instead attempted to identify transcription factors (TFs) that regulate *TXNRD1* and are also targets of miR526b and miR655. In miR526b/miR655 overexpressing cells, we observed that *TXNRD1* expression is high, which indicates that these miRNAs might be targeting negative regulators of *TXNRD1*. To identify these TFs, we used Enrichr, a tool that consists of both a validated user-submitted gene list and a search engine for further analysis [27]. By comparing miRNA target genes and *TXNRD1* regulatory TFs, we identified eight TFs as direct targets of miR526b (blue down arrows in the yellow circle) and eleven TFs as direct targets of miR655 (blue down arrows in the pink circle) (Figure 8A). Finally, we identified two TFs (PBRM1 and TCF21) as common targets of both miRNAs (blue down arrows in the green circle), which negatively regulate *TXNRD1* (Figure 8A). Both *PBRM1* and *TCF21* have been shown to have tumor suppressor-like functions in breast cancer [28,29]. Therefore, we hypothesize that when miR526b and miR655 are upregulated, their targets *PBRM1* and *TCF21* are downregulated, leading to the upregulation of *TXNRD1*. This result justifies the abundance of *TXNRD1* in miRNA-high cells.

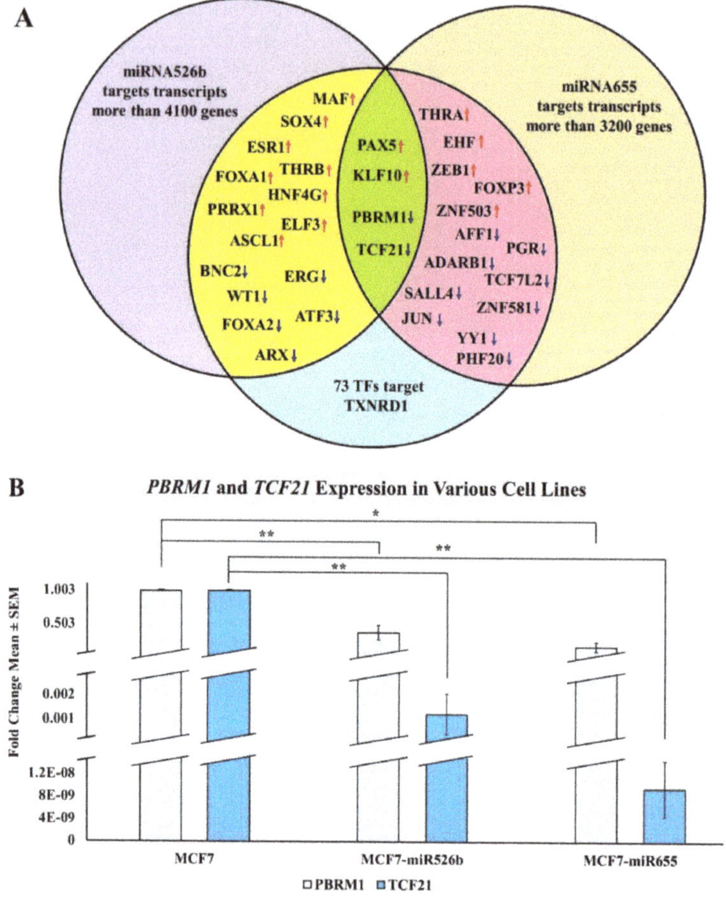

Figure 8. (A) Overlap of TFs regulating *TXNRD1*, and miR526b, miR655 target genes. The purple area represents the list of 4133 miR526b target genes, the brown area represents the list of 3264 miR655 target genes, and the blue area represents all 155 TFs regulating *TXNRD1*. The yellow area shows the miR526b target genes which are also TFs regulating *TXNRD1*. The pink area indicates miR655 targets which are also TFs regulating *TXNRD1*. The green center represents the overlap of all three criteria, which shows the four TFs of *TXNRD1* which are common targets of both miRNAs. The red up arrow symbolizes that the TF upregulates *TXNRD1* expression and the blue down arrow signifies that the TFs downregulates *TXNRD1* expression. Because we observed that *TXNRD1* is upregulated in miRNA-high cells, we considered both miRNAs targeting the two TFs, *PBRM1*, and *TCF21*, which are the negative regulators of *TXNRD1*. (B) *PBRM1* and *TCF21* expression in MCF7, MCF7-miR526b, and MCF7-miR655 cell lines. miR526b/miR655-high cell lines show significantly lower expression of both *PBRM1* and *TCF21*. This indicates both miRNAs target the negative regulator of *TXNRD1*. Data presented as the mean of quadruplicate replicates; * $p < 0.01$, ** $p < 0.001$.

2.3.4. miRNA Overexpression Indirectly Upregulates TXNRD1 by Targeting Negative Regulator of the Gene

The expression of *PBRM1* and *TCF21* was measured in MCF7, MCF7-miR526b, and MCF7-miR655 cell lines to further confirm that miR526b and miR655 target these TFs to regulate the expression of *TXNRD1*. Results show that both *PBRM1* and *TCF21* are significantly downregulated in both miR526b/miR655-high cell lines as compared to MCF7 cells, validating the in silico analysis (Figure 8B).

2.3.5. MCF7 Cells Treated with miR526b and miR655-High Cell-Free Conditioned Media Show Upregulation of TXNRD1

MCF7 cells were treated with basal media or miR526b/miR655-high cell-free conditioned media for 12–18 h as mentioned before. RNA extraction and gene expression assays were carried out to quantify the expression of *TXNRD1*. It was found that *TXNRD1* expression in MCF7-miR655 conditioned media-treated cells was significantly higher compared to the basal media treated MCF7 cells. MCF7 cells treated with MCF7-miR526b conditioned media had marginally higher, but statistically non-significant *TXNRD1* expression compared to basal media treated MCF7 cells (Figure 9A). These results piqued our interest in the paracrine effect of miRNA-overexpressing cells. In the tumor microenvironment, oxidative stress in neighboring normal, immune, and endothelial cells would also be increased. Thus, we wanted to investigate this principle in non-cancerous cells, using a primary endothelial cell line (HUVECs).

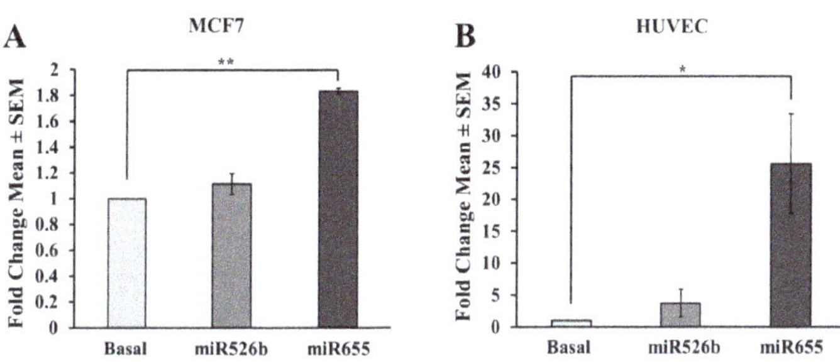

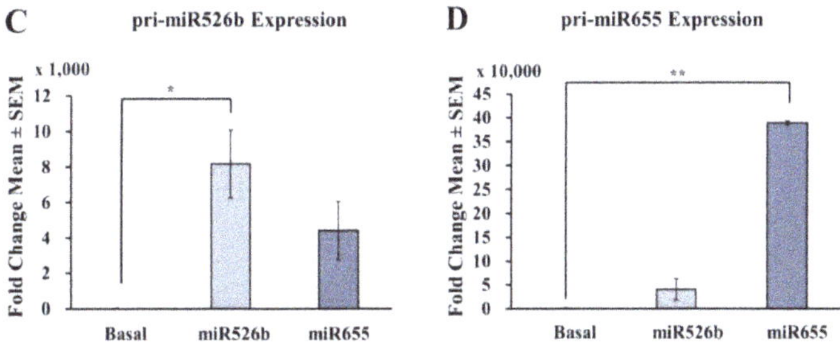

Figure 9. Indirect effects of miRNA overexpression on *TXNRD1*, pri-miR526b, and pri-miR655 expression. (**A**) *TXNRD1* expression in MCF7 cells treated with MCF7-miR526b or MCF7-miR655 conditioned media compared to non-treated MCF7 cells. (**B**) *TXNRD1* expression in HUVEC cells treated with MCF7-miR526b or MCF7-miR655 conditioned media compared to non-treated MCF7 cells. (**C**) pri-miR526b expression in MCF7 cells treated with MCF7-miR526b or MCF7-miR655 conditioned media compared to non-treated MCF7 cells. (**D**) pri-miR655 expression in MCF7 cells treated with MCF7-miR526b or MCF7-miR655 conditioned media compared to non-treated MCF7 cells. Data presented as the mean ± SEM of triplicate replicates; * $p < 0.05$, ** $p < 0.001$.

2.3.6. HUVECs Treated with Cell-Free miR526b and miR655 Conditioned Media Show Upregulation of TXNRD1

HUVECs were treated with basal media, MCF7-miR526b conditioned media, or MCF7-miR655 conditioned media for 12 h. Using qRT-PCR, quantification for the expression of *TXNRD1* in treated and non-treated HUVECs was performed. Results show that HUVECs treated with MCF7-miR526b conditioned media containing secretory proteins and metabolites show a marginal upregulation of *TXNRD1* compared to HUVECs treated with basal media, but was not statistically significant (Figure 9B). However, HUVECs treated with MCF7-miR655 conditioned media show a significant overexpression of *TXNRD1* when compared to HUVECs treated with basal media (Figure 9B).

2.4. Cell-Free miRNA Conditioned Media Indirectly Induces miRNA Overexpression in MCF7 Cells

Since we have shown that cell-free conditioned media from miR526b/miR655-high cell lines induces ROS production and *TXNRD1* expression in MCF7 cells, we wanted to test if cell-metabolites and secretory proteins could also induce oncogenic miRNA upregulation in poorly metastatic MCF7 cells. MCF7 cells were treated with serum-free basal media, MCF7-miR526b conditioned media, or MCF7-miR655 conditioned media for 21 h. RNA was extracted and reverse transcribed into cDNA to quantify pri-miR526b and pri-miR655 expressions. After relative gene expression analysis, the results showed that all miRNA conditioned media-treated MCF7 cells had a significant increase in expression of pri-miR526b and pri-miR655 compared to basal control MCF7 cells (Figure 9C,D). These results establish the dynamic roles of miR526b/miR655 and ROS in the tumor microenvironment where a complex interplay between the tumor cell, tumor cell secretions, and endothelial cells is ongoing, thus promoting tumor growth.

2.5. Induction of Oxidative Stress Upregulates miR526b and miR655 Expression in MCF7 Cells

Finally, we wanted to investigate if miR526b and miR655 are key responders to oxidative stress. Thus, we induced oxidative stress in MCF7 cells using a chemical inducer, H_2O_2, and measured its effects on miR526b and miR655 expression. MCF7 cells were grown until 80% confluent, and then treated with either, 25 μM or 50 μM of H_2O_2 for 24 h. Following treatment, RNA was extracted and reverse transcribed into cDNA. qRT-PCR was then carried out to quantify the expression of pri-miR526b and pri-miR655 in the H_2O_2-treated and non-treated MCF7 cells. Results show a significant dose-dependent increase in the expression of pri-miR655 in H_2O_2-treated MCF7 cells at both 25 μM and 50 μM concentrations (Figure 10). Expression of pri-miR526b following treatment with 50 μM of H_2O_2 showed marginal upregulation; however, this was not statistically significant (Figure 10). These results support the notion that these two miRNAs are immediate responders to oxidative stress in breast cancer.

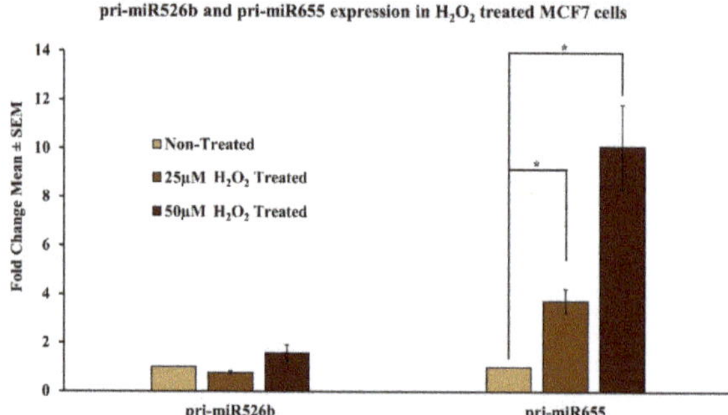

Figure 10. Expression of miR526b and miR655 in H_2O_2 treated MCF7 cells. pri-miR526b and pri-miR655 expression quantified in MCF7 cells, MCF7 cells treated with 25 μM H_2O_2, or 50 μM H_2O_2 using qRT-PCR. Data presented as the mean ± SEM of triplicate replicates; * $p < 0.05$.

3. Discussion

Previously, we have established the roles of oncogenic miR526b and miR655 in breast cancer disease progression, angiogenesis, cancer stem cell regulation, and metastasis [12–15]. We have also previously shown that overexpression of miR526b and miR655 is associated with poor breast cancer patient survival and found that miRNA expression was elevated in advanced grades of breast cancer [12,13], suggesting these two miRNAs are oncogenic and metastasis-promoting miRNAs. Here, we tested the potential roles of these miRNAs in the induction of oxidative stress and the effects of this potential regulation within the tumor microenvironment. ROS including SO, free radicals, and charged ions are the byproducts of cellular metabolism. Under normal physiological conditions, cells keep a balance of ROS production and neutralization to maintain tissue homeostasis [4,5]. However, overproduction of ROS induces oxidative stress, which is associated with cancer development and progression. Production of ROS causes DNA mutation, oncogenic miRNA expression, protein malfunction, apoptosis, and the induction of oxidative stress, which has been identified as a major cause of breast cancer [30]. Superoxide (SO) serves as a growth-stimulating molecule that regulates signaling cascades, which leads to cell survival and proliferation [4]. Moreover, it has been shown that ER-positive breast cancer tumor samples exhibit higher SO levels compared to matched normal tissues, and that SO levels are higher in the blood of breast cancer patients [31,32]. In this study we used the ER-positive MCF7 breast cancer cell line as an in vitro tumor model to establish the link between miRNA and ROS/SO production in breast cancer. The link between various miRNAs and oxidative stress has also been previously reported, such as the expression of miR155 shown to regulate oxidative stress in endothelial cells, and the in vitro induction of oxidative stress being shown to regulate the expression of miR146a and miR34a [33,34]. Oxidative stress is the result of excess ROS, which is due to an imbalance between the generation of ROS and the cell's ability to neutralize and eliminate them. Previous studies have linked the roles of miRNA with ROS production; for example, Zhang et al. showed that miR21 modulates oxidative stress by measuring ROS production in cells through ROS detection [35].

We wanted to investigate if miRNA expression can regulate ROS production and induce oxidative stress, while oxidative stress can also regulate miRNA expression in breast cancer. In this study, we used cell-permeable dyes which interact with cellular ROS and SO to detect and quantify ROS/SO production in cells using fluorescence assays. First, we measured and compared ROS/SO production in MCF7, MCF7-miR526b, and MCF7-miR655 cell lines, to test for the direct regulation of oxidative stress in breast cancer cells by these miRNAs. We have shown that MCF7-miR526b and MCF7-miR655 cell lines, especially the MCF7-miR655 cell line, have higher production of ROS/SO than MCF7 cells, showing that miR526b and miR655 have a role in the endogenous or "direct" induction of oxidative stress.

Breast tumors consist of heterogeneous cells and interactions between tumor cells and cells within the tumor microenvironment to promote tumor sustenance and metastasis. Specifically, cell metabolites and secretions from tumor cells into the tumor microenvironment function to communicate between tumor cells with nearby non-tumor cells, which can regulate many different pathways and networks to promote tumor metastasis [5,36]. Therefore, we investigated the paracrine or "indirect" induction of oxidative stress by treating MCF7 cells and HUVECs with tumor cell metabolites and secretions from miR526b/miR655-high cell lines. We observed that cell-free conditioned media collected from miR526b/miR655-high cells induced ROS/SO production in both MCF7 cells and HUVECs, which suggests that miR526b and miR655 secretory proteins and metabolites indirectly induce oxidative stress in the tumor microenvironment.

The roles of extracellular or cell-free miR526b and miR655 in the complexity of breast tumor metastasis has not been well investigated. Although in recent years many reports studied the detection of miRNAs in the blood of cancer patients, it was only recently shown by a group that extracellular miRNAs can be found in the media of *Drosophila* cell lines growth in petri dish [37]; giving an excellent model to test cell-free miRNA in vitro. We previously have shown that miR526b/miR655 cell-free conditioned media contain stimulatory proteins which induce angiogenesis in the tumor microenvironment [15]. However, we never measured the presence of miR526b/miR655 themselves in the cell-free conditioned media. Here, for the first time, we showed that cell-free supernatant (cell-free conditioned media) collected from MCF7, MCF7-miR526b, and MCF7-miR655 serum-starved cells media contain miR5256b/miR655. Moreover, we found that both MCF7-miR526b and MCF7-miR655 conditioned media had a higher expression of both pri-miR526b and pri-miR655 compared to the MCF7 conditioned media. These results prove that cell secretions from miR526b/miR655-high cell lines also contain miRNAs and indirectly play a role in oxidative stress induction in the tumor microenvironment.

Since ROS activates signaling cascades that promote cell survival and tumor growth, it is expected that highly metastatic and aggressive breast cancer cell lines will be under higher oxidative stress than poorly metastatic breast cancer cell lines [38]. Here, we observed that a key regulatory protein of oxidative stress, TXNRD1, is upregulated in highly metastatic and aggressive breast cancer cell lines, which is supported by other studies showing a link between oxidative stress and breast cancer [10,39]. Next, it was found that miRNA overexpression induced *TXNRD1* expression in MCF7-miR526b and MCF7-miR655 cell lines. These results led us to investigate potential targets of miR526b and miR655 to explain the upregulation of *TXNRD1* in miRNA-overexpressing cell lines. It was found that two transcription factors, PBRM1 (polybromo 1) and TCF21 (Transcription Factor 21), which are negative regulators of *TXNRD1*, are both targets of miR526b and miR655. *PBRM1* has been described as a tumor suppressor gene that is responsible for the control of the cell cycle [28]. Low *PBRM1* expression has been shown to predict poor prognosis in breast cancer and mutations in *PBRM1* have been reported in many tumor types such as renal cell carcinoma, biliary carcinoma, gallbladder carcinoma, and intrahepatic cholangiocarcinoma [28,40]. *TCF21* has also been reported as a tumor suppressor gene in gastric cancer, colorectal cancer, head and neck carcinomas, and breast cancer [29,41–43]. Following the Bioinformatics analysis, we validated this observation by measuring the expression of these two TFs in MCF7, MCF7-miR526b, and MCF7-miR655 cell lines. Our results showed that *PBRM1* and *TCF21* are indeed downregulated in miR526b/miR655-high cell lines, proving that miR526b and miR655 upregulate *TXNRD1* by targeting these two negative regulators of *TXNRD1*.

In the tumor microenvironment, dynamics between tumor cell secretion of inflammatory molecules and growth factors, communication with endothelial cells, and activation of immune cells are well established [15,44]. We have previously shown that treatment of HUVECs with MCF7-miR526b or MCF7-miR655 conditioned media induced cancer related phenotypes, such as angiogenesis and lymphangiogenesis via paracrine regulation [15]. In addition, here we showed that even cell-free conditioned media contain miR526b/miR655. To further investigate the roles of miR526b and miR655 in the indirect induction of oxidative stress, *TXNRD1* expression was quantified and compared in MCF7 cells and HUVECs treated with MCF7-miR526b or MCF7-miR655 cell-free conditioned media. Here, we have found that in both MCF7 and HUVECs treated with MCF7-miR526b or MCF7-miR655 cell-conditioned media, there is an upregulation of *TXNRD1*, which supports our findings of miR526b and miR655 indirectly regulating the production of ROS and induction of oxidative stress.

While ROS production is a component of the cell's physiological process, high concentrations of ROS are detrimental for the cell, which induces apoptosis. However, epigenetic changes, such as miRNA overexpression by tumor cells, protect cellular death and promote cell proliferation. It has been shown that the induction of oxidative stress can alter the expression of specific miRNAs by inhibiting or inducing their expression [16]. Similarly, we have shown that conditioned media collected from MCF7-miR526b and MCF7-miR655 cell lines induce ROS production in MCF7 cells, thus miR526b and miR655 are involved in the regulation of oxidative stress both directly, and indirectly.

Next, we tested to see if miR526b and miR655 are immediate responders to cellular oxidative stress. In this study, we have shown that cell-free conditioned media collected from MCF7-miR526b and MCF7-miR655 cell lines induce oxidative stress; thus, we again used conditioned media from miRNA-overexpressing cells to induce oxidative stress in MCF7 cells and examined miR526b and miR655 expression. We observed that MCF7 cells treated with MCF7-miR526b or MCF7-miR655 conditioned media had increased expressions of both pri-miR526b and pri-miR655. Interestingly, we observed that MCF7 cells treated with miR526b conditioned media and metabolites showed a higher expression of pri-miR526b than pri-miR655, and MCF7 cells treated with conditioned media from MCF7-miR655 showed a higher expression of pri-miR655 than pri-miR526b. It has previously been shown by other groups that H_2O_2 treatment induces oxidative stress in MCF7 cells [45,46]. Therefore, to further validate that miR526b and miR655 are immediate responders to cellular oxidative stress, we tested the effects of H_2O_2 treatment on MCF7 cells. Interestingly, H_2O_2 treatment significantly increased the expression of pri-miR655 in MCF7 cells, and marginally increased pri-miR526b expression in a dose dependent manner. Taken together, these results suggest that a positive feedback loop exists between oxidative stress and miRNA in breast cancer, which is driven by miRNA-high cell line secretions.

Interestingly, we noticed a common trend in which miR655 appeared to have a stronger role in both the direct and indirect induction of oxidative stress than miR526b. MCF7-miR655 was shown to have the greatest expression of *TXNRD1*, and the greatest production of ROS/SO as compared to MCF7 cells. Furthermore, MCF7 and HUVECs treated with cell-free conditioned media from MCF7-miR655 showed the greatest expression of *TXNRD1* and the greatest amount of ROS/SO production. This shows that while miR526b still appeared to be involved in oxidative stress and the *TXNRD1* pathway, miR655 has a stronger role in oxidative stress pathways in breast cancer. Differential roles of miRNAs in regulating oxidative stress may be due to various targets of miRNAs (Figure 8A). In the future, it would be interesting to investigate the signaling pathways involved in miR526b and miR655's regulation of oxidative stress in breast cancer.

In this study, we identified the novel roles of miR526b and miR655 in oxidative stress in breast cancer. Specifically, this is the first time that miR526b and miR655 has been linked to oxidative stress, as we show that miR526b and miR655 regulate ROS production, as well as show greater expression of miRNAs during cellular oxidative stress. Furthermore, we suggest a positive feedback loop exists between miR526b/miR655 and oxidative stress in breast cancer. Here we also show that miR526b and miR655 are present in the extracellular tumor microenvironment, which suggests that these cell free miRNAs might also be regulating extracellular signaling and regulating oxidative stress hence promoting tumor growth and metastasis. These discoveries add to the accumulation of evidence that miR526b and miR655 are strong candidates for potential biomarkers in breast cancer. Future studies require a complete analysis of miRNA cell metabolites and cell secretome to discover new functions of miR526b and miR655. This will allow us to discover complex mechanisms behind oxidative stress induction in breast cancer and the possibility of these miRNAs as therapeutic targets to abrogate oxidative stress.

4. Materials and Methods

We conducted all experiments at Brandon University, following the regulations of Brandon University Research Ethics (#21986, approved on April 21, 2017) and Biohazard Committee (#2017-BIO-02, approved on September 13, 2017). An overview of the methods workflow is presented in Figure 11.

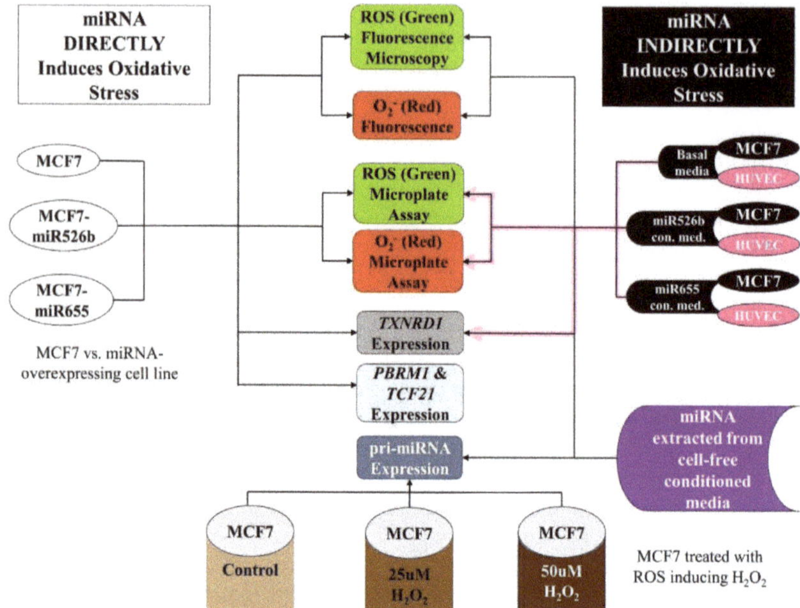

Figure 11. Outline of the in vitro approaches taken in establishing the direct and indirect induction of oxidative stress by miR526b and miR655, as well as the effect of ROS induction on the regulation of miRNA.

4.1. Cell Culture

All human breast cancer cell lines MCF7, SKBR3, T47D, MDAMB231, and Hs578T were purchased from American Type Culture Collection (ATCC, Rockville, MD, USA). All breast cancer cell lines were grown in Dulbecco's Modified Eagle Medium (DMEM) (Gibco, Mississauga, ON, Canada) supplemented with 10% fetal bovine serum (FBS) and 1% Penstrep as described before following manufacturer protocols [12,13,47]. Stable miRNA-overexpressing MCF7-miR526b and MCF7-miR655 cell lines were established as previously described [12,13]. MCF7-miR526b and MCF7-miR655 cells were grown in Roswell Park Memorial Institute (RPMI) 1640 medium (Gibco, Mississauga, ON, Canada) supplemented with 10% FBS and 1% Penstrep. Furthermore, MCF7-miR526b and MCF7-miR655 cell lines were sustained with Geneticin (Gibco, Mississauga, ON, CAN) at 40 mg/mL. An immortalized non-tumorigenic mammary epithelial cell line MCF10A was cultured and maintained by Ling Liu at the University of Western Ontario in Professor Peeyush K Lala's laboratory as described earlier [47] and they kindly shared an aliquot of MCF10A cDNA.

HUVECs were purchased from Life Technologies (NY, USA) and grown in Medium 200 (Gibco, Mississauga, ON, Canada), supplemented with Low Serum Growth Supplement Kit containing 2% FBS, hydrocortisone (1 μg/mL), human epidermal growth factor (10ng/mL), basic fibroblast growth factor (3 ng/mL), and heparin (10 μg/mL). All cell lines were maintained in a humidified incubator at 37 °C with 5% CO_2.

4.2. Collection of Conditioned Media

MCF7, MCF7-miR526b, and MCF7-miR655 cell lines were grown in complete RPMI 1640 until 90% confluent. Cells were then washed with phosphate buffered saline (PBS) to remove any trace of the complete media. The cells were then starved with basal RPMI 1640 medium (serum-free) for 12–16 h prior to collection of media, and then centrifuged. Cell-free supernatant was then collected for assays testing the indirect induction of oxidative stress by miR526b and miR655. We hypothesized that these cell supernatants contain cell metabolites and secretory proteins with unknown function.

4.3. RNA Extraction and Quantitative Real-Time PCR

Total RNA was extracted from all cell lines using the miRNeasy Mini Kit (Qiagen, Toronto, ON, Canada) and reverse transcribed using the microRNA and mRNA cDNA Reverse Transcription Kit (Applied Biosystems, Waltham, MA, USA). For conditioned media miRNA extraction, MCF7, MCF7-miR526b, and MCF7-miR655 conditioned media were centrifuged at 3000 RPM for 5 min, and the supernatants were collected for RNA extraction following the miRNeasy Mini Kit protocol (Qiagen, Toronto, ON, Canada). The TaqMan miRNA or Gene Expression Assays was used for qRT-PCR. The expressions of two endogenous control genes, *Beta-actin* (Hs01060665_g1) and *RPL5* (Hs03044958_g1), were quantified using qRT-PCR and were used to normalize the expression of *TXNRD1* (Hs00917067), *PBRM1* (Hs01015916_m1), *TCF21* (Hs00162646_m1), pri-miR526b (Hs03296227), and pri-miR655 (Hs03304873) markers using relative analysis. Gene expression was measured using CT values from each curve, which are obtained from the point at which each curve reaches the threshold. To determine the relative levels of gene expression, the comparative threshold cycle method (ΔCt) was used [15,47].

4.4. Fluorescence Microplate Assay

MCF7, MCF7-miR526b, and MCF7-miR655 cells were seeded in a 96-well plate as shown in Figure S1A and were grown until 70% confluent. Total ROS and SO levels were detected using the ROS-ID Total ROS/SO detection kit (Enzo Life Sciences, Farmingdale, NY, USA) according to manufacturer's instructions. Negative controls and test groups were prepared for each cell line. The negative controls were treated with 5 μM of *N*-acetyl-L-cysteine (ROS inhibitor) for 30 min to eliminate all ROS present in the cells. Following this, 200 μM of Pyocyanin (ROS inducer) was added to induce ROS production in all wells. The test groups were treated with only the ROS inducer. Detection reagents from the ROS-ID kit were used to measure ROS/SO production. Microplate readings were done at 1 and 21 h following the addition of detection dyes, using the standard Fluorescein filter (Ex/Em: 485/535 nm) and Rhodamine filter (Ex/Em: 550/625 nm). Data was collected using the SoftMax Pro 6 Microplate Data Acquisition and Analysis software (Molecular Devices, San Jose, CA, USA). Concentrations of the ROS inhibitor, inducer, and detection reagents were determined based on a known standard curve. For normalization, negative control emissions were subtracted from the test group emissions to show the total production ROS in each cell line (Figure S1A).

Two more plate-reading experiments were done using MCF7 (Figure S1B) or HUVECs (Figure S1C) treated with basal media (no serum added) or MCF7-miR526b/miR655 conditioned media. MCF7 cells/HUVECs were seeded as shown in Figure S1B/C, and when 70% confluent, they were washed with PBS to remove traces of the serum and growth factors. They were then treated with basal media or MCF7-miR526b/miR655 supernatant for 12–18 h. The assay was then performed as described above.

4.5. Fluorescence Microscopy Assay

We used the same ROS/SO detection kit to determine the number of cells producing ROS and SO following the manufacturer's protocol. Test groups and negative controls were prepared for the MCF7, MCF7-miR526b, and MCF7-miR655 cell lines and seeded as described above. When 70% confluent, the cells were washed PBS and treated as described above. The assay was performed on the NIS Elements Advanced Research software (Nikon, Melville, NY, USA), using a Nikon Ds-Ri1 microscopy camera. The fluorescent cells in each experiment were quantified using the ImageJ software (National Institute of Health, Bethesda, MD, USA). Fluorescent images were converted to 8-bit and adjustments were made. Particle analysis was then done on ImageJ to quantify the number of fluorescing cells (Figure S2B,D,F,H,J,L,N,S,X) and (Figure S3B,D,F,H,J,L,N,S,X) For each condition, the negative control was used as a threshold for quantification (Figures S2 and S3). Negative control quantifications were subtracted from test group quantifications, and then divided by the total number of cells to present the total ROS/SO production in each cell line as ratios.

A second experiment was conducted using the same ROS-ID kit, following the same protocol as described above. MCF7 cells were seeded in a 96-well plate as shown in Figure S1B/C, and once they have reached 70% confluency, washed with PBS and treated with miRNA-conditioned media. The assay was performed and fluorescent cells were quantified and presented using the same methods outlined above. The assay was then performed as described above.

4.6. Bioinformatics Analysis

A total of 4133 target transcript genes for human miR526b (hsa-miR-526b) and 3264 target transcription genes for human miR655 (hsa-miR-655) were found using TargetScan (analysis tool which can predict miRNA target genes in mammalian mRNA pool) and miRBase database [22–26]. Finding the TFs of the *TXNRD1* gene allows us to distinguish what up/down regulates *TXNRD1* expression within the human system.

We used the Enrichr (a tool that consists of both a validated user-submitted gene list and a search engine for further analysis) and found 155 TFs perturbations followed by gene expression [27]. These 155 TFs upregulate or downregulate *TXNRD1* gene expression. We then compared the two data sets to find common genes between miR526b/miR655 target genes and *TXNRD1* regulatory genes (TFs). We observed 19 genes that are common between miR526b targets and *TXNRD1* regulators, of which 8 down-regulated the *TXNRD1* gene expression. We then compared the gene list to find common genes between miR655 targets and *TXNRD1* regulators, and observed 18 genes that are common in both gene sets, of which 11 down-regulated the *TXNRD1* gene. Finally, we found two TFs as common targets of both miRNAs that are negative regulators of *TXNRD1*. To determine the target a nominal $p < 0.05$ was used and the p value was calculated with Fisher exact test, which is a proportion test that assumes a binomial distribution and independence for the probability of any gene belonging to any set.

4.7. Treatment of MCF7 Cells with H_2O_2

MCF7 cells were grown and maintained until 90% confluent. H2O2 at a concentration of either 25 μM or 50 μM was added to confluent MCF7 cells for 24 h. H_2O_2 was used instead of pyocyanin to test the effects of a different ROS inducer. These concentrations of H_2O_2 have been previously reported to induce oxidative stress in the MCF7 cell line [45,46]. Following the addition of H_2O_2 for 24 h, MCF7 cells were collected for RNA extraction, carried out with miRNeasy Mini Kit (Qiagen, Toronto, ON, Canada) and reverse transcribed using the TaqMan microRNA and mRNA cDNA Reverse Transcription Kit (Applied Biosystems, Waltham, MA, USA). qRT-PCR was carried out as mentioned above to measure pri-miR526b and pri-miR655 expression, and were normalized to *Beta-actin* and *RPL5*.

4.8. Statistical Analysis

Statistical calculations were performed using GraphPad Prism software version 8 (https://www.graphpad.com/quickcalcs/ttest1/?Format=SEM). All parametric data were analyzed with one-way ANOVA followed by Tukey–Kramer or Dunnett post-hoc comparisons. Student's t-test was used when comparing two datasets. Statistically relevant differences between means were accepted at $p < 0.05$. Fisher exact test was performed for miRNA database and target TFs analysis followed by false positive rate (FDR) correction to identify significant changes in target gene expression ($p < 0.05$).

Supplementary Materials: Supplementary materials can be found at http://www.mdpi.com/1422-0067/20/16/4039/s1.

Author Contributions: Concept, project design, and supervision: M.M.; Experiments: B.S., R.F., B.N., S.H., S.M., K.C.U.; Data Analysis: B.S., R.F., B.N., S.M.; Figures and Image Data Processing: B.S., R.F., B.N., S.M.; Manuscript writing: B.S., R.F., S.H., B.N., and M.M.

Funding: This project is funded by NSERC-Discovery grant and Brandon University Research Committee grants to M.M., B.S., S.H., B.N. are recipients of NSERC-USRA scholarships. R.F., S.H. were partly supported by Canadian Summer Job grant to M.M.

Acknowledgments: The authors of this article would like to thank Danielle Laroque at Brandon University for her help with cell number quantification with ImageJ. We sincerely thank Peeyush K Lala at the University of Western Ontario to share MCF10A cDNA with us. Thanks to Bernadette Ardelli at Brandon University to give us access to Fluorescence Microscope, Microplate Reader, and the Rotor Gene PCR machine in her laboratory. We also want to thank Vincent Chen at Brandon University to help us with the analysis of cell metabolites.

Conflicts of Interest: The authors declare no conflict of interest.

Abbreviations

ROS	Reactive Oxygen Species
SO	Superoxide
miRNA	microRNA
TXNRD1	Thioredoxin Reductase 1
TXN	Thioredoxin
TF	Transcription Factor
ER	Estrogen Receptor

References

1. Breast Cancer—Early Diagnosis and Screening. Available online: https://www.who.int/cancer/prevention/diagnosis-screening/breast-cancer/en/ (accessed on 19 July 2019).
2. Hecht, F.; Pessoa, C.F.; Gentile, L.B.; Rosenthal, D.; Carvalho, D.P.; Fortunato, R.S. The role of oxidative stress on breast cancer development and therapy. *Tumor Biol.* **2016**, *37*, 4281–4291. [CrossRef] [PubMed]
3. Federico, A.; Morgillo, F.; Tuccillo, C.; Ciardiello, F.; Loguercio, C. Chronic inflammation and oxidative stress in human carcinogenesis. *Int. J. Cancer* **2007**, *121*, 2381–2386. [CrossRef] [PubMed]
4. Buetler, T.M.; Krauskopf, A.; Ruegg, U.T. Role of superoxide as a signaling molecule. *Physiology* **2004**, *19*, 120–123. [CrossRef]
5. Calaf, G.M.; Urzua, U.; Termini, L.; Aguayo, F. Oxidative stress in female cancers. *Oncotarget* **2018**, *9*, 23824. [CrossRef] [PubMed]
6. Lobo, V.; Patil, A.; Phatak, A.; Chandra, N. Free radicals, antioxidants and functional foods: Impact on human health. *Pharmacogn. Rev.* **2010**, *4*, 118–126. [CrossRef]
7. Arnér, E.S.; Holmgren, A. Physiological functions of thioredoxin and thioredoxin reductase. *Eur. J. Biochem.* **2000**, *267*, 6102–6109. [CrossRef] [PubMed]
8. Lu, J.; Vlamis-Gardikas, A.; Kandasamy, K.; Zhao, R.; Gustafsson, T.N.; Engstrand, L.; Hoffner, S.; Engman, L.; Holmgren, A. Inhibition of bacterial thioredoxin reductase: An antibiotic mechanism targeting bacteria lacking glutathione. *FASEB J.* **2013**, *7*, 1394–1403. [CrossRef]
9. Urig, S.; Lieske, J.; Fritz-Wolf, K.; Irmler, A.; Becker, K. Truncated mutants of human thioredoxin reductase 1 do not exhibit glutathione reductase activity. *FEBS Lett.* **2006**, *580*, 3595–3600. [CrossRef]
10. Leone, A.; Roca, M.S.; Ciardiello, C.; Costantini, S.; Budillon, A. Oxidative stress gene expression profile correlates with cancer patient poor prognosis: Identification of crucial pathways might select novel therapeutic approaches. *Oxid. Med. Cell. Longevity* **2017**, *2017*. [CrossRef]
11. Singh, R.P.; Massachi, I.; Manickavel, S.; Singh, S.; Rao, N.P.; Hasan, S.; Mc Curdy, D.K.; Sharma, S.; Wong, D.; Hahn, B.H.; et al. The role of miRNA in inflammation and autoimmunity. *Autoimmun. Rev.* **2013**, *12*, 1160–1165. [CrossRef]
12. Majumder, M.; Landman, E.; Liu, L.; Hess, D.; Lala, P.K. COX-2 elevates oncogenic miR-526b in breast cancer by EP4 activation. *Mol. Cancer Res.* **2015**, *13*, 1022–1033. [CrossRef]
13. Majumder, M.; Dunn, L.; Liu, L.; Hasan, A.; Vincent, K.; Brackstone, M.; Hess, D.; Lala, P.K. COX-2 induces oncogenic microRNA miR655 in human breast cancer. *Sci. Rep.* **2018**, *8*, 327. [CrossRef]
14. Majumder, M.; Nandi, P.; Omar, A.; Ugwuagbo, K.; Lala, P.K. EP4 as a therapeutic target for aggressive human breast cancer. *Int. J. Mol. Sci.* **2018**, *19*, 1019. [CrossRef]
15. Hunter, S.; Nault, B.; Ugwuagbo, K.; Maiti, S.; Majumder, M. Mir526b and Mir655 Promote Tumour Associated Angiogenesis and Lymphangiogenesis in Breast Cancer. *Cancers* **2019**, *11*, 938. [CrossRef]

16. He, J.; Jiang, B.H. Interplay between reactive oxygen species and microRNAs in cancer. *Curr. Pharmacol. Rep.* **2016**, *2*, 82–90. [CrossRef]

17. Longo, A.; Librizzi, M.; Chuckowree, I.; Baltus, C.; Spencer, J.; Luparello, C. Cytotoxicity of the urokinase-plasminogen activator inhibitor carbamimidothioic acid (4-boronophenyl) methyl ester hydrobromide (BC-11) on triple-negative MDA-MB231 breast cancer cells. *Molecules* **2015**, *20*, 9879–9889. [CrossRef]

18. Librizzi, M.; Longo, A.; Chiarelli, R.; Amin, J.; Spencer, J.; Luparello, C. Cytotoxic effects of Jay Amin hydroxamic acid (JAHA), a ferrocene-based class I histone deacetylase inhibitor, on triple-negative MDA-MB231 breast cancer cells. *Chem. Res. Toxicol.* **2012**, *25*, 2608–2616. [CrossRef]

19. Liang, H.H.; Huang, C.Y.; Chou, C.W.; Makondi, P.T.; Huang, M.T.; Wei, P.L.; Chang, Y.J. Heat shock protein 27 influences the anti-cancer effect of curcumin in colon cancer cells through ROS production and autophagy activation. *Life Sci.* **2018**, *209*, 43–51. [CrossRef]

20. Chang, T.C.; Wei, P.L.; Makondi, P.T.; Chen, W.T.; Huang, C.Y.; Chang, Y.J. Bromelain inhibits the ability of colorectal cancer cells to proliferate via activation of ROS production and autophagy. *PLoS ONE* **2019**, *14*, e0210274. [CrossRef]

21. Wei, P.L.; Huang, C.Y.; Chang, Y.J. Propyl gallate inhibits hepatocellular carcinoma cell growth through the induction of ROS and the activation of autophagy. *PloS ONE* **2019**, *14*, e0210513. [CrossRef]

22. TargetScanHuman 7.1—Predicted miRNA targets of miR-526b-5p. Available online: http://www.targetscan.org/cgi-bin/targetscan/vert_71/targetscan.cgi?mirg=hsa-miR-526b-5p (accessed on 19 July 2019).

23. TargetScanHuman 7.1—Predicted miRNA targets of miR-655-5p. Available online: http://www.targetscan.org/cgi-bin/targetscan/vert_71/targetscan.cgi?mirg=hsa-miR-655-5p (accessed on 19 July 2019).

24. Agarwal, V.; Bell, G.W.; Nam, J.; Bartel, D.P. Predicting effective microRNA target sites in mammalian mRNAs. *eLife* **2015**, *4*, e05005. [CrossRef]

25. miRBase-Release 22.1.—miRNA Entry for MI0003150. Available online: http://www.mirbase.org/cgi-bin/mirna_entry.pl?acc=MI0003150 (accessed on 19 July 2019).

26. miRBase-Release 22.1.—miRNA Entry for MI0003677. Available online: http://www.mirbase.org/cgi-bin/mirna_entry.pl?acc=MI0003677 (accessed on 19 July 2019).

27. Keenan, A.B.; Torre, D.; Lachmann, A.; Leong, A.K.; Wojciechowicz, M.L.; Utti, V.; Jagodnik, K.M.; Kropiwnicki, E.; Wang, Z.; Ma'ayan, A. ChEA3: Transcription factor enrichment analysis by orthogonal omics integration. *Nucleic Acids Res.* **2019**, *2*, 212–224. [CrossRef]

28. Wang, H.; Qu, Y.; Dai, B.; Zhu, Y.; Shi, G.; Zhu, Y.; Shen, Y.; Zhang, H.; Ye, D. PBRM1 regulates proliferation and the cell cycle in renal cell carcinoma through a chemokine/chemokine receptor interaction pathway. *PLoS ONE* **2017**, *12*, e0180862. [CrossRef]

29. Wang, J.; Gao, X.; Wang, M.; Zhang, J. Clinicopathological significance and biological role of TCF21 mRNA in breast cancer. *Tumor Biol.* **2015**, *36*, 8679–8683. [CrossRef]

30. Nourazarian, A.R.; Kangari, P.; Salmaninejad, A. Roles of oxidative stress in the development and progression of breast cancer. *Asian Pac. J. Cancer Prev.* **2014**, *15*, 4745–4751. [CrossRef]

31. Kanchan, R.K.; Tripathi, C.; Baghel, K.S.; Dwivedi, S.K.; Kumar, B.; Sanyal, S.; Sharma, S.; Mitra, K.; Garg, V.; Singh, K.; et al. Estrogen receptor potentiates mTORC2 signaling in breast cancer cells by upregulating superoxide anions. *Free Radical Bio. Med.* **2012**, *53*, 1929–1941. [CrossRef]

32. Yeh, C.C.; Hou, M.F.; Tsai, S.M.; Lin, S.K.; Hsiao, J.K.; Huang, J.C.; Wang, L.H.; Wu, S.H.; Hou, L.A.; Ma, H.; et al. Superoxide anion radical, lipid peroxides and antioxidant status in the blood of patients with breast cancer. *Clin. Chim. Acta* **2005**, *361*, 104–111. [CrossRef]

33. Chen, H.; Gao, L.; Yang, M.; Zhang, L.; He, F.L.; Shi, Y.K.; Pan, X.H.; Wang, H. MicroRNA-155 affects oxidative damage through regulating autophagy in endothelial cells. *Oncol. Lett.* **2019**, *17*, 2237–2243. [CrossRef]

34. Cheleschi, S.; de Palma, A.; Pascarelli, N.; Giordano, N.; Galeazzi, M.; Tenti, S.; Fioravanti, A. Could oxidative stress regulate the expression of microRNA-146a and microRNA-34a in human osteoarthritic chondrocyte cultures? *Int. J. Mol. Sci.* **2017**, *18*, 2660. [CrossRef]

35. Zhang, X.; Ng, W.L.; Wang, P.; Tian, L.; Werner, E.; Wang, H.; Doetsch, P.; Wang, Y. MicroRNA-21 modulates the levels of reactive oxygen species by targeting SOD3 and TNFα. *Cancer Res.* **2012**, *72*, 4707–4713. [CrossRef]

36. Huang, W.; Luo, S.; Burgess, R.; Yi, Y.H.; Huang, G.; Huang, R.P. New insights into the tumor microenvironment utilizing protein array technology. *Int. J. Mol. Sci.* **2018**, *19*, 559. [CrossRef]

37. Van den Brande, S.; Gijbels, M.; Wynant, N.; Santos, D.; Mingels, L.; Gansemans, Y.; van Nieuwerburgh, F.; Vanden Broeck, J. The presence of extracellular microRNAs in the media of cultured Drosophila cells. *Sci. Rep.* **2018**, *8*, 17312. [CrossRef]

38. Reuter, S.; Gupta, S.C.; Chaturvedi, M.M.; Aggarwal, B.B. Oxidative stress, inflammation, and cancer: How are they linked? *Free Radicals Biol. Med.* **2010**, *49*, 1603–1616. [CrossRef]

39. Zajchowski, D.A.; Bartholdi, M.F.; Gong, Y.; Webster, L.; Liu, H.L.; Munishkin, A.; Beauheim, C.; Harvey, S.; Ethier, S.P.; Johnson, P.H. Identification of gene expression profiles that predict the aggressive behavior of breast cancer cells. *Cancer Res.* **2001**, *61*, 5168–5178.

40. Mo, D.; Li, C.; Liang, J.; Shi, Q.; Su, N.; Luo, S.; Zeng, T.; Li, X. Low PBRM1 identifies tumor progression and poor prognosis in breast cancer. *Int. J. Clin. Exp. Pathol.* **2015**, *8*, 9307–9313.

41. Yang, Z.; Li, D.M.; Xie, Q.; Dai, D.Q. Protein expression and promoter methylation of the candidate biomarker TCF21 in gastric cancer. *J. Cancer Res. Clin. Oncol.* **2015**, *141*, 211–220. [CrossRef]

42. Dai, Y.; Duan, H.; Duan, C.; Zhu, H.; Zhou, R.; Pei, H.; Shen, L. TCF21 functions as a tumor suppressor in colorectal cancer through inactivation of PI3K/AKT signaling. *Onco. Targets Ther.* **2017**, *10*, 1603–1611. [CrossRef]

43. Smith, L.T.; Lin, M.; Brena, R.M.; Lang, J.C.; Schuller, D.E.; Otterson, G.A.; Morrison, C.D.; Smiraglia, D.J.; Plass, C. Epigenetic regulation of the tumor suppressor gene *TCF21* on 6q23-q24 in lung and head and neck cancer. *Proc. Natl. Acad. Sci. USA* **2006**, *103*, 982–987. [CrossRef]

44. Majumder, M.; Xin, X.; Liu, L.; Girish, G.V.; Lala, P.K. Prostaglandin E2 receptor EP4 as the common target on cancer cells and macrophages to abolish angiogenesis, lymphangiogenesis, metastasis and stem-like cell functions. *Cancer Sci.* **2014**, *105*, 1142–1151. [CrossRef]

45. Chua, P.J.; Yip, G.W.C.; Bay, B.H. Cell cycle arrest induced by hydrogen peroxide is associated with modulation of oxidative stress related genes in breast cancer cells. *Exp. Biol. Med.* **2009**, *234*, 1086–1094. [CrossRef]

46. Mahalingaiah, P.K.S.; Ponnusamy, L.; Singh, K.P. Chronic oxidative stress causes estrogen-independent aggressive phenotype, and epigenetic inactivation of estrogen receptor alpha in MCF-7 breast cancer cells. *Breast Cancer Res. Treat.* **2015**, *153*, 41–56. [CrossRef]

47. Majumder, M.; Xin, X.; Liu, L.; Tutunea-Fatan, E.; Rodriguez-Torres, M.; Vincent, K.; Postovit, L.M.; Hess, D.; Lala, P.K. COX-2 induces breast cancer stem cells via EP4/PI3K/AKT/NOTCH/WNT axis. *Cancer Stem Cells* **2016**, *34*, 2290–2305. [CrossRef]

International Journal of
Molecular Sciences

MDPI

Article

miR-27a-5p Attenuates Hypoxia-induced Rat Cardiomyocyte Injury by Inhibiting *Atg7*

Jinwei Zhang [1,2,†], Wanling Qiu [1,2,†], Jideng Ma [1,2,†], Yujie Wang [1,2], Zihui Hu [1,2], Keren Long [1,2], Xun Wang [1,2], Long Jin [1,2], Qianzi Tang [1,2], Guoqing Tang [1,2], Li Zhu [1,2], Xuewei Li [1,2], Surong Shuai [1,2,*] and Mingzhou Li [1,2,*]

[1]   Institute of Animal Genetics and Breeding, College of Animal Science and Technology, Sichuan Agricultural University, Chengdu 611130, Sichuan, China; Jinweizhang50@163.com (J.Z.); qiuwanling2016@163.com (W.Q.); jideng.ma@sicau.edu.cn (J.M.); wangyujie715@163.com (Y.W.); Huzihui2016@163.com (Z.H.); keren.long@sicau.edu.cn (K.L.); xun_wang007@163.com (X.W.); longjin8806@163.com (L.J.); wupie@163.com (Q.T.); tyq003@163.com (G.T.); zhuli7508@163.com (L.Z.); xuewei.li@sicau.edu.cn (X.L.)

[2]   Farm Animal Genetic Resource Exploration and Innovation Key Laboratory of Sichuan Province, Sichuan Agricultural University, Chengdu 611130, Sichuan, China

*   Correspondence: srshuai@sohu.com (S.S.); mingzhou.li@sicau.edu.cn (M.L.); Tel.: +86-28-8629-0998 (S.S.); +86-28-8629-0962 (M.L.)

†   These authors contributed equally to this work.

Received: 14 April 2019; Accepted: 13 May 2019; Published: 16 May 2019

Abstract: Acute myocardial infarction (AMI) is an ischemic heart disease with high mortality worldwide. AMI triggers a hypoxic microenvironment and induces extensive myocardial injury, including autophagy and apoptosis. MiRNAs, which are a class of posttranscriptional regulators, have been shown to be involved in the development of ischemic heart diseases. We have previously reported that hypoxia significantly alters the miRNA transcriptome in rat cardiomyoblast cells (H9c2), including miR-27a-5p. In the present study, we further investigated the potential function of miR-27a-5p in the cardiomyocyte response to hypoxia, and showed that miR-27a-5p expression was downregulated in the H9c2 cells at different hypoxia-exposed timepoints and the myocardium of a rat AMI model. Follow-up experiments revealed that miR-27a-5p attenuated hypoxia-induced cardiomyocyte injury by regulating autophagy and apoptosis via *Atg7*, which partly elucidated the anti-hypoxic injury effects of miR-27a-5p. Taken together, this study shows that miR-27a-5p has a cardioprotective effect on hypoxia-induced H9c2 cell injury, suggesting it may be a novel target for the treatment of hypoxia-related heart diseases.

Keywords: miR-27a-5p; acute myocardial infarction; autophagy; apoptosis; hypoxia

---

1. Introduction

Acute myocardial infarction (AMI) is often the primary pathological cause of death and disability worldwide [1]. During AMI, acute occlusion of the coronary artery deprives the oxygen and nutrients in myocardium and will contributes to cardiac dysfunction, including hypertrophy and remodeling, eventually leads to heart failure [2]. Since cardiomyocytes are terminally differentiated cells that have no or little regenerative potentialities, thus preventing cardiomyocytes loss after AMI injury is clinically a vital therapeutic strategy. Cardiomyocytes death and survival are affected predominantly via three cellular pathways: apoptosis, necrosis and autophagy [3]. Out of these three cellular pathways, apoptosis and necrosis have been extensively researched in AMI, but the effect of autophagy underlying AMI is still controversial to date [4]. Autophagy is an evolutionarily conserved process that maintains homeostasis in a cellular response to stresses by degrading abnormal protein and damaged organelles, which is considered to be closely associated with many heart diseases such as AMI [5]. Recently, autophagy has been considered a double-edged sword in the context of AMI, i.e., autophagy in early stage of AMI is beneficial to cardiomyocytes survival but excessive autophagy after AMI will induce autophagic cell death [6]. Thus, it is indispensable to further elucidate the autophagy regulation mechanism in cardiomyocytes survival after AMI.

MicroRNAs (miRNAs), a class of highly conserved non-coding RNAs, are major posttranscriptional regulators that involving in almost all cellular processes [7]. Currently, accumulating evidence has shown that miRNAs play essential roles in some heart diseases by regulating autophagy-related genes [4]. miRNA-212/132 family induce both cardiac hypertrophy and heart failure by activating pro-hypertrophic calcineurin/NFAT signaling, while inhibiting autophagic response upon starvation by directly targeting the anti-hypertrophic and pro-autophagic FoxO3 transcription factor [8]. miR-188-3p inhibits autophagy and autophagic cell death in the heart by targeting *Atg7* expression, meanwhile this effect can be suppressed by lncRNA APF (autophagy promoting factor) [9]. miR-21 alleviates hypoxia/reoxygenation-induced injury in H9c2 cells through weakening excessive autophagy and apoptosis via the Akt/mTOR pathway [10]. miR-204 has a protective effect against H9c2 cells hypoxia/reoxygenation-induced injury by regulating SIRT1-mediated autophagy [11]. Moreover, a recent study reports that miR-223 alleviates hypoxia-induced excessive autophagy and apoptosis in rat cardiomyocytes via the Akt/mTOR pathway by targeting *PARP-1* [12]. These miRNA may be serve as a potential target for ischemic heart disease treatment. In our previous study, we noted that the expression of miR-27a-5p decreased in acute hypoxia-exposed H9c2 cells using a small RNA-seq [13]. However, whether miR-27a-5p affects hypoxia-induced cardiomyocyte survival through regulating cell autophagy after AMI are still unknown.

In this study, we established a model of hypoxia in H9c2 cells and developed an AMI model in the rat to investigate the miR-27a-5p expression pattern in H9c2 cells and the main visceral tissues of rats (Figure 1). We found that hypoxia induced cell injury in vivo and in vitro and was accompanied by downregulation of miR-27a-5p expression. miR-27a-5p upregulation attenuated hypoxia-induced cardiomyocyte injury by regulating autophagy and apoptosis via *Atg7*, suggesting that miR-27a-5p may be a novel treatment strategy for hypoxia-related heart diseases.

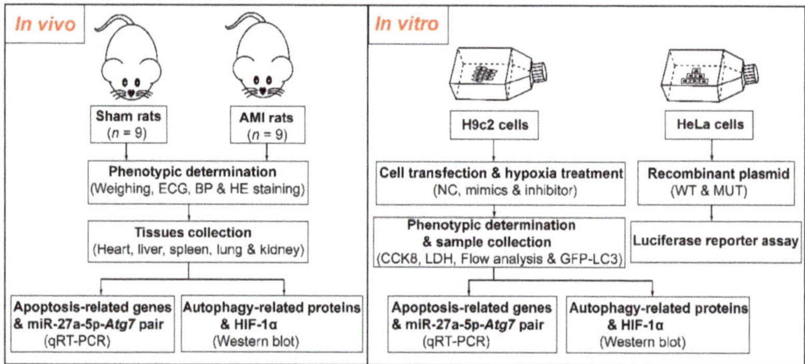

Figure 1. The flow chart of this study. ECG, electrocardiogram; BP, blood pressure; HE staining, hematoxylin & eosin staining; qRT-PCR, quantitative reverse-transcription polymerase chain reaction NC, negative control; CCK8, cell counting kit-8; LDH, lactate dehydrogenase; WT/MUT, wild-type/mutant.

2. Results

2.1. Hypoxia Induces H9c2 Cells Injury and Reduces miR-27a-5p Expression

In this study, we first cultured H9c2 cells in hypoxic condition for 24 h to simulate hypoxia induced by AMI *in vitro*. We found that hypoxia increased HIF-1α protein expression (Figure 2A) and triggered cell injury, including a decrease in cell viability ($p < 0.01$; Figure 2B), increased cell membrane damage ($p < 0.01$; Figure 2C) and apoptosis and necrosis ($p < 0.01$; Figure 2D,E). Meanwhile, hypoxia significantly increased the expression of proapoptotic genes (*Caspase-3*, *BAX*, *Faslg* and *P53*, $p < 0.01$; Figure 2F), but decreased expression of the antiapoptotic gene *Bcl-2* ($p < 0.05$; Figure 2F). Autophagy has previously been observed in ischemic heart disease [14,15] and autophagy levels were assessed in hypoxia-exposed H9c2 cells by western blot and autophagosome formation. These data showed that hypoxia increased autophagosome formation (Figure 2G) and promoted the switch of LC3-I to LC3-II. It also resulted in a reduction in P62 protein expression ($p < 0.01$; Figure 2H). Next, miR-27a-5p expression pattern was assessed in hypoxia-exposed H9c2 cells using qRT-PCR. miR-27a-5p expression decreased in a time-dependent manner (Figure 2I). These results indicate that hypoxia induced cell injury and reduced miR-27a-5p expression levels in H9c2 cells.

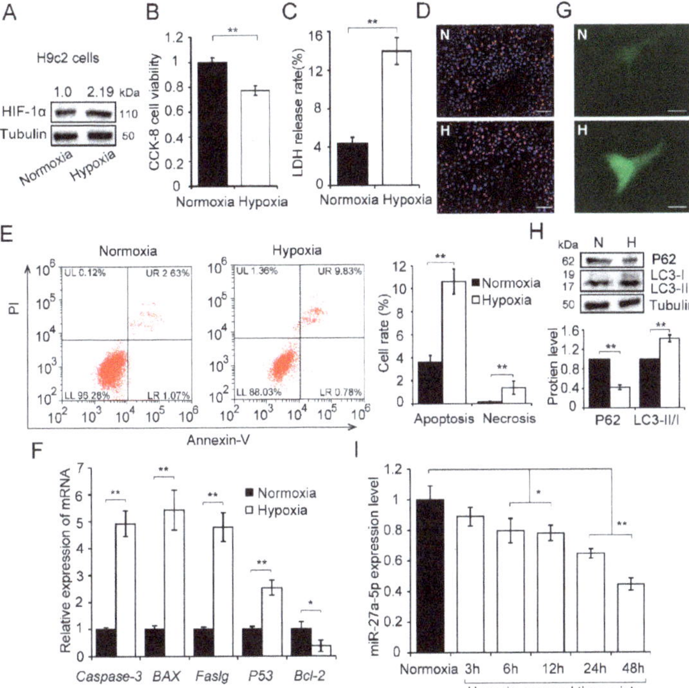

Figure 2. Hypoxia induces H9c2 cell injury and downregulation of miR-27a-5p. H9c2 cells were cultured under hypoxia or normoxia for 24 h. HIF-1α protein increased in H9c2 cells after hypoxia (**A**). Cell viability (**B**), membrane damage (**C**), and cell apoptosis (**D–F**) were evaluated by CCK8 assay, LDH release assays, apoptosis staining (scale bar: 50 μm), flow cytometry, and qRT-PCR analysis, respectively. H9c2 cells were transfected with GFP-LC3 plasmids and exposed to hypoxia for 24 h, fluorescence was observed by confocal fluorescence microscopy (**G**); scale bar: 5 μm. The autophagy-related proteins were detected by western blot (**H**). The expression of miR-27a-5p was tested using qRT-PCR at different hypoxia-exposed timepoints (**I**). Three independent experiments were performed in triplicate. Data are expressed as the mean ± SD. * $p < 0.05$, ** $p < 0.01$. N: normoxia; H: hypoxia.

2.2. AMI Triggers Widespread Injury Accompanied by Downregulation of miR-27a-5p in Rats

To investigate whether the miR-27a-5p expression under hypoxia induced by AMI in vivo was similar to that in hypoxia-exposed cardiomyocytes in vitro, an AMI rat model was established by ligating the coronary artery [16]. We observed S-T segment elevation in the electrocardiogram (ECG) and a reduction in blood pressure (BP) in the AMI group compared with sham, which confirmed successful AMI (Figure 3A,B). A *post hoc* power analysis of Δ BP obtained a power of > 0.90 with $p = 0.05$ in every LAD ligation timepoint (see "Statistical Analysis" for details on power analysis) (Table S1). We also found that the organ index in several main visceral tissues (except lung) was reduced (Table S2), which may be associated with the decreased left ventricular ejection fraction commonly observed after AMI [12]. Meanwhile, HE staining of the left ventricle showed that the cells in sham rat hearts were arranged uniformly with a normal gap, but local necrosis (indicated by arrowhead) and intercellular gaps (indicated by asterisk) were observed in AMI rats (Figure 3C). These data indicate that AMI induced severe damage in the rat myocardium.

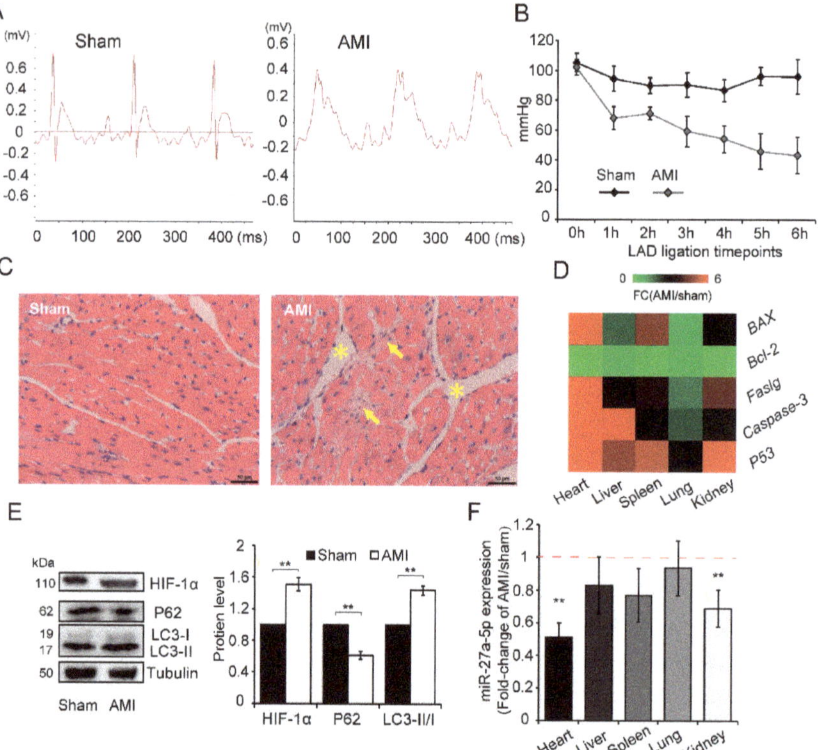

Figure 3. AMI widely induces injury and reduces miR-27a-5p expression in rats. A model of AMI was established in rats by ligating the coronary artery, and confirmed by analyzing ECG (**A**) and BP (**B**). (**C**) HE staining showed morphological differences between sham and AMI rats in coronal sections of the left ventricle; Yellow arrowheads and asterisks highlight local necrosis and intercellular gaps, respectively; scale bar: 50 μm. Expression patterns of apoptosis-related genes (**D**) in main visceral tissues (including heart, liver, spleen, lung and kidney) were determined by qRT-PCR (AMI *vs* Sham). AMI increased HIF-1α expression and promoted the conversion of LC3-I to LC3-II, but decreased P62 expression (**E**). miR-27a-5p expression (**F**) in main visceral tissues by qRT-PCR analysis (AMI *vs* Sham). Data are presented as the means ± SD of three independent experiments. ** $p < 0.01$. LAD: left anterior descending coronary artery.

The expression pattern of apoptosis-related genes showed that AMI triggers widespread apoptosis in the main visceral tissues, especially heart ($p < 0.01$), compared with sham (Figure 3D & Figure S1A). AMI also increased HIF-1α expression, shifted the expression of LC3-I to LC3-II, and decreased the expression of P62 protein (Figure 3E), which indicates that AMI synchronously promotes autophagy and apoptosis in the rat myocardium. In addition, AMI caused a reduction in miR-27a-5p expression in several visceral tissues, in particular the heart and kidney ($p < 0.01$; Figure 3F) when assessed by qRT-PCR analysis. The above results indicate that, similar to the in vitro results, hypoxic injury is widely induced in AMI rats and is accompanied by widespread downregulation of miR-27a-5p. Thus, miR-27a-5p may play a role in AMI-induced hypoxic injury.

2.3. Upregulation of miR-27a-5p Attenuates Hypoxia-Induced Excessive Autophagy and Apoptosis

Several studies have previously reported that autophagy and apoptosis successively appear in the cardiovascular diseases and the crosstalk between them plays an important role in the development of ischemic heart disease [17,18]. Autophagy have bidirectional effects in AMI, as autophagy may have both damaging and protective roles depending on the hypoxic conditions, such as duration or severity [19]. In the present study, assessment of autophagic flux showed that hypoxia-exposed H9c2 cells increased the level of autophagy in a time-dependent manner (Figure S1B). Next, cell viability and membrane damage were assessed after hypoxia in H9c2 cells pretreated with 10 mM 3-MA (a widely-used autophagy inhibitor). Cell viability was decreased in 3-MA-treated cells compared with control at the early stages of hypoxia exposure (within first 12 h); however, cell viability was higher in 3-MA-treated cells than control after hypoxia for 24 h (Figure S1C). Conversely, 3-MA pretreatment increased membrane damage in early stages of hypoxia but then this damage was alleviated after hypoxia for 24 h (Figure S1D). These results indicate that autophagy plays different roles in hypoxia-induced H9c2 cell injury over time and is beneficial in early stage of hypoxia but detrimental after 24 h of hypoxia (excessive autophagy), in keeping with previous reports [12]. Hypoxia for 24 h was used in subsequent experiments.

Based on the miR-27a-5p expression pattern and cell injury in hypoxia-exposed H9c2 cells and AMI rat myocardium, we hypothesized that miR-27a-5p is involved in mediating this biological process. To test this hypothesis, gain and loss of function analyses were performed. Effective overexpression and downregulation of miR-27a-5p was achieved in H9c2 cells by transfecting cells with a miR-27a-5p mimics or inhibitor, respectively, after exposure to hypoxia for 24 h (Figure 4A). Overexpression of miR-27a-5p significantly mitigated hypoxic injury, including improved cell viability ($p < 0.01$; Figure 4B), alleviated cell membrane damage ($p < 0.01$; Figure 4C), and reduced cell apoptosis (Figure 4D–F). Meanwhile, miR-27a-5p downregulation yielded the opposite effects (Figure 4D–F). These results demonstrate that miR-27a-5p can reduce hypoxia-induced H9c2 cell injury by inhibiting apoptosis. To further assess the impact of miR-27a-5p on autophagy, the autophagic flux and autophagy-related proteins were assessed in cells exposed to hypoxia for 24 h after transfection. As shown in Figure 4G,H, miR-27a-5p overexpression decreased the level of autophagy, shifted LC3-I expression to LC3-II expression ($p < 0.05$), and increased P62 protein expression compared with control ($p < 0.01$). However, miR-27a-5p downregulation resulted in a more severe autophagy (Figure 4G,H). Altogether, these results indicate that miR-27a-5p has a negative effect on hypoxia-induced autophagy and that miR-27a-5p protects against hypoxia-induced cardiomyocyte injury by reducing apoptosis and excessive autophagy.

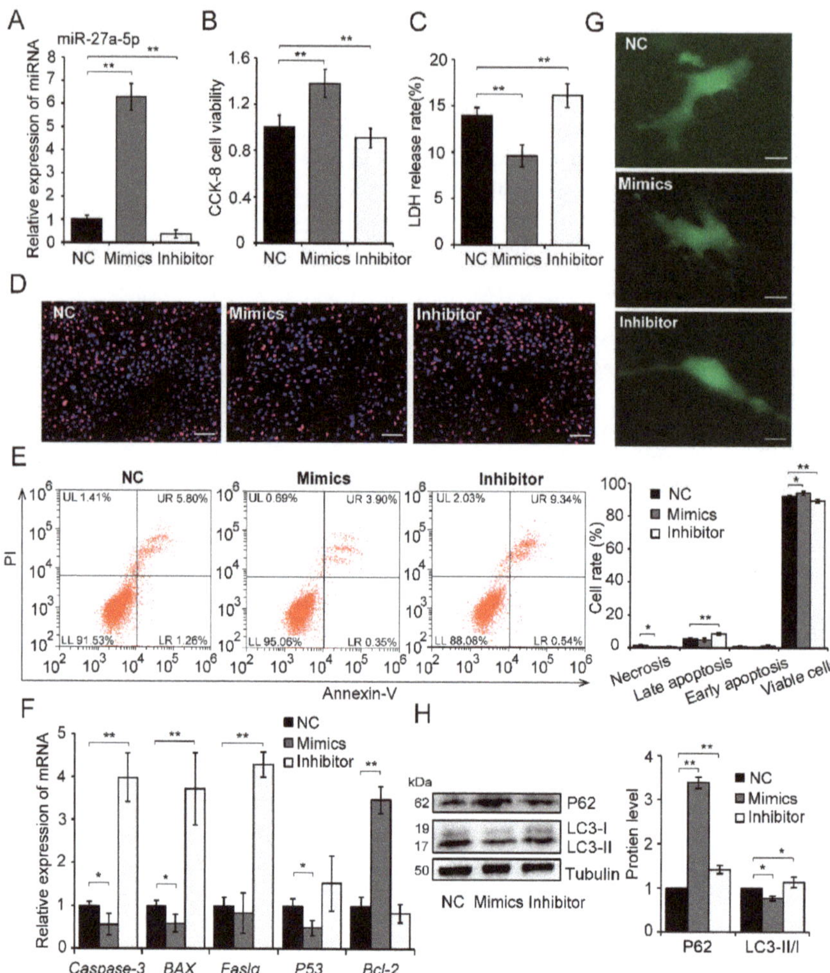

Figure 4. miR-27a-5p attenuates hypoxia-induced excessive autophagy and apoptosis in H9c2 cells. H9c2 cells were exposed to hypoxia for 24 h after transfection of a specific miR-27a -5p mimics or inhibitor. Transfection efficiency was analyzed by qRT-PCR (**A**). Cell viability (**B**), membrane damage (**C**), and cell apoptosis (**D–F**) were assessed by CCK8 assays, LDH release assays, apoptosis staining (scale bar: 50 μm), flow cytometry and qRT-PCR analysis, respectively. The level of autophagy was evaluated by GFP-LC3 fluorescence after hypoxia for 24 h (**G**); scale bar: 5 μm. Autophagy-related proteins were detected by western blot (**H**). Three independent experiments were performed in triplicate. Data are expressed as the mean ± SD. * $p < 0.05$, ** $p < 0.01$. NC: negative control.

2.4. Atg7 is The Target of miR-27a-5p

To explore the mechanism underlying miR-27a-5p regulation of excessive autophagy and inhibition of apoptosis, we analyzed candidate target genes of miR-27a-5p using TargetScan (release 7.2) [20] and RNAhybrid 2.2 prediction [21]. The prediction results showed that the 3′-UTR region of *Atg7* mRNA contained a target site for miR-27a-5p, and *Atg7* has been linked to autophagy [22]. We tested the expression of *Atg7* and miR-27a-5p in hypoxia-exposed H9c2 cells and in the main visceral tissues of AMI rat, and then performed a correlation analysis. We found a strongly negative correlation between the expression of *Atg7* and miR-27a-5p in hypoxia-exposed H9c2 cells at different timepoints ($r = -0.807$; Figure 5A). Meanwhile, a moderate negative correlation was observed in AMI rat visceral tissue ($r = -0.569$; Figure 5B). Additionally, overexpression of miR-27a-5p in hypoxia-exposed H9c2 cells significantly reduced *Atg7* mRNA and protein expression, while miR-27a-5p downregulation showed an opposite trend ($p < 0.01$; Figure 5C,D). The aforementioned results suggest that miR-27a-5p alleviates hypoxia-induced cardiomyocyte injury by targeting *Atg7*.

We subsequently performed a dual-luciferase reporter assay to confirm the potential relationship between *Atg7* and miR-27a-5p. The sequence alignment of miR-27a-5p showed high similarity, and likewise miR-27a-5p-binding site in *Atg7* 3′-UTR among several representative species were also conserved, which suggested the conservative interaction mechanism of miR-27a-5p-*Atg7* pair among species (Figure 5E). *Atg7* 3′-UTR containing the miR-27a-5p binding site (WT or MUT) was inserted into dual luciferase plasmid (pmirGLO-*Atg7*-3′-UTR) (Figure 5E). HeLa cells were co-transfected with the WT or MUT recombinant plasmid and miR-27a-5p mimics. Luciferase activity was detected 48 h after transfection. As shown in Figure 5F, co-transfection of miR-27a-5p and WT pmirGLO reporter significantly inhibited luciferase activity compared with the negative control (0.462 fold-change, $p < 0.01$). This effect was eliminated with the MUT pmirGLO reporter, which indicates that *Atg7* is a direct target for miR-27a-5p. A standard validation reporting for miR-27a-5p-*Atg7* interaction in this study is shown in Table S3 [23,24].

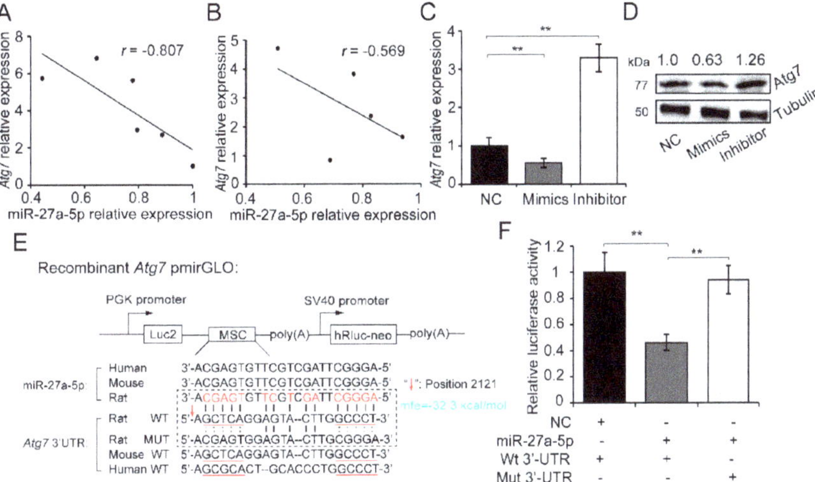

Figure 5. *Atg7* is a direct target of miR-27a-5p. Relative expression correlation analysis between miR-27a-5p and *Atg7* during hypoxia at different timepoints (0, 3, 6, 12, 24 and 48 h after hypoxia) in H9c2 cells (**A**), and in AMI/sham rat visceral tissues (**B**). mRNA (**C**) and protein (**D**) expression of *Atg7* was tested by qRT-PCR and western blotting after miR-27a-5p gain and loss of function in hypoxia-exposed H9c2 cells. (**E**) Schematic diagram showing the structure of dual-luciferase reporter plasmid pmirGLO and the sequence alignment of miR-27a-5p and *Atg7* 3′-UTR among several representative species (human, mouse and rat). *Atg7* 3′-UTR containing the miR-27a-5p binding site (WT or MUT) was inserted into the multiple cloning site (MSC) of pmirGLO plasmid. (**F**) Luciferase activity was analyzed after co-transfection of recombinant plasmid (WT or MUT) with miR-27a-5p mimic or control into HeLa cells. Three independent experiments were performed in triplicate. Data are expressed as the mean ± SD. ** $p < 0.01$. NC: negative control; mfe: minimum free energy.

3. Discussion

In recent years, miRNAs have frequently been reported in cardiovascular disease and play important roles in ischemic heart diseases by regulating the process of autophagy and apoptosis [4]. miR-27a-5p belongs to the miRNA-23a-27a-24 cluster that is reported to be involved in many cardiac diseases [25]. miR-24 has been shown to attenuate mouse AMI and reduces cardiac dysfunction by inhibiting cardiomyocyte apoptosis [26]; miR-23a has been shown to positively regulate cardiac hypertrophy by targeting anti-hypertrophic factor *MuRF1* [27] and *Foxo3a* [28]. Although miRNA-27a has been shown to be involved in the regulation of cardiomyocyte apoptosis, during cardioplegia-induced cardiac arrest through IL10-related pathways [29]; whether it regulates cardiomyocyte survival under hypoxic stress caused by ischemic heart diseases such as AMI, remains to be investigated. Based on previous report that the expression of miR-27a-5p decreased in hypoxia-exposed H9c2 cells, we found in the present study that miR-27a-5p expression likewise decreased in AMI rat myocardium (Figure 3F). More deeply, we revealed the miR-27a-5p-*Atg7* interaction in vivo and in vitro, and functionally, miR-27a-5p attenuated hypoxia-induced cardiomyocyte injury by regulating autophagy and apoptosis via *Atg7*, which further confirmed the crucial roles of miRNA-23a-27a-24 cluster in heart diseases.

Autophagy is an evolutionarily conserved and tightly regulated process that maintains cellular homeostasis in response to stresses, such as hypoxia, by degrading abnormal protein and damaged organelles [30,31]. Nevertheless, autophagy is considered a double-edged sword in the context of AMI, i.e., autophagy may have both damaging and protective roles depending on the hypoxic conditions, such as duration or severity [6,19]. In this study, we found that the degree of autophagy in hypoxia-exposed H9c2 cells increased in a time-dependent manner (Figure S1B). Inhibition of autophagy (hypoxia + 3-MA pretreatment) decreased cell viability and increased hypoxia-induced membrane damage compared with control (hypoxia) at the early stages of hypoxia exposure (within first 12 h), however these effects were alleviated after hypoxia for 24 h (Figure S1C,D). These results indicate that autophagy plays different roles in hypoxia-induced H9c2 cell injury over time and is beneficial in early stage of hypoxia but detrimental after 24 h of hypoxia (excessive autophagy), in keeping with previous reports [12]. Thus, elucidating and manipulating the development of cardiomyocyte autophagy under hypoxia may be beneficial to the clinical treatment of ischemic heart diseases.

Acting as the only E1-like enzyme, *Atg7* is located in the hub of the LC3 and Atg12 ubiquitin-like systems and is essential for the expansion of autophagosomal membranes [22]. Accumulating evidence suggests that *Atg7* is not only a crucial marker of autophagy, but also participates in the regulation of cell death and survival [32,33], including in cardiac progenitor cells [34]. Previously, we noted that the expression of miR-27a-5p decreased in acute hypoxia-exposed H9c2 cells using a small RNA-seq, as a known hypoxamiR, however, its underlying function in the cardiomyocyte hypoxic response is unclear [13]. In this study, we showed for the first time, to our knowledge, the negative correlation of miR-27a-5p-*Atg7* pair in vivo and in vitro, and that miR-27a-5p alleviated hypoxia-induced cardiomyocyte injury through regulation of excessive autophagy and apoptosis by inhibiting *Atg7* in vitro. This further highlights miRNA regulation in hypoxia-related heart diseases and may have potential implications for the treatment of ischemic cardiomyopathy in the future. However, the function of miR-27a-5p in hypoxia-induced cardiomyocyte injury is mainly focused on the cell-based experiments in vitro. Thus, animal studies on miR-27a-5p knock in/out, such as CRISP-Cas9-mediated gene editing, may better demonstrate the function of miR-27a-5p in hypoxia-induced cardiomyocyte injury after AMI and this should be performed in future research. In addition, although the sequence in miR-27a-5p and *Atg7* 3′-UTR has high similarity among several representative species, the function and strength of miR-27a-5p and its clinical application in human remain to be further elucidated.

4. Materials and Methods

4.1. Rat AMI Model

Healthy male Sprague Dawley (SD) rats (308 ± 14 g) were bought from Dashuo Laboratory Animal Center (Chengdu, Sichuan, China) and housed in a standard environment (20 ± 2 °C and 58% ± 2% humidity), with free choice feeding for 1 week before experiment. All animal procedures complied with the Ethics Committee of Sichuan Agricultural University rules (Approval Number DKY-B20171903, 15 February 2018). Coronary artery ligation was performed as previously described, to establish the rat AMI model [35]. Arterial BP and ECG were measured throughout the experiment. A clear elevation of the S-T segment of the ECG indicated successful AMI in the rat ($n = 9$). The same procedure was carried out without coronary artery ligation as sham control ($n = 9$). All rats were anesthetized and euthenized 6 h after coronary artery ligation. Several main visceral tissues were collected and immediately immersed in liquid nitrogen before storing at $-80\ °C$ for further experimentation.

4.2. H9c2 Cell Culture and Hypoxia Treatment

H9c2 cells (an embryonic rat heart-derived cell line) were routinely maintained in Dulbecco's Modified Eagle Medium (DMEM) (Hyclone, Logan, UT, USA) with 10% fetal bovine serum (FBS) (GIBCO, Grand Island, NY, USA) at 37 °C in a humidified atmosphere containing 5% CO_2 and 95% air. To establish hypoxia in vitro, cells with 50% confluency received hypoxia treatment for 24 h in a modular incubator chamber with 5% CO_2, 1% O_2 and 95% N_2 (MIC-101, Billups-Rothenberg, Del Mar, CA). Cells in the normoxic group were placed in conventional conditions (5% CO_2 and 95% air) and served as the control.

4.3. H9c2 Cell Transfection

Specific mimics and inhibitor of miR-27a-5p (RIBOBIO, Guangzhou, Guangdong, China) were transfected in cells at 50% confluency to facilitate gain and loss of function. Three groups of cells were designed; a mimic, an inhibitor and a negative control. Transfection solutions were premixed and added to the medium at a final concentration of 50 nM (or 100 nM for the inhibitor) using Lipofectamine 2000 (Invitrogen, Grand Island, NY, USA) in accordance with the manufacturer's protocol. After 6 h in the transfection medium, all groups were replaced with new medium before receiving hypoxia treatment for 24 h for subsequent experimentation.

4.4. Cell Counting Kit-8 (CCK8) and Lactate Dehydrogenase (LDH) Release Assay

To evaluate hypoxia-induced cell injury, cell viability and LDH release were analyzed using a CCK8 and a LDH Cytotoxicity Assay Kit (Beyotime, Shanghai, China), respectively. H9c2 cells were cultured in 96-well plate and received the relevant treatments (such as hypoxia, transfection) at the given time. For CCK8 detection, 10 μL CCK8 reagent was added to the culture medium 4 h before analysis. Optical density $(OD)_{450}$ values were measured using a microplate reader (Thermo Fisher Scientific, Madrid, Spain). For LDH release analysis, the culture medium in each group was premixed with the relevant reagent and incubated in accordance with the manufacturer's protocol. OD_{490} values were measured and LDH release rate presented as the percentage of the maximum enzymatic activity. At least three independent experiments were repeated three times. All values are presented as mean ± standard deviation (SD).

4.5. Cell Apoptosis Analysis

Cell apoptosis was assessed using an Annexin V-FITC and propidium iodide (PI) detection kit (BD Pharmingen, San Diego, CA, USA), in accordance with the manufacturer's protocols. Briefly, cells were digested by trypsin and gently washed with phosphate buffered saline (PBS). Cells were then incubated with Annexin V and PI for 10 min at room temperature and assessed by flow cytometry (Beckman Coulter, Brea, USA). The raw data were analyzed using CytExpert 2.0 software and more than 10,000 cells in each group were used for statistical analysis. All values are presented as mean ± SD.

4.6. HE Staining and Fluorescence Staining of Apoptosis

Tissue sections of the left ventricle were assessed using HE staining. In brief, the rat myocardium was fixed with 4% paraformaldehyde at room temperature, followed by dehydration and embedding in paraffin. The sections were prepared and successively stained using eosin and hematoxylin (Beyotime, Shanghai, China). To observe cell apoptosis, fluorescence staining of H9c2 cells was performed using an apoptosis and necrosis assay kit (Beyotime, Shanghai, China) in accordance with the manufacturer's instructions. Stained tissue sections and cells were imaged using an Olympus IX53 microscope (Olympus, Tokyo, Japan).

4.7. Detection of Autophagosome Formation

H9c2 cells were plated on coverslips. GFP-LC3 plasmids (Beyotime, Shanghai, China) were transfected into H9c2 cells at 50% confluency, before miRNA transfection and exposure to hypoxia. Afterwards, the cells were fixed with 10% formalin and GFP-LC3 fluorescence punctae were imaged using a confocal fluorescence microscope (Olympus, Tokyo, Japan).

4.8. Luciferase Reporter Assay

Luciferase activity assays were performed to validate the potential relationship between miR-27a-5p and *Atg7*. Briefly, HeLa cells were routinely maintained in DMEM with 10% FBS at 37 °C. We synthesized the *Atg7* 3′-UTR sequence containing the miR-27a-5p binding site (WT or MUT) and then cloned into the MCS of pmirGLO plasmid (Figure 5E). The WT or MUT recombinant pmirGLO vector was cotransfected with miR-27a-5p mimic or negative control into HeLa cells using Lipofectamine 3000 (Invitrogen, Grand Island, NY, USA), in accordance with the manufacturer's instructions. Dual luciferase activity was tested by Luciferase Dual Assay Kit (Promega, Madison, WI, USA) 48 h after transfection. Luciferase activity is expressed as an adjusted value (firefly normalized to renilla).

4.9. Total RNA Extraction and qRT-PCR

Total RNA was extracted from tissue or cultured cells using HiPure Total RNA Mini Kit (Magen, Guangzhou, China). The quality of total RNA was assessed by NanoDrop 2000 (Thermo Fisher Scientific, Wilmington, DE) and gel electrophoresis. The reverse transcription of mRNA and miRNA from the qualified total RNA was performed using PrimeScript™ RT Reagent Kit (Takara, Beijing, China) and Mir-X™ miRNA First Strand Synthesis Kit (Clontech, Mountain View, USA), respectively, according to the manufacturers' protocols. qPCR reactions were prepared using an SYBR Premix Ex Taq kit (Takara, Beijing, China) and performed in a Bio-Rad CFX96 PCR System (Bio-Rad, Hercules, USA). The relative expression of mRNA and miRNA was calculated using the $2^{-\Delta\Delta Ct}$ method and expressed as fold-change relative to the corresponding control. *GAPDH* and *U6* served as the reference genes for miRNA and mRNA, respectively. All primers used for qPCR are listed in Table S4.

4.10. Western Blot Analysis

Western blot analysis was performed as previously described [36]. Total protein was extracted from the H9c2 cells and rat myocardium using radioimmunoprecipitation assay lysis buffer containing protease and phosphatase inhibitors (Beyotime, Beijing, China) and quantified using a BCA protein assay. Approximately 30 g of protein was loaded and separated on an 8% SDS-PAGE gel, and then transferred to polyvinylidene difluoride membranes (BIO-RAD, Hercules, USA). The membranes were blocked with nonfat milk for 2 h at room temperature, and then incubated with primary antibodies at 4 °C overnight. Subsequently, the membranes were washed in PBS with Tween-20 before incubating with secondary antibodies for 2 h at room temperature. The antigen–antibody bands were visualized and quantified using ImageJ software (Bethesda, MA, USA). The primary antibodies used in this study and corresponding dilution ratios were as follows: anti-alpha Tubulin (1:1000), anti-Atg7 (1:500), anti-LC3 (1:1000), anti-P62 (1:1000), anti-HIF-1α (1:1000) (Abcam, Cambridge, USA).

4.11. Statistical Analysis

All experiments were performed as at least three independent experiments with three technical repetitions. The data are expressed as mean ± SD. Significance tests were performed using SPSS 22.0 software (SPSS, Chicago, USA). Unpaired Student's *t*-test and one-way ANOVA with Tukey's post-hoc test were used to evaluate the differences between two groups or three or more groups, respectively. The Δ BP were used as a surrogate measure of effect to perform a *post hoc* power analysis. The parameters "($n = 9$, d $= \frac{|\mu_1 - \mu_2|}{\rho}$, sig.level = 0.05, power = , type = "two.sample", alternative = "two.sided")" were performed with R (Version 3.2.0) computed by the *pwr* package [37]. $p < 0.05$ was considered as statistically significant (* $p < 0.05$, ** $p < 0.01$).

5. Conclusions

We have shown that AMI-induced hypoxia causes cell injury and the expression of miR-27a-5p is decreased in hypoxia-exposed H9c2 cells and AMI rat myocardium. miR-27a-5p attenuates hypoxia-induced cardiomyocyte injury by inhibiting excessive autophagy and apoptosis via *Atg7*. Our findings show that miR-27a-5p has a cardioprotective effect on hypoxia-induced H9c2 injury, and may serve as a novel target for the treatment of hypoxia-related heart diseases.

Supplementary Materials: Supplementary materials can be found at http://www.mdpi.com/1422-0067/20/10/2418/s1.

Author Contributions: J.Z., W.Q., S.S. and M.L. conceived and designed the study and drafted the manuscript. J.Z., J.M., Y.W., Z.H., and K.L. performed the experiments. X.W., L.J., Q.T., G.T., and L.Z. analyzed the experiment data. X.L., S.S. and M.L. revised the manuscript. All authors read and approved the final manuscript.

Funding: This work was supported by grants from the National Key R & D Program of China (2018YFD0501204 and 2018YFD0500403), the National Natural Science Foundation of China (31601918, 31530073, 31872335 and 31772576), the Sichuan Province & Chinese Academy of Science of Science & Technology Cooperation Project (2017JZ0025), the Earmarked Fund for China Agriculture Research System (CARS-35-01A), the Key Project of Sichuan Education Department (16ZA0025), the Tibet Municipal Natural Science Foundation Project (XZ2018ZRG-132), the Science & Technology Support Program of Sichuan (2016NYZ0042 and 2017NZDZX0002).

Conflicts of Interest: The authors declare no conflict of interest.

Abbreviations

AMI	acute myocardial infarction
miRNA	microRNA
LAD	left anterior descending
CCK8	cell counting kit-8
LDH	lactate dehydrogenase
ECG	electrocardiogram
BP	blood pressure
NC	negative control
FBS	fetal bovine serum
PBS	phosphate buffered saline
PI	propidium iodide
OD	optical density
HE staning	hematoxylin & eosin staning
MSC	multiple cloning site
kDa	kilodalton
UL	upper left
UR	upper right
LL	lower left
LR	lower right
SD	standard deviation
mfe	minimum free energy
WT/Mut	wild-type/mutant

qRT-PCR   quantitative reverse-transcription polymerase chain reaction
ANOVA   analysis of variance

References

1. Benjamin, E.J.; Virani, S.S.; Callaway, C.W.; Chamberlain, A.M.; Chang, A.R.; Cheng, S.; de Ferranti, S.D. Heart disease and stroke statistics—2018 update: A report from the American Heart Association. *Circulation* **2018**, *137*, e67–e492. [CrossRef]
2. Anderson, J.L.; Morrow, D.A. Acute Myocardial Infarction. *New Engl. J. Med.* **2017**, *376*, 2053. [CrossRef]
3. Nikoletopoulou, V.; Markaki, M.; Palikaras, K.; Tavernarakis, N. Crosstalk between apoptosis, necrosis and autophagy. *BBA-Mol. Cell Res.* **2013**, *1833*, 3448–3459. [CrossRef]
4. Sermersheim, M.A.; Park, K.H.; Gumpper, K.; Adesanya, T.M.; Song, K.; Tan, T.; Ren, X.; Yang, J.M.; Zhu, H. MicroRNA regulation of autophagy in cardiovascular disease. *Front Biosci.* **2017**, *22*, 48.
5. Nishida, K.; Kyoi, S.; Yamaguchi, O.; Sadoshima, J.; Otsu, K. The role of autophagy in the heart. *Cell Death Differ.* **2009**, *16*, 31–38. [CrossRef]
6. Hongxin, Z.; Paul, T.; Johnstone, J.L.; Yongli, K.; Shelton, J.M.; Richardson, J.A.; Vien, L.; Beth, L.; Rothermel, B.A.; Hill, J.A. Cardiac autophagy is a maladaptive response to hemodynamic stress. *J. Clin. Invest.* **2007**, *117*, 1782–1793.
7. Krol, J.; Loedige, I.W. The widespread regulation of microRNA biogenesis, function and decay. *Nat. Rev. Genet.* **2010**, *11*, 597–610. [CrossRef]
8. Ucar, A.; Gupta, S.K.; Fiedler, J.; Erikci, E.; Kardasinski, M.; Batkai, S.; Dangwal, S.; Kumarswamy, R.; Bang, C.; Holzmann, A. The miRNA-212/132 family regulates both cardiac hypertrophy and cardiomyocyte autophagy. *Nat. Commun.* **2012**, *3*, 1078. [CrossRef]
9. Wang, K.; Liu, C.Y.; Zhou, L.Y.; Wang, J.X.; Wang, M.; Zhao, B.; Zhao, W.K.; Xu, S.J.; Fan, L.H.; Zhang, X.J. APF lncRNA regulates autophagy and myocardial infarction by targeting miR-188-3p. *Nat. Commun.* **2015**, *6*, 6779. [CrossRef]
10. Huang, Z.; Wu, S.; Kong, F.; Cai, X.; Ye, B.; Shan, P.; Huang, W. MicroRNA-21 protects against cardiac hypoxia/reoxygenation injury by inhibiting excessive autophagy in H9c2 cells via the Akt/mTOR pathway. *J. Cell Mol. Med.* **2017**, *21*, 467–474. [CrossRef]
11. Qiu, R.; Wen, L.; Liu, Y. MicroRNA-204 protects H9C2 cells against hypoxia/reoxygenation-induced injury through regulating SIRT1-mediated autophagy. *Biomed. Pharmacol.* **2018**, *100*, 15–19. [CrossRef]
12. Liu, X.; Deng, Y.; Xu, Y.; Jin, W.; Li, H. MicroRNA-223 protects neonatal rat cardiomyocytes and H9c2 cells from hypoxia-induced apoptosis and excessive autophagy via the Akt/mTOR pathway by targeting PARP-1. *J. Mol. Cell Cardiol.* **2018**, *118*, 133–146. [CrossRef]
13. Zhang, J.; Ma, J.; Long, K.; Qiu, W.; Wang, Y.; Hu, Z.; Liu, C.; Luo, Y.; Jiang, A.; Jin, L. Overexpression of Exosomal Cardioprotective miRNAs Mitigates Hypoxia-Induced H9c2 Cells Apoptosis. *Int. J. Mol. Sci.* **2017**, *18*, 711. [CrossRef] [PubMed]
14. Gao, Y.H.; Qian, J.Y.; Chen, Z.W.; Fu, M.Q.; Xu, J.F.; Xia, Y.; Ding, X.F.; Yang, X.D.; Cao, Y.Y.; Zou, Y.Z. Suppression of Bim by microRNA-19a may protect cardiomyocytes against hypoxia-induced cell death via autophagy activation. *Toxicol. Lett.* **2016**, *257*, 72–83. [CrossRef]
15. Yutaka, M.; Hiromitsu, T.; Xueping, Q.; Maha, A.; Hideyuki, S.; Tomoichiro, A.; Beth, L.; Junichi, S. Distinct roles of autophagy in the heart during ischemia and reperfusion: Roles of AMP-activated protein kinase and Beclin 1 in mediating autophagy. *Circ. Res.* **2007**, *100*, 914–922.
16. Boon, R.A.; Kazuma, I.; Stefanie, L.; Timon, S.; Ariane, F.; Susanne, H.; David, K.; Karine, T.; Guillaume, C.; Angelika, B. MicroRNA-34a regulates cardiac ageing and function. *Nature* **2013**, *495*, 107–110. [CrossRef] [PubMed]
17. Nishida, K.; Yamaguchi, O.; Otsu, K. Crosstalk between autophagy and apoptosis in heart disease. *Circ. Res.* **2008**, *103*, 343–351. [CrossRef] [PubMed]
18. Li, M.; Gao, P.; Zhang, J. Crosstalk between Autophagy and Apoptosis: Potential and Emerging Therapeutic Targets for Cardiac Diseases. *Int. J. Mol. Sci.* **2016**, *17*, 332. [CrossRef] [PubMed]
19. Sai, M.; Yabin, W.; Yundai, C.; Feng, C. The role of the autophagy in myocardial ischemia/reperfusion injury. *BBA-Mol. Basis Dis.* **2015**, *1852*, 271–276.
20. Maziere, P.; Enright, A.J. Prediction of microRNA targets. *Drug Discov. Today* **2007**, *12*, 452–458. [CrossRef]

Int. J. Mol. Sci. **2019**, *20*, 2418

21. Jan, K.; Marc, R. RNAhybrid: MicroRNA target prediction easy, fast and flexible. *Nucleic Acids Res.* **2006**, *34*, W451–W454.
22. Xiong, J. Atg7 in development and disease: Panacea or pandora's box? *Protein Cell* **2015**, *6*, 722–734. [CrossRef]
23. Piletič, K.; Kunej, T. Minimal standards for reporting microRNA: Target interactions. *Omics* **2017**, *21*, 197–206. [CrossRef]
24. Desvignes, T.; Batzel, P.; Berezikov, E.; Eilbeck, K.; Eppig, J.T.; McAndrews, M.S.; Singer, A.; Postlethwait, J. miRNA nomenclature: A view incorporating genetic origins, biosynthetic pathways, and sequence variants. *Trends Genet.* **2015**, *31*, 613–626. [CrossRef]
25. Chhabra, R.; Dubey, R.; Saini, N. Cooperative and individualistic functions of the microRNAs in the miR-23a~27a~24-2 cluster and its implication in human diseases. *Mol. Cancer* **2010**, *9*, 232. [CrossRef]
26. Li, Q.; Van Laake, L.W.; Yu, H.; Siyuan, L.; Wendland, M.F.; Deepak, S. miR-24 inhibits apoptosis and represses Bim in mouse cardiomyocytes. *J. Exp. Med.* **2011**, *208*, 549–560.
27. Zhiqiang, L.; Iram, M.; Kun, W.; Jianqin, J.; Jie, G.; Pei-Feng, L. miR-23a functions downstream of NFATc3 to regulate cardiac hypertrophy. *Proc. Natl. Acad. Sci. USA* **2009**, *106*, 12103–12108.
28. Kun, W.; Zhi-Qiang, L.; Bo, L.; Jian-Hui, L.; Jing, Z.; Pei-Feng, L. Cardiac hypertrophy is positively regulated by MicroRNA miR-23a. *J. Biol. Chem.* **2012**, *287*, 589–599.
29. Yeh, C.H.; Chen, T.P.; Wang, Y.C.; Lin, Y.M.; Fang, S.W. MicroRNA-27a regulates cardiomyocytic apoptosis during cardioplegia-induced cardiac arrest by targeting interleukin 10-related pathways. *Shock* **2012**, *38*, 607–614. [CrossRef] [PubMed]
30. Kroemer, G.; Mariño, G.; Levine, B. Autophagy and the integrated stress response. *Mol. Cell* **2010**, *40*, 280–293. [CrossRef]
31. Chun, Y.; Kim, J. Autophagy: An Essential Degradation Program for Cellular Homeostasis and Life. *Cells* **2018**, *7*, 278. [CrossRef]
32. In Hye, L.; Yoshichika, K.; Fergusson, M.M.; Rovira, I.I.; Bishop, A.J.R.; Noboru, M.; Liu, C.; Toren, F. Atg7 modulates p53 activity to regulate cell cycle and survival during metabolic stress. *Science* **2012**, *336*, 225–228.
33. Li, Y.; Ajjai, A.; Helen, S.; Parmesh, D.; Eric, F.; Sarah, W.; Baehrecke, E.H.; Lenardo, M.J. Regulation of an ATG7-beclin 1 program of autophagic cell death by caspase-8. *Science* **2004**, *304*, 1500–1502.
34. Ma, W.; Ding, F.; Wang, X.; Huang, Q.; Zhang, L.; Bi, C.; Hua, B.; Yuan, Y.; Han, Z.; Jin, M. By Targeting Atg7 MicroRNA-143 Mediates Oxidative Stress-Induced Autophagy of c-Kit + Mouse Cardiac Progenitor Cells. *EBioMedicine* **2018**, *32*, 182–191. [CrossRef] [PubMed]
35. Xin, L.; Baoqiu, W.; Hairong, C.; Yue, D.; Yang, S.; Lei, Y.; Qi, Z.; Fei, S.; Dan, L.; Chaoqian, X. Let-7e replacement yields potent anti-arrhythmic efficacy via targeting beta 1-adrenergic receptor in rat heart. *J. Cell Mol. Med.* **2014**, *18*, 1334–1343.
36. Ma, J.; Zhang, J.; Wang, Y.; Long, K.; Wang, X.; Jin, L.; Tang, Q.; Zhu, L.; Tang, G.; Li, X. MiR-532-5p alleviates hypoxia-induced cardiomyocyte apoptosis by targeting *PDCD4*. *Gene* **2018**, *675*, 36–43. [CrossRef]
37. Basic Functions for Power Analysis. Available online: https://cran.r-project.org/web/packages/pwr/index.html (accessed on 5 May 2019).

International Journal of
Molecular Sciences

Review

microRNAs Tune Oxidative Stress in Cancer Therapeutic Tolerance and Resistance

Wen Cai Zhang

Burnett School of Biomedical Sciences, College of Medicine, University of Central Florida, 6900 Lake Nona Blvd, Orlando, FL 32827, USA; wencai.zhang@ucf.edu; Tel.: +1-(407) 266-7178

Received: 30 October 2019; Accepted: 27 November 2019; Published: 3 December 2019

Abstract: Relapsed disease following first-line therapy remains one of the central problems in cancer management, including chemotherapy, radiotherapy, growth factor receptor-based targeted therapy, and immune checkpoint-based immunotherapy. Cancer cells develop therapeutic resistance through both intrinsic and extrinsic mechanisms including cellular heterogeneity, drug tolerance, bypassing alternative signaling pathways, as well as the acquisition of new genetic mutations. Reactive oxygen species (ROSs) are byproducts originated from cellular oxidative metabolism. Recent discoveries have shown that a disabled antioxidant program leads to therapeutic resistance in several types of cancers. ROSs are finely tuned by dysregulated microRNAs, and vice versa. However, mechanisms of a crosstalk between ROSs and microRNAs in regulating therapeutic resistance are not clear. Here, we summarize how the microRNA–ROS network modulates cancer therapeutic tolerance and resistance and direct new vulnerable targets against drug tolerance and resistance for future applications.

Keywords: microRNA; cancer; oxidative stress; reactive oxygen species; redox signaling; hypoxia; therapeutic tolerance; therapeutic resistance

---

1. Reactive Oxygen Species (ROSs)

There are many types of free radicals including oxygen- and nitrogen-based species. ROSs or reactive oxygen metabolites are free radicals containing oxygen metabolites such as single oxygen, the superoxide anion, hydrogen peroxide, and the hydroxyl radical [1]. ROSs are generated from cellular oxidative metabolism, including mitochondrial oxidative phosphorylation and electron transfer reactions, and optimal levels of ROSs play a pivotal role in many cellular functions [2]. At physiological levels, ROSs are considered signaling molecules or secondary messengers that participate in cell signal transduction, a process known as redox signaling [3]. In addition, the production of ROSs by phagocytic cells is recognized as an important part of innate immunity that kills invading pathogens [4].

The coordination between ROS generation and scavenging ensures that ROS levels are tightly controlled and fine-tuned so as to act as secondary messengers for cell signaling [5]. However, the aberrant production of ROSs, or the failure of the capacity to scavenge excessive ROSs, results in an imbalance in the redox environment of the cell [6]. High levels of ROSs have deleterious effects including nucleic acid (DNA and RNA), lipid, and protein oxidation, as well as membrane destruction by lipid peroxide formation, leading to the development of various diseases such as cancer [7]. Using antioxidant-based strategies [8] to decrease ROS levels or inhibit oxidative damage may prevent ROS-induced cell damage. For example, peroxisome proliferator-activated receptor-gamma coactivator 1 alpha upregulates expression levels of superoxide dismutase enzymes (SOD2/SOD3) and catalase to protect cells from oxidative damage via detoxification and DNA repair [9].

Aberrantly regulated metabolic pathways lead to tumorigenesis [10] and preferential survival of tumor cells [11]. Accumulating evidence suggests that tumorigenesis is dependent on mitochondrial metabolism [12], especially the tricarboxylic acid (TCA) cycle [13]. The TCA cycle is a central pathway in the metabolism of sugars, lipids, and amino acids [14]. Dysregulation of the TCA cycle can induce oncogenesis by activating pseudohypoxia responses, which result in the expression of hypoxia-associated proteins irrespective of oxygen status [15]. For example, succinate accumulation caused by functional loss of the TCA cycle enzyme succinate dehydrogenase complex stabilizes hypoxia-inducible factor (HIF)-1α via inhibition of prolyl hydroxylase (PHD) [16]. In addition, loss of function of the von Hippel–Lindau (VHL) protein [17] also induces pseudohypoxia responses through decreased ubiquitination and proteasomal degradation of HIF-1α [18]. Among the 1158 mitochondrial genes discovered in MitoCarta2.0 (Broad Institute) [19,20], the succinate dehydrogenase complex [21] inclusive of succinate dehydrogenase A [22], succinate dehydrogenase B [23], succinate dehydrogenase C [24], and succinate dehydrogenase D [25], as well as glycine decarboxylase [26–29] and glutaminase [30], is especially critical for tumorigenesis. Hypoxia, acting through HIF-1α, results in a low production of ROSs and high antioxidant defense in cancers such as leukemia [31]. It suggests that targeting key enzymes of hypoxia metabolism pathways might provide a new way to eradicate tumor formation [32].

2. microRNAs (miRNAs)

miRNAs are important regulators of mRNA expression [33] and play critical roles in regulating tumor initiation and progression [34]. Importantly, single miRNAs have been shown to regulate entire cell signaling networks in a cell-context dependent manner [35] and may also be utilized as biomarkers [36–38] for both invasive [39,40] and non-invasive [41–43] detection. Dysregulated expressions of miRNAs may function as oncogenes (oncomiRs) [44] such as miR-21 [45], miR-31 [46], miR-155 [47,48], and miR-10b [49] or as tumor suppressors such as *let-7* [50] and miR-34 [51,52] in many cancers.

ROSs are finely tuned by dysregulated miRNAs, and vice versa. Many studies are focused on regulatory interactions between miRNAs and ROSs attributing to oxidative stress-related tissue [53]. It is important for a well-regulated cellular ROS level, and miRNAs fill in the role of maintaining this homeostasis. A dysregulation of normal physiological miRNA levels can thus lead to oxidative damage and the development of diseases such as cancer. For example, oncogenic miR-21 enhances both KRAS [54] and epidermal growth factor receptor (EGFR) signaling [55] and promotes tumorigenesis through stimulation of mitogen-activated protein kinase (MAPK)-mediated ROS production by downregulation of SOD2/SOD3 [56]. On the other hand, oxidative stress can alter the expression level of many miRNAs [57–59]. For instance, oxidative stress such as hydrogen peroxide elevates miR-34a with concomitant reduction of sirtuin-1 and sirtuin-6 in bronchial epithelial cells [60], which is associated with chronic obstructive pulmonary disease and tumorigenesis [61]. However, oxidative stress decreases expression levels of the *let-7* family [62] in a p53-dependent manner in a variety of tumor cells [63]. These findings suggest that ROSs may exert a pivotal role in the regulation of microRNA expression in a cell-context-dependent manner.

miRNA-based monotherapy has not been developed well in clinical settings [64–66]. For example, a first-in-man, phase 1 clinical trial of miR-16-loaded nanoparticles as a treatment for recurrent malignant pleural mesothelioma patients has been completed [67]. Delivery of tumor suppressive miR-16 in 22 patients led to 5% objective response, 68% stable disease, and 27% progressive disease. Possible mechanisms of low objective response include miRNA sequestration through leaky cancer blood vessels as well as endocytosis by cancer cells [68]. Nevertheless, miR-16 expression levels in patients should be detected prior to receiving miR-16 treatment in future clinical trials [69]. Furthermore, miRNA-based treatment may combine with other current or potential therapeutics in combating cancer [70,71]. In addition, increasing evidence has revealed that miRNAs can be directly linked to therapeutic resistance in some cancers. For instance, overexpressing miR-205 sensitizes radioresistant breast cancer cells to radiation in a xenograft model [72]. Similarly, administration of miR-24 sensitizes radioresistant nasopharyngeal carcinoma cells to radiation in vitro [73]. miRNA-mediated regulation of signaling pathways involved in tumorigenesis as well as therapeutic tolerance and resistance is summarized in Table 1. It is revealed that miRNAs may serve both as drug targets and as therapeutic agents to eradicate cancer cells and sensitize therapeutic resistant cells [74].

Table 1. miRNA-mediated regulation of signaling pathways involved in tumorigenesis as well as therapeutic tolerance and resistance.

miRNA	Signaling Involved in Tumorigenesis	Signaling Involved in Therapeutic Tolerance and Resistance
miR-1246 and miR-1290 ↑	(+) tumorigenesis via repressing metallothioneins in human non-small cell lung cancer [75]	(+) resistance to EGFR tyrosine kinase inhibitor gefitinib via repressing metallothioneins in human non-small cell lung cancer [75]
miR-147b ↑	N.A.	(+) tolerance to EGFR tyrosine kinase inhibitor osimertinib through activating pseudohypoxia signaling pathways via repressing VHL and succinate dehydrogenase in human non-small cell lung cancer [76]
miR-155 ↑	(+) tumorigenesis in mouse miR155 transgenic B cell lymphomas [77]	(+) chemoresistance to gemcitabine through decreasing apoptosis in human pancreatic cancer [78]
miR-21 ↑	(+) Ras/MEK/ERK signaling via repressing negative regulators of the Ras/MEK/ERK pathway and inhibition of apoptosis in mouse KRAS transgenic non-small cell lung cancer [54]	(+) chemoresistance to gemcitabine through decreasing apoptosis and activating Akt phosphorylation in human pancreatic cancer [79,80] (+) radioresistance through upregulation of hypoxia-inducible factor 1α in human non-small cell lung cancer [81] (+) resistance to EGFR tyrosine kinase inhibitors through activating PI3K-AKT signaling pathway in human non-small cell lung cancer [82]
miR-31 ↑	(+) tumorigenesis through activating RAS/MAPK signaling via repressing negative regulators of RAS/MAPK signaling in mouse *KRAS* transgenic non-small cell lung cancer [46]	N.A.
let-7 family ↓	(+) tumorigenesis in human breast cancer through repressing H-RAS and high mobility group AT-hook 2 [83]	(+) resistance to EGFR tyrosine kinase inhibitor gefitinib through upregulation of MYC in human non-small cell lung cancer [84]
miR-30 ↓	• (+) tumor initiation and (−) apoptosis by repressing ubiquitin-conjugating enzyme 9 and integrin beta3, respectively, in human breast cancer [85] • (+) mTOR/AKT-signaling pathway through repressing transmembrane 4 super family member 1 in human non-small cell lung cancer [86]	• (−) resistance to EGFR tyrosine kinase inhibitor gefitinib through repressing BCL2-like 11 and apoptotic peptidase activating factor 1 in human non-small cell lung cancer [87] • (+) chemoresistance to cisplatin through activating autophagy in human gastric cancer [88]
miR-34a/b/c ↓	• (+) tumor initiation in mouse *Kras; Trp53* transgenic lung cancer [51] • (+) tumor initiation by repressing inhibin subunit beta B and AXL in mouse *Apc* transgenic colorectal cancer [89]	(+) chemoresistance to fludarabine through p53 inactivation and apoptosis resistance in human chronic lymphocytic leukemia [90]

EGFR: epidermal growth factor receptor; Akt: Akt Serine/Threonine Kinase; MAPK: mitogen-activated protein kinase; MEK: Mitogen-activated protein kinase kinase; ERK: extracellular-signal-regulated kinase; PI3K: phosphatidylinositol 3-kinase; AXL: AXL receptor tyrosine kinase; Apc: adenomatous polyposis coli; VHL: Von Hippel–Lindau; mTOR: mammalian target of rapamycin; ↑: upregulation; ↓: downregulation; (+): promotion; (−): repression; N.A.: not available.

3. Therapeutic Tolerance and Resistance

The discovery of genetic mutation on tyrosine kinase, such as *EGFR* mutations including exon 19 deletion (*Del19*) and exon 21 Leu858Arg substitution (*L858R*), that confer sensitivity to EGFR-targeted tyrosine kinase inhibitors in lung adenocarcinomas heralded the beginning of the era of precision medicine for lung cancer [91,92]. However, the success of EGFR-based therapy was compromised by therapeutic resistance following initial treatment response in most cancer patients [93]. Exon 20 Thr790Met substitution (*T790M*), affecting the ATP binding pocket of the EGFR kinase domain, accounts for approximately half of all lung cancer cases with acquired resistance to the current first generation EGFR tyrosine kinase inhibitors, erlotinib and gefitinib [94]. In erlotinib- and gefitinib-resistant lung tumors with $EGFR^{T790M}$, rociletinib and osimertinib are highly active [95]. However, resistance to the third generation EGFR tyrosine kinase inhibitor osimertinib is now emerging clinically [96]. In addition to genetic mutations, intratumor heterogeneity also drives neoplastic progression and therapeutic resistance [97]. Recently, it has been found that $EGFR^{T790M}$-positive drug-resistant cells are derived from $EGFR^{T790M}$-negative drug-tolerant persister cells that survive initial EGFR tyrosine kinase inhibitors treatment [98,99]. It is therefore crucial to identify molecular changes that drive drug tolerance.

Consistently, Zhang et al. have revealed that lung tumor cells protect themselves with a drug-tolerance mechanism when the cells are treated with osimertinib [76]. These findings align with previous data showing that tumor cells enter into a tolerant state when they are treated with tyrosine kinase inhibitors in lung and other cancers [100–102]. These tolerant persister cells precede and evolve into resistant cells over time by acquiring *EGFR*-resistant mutations [98,99]. These tolerant cells are slow cycling and are enriched in the expression of stem-associated genes in the WNT/planar cell polarity signaling pathway, such as *WNT5A*, *FZD2*, and *FZD7*. These findings are conceptually similar to a recent report that post-drug transition to stable resistance consists of dedifferentiation [103].

Excessive ROSs produced by damaged mitochondria can trigger mitophagy, a process that can scavenge impaired mitochondria and reduce ROS levels to maintain a stable mitochondrial function in cells [104]. Therefore, mitophagy helps maintain cellular homeostasis under oxidative stress. For example, protein kinase inhibitor sorafenib shows activities against many protein kinases, including vascular endothelial growth factor receptor (VEGFR), platelet-derived growth factor receptor (PDGFR), and rapidly accelerated fibrosarcoma (RAF) kinases [105]. Resistance to sorafenib in cancers such as hepatocellular carcinoma is frequent [106] partially due to antiangiogenic effects-mediated hypoxia [107]. Administration of tryptophan-derived metabolites such as melatonin [108] increased ROS production and mitophagy, resulting in increased sensitivity to sorafenib in hepatocellular carcinoma cells [109]. Additionally, melatonin downregulated the HIF-1α protein synthesis through inhibition of the mammalian target of rapamycin complex 1 (mTORC1)-mediated pathway [110]. Most recently, it was shown that drug-tolerant persister cancer cells were vulnerable to inhibition of the glutathione peroxidase 4, owing to a disabled antioxidant program [102]. It suggests that increasing ROS levels may re-sensitize therapeutic resistant cancer cells to current treatments.

4. miRNA–ROS Interaction Regulates Therapeutic Tolerance/Resistance at the Phenotypic Level

The miRNA–ROS network in a scenario of therapeutic tolerance/resistance is grouped at three levels including phenotype, signaling/metabolism, and genetics/epigenetics (Figure 1). Phenotypic changes include the enrichment of tumor-initiating cells, the histological transformation from *EGFR*-mutant non-small cell lung cancer to small cell lung cancer, and epithelial–mesenchymal transition resulting in therapeutic tolerance/resistance.

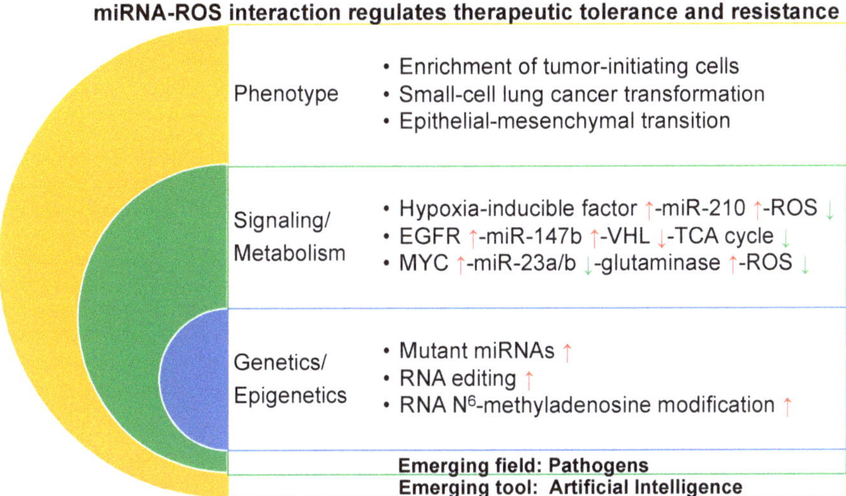

Figure 1. miRNA–ROS interaction regulates cancer therapeutic tolerance and resistance through heterogeneous mechanisms. The mechanisms at hierarchy levels include phenotypic, signaling/ metabolic, and genetic/epigenetic changes. ROS: reactive oxygen species; HIF: hypoxia-inducible factor; EGFR: epidermal growth factor receptor; VHL: von Hippel–Lindau; TCA: tricarboxylic acid; ↑: upregulation; ↓: downregulation.

4.1. Enrichment of Tumor-Initiating Cells

Therapeutic resistance is frequent after primary and adjuvant cancer therapy, often evolving into a lethal relapse disease [111]. These observations may be attributed to the highly heterogeneous nature of tumors that contain distinct tumoral and microenvironment cells, all of which contribute in varying degrees toward self-renewal, drug resistance, and relapse [112]. The tumor-initiating cell or cancer stem cell model provides one explanation for the phenotypic and functional diversity among cancer cells in some tumors [113]. Tumor-initiating cells have been demonstrated to be more resistant to conventional therapeutic interventions [114] and are key drivers of relapse in many types of cancers including leukemia [115], lung cancer [116], breast cancer [117], brain cancer [118], colon cancer [119], and nasopharyngeal carcinoma [120]. There is, therefore, increasing interest in developing strategies that can specifically target tumor-initiating cells with novel and emerging therapeutic modalities, thereby halting cancer progression and improving disease outcome [121]. Tumor-initiating cells protect their genomes from ROS-mediated damage [122] via increased production of free radical scavengers [123] leading to low ROS levels [124]. Thus, heterogeneity of ROS levels in cancers such as glioma may influence the extent to which tumor-initiating cell-enriched populations are resistant to therapies such as ionizing radiation [125]. Tumor-initiating cells display heterogeneous phenotypes due to different genotypes in tumors [126]. Thus, the genetic backgrounds, such as mutant *EGFR* and *RAS*, need to be taken into consideration to better understand the association between tumor-initiating cells and therapeutic resistance in the future.

In non-small cell lung cancer, a panel of tumor-initiating cell-relevant miRNAs is enriched when assessed by a miRNA microarray [75]. Those top upregulated miRNAs include miR-1290 and miR-1246 (Table 1). The top downregulated miRNAs comprise miR-23a and *let-7b/c/d/i*. Further analysis showed that miR-1246 and miR-1290 regulate tumor-initiating cells via repressing cysteine-rich metal-binding proteins (metallothioneins) [75]. The reduced expression of metallothioneins has been implicated as biomarkers of low ROSs, which is consistent with the previous finding that pharmacological anti-oxidants such as N-acetyl cysteine or the knock-down of *nuclear respiratory factor 2* (*NRF2*) prevented the induction of metallothionein-1 induced by tyrosine kinase inhibitor sorafenib [127].

Another direct target of miR-1290, glioma pathogenesis-related protein 1, promotes apoptosis through upregulating ROS production by activating the c-Jun-NH(2) kinase signaling cascade in cancer cells [128]. Other evidence has shown that extracellular miR-1246 could enhance radioresistance of lung cancer cells [129]. In addition, miR-21 is enriched in tumor-initiating cells in many types of cancers such as gastric and breast cancers [130]. Functional loss of miR-21 reduces a frequency of tumor-initiating cells, consistently with decreased capacity of therapeutic resistance against EGFR tyrosine kinase inhibitors [82] (Table 1). Whether these miRNAs regulate ROSs resulting in therapeutic tolerance and resistance still needs further study. Thus, targeting enriched tumor-initiating cells might overcome miRNA–ROS-mediated therapeutic tolerance/resistance.

4.2. Small Cell Lung Cancer Transformation

Small cell lung cancer is a highly aggressive disease that exhibits rapid growth and genetic instability including inactivated *tumor suppressor retinoblastoma 1* (*RB1*) and amplified *MYC proto-oncogene* (*MYC*) [131]. Histologic transformation of *EGFR* mutant non-small cell lung cancer to small cell lung cancer is an important mechanism of resistance to EGFR tyrosine kinase inhibitors that occurs in approximately 3–10% of *EGFR* mutant non-small cell lung cancers [132]. Transformation to small cell lung cancer occurs in a subpopulation of *EGFR* mutant non-small cell lung cancer patients and is frequently associated with mutant *RB1*, *TP53*, and *PIK3CA* [133,134]. Future studies might help define which subsets of non-small cell lung cancer are most prone to small cell lung cancer transformation.

Frequent overexpression of the miR-17~92 cluster in small cell lung cancer [135] is a fine-tuner to reduce excessive ROS-induced DNA damage in RB1-inactivated small cell lung cancer cells [136]. Therefore, miR-17~92 may be excellent therapeutic target candidates to overcome small cell lung cancer transformation.

4.3. Epithelial–Mesenchymal Transition

An epithelial–mesenchymal transition is a biologic process that allows a polarized epithelial cell to undergo multiple biochemical changes that enable it to assume a mesenchymal cell phenotype, which includes increased resistance to apoptosis [137]. Epithelial–mesenchymal transition is tightly regulated by microRNAs. For example, downregulation of miR-200 family members is linked to enhanced epithelial–mesenchymal transition and tumor-initiating cell acquisition [138,139] in many cancers [140]. Reduced miR-200s directly increase p38α [141], leading to decreased levels of ROSs and subsequent inactivation of the NRF2 oxidative stress response pathway [142]. The decreased ROSs, in turn, inhibit expression of the miR-200s [143], thus establishing a miR-200s-activated stress signature, which strongly correlates with shorter patient survival caused by chemotherapeutic resistance. In addition, miR-30b/c and miR-222 mediate gefitinib-induced apoptosis and the epithelial–mesenchymal transition leading to therapeutic resistance in non-small cell lung cancer [87]. These discoveries collectively indicate potential roles of the miRNA family in the regulation of ROS homeostasis in tumor-initiating cells and therapeutic resistance.

5. miRNA–ROS Interaction Regulates Therapeutic Tolerance/Resistance at a Signaling/Metabolic Level

5.1. HIF-miR-210-ROS

Under hypoxic conditions, upregulated HIF-1α directly binds to a hypoxia-responsive element on the proximal miR-210 promoter and induces miR-210 expression in cancer cells [144]. miR-210 activates generation of ROSs [145] via suppressing iron–sulfur cluster assembly enzyme [146,147] and cytochrome c oxidase assembly protein [148] in the mitochondria electron transport chain and the TCA cycle. miR-210 knockdown decreased resistance to radiotherapy in hypoxic glioma stem cells and hepatoma cells [149,150]. These discoveries suggest that the HIF-miR-210-ROS [151] pathway might be a target to overcome therapeutic resistance (Figure 1).

5.2. EGFR-miR-147b-VHL-TCA Cycle

Increasing evidence suggests that the metabolic enzymes and the catalyzed metabolites, such as isocitrate dehydrogenase, succinate dehydrogenase, and succinate [16,152] in the TCA cycle, are involved in not only tumorigenesis but also therapeutic resistance. A hypoxia response is linked to tumor cell survival and drug-resistance in many cancers [153,154]. Dysregulated cancer metabolism has recently gained attention for its potential role in promoting therapeutic resistance by a therapeutic tolerance strategy in a novel manner [102]. Furthermore, Zhang et al. discovered that lung cancer cells adopt a tolerance strategy to protect from EGFR tyrosine kinase inhibitors by modulating miR-147b-dependent pseudohypoxia signaling pathways [76]. The study revealed that VHL [155] and succinate dehydrogenase play roles in tolerance-mediated cancer progression. Decreasing miR-147b and reactivation of the TCA cycle pathway provides a promising strategy to prevent therapeutic tolerance-mediated tumor relapse (Figure 1).

In addition, VHL regulates Akt activity [156], suggesting that miR-147b-VHL axis might confer therapeutic tolerance through activating Akt activity. In addition, other upstream transcription factors such as the inhibitor of DNA binding 2 might regulate VHL levels [157]. The interaction between miR-147b and other transcription factors controlling VHL needs to be investigated in the future.

Furthermore, the reciprocal changes of metabolites in the TCA cycle such as increased levels of succinate and 2-oxoglutarate (also known as α-ketoglutarate) [158] as well as decreased levels of malate and fumarate in osimertinib-tolerant cells indicate that silenced activity for succinate dehydrogenase is linked to therapeutic tolerance. In addition, small molecule inhibitor R59949 silencing succinate dehydrogenase activity enhances therapeutic tolerance, which is comparable to the function of miR-147b overexpression in tolerant persister cells. It is not surprising that accumulated succinate due to a loss of function of succinate dehydrogenase could activate the pseudohypoxia signaling pathway by repressing PHD2 as reported previously [16]. This is consistent with the findings that the miR-147b/succinate dehydrogenase axis could increase the gene expression for pseudohypoxia signaling pathways. In addition to inactivated VHL and succinate dehydrogenase, other factors such as reduced nicotinamide adenine dinucleotide (NAD^+) and decreased glutathione [159] might also activate pseudohypoxia responses leading to therapeutic tolerance. In addition, these pseudohypoxia responses may further perturb the TCA cycle and cooperatively regulate therapeutic tolerance.

These discoveries suggests that miR-147b may promote drug-tolerance to EGFR tyrosine kinase inhibitors either through reactivation of the EGFR downstream signaling pathway or through bypass by another receptor tyrosine kinase that sustains downstream signaling despite inhibition of EGFR [160,161].

5.3. Myc-miR-23a/b-Glutaminase-ROS

Cancer cells depend on both glycolysis and glucose oxidation to support their growth [162,163] as well as glutaminolysis that catabolizes glutamine to generate ATP and lactate [164]. Oncogenic c-Myc represses miR-23a and miR-23b, resulting in increased levels of mitochondrial glutaminase in cancer cells [30]. Glutaminase converts glutamine to glutamate, which is further catabolized through the TCA cycle for the production of adenosine triphosphate (ATP) or serves as substrate for glutathione synthesis [165]. Glutamine withdrawal or glutaminase knockdown resulted in increased levels of ROSs. Thus, the Myc-miR-23-glutaminase axis provides a new mechanism for regulating ROS homeostasis in cancer cells. Considering that downregulated miR-23a is enriched in tumor-initiating cells [75], it is of great interest to explore a link between miR-23 and ROSs in therapeutic tolerance/resistance (Figure 1).

6. miRNA–ROS Interaction Regulates Therapeutic Tolerance/Resistance at a Genetic/Epigenetic Level

6.1. Mutant miRNAs

The whole genome sequencing analysis of lung adenocarcinomas showed noncoding somatic mutational hotspots near *vacuolar membrane protein 1/MIR21* [166]. Samples harboring indels or single nucleotide variants in this locus demonstrated significantly higher levels of *MIR21* expression. miR-21 high levels are linked to therapeutic resistance to several treatments, including EGFR tyrosine kinase inhibitors [167] and chemotherapeutic agents [168]. Thus, it is valuable to predict therapeutic response by detecting the sequence of miR-21 in biopsies from cancer patients before they receive treatments such as EGFR tyrosine kinase inhibitors (Figure 1).

6.2. RNA Editing

Adenosine deaminases acting on RNA (ADARs) convert adenosine to inosine in double-stranded RNA including both protein-coding [169] non-coding RNAs [170]. ADAR editase activation has been associated with progression of a broad array of malignancies including therapeutic resistance [171]. ADAR1 promotes tumor-initiating cell activity [172] and resistance to BCR-ABL1 inhibitor or janus kinase 2 inhibitor in chronic myeloid leukemia through inactivating biogenesis of the *let-7* [173] or pri-miR-26a maturation [174]. In addition, most cancer patients either do not respond to the immune checkpoint blockade or develop resistance to it, often because of acquired mutations [175] that impair antigen presentation [176]. Loss of function of ADAR1 in tumor cells profoundly sensitizes tumors to immunotherapy and overcomes resistance to the programmed cell death protein 1 (PD-1) checkpoint blockade [177]. It is of interest to further study how the ADAR-miRNA axis regulates therapeutic tolerance/resistance through controlling potential genes encoding ROS scavengers [178] such as *Drosophila homolog of the mammalian protein thioredoxin-1* and *cytochrome P450 4g1* (Figure 1).

6.3. RNA m^{6}A Modification

N^6-methyladenosine (m^{6}A) modification of mRNA (RNA m^{6}A modification) is the most abundant RNA modification in eukaryotes and highly conserved among multiple species [179]. RNA m^{6}A modification is emerging as an important regulator of gene expression that affects different developmental and biological processes [180], and altered m^{6}A homeostasis is linked to cancer [181–183]. RNA m^{6}A modification is catalyzed by the dynamic regulation of methyltransferases and demethylases. Methyltransferase include methyltransferase-like 3 (METTL3), METTL14, and Wilms' tumor 1-associating protein, and the demethylases include fat mass- and obesity-associated protein and ALKB homolog 5 [184]. Upregulation of METTL3 is associated with poor prognosis in tumorigenesis and increased chemo- and radio-resistance in cancers such as glioblastomas [185] and pancreatic cancer [186]. Developing resistant phenotypes during tyrosine kinase inhibitor therapy is controlled by m^{6}A modification [187]. Leukemia cells with mRNA m^{6}A demethylation are more tolerant to tyrosine kinase inhibitor treatment. Recovery of m^{6}A methylation re-sensitizes therapeutic resistant cells towards tyrosine kinase inhibitors. The findings identify a novel function for the m^{6}A methylation in regulating reversible tyrosine kinase inhibitor-tolerance state, providing a mechanistic paradigm for drug resistance in cancer. In addition, METTL3 plays roles in the maturation process of miRNAs against ROSs in an m^{6}A-dependent manner [188]. For example, METTL3-mediated miR-873 upregulation controls the kelch-like ECH associated protein 1 (KEAP1)-NRF2 [142] pathway against ROSs. These studies revealed that RNA m^{6}A might regulate therapeutic tolerance/resistance through miRNA–ROS pathways (Figure 1).

7. Emerging Fields and Tools in Preventing and Overcoming Therapeutic Tolerance/Resistance

7.1. Artificial Intelligence (AI)

AI is an area of computer science that emphasizes the creation of intelligent machines that work and react like humans and that uses labeled big data along with markedly enhanced computing power and cloud storage [189]. The most common applications of AI in drug treatment have to do with matching patients to their optimal drug or combination of drugs, predicting drug–target or drug–drug interactions and optimizing treatment protocols [190]. AI-based models have been developed for predicting synergistic treatment combinations in many diseases such as infectious diseases [191] and cancers [192,193]. One challenge is determining how AI-based technology may design tools which improve identification of therapeutic tolerance and resistance and develop new treatment combinations against tolerant and resistant cancers. The success of this AI-based approach may provide earlier and targeted anticancer treatment, which would prevent therapeutic tolerance/resistance emerging and cure cancer patients more effectively (Figure 1).

7.2. Pathogens

Pathogens such as microbiomes and viruses are becoming increasingly recognized for their effects on tumorigenesis and therapeutic resistance to cancer treatment [194]. Bacterial dysbiosis accompanies carcinogenesis in several malignancies such as gastric [195], colon [196], liver [197], and pancreatic [198] cancers by affecting metabolism and impairing immune functions [199]. Additionally, fungi [200] and viruses [201] also induce carcinogenesis in several cancers. Furthermore, intratumoral bacteria induced therapeutic resistance through breakdown of chemotherapy gemcitabine into inactive metabolites via bacterial enzymes such as cytidine deaminase [202] and via impairing response to immune checkpoint blockade [198]. Gut microbiota plays a critical role in mediating colorectal cancer chemoresistance in response to chemotherapeutics via a selective target loss of miR-18a* (miR-18a-3p) and miR-4802, and via activation of the autophagy pathway [203]. In addition, miR-18a* is a tumor suppressor that inhibits KRAS expression [204]. Activating *KRAS* mutations confer both primary [205] and acquired [206] resistance to anti-EGFR cetuximab therapy in colorectal cancer. Thus, targeting intratumoral pathogens provide a new angle in cancer treatment to overcome therapeutic tolerance/resistance. Some intracellular pathogens interact directly with receptor tyrosine kinases, and this interaction is critical for pathogen entry [207]. This establishes that pathogen-encoded receptor tyrosine kinase-interacting epitopes represent promising candidates for the development of novel therapeutic and prophylactic vaccines and of small-molecule interaction disruptors [208]. It would be of great interest to investigate whether those pathogens will confer therapeutic tolerance/resistance in host tumor cells by regulating miRNA–ROS interaction (Figure 1).

8. Concluding Remarks and Future Directions

Therapeutic tolerance/resistance raise major problems for the successful treatment of cancer, including conventional therapy and recent molecular therapy. There is an increasing importance of studying the role of ROS-relevant miRNAs to identify more effective biomarkers and develop better therapeutic targets against therapeutic tolerance/resistance. The interaction between miRNAs and ROSs fits in with the opportunities and challenges of studying mechanisms by which cancer cells resist therapy and ways by which therapeutic tolerance/resistance can be overcome. New concepts and emerging research tools bring potential to overcome therapeutic tolerance/resistance. However, some major challenges should be addressed properly. First, cancer relapse is driven by a small subpopulation of drug-tolerant persister cells, known as minimal residual disease in clinic. Single cell-relevant technologies, such as single-cell sequencing [209] might be applied to track single tolerant persister cells to gain insights into drug tolerance dynamics and heterogeneity [210]. In addition, preventative strategies using potential agents targeting those therapeutic tolerant cells at early stages in combination with molecular therapeutics will help prevent therapeutic tolerance and the resulting therapeutic resistance [211]. Second, new ex vivo models such as the organoid have been widely applied in cancer treatment response and therapeutic tolerance/resistance [212,213]. One of the advantages of the three-dimensional organoid model compared to a conventional two-dimensional monolayer is that tumor microenvironments established in organoids are similar to those found in vivo. For example, cancer organoids show heterogeneous hypoxic regions and show their enriched tumor-initiating cells and relevant metabolism pathway [214]. The organoid model may be used for large-scale screening, especially when incorporated with AI-based technology, to optimize the best drug combinations and thus reduce therapeutic tolerance/resistance. However, lacking immune cells and other types of cells has challenged this model [215]. Thus, incorporating immune cells will help better understand tolerance and resistance to immunotherapy [216]. Third, applications of non-invasive biomarkers to predict drug response represents a future direction in clinical settings. For example, cell-free circulating miRNAs have been successfully combined with low dose computed tomography scanning for diagnoses of early-stage lung cancer patients [217]. It is reasonable to incorporate cell-free circulating miRNAs signature together with cell-free DNAs signature [218] to predict and track the emergence of therapeutic tolerance/resistance. However, microRNAs predicting therapeutic tolerance/resistance might be dependent on specific mutant driver genes. For instance, increased miR-147b is relevant to mutant *EGFR* [76], and downregulated miR-23a is relevant to mutant *MYC* [30]. Thus, genetic mutation background and specific treatment agents should be considered comprehensively. Ultimately, early intervention on genetic/epigenetic, signaling/metabolic, and phenotypic changes in the miRNA–ROS network should be considered comprehensively to prevent and overcome therapeutic tolerance/resistance.

Acknowledgments: This work is supported by the Burnett School of Biomedical Sciences, College of Medicine, University of Central Florida grant 25400714, the NIH-Yale SPORE in Lung Cancer Career Development Program Award, and NRSA grant 5T32HL007893 awarded to W.C.Z. We thank Joshua Roney for critical reading and comments. We apologize to all researches whose work could not be cited due to reference limitations.

Conflicts of Interest: The author declares no conflict of interest.

References

1. Harman, S.M.; Liang, L.; Tsitouras, P.D.; Gucciardo, F.; Heward, C.B.; Reaven, P.D.; Ping, W.; Ahmed, A.; Cutler, R.G. Urinary excretion of three nucleic acid oxidation adducts and isoprostane $F_2\alpha$ measured by liquid chromatography-mass spectrometry in smokers, ex-smokers, and nonsmokers. *Free Radic. Biol Med.* **2003**, *1301*, 35–1309. [CrossRef]

2. Apel, K.; Hirt, H. Reactive oxygen species: Metabolism, oxidative stress, and signal transduction. *Annu. Rev. Plant. Biol.* **2004**, *55*, 373–399. [CrossRef]

3. Finkel, T. Signal transduction by reactive oxygen species. *J. Cell Biol.* **2011**, *194*, 7–15. [CrossRef]

4. Dupré-Crochet, S.; Erard, M.; Nüße, O. ROS production in phagocytes: Why, when, and where? *J. Leukoc. Biol.* **2013**, *94*, 657–670. [CrossRef]

5. Li, R.; Jia, Z.; Trush, M.A. Defining ROS in biology and medicine. *React. Oxyg. Species* **2016**, *1*, 9–21.

6. Sabharwal, S.S.; Schumacker, P.T. Mitochondrial ROS in cancer: Initiators, amplifiers or an Achilles' heel? *Nat. Rev. Cancer* **2014**, *14*, 709–721. [CrossRef]

7. Zorov, D.B.; Juhaszova, M.; Sollott, S.J. Mitochondrial Reactive Oxygen Species (ROS) and ROS-Induced ROS Release. *Physiol. Rev.* **2014**, *94*, 909–950. [CrossRef]

8. Yeung, A.W.K.; Tzvetkov, N.T.; El-Tawil, O.S.; Bungau, S.G.; Abdel-Daim, M.M.; Atanasov, A.G. Antioxidants: Scientific literature landscape analysis. *Oxid. Med. Cell Longev.* **2019**, *8278*, 2019454. [CrossRef]

9. Valle, I.; Alvarez-Barrientos, A.; Arza, E.; Lamas, S.; Monsalve, M. Pgc-1α regulates the mitochondrial antioxidant defense system in vascular endothelial cells. *Cardiovasc. Res.* **2005**, *66*, 562–573. [CrossRef]

10. Ward, P.S.; Thompson, C.B. Metabolic reprogramming: A cancer hallmark even Warburg did not anticipate. *Cancer Cell* **2012**, *21*, 297–308. [CrossRef]

11. Vander Heiden, M.G.; Cantley, L.C.; Thompson, C.B. Understanding the Warburg effect: The metabolic requirements of cell proliferation. *Science* **2009**, *324*, 1029–1033. [CrossRef]

12. Vyas, S.; Zaganjor, E.; Haigis, M.C. Mitochondria and cancer. *Cell* **2016**, *166*, 555–566. [CrossRef]

13. Raimundo, N.; Baysal, B.E.; Shadel, G.S. Revisiting the TCA cycle: Signaling to tumor formation. *Trends Mol. Med.* **2011**, *17*, 641–649. [CrossRef]

14. Fernie, A.R.; Carrari, F.; Sweetlove, L.J. Respiratory metabolism: Glycolysis, the TCA cycle and mitochondrial electron transport. *Curr. Opin. Plant Biol.* **2004**, *7*, 254–261. [CrossRef]

15. Dahia, P.L. Pheochromocytoma and paraganglioma pathogenesis: Learning from genetic heterogeneity. *Nat. Rev. Cancer* **2014**, *14*, 108–119. [CrossRef]

16. Selak, M.A.; Armour, S.M.; MacKenzie, E.D.; Boulahbel, H.; Watson, D.G.; Mansfield, K.D.; Pan, Y.; Simon, M.C.; Thompson, C.B.; Gottlieb, E. Succinate links TCA cycle dysfunction to oncogenesis by inhibiting HIF-α prolyl hydroxylase. *Cancer Cell* **2005**, *7*, 77–85. [CrossRef]

17. Kaelin, W.G., Jr. Molecular basis of the VHL hereditary cancer syndrome. *Nat. Rev. Cancer* **2002**, *2*, 673–682. [CrossRef]

18. Krieg, M.; Haas, R.; Brauch, H.; Acker, T.; Flamme, I.; Plate, K.H. Up-regulation of hypoxia-inducible factors HIF-1α and HIF-2α under normoxic conditions in renal carcinoma cells by von Hippel-Lindau tumor suppressor gene loss of function. *Oncogene* **2000**, *19*, 5435–5443. [CrossRef]

19. Pagliarini, D.J.; Calvo, S.E.; Chang, B.; Sheth, S.A.; Vafai, S.B.; Ong, S.E.; Walford, G.A.; Sugiana, C.; Boneh, A.; Chen, W.K.; et al. A mitochondrial protein compendium elucidates complex I disease biology. *Cell* **2008**, *134*, 112–123. [CrossRef]

20. Calvo, S.E.; Clauser, K.R.; Mootha, V.K. MitoCarta2.0: An updated inventory of mammalian mitochondrial proteins. *Nucleic Acids Res.* **2016**, *44*, D1251–D1257. [CrossRef]

21. Gottlieb, E.; Tomlinson, I.P.M. Mitochondrial tumour suppressors: A genetic and biochemical update. *Nat. Rev. Cancer* **2005**, *5*, 857–866. [CrossRef]

22. Wagner, A.J.; Remillard, S.P.; Zhang, Y.X.; Doyle, L.A.; George, S.; Hornick, J.L. Loss of expression of SDHA predicts *SDHA* mutations in gastrointestinal stromal tumors. *Mod. Pathol.* **2013**, *26*, 289–294. [CrossRef] [PubMed]

23. Astuti, D.; Latif, F.; Dallol, A.; Dahia, P.L.M.; Douglas, F.; George, E.; Skoldberg, F.; Husebye, E.S.; Eng, C.; Maher, E.R. Gene mutations in the succinate dehydrogenase subunit SDHB cause susceptibility to familial pheochromocytoma and to familial paraganglioma. *Am. J. Hum. Genet.* **2001**, *69*, 49–54. [CrossRef] [PubMed]

24. Niemann, S.; Muller, U. Mutations in *SDHC* cause autosomal dominant paraganglioma, type 3. *Nat. Genet.* **2000**, *26*, 268–270. [CrossRef]

25. Baysal, B.E.; Ferrell, R.E.; Willett-Brozick, J.E.; Lawrence, E.C.; Myssiorek, D.; Bosch, A.; van der Mey, A.; Taschner, P.E.M.; Rubinstein, W.S.; Myers, E.N.; et al. Mutations in *SDHD*, a mitochondrial complex II gene, in hereditary paraganglioma. *Science* **2000**, *287*, 848–851. [CrossRef]

26. Zhang, W.C.; Shyh-Chang, N.; Yang, H.; Rai, A.; Umashankar, S.; Ma, S.; Soh, B.S.; Sun, L.L.; Tai, B.C.; Nga, M.E.; et al. Glycine decarboxylase activity drives non-small cell lung cancer tumor-initiating cells and tumorigenesis. *Cell* **2012**, *148*, 259–272. [CrossRef]

27. Zhang, W.C.; Lim, B. Targeting metabolic enzyme with locked nucleic acids in non-small cell lung cancer. *Cancer Res.* **2014**, *74*, 1438.

28. Lim, B.; Zhang, W. Targeting Metabolic Enzymes in Human Cancer. U.S. Patent 9,297,813,B2, 11 November 2011.

29. Go, M.K.; Zhang, C W.; Lim, B.; Yew, W.S. Glycine decarboxylase is an unusual amino acid decarboxylase involved in tumorigenesis. *Biochemistry* **2014**, *53*, 947–956. [CrossRef]

30. Gao, P.; Tchernyshyov, I.; Chang, T.C.; Lee, Y.S.; Kita, K.; Ochi, T.; Zeller, K.I.; De Marzo, A.M.; Van Eyk, J.E.; Mendell, J.T.; et al. C-Myc suppression of miR-23a/B enhances mitochondrial glutaminase expression and glutamine metabolism. *Nature* **2009**, *458*, 762–765. [CrossRef]

31. Testa, U.; Labbaye, C.; Castelli, G.; Pelosi, E. Oxidative stress and hypoxia in normal and leukemic stem cells. *Exp. Hematol.* **2016**, *44*, 540–560. [CrossRef]

32. Hanahan, D.; Weinberg, R.A. Hallmarks of cancer: The next generation. *Cell* **2011**, *144*, 646–674. [CrossRef]

33. Kasinski, A.L.; Slack, F.J. MicroRNAs en route to the clinic: Progress in validating and targeting microRNAs for cancer therapy. *Nat. Rev. Cancer* **2011**, *11*, 849–864. [CrossRef]

34. Anastasiadou, E.; Jacob, L.S.; Slack, F.J. Non-coding RNA networks in cancer. *Nat. Rev. Cancer* **2018**, *18*, 5–18. [CrossRef]

35. Fan, M.; Krutilina, R.; Sun, J.; Sethuraman, A.; Yang, H.C.; Wu, H.Z.; Yue, J.; Pfeffer, L.M. Comprehensive analysis of microRNA (miRNA) targets in breast cancer cells. *J. Biol. Chem.* **2013**, *288*, 27480–27493. [CrossRef]

36. Calin, G.A.; Croce, C.M. MicroRNA signatures in human cancers. *Nat. Rev. Cancer* **2006**, *6*, 857–866. [CrossRef]

37. Lu, J.; Getz, G.; Miska, A.E.; Alvarez-Saavedra, E.; Lamb, J.; Peck, D.; Sweet-Cordero, A.; Ebert, B.L.; Mak, R.H.; Ferrando, A.A.; et al. MicroRNA expression profiles classify human cancers. *Nature* **2005**, *435*, 834–838. [CrossRef]

38. Nouraee, N.; Calin, G.A. MicroRNAs as cancer biomarkers. *MicroRNA* **2013**, *2*, 102–117. [CrossRef]

39. Tentori, A.M.; Nagarajan, M.B.; Kim, J.J.; Zhang, W.C.; Slack, F.J.; Doyle, P.S. Quantitative and multiplex microRNA assays from unprocessed cells in isolated nanoliter well arrays. *Lab Chip* **2018**, *18*, 2410–2424. [CrossRef]

40. Nagarajan, M.B.; Tentori, A.M.; Zhang, W.C.; Slack, F.J.; Doyle, P.S. Nonfouling, encoded hydrogel microparticles for multiplex microRNA profiling directly from formalin-fixed, paraffin-embedded tissue. *Anal. Chem.* **2018**, *90*, 10279–10285. [CrossRef]

41. Anfossi, S.; Babayan, A.; Pantel, K.; Calin, G.A. Clinical utility of circulating non-coding RNAs—An update. *Nat. Rev. Clin. Oncol.* **2018**, *15*, 541–563. [CrossRef]

42. Schwarzenbach, H.; Nishida, N.; Calin, G.A.; Pantel, K. clinical relevance of circulating cell-free microRNAs in cancer. *Nat. Rev. Clin. Oncol.* **2014**, *11*, 145–156. [CrossRef]

43. Cortez, M.A.; Bueso-Ramos, C.; Ferdin, J.; Lopez-Berestein, G.; Sood, A.K.; Calin, G.A. MicroRNAs in body fluids—The mix of hormones and biomarkers. *Nat. Rev. Clin. Oncol.* **2011**, *8*, 467–477. [CrossRef]

44. Esquela-Kerscher, A.; Slack, F.J. Oncomirs—MicroRNAs with a role in cancer. *Nat. Rev. Cancer* **2006**, *6*, 259–269. [CrossRef]

45. Medina, P.P.; Nolde, M.; Slack, F.J. Oncomir addiction in an in vivo model of microRNA-21-induced pre-B-cell lymphoma. *Nature* **2010**, *467*, 86–90. [CrossRef]

46. Edmonds, M.D.; Boyd, K.L.; Moyo, T.; Mitra, R.; Duszynski, R.; Arrate, M.P.; Chen, X.; Zhao, Z.; Blackwell, T.S.; Andl, T.; et al. MicroRNA-31 initiates lung tumorigenesis and promotes mutant KRAS-driven lung cancer. *J. Clin. Investig.* **2016**, *126*, 349–364. [CrossRef] [PubMed]

47. Cheng, C.J.; Bahal, R.; Babar, I.A.; Pincus, Z.; Barrera, F.; Liu, C.; Svoronos, A.; Braddock, D.T.; Glazer, P.M.; Engelman, D.M.; et al. MicroRNA silencing for cancer therapy targeted to the tumour microenvironment. *Nature* **2015**, *518*, 107–110. [CrossRef] [PubMed]

48. Babar, I.A.; Cheng, C.J.; Booth, C.J.; Liang, X.; Weidhaas, J.B.; Saltzman, W.M.; Slack, F.J. Nanoparticle-based therapy in an in vivo microRNA-155 (miR-155)-dependent mouse model of lymphoma. *Proc. Natl. Acad. Sci. USA* **2012**, *109*, E1695–E1704. [CrossRef]

49. Ma, L.; Reinhardt, F.; Pan, E.; Soutschek, J.; Bhat, B.; Marcusson, E.G.; Teruya-Feldstein, J.; Bell, G.W.; Weinberg, R.A. Therapeutic silencing of miR-10b inhibits metastasis in a mouse mammary tumor model. *Nat. Biotechnol.* **2010**, *28*, 341–347. [CrossRef]

50. Johnson, S.M.; Grosshans, H.; Shingara, J.; Byrom, M.; Jarvis, R.; Cheng, A.; Labourier, E.; Reinert, K.L.; Brown, D.; Slack, F.J. RAS is regulated by the let-7 microRNA family. *Cell* **2005**, *120*, 635–647. [CrossRef]

51. Kasinski, A.L.; Slack, F.J. MiRNA-34 prevents cancer initiation and progression in a therapeutically resistant k-RAS and P53-induced mouse model of lung adenocarcinoma. *Cancer Res.* **2012**, *72*, 5576–5587. [CrossRef]

52. Krzeszinski, J.Y.; Wei, W.; Huynh, H.; Jin, Z.; Wang, X.; Chang, T.C.; Xie, X.J.; He, L.; Mangala, L.S.; Lopez-Berestein, G.; et al. MiR-34a blocks osteoporosis and bone metastasis by inhibiting osteoclastogenesis and Tgif2. *Nature* **2014**, *512*, 431–435. [CrossRef]

53. Banerjee, J.; Khanna, S.; Bhattacharya, A. MicroRNA regulation of oxidative stress. *Oxid. Med. Cell Longev.* **2017**, *2872*, 2017156. [CrossRef]

54. Hatley, M.E.; Patrick, D.M.; Garcia, M.R.; Richardson, J.A.; Bassel-Duby, R.; van Rooij, E.; Olson, E.N. Modulation of K-Ras-dependent lung tumorigenesis by microRNA-21. *Cancer Cell* **2010**, *18*, 282–293. [CrossRef]

55. Seike, M.; Goto, A.; Okano, T.; Bowman, D.E.; Schetter, J.A.; Horikawa, I.; Mathe, A.E.; Jen, J.; Yang, P.; Sugimura, H.; et al. MiR-21 is an EGFR-regulated anti-apoptotic factor in lung cancer in never-smokers. *Proc. Natl. Acad. Sci. USA* **2009**, *106*, 12085–12090. [CrossRef]

56. Zhang, X.; Ng, W.L.; Wang, P.; Tian, L.; Werner, E.; Wang, H.; Doetsch, P.; Wang, Y. MicroRNA-21 modulates the levels of reactive oxygen species by targeting Sod3 Tnfα. *Cancer Res.* **2012**, *72*, 4707–4713. [CrossRef]

57. Engedal, N.; Zerovnik, E.; Rudov, A.; Galli, F.; Olivieri, F.; Procopio, A.D.; Rippo, M.R.; Monsurro, V.; Betti, M.; Albertini, M.C. From oxidative stress damage to pathways, networks, and autophagy via microRNAs. *Oxid. Med. Cell Longev.* **2018**, *2018*, 4968321. [CrossRef]

58. Poyil, P.; Son, Y.-O.; Divya, S.P.; Wang, L.; Zhang, Z.; Shi, H. Oncogenic transformation of human lung bronchial epithelial cells induced by arsenic involves ROS-dependent activation of STAT3-miR-21-PDCD4 mechanism. *Sci. Rep.* **2016**, *6*, 37227.

59. Thulasingam, S.; Massilamany, C.; Gangaplara, A.; Dai, H.; Yarbaeva, S.; Subramaniam, S.; Riethoven, J.J.; Eudy, J.; Lou, M.; Reddy, J. MiR-27b*, an oxidative stress-responsive microRNA modulates nuclear factor-kB pathway in Raw 264.7 cells. *Mol. Cell Biochem.* **2011**, *352*, 181–188. [CrossRef]

60. Baker, J.R.; Vuppusetty, C.; Colley, T.; Papaioannou, A.I.; Fenwick, P.; Donnelly, L.; Ito, K.; Barnes, P.J. Oxidative stress dependent microRNA-34a activation via PI3Kα reduces the expression of sirtuin-1 and sirtuin-6 in epithelial cells. *Sci. Rep.* **2016**, *6*, 35871. [CrossRef]

61. Lin, Z.; Fang, D. The roles of SIRT1 in cancer. *Genes Cancer* **2013**, *4*, 97–104. [CrossRef] [PubMed]

62. Bussing, I.; Slack, F.J.; Grosshans, H. Let-7 microRNAs in development, stem cells and cancer. *Trends Mol. Med.* **2008**, *14*, 400–409. [CrossRef] [PubMed]

63. Saleh, A.D.; Savage, J.E.; Cao, L.; Soule, B.P.; Ly, D.; DeGraff, W.; Harris, C.C.; Mitchell, J.B.; Simone, N.L. Cellular stress induced alterations in microRNA let-7a and let-7b expression are dependent on P53. *PLoS ONE* **2011**, *6*, e24429. [CrossRef] [PubMed]

64. Ling, H.; Fabbri, M.; Calin, G.A. MicroRNAs and other non-coding RNAs as targets for anticancer drug development. *Nat. Rev. Drug Discov.* **2013**, *12*, 847–865. [CrossRef]

65. Adams, B.D.; Parsons, C.; Walker, L.; Zhang, W.C.; Slack, F.J. Targeting noncoding RNAs in disease. *J. Clin. Investig.* **2017**, *127*, 761–771. [CrossRef]

66. Rupaimoole, R.; Slack, F.J. MicroRNA therapeutics: Towards a new era for the management of cancer and other diseases. *Nat. Rev. Drug Discov.* **2017**, *16*, 203–222. [CrossRef]

67. Van Zandwijk, N.; Pavlakis, N.; Kao, S.C.; Linton, A.; Boyer, M.J.; Clarke, S.; Huynh, Y.; Chrzanowska, A.; Fulham, M.J.; Bailey, D.L.; et al. Safety and activity of microRNA-loaded minicells in patients with recurrent malignant pleural mesothelioma: A first-in-man, phase 1, open-label, dose-escalation study. *Lancet Oncol.* **2017**, *18*, 1386–1396. [CrossRef]

68. Van Zandwijk, N.; McDiarmid, J.; Brahmbhatt, H.; Reid, G. Response to an innovative mesothelioma treatment based on miR-16 mimic loaded EGFR targeted minicells (TargomiRs). *Transl. Lung Cancer Res.* **2018**, *7*, S60–S61. [CrossRef]

69. Fennell, D. MiR-16: Expanding the range of molecular targets in mesothelioma. *Lancet Oncol.* **2017**, *18*, 1296–1297. [CrossRef]

70. Slack, F.J.; Chinnaiyan, A.M. The role of non-coding RNAs in oncology. *Cell* **2019**, *179*, 1033–1055. [CrossRef]

71. Garzon, R.; Marcucci, G.; Croce, C.M. Targeting microRNAs in cancer: Rationale, strategies and challenges. *Nat. Rev. Drug Discov.* **2010**, *9*, 775–789. [CrossRef]

72. Zhang, P.; Wang, L.; Rodriguez-Aguayo, C.; Yuan, Y.; Debeb, B.G.; Chen, D.; Sun, Y.; You, M.J.; Liu, Y.; Dean, C.D.; et al. MiR-205 acts as a tumour radiosensitizer by targeting ZEB1 and Ubc13. *Nat. Commun.* **2014**, *5*, 5671. [CrossRef] [PubMed]

73. Wang, S.; Zhang, R.; Claret, F.X.; Yang, H. Involvement of microRNA-24 and DNA methylation in resistance of nasopharyngeal carcinoma to ionizing radiation. *Mol. Cancer Ther.* **2014**, *13*, 3163–3174. [CrossRef] [PubMed]

74. Babar, I.A.; Czochor, J.; Steinmetz, A.; Weidhaas, J.B.; Glazer, P.M.; Slack, F.J. Inhibition of hypoxia-induced miR-155 radiosensitizes hypoxic lung cancer cells. *Cancer Biol. Ther.* **2011**, *12*, 908–914. [CrossRef]

75. Zhang, W.C.; Chin, T.M.; Yang, H.; Nga, M.E.; Lunny, D.P.; Lim, E.K.; Sun, L.L.; Pang, Y.H.; Leow, Y.N.; Malusay, S.R.; et al. Tumour-initiating cell-specific miR-1246 and miR-1290 expression converge to promote non-small cell lung cancer progression. *Nat. Commun.* **2016**, *7*, 11702. [CrossRef]

76. Zhang, W.C.; Wells, J.M.; Chow, K.H.; Huang, H.; Yuan, M.; Saxena, T.; Melnick, M.A.; Politi, K.; Asara, J.M.; Costa, D.B.; et al. MiR-147b-mediated TCA cycle dysfunction and pseudohypoxia initiate drug tolerance to EGFR inhibitors in lung adenocarcinoma. *Nat. Metab.* **2019**, *1*, 460–474. [CrossRef]

77. Costinean, S.; Zanesi, N.; Pekarsky, Y.; Tili, E.; Volinia, S.; Heerema, N.; Croce, M.C. Pre-B Cell Proliferation and Lymphoblastic Leukemia/High-Grade Lymphoma in E(Mu)-Mir155 Transgenic Mice. *Proc. Natl. Acad. Sci. USA* **2006**, *103*, 7024–7029. [CrossRef]

78. Mikamori, M.; Yamada, D.; Eguchi, H.; Hasegawa, S.; Kishimoto, T.; Tomimaru, Y.; Asaoka, T.; Noda, T.; Wada, H.; Kawamoto, K.; et al. MicroRNA-155 controls exosome synthesis and promotes gemcitabine resistance in pancreatic ductal adenocarcinoma. *Sci. Rep.* **2017**, *7*, 42339. [CrossRef]

79. Moriyama, T.; Ohuchida, K.; Mizumoto, K.; Yu, J.; Sato, N.; Nabae, T.; Takahata, S.; Toma, H.; Nagai, E.; Tanaka, M. MicroRNA-21 modulates biological functions of pancreatic cancer cells including their proliferation, invasion, and chemoresistance. *Mol. Cancer Ther.* **2009**, *8*, 1067–1074. [CrossRef]

80. Giovannetti, E.; Funel, N.; Peters, J.G.; Del Chiaro, M.; Erozenci, L.A.; Vasile, E.; Leon, L.G.; Pollina, L.E.; Groen, A.; Falcone, A.; et al. MicroRNA-21 in pancreatic cancer: correlation with clinical outcome and pharmacologic aspects underlying its role in the modulation of gemcitabine activity. *Cancer Res.* **2010**, *70*, 4528–4538. [CrossRef]

81. Jiang, S.; Wang, R.; Yan, H.; Jin, L.; Dou, X.; Chen, D. MicroRNA-21 modulates radiation resistance through upregulation of hypoxia-inducible factor-1α-promoted glycolysis in non-small cell lung cancer cells. *Mol. Med. Rep.* **2016**, *13*, 4101–4107. [CrossRef]

82. Zhang, W.C.; Slack, F.J. MicroRNA-21 mediates resistance to EGFR tyrosine kinase inhibitors in lung cancer. *J. Thorac. Oncol.* **2017**, *12*, S1536. [CrossRef]

83. Yu, F.; Yao, H.; Zhu, P.; Zhang, X.; Pan, Q.; Gong, C.; Huang, Y.; Hu, X.; Su, F.; Lieberman, J.; et al. Let-7 regulates self renewal and tumorigenicity of breast cancer cells. *Cell* **2007**, *131*, 1109–1123. [CrossRef] [PubMed]

84. Yin, J.; Hu, W.; Pan, L.; Fu, W.; Dai, L.; Jiang, Z.; Zhang, F.; Zhao, J. Let-7 and miR-17 promote self-renewal and drive gefitinib resistance in non-small cell lung cancer. *Oncol. Rep.* **2019**, *42*, 495–508. [CrossRef] [PubMed]

85. Yu, F.; Deng, H.; Yao, H.; Liu, Q.; Su, F.; Song, E. MiR-30 reduction maintains self-renewal and inhibits apoptosis in breast tumor-initiating cells. *Oncogene* **2010**, *29*, 4194–4204. [CrossRef] [PubMed]

86. Ma, Y.S.; Yu, F.; Zhong, X.M.; Lu, G.X.; Cong, X.L.; Xue, S.B.; Xie, W.T.; Hou, L.K.; Pang, L.J.; Wu, W.; et al. MiR-30 family reduction maintains self-renewal and promotes tumorigenesis in NSCLC-initiating cells by targeting oncogene TM4SF1. *Mol. Ther.* **2018**, *26*, 2751–2765. [CrossRef] [PubMed]

87. Garofalo, M.; Romano, G.; Di Leva, G.; Nuovo, G.; Jeon, Y.J.; Ngankeu, A.; Sun, J.; Lovat, F.; Alder, H.; Condorelli, G.; et al. EGFR and MET receptor tyrosine kinase-altered microRNA expression induces tumorigenesis and gefitinib resistance in lung cancers. *Nat. Med.* **2012**, *18*, 74–82. [CrossRef]

88. Du, X.; Liu, B.; Luan, X.; Cui, Q.; Li, L. MiR-30 decreases multidrug resistance in human gastric cancer cells by modulating cell autophagy. *Exp. Ther. Med.* **2018**, *15*, 599–605. [CrossRef]

89. Jiang, L.; Hermeking, H. MiR-34a and miR-34b/c suppress intestinal tumorigenesis. *Cancer Res.* **2017**, *77*, 2746–2758. [CrossRef]

90. Zenz, T.; Mohr, J.; Eldering, E.; Kater, A.P.; Bühler, A.; Kienle, D.; Winkler, D.; Dürig, J.; van Oers, M.H.; Mertens, D.; et al. MiR-34a as part of the resistance network in chronic lymphocytic leukemia. *Blood* **2009**, *113*, 3801–3808. [CrossRef]

91. Pao, W.; Miller, V.; Zakowski, M.; Doherty, J.; Politi, K.; Sarkaria, I.; Singh, B.; Heelan, R.; Rusch, V.; Fulton, L.; et al. EGF receptor gene mutations are common in lung cancers from "never smokers" and are associated with sensitivity of tumors to gefitinib and erlotinib. *Proc. Natl. Acad. Sci. USA* **2004**, *101*, 13306–13311. [CrossRef]

92. Sordella, R.; Bell, D.W.; Haber, D.A.; Settleman, J. Gefitinib-sensitizing EGFR mutations in lung cancer activate anti-apoptotic pathways. *Science* **2004**, *305*, 1163–1167. [CrossRef]

93. Politi, K.; Herbst, R.S. Lung cancer in the era of precision medicine. *Clin. Cancer Res.* **2015**, *21*, 2213–2220. [CrossRef]

94. Kobayashi, S.; Boggon, T.J.; Dayaram, T.; Janne, P.A.; Kocher, O.; Meyerson, M.; Johnson, B.E.; Eck, M.J.; Tenen, D.G.; Halmos, B. EGFR mutation and resistance of non-small-cell lung cancer to gefitinib. *N. Engl. J. Med.* **2005**, *352*, 786–792. [CrossRef] [PubMed]

95. Politi, K.; Ayeni, D.; Lynch, T. The next wave of EGFR tyrosine kinase inhibitors enter the clinic. *Cancer Cell* **2015**, *27*, 751–753. [CrossRef] [PubMed]

96. Thress, K.S.; Paweletz, C.P.; Felip, E.; Cho, B.C.; Stetson, D.; Dougherty, B.; Lai, Z.; Markovets, A.; Vivancos, A.; Kuang, Y.; et al. Acquired EGFR C797S mutation mediates resistance to AZD9291 in non-small cell lung cancer harboring EGFR T790M. *Nat. Med.* **2015**, *21*, 560–562. [CrossRef]

97. Greaves, M.; Maley, C.C. Clonal evolution in cancer. *Nature* **2012**, *481*, 306–313. [CrossRef]

98. Hata, A.N.; Niederst, M.J.; Archibald, H.L.; Gomez-Caraballo, M.; Siddiqui, F.M.; Mulvey, H.E.; Maruvka, Y.E.; Ji, F.; Bhang, H.E.; Krishnamurthy Radhakrishna, V.; et al. Tumor cells can follow distinct evolutionary paths to become resistant to epidermal growth factor receptor inhibition. *Nat. Med.* **2016**, *22*, 262–269. [CrossRef]

99. Ramirez, M.; Rajaram, S.; Steininger, R.J.; Osipchuk, D.; Roth, M.A.; Morinishi, L.S.; Evans, L.; Ji, W.; Hsu, C.H.; Thurley, K.; et al. Diverse drug-resistance mechanisms can emerge from drug-tolerant cancer persister cells. *Nat. Commun.* **2016**, *7*, 10690. [CrossRef]

100. Sharma, S.V.; Lee, D.Y.; Li, B.; Quinlan, M.P.; Takahashi, F.; Maheswaran, S.; McDermott, U.; Azizian, N.; Zou, L.; Fischbach, M.A.; et al. A chromatin-mediated reversible drug-tolerant state in cancer cell subpopulations. *Cell* **2010**, *141*, 69–80. [CrossRef]

101. Smith, M.P.; Brunton, H.; Rowling, E.J.; Ferguson, J.; Arozarena, I.; Miskolczi, Z.; Lee, J.L.; Girotti, M.R.; Marais, R.; Levesque, M.P.; et al. Inhibiting drivers of non-mutational drug tolerance is a salvage strategy for targeted melanoma therapy. *Cancer Cell* **2016**, *29*, 270–284. [CrossRef]

102. Hangauer, M.J.; Viswanathan, V.S.; Ryan, J.M.; Bole, D.; Eaton, J.K.; Matov, A.; Galeas, J.; Dhruv, H.D.; Berens, M.E.; Schreiber, L.S.; et al. Drug-tolerant persister cancer cells are vulnerable to GPX4 inhibition. *Nature* **2017**, *551*, 247–250. [CrossRef] [PubMed]

103. Shaffer, S.M.; Dunagin, M.C.; Torborg, S.R.; Torre, E.A.; Emert, B.; Krepler, C.; Beqiri, M.; Sproesser, K.; Brafford, P.A.; Xiao, M.; et al. Rare cell variability and drug-induced reprogramming as a mode of cancer drug resistance. *Nature* **2017**, *546*, 431–435. [CrossRef] [PubMed]

104. Pavlides, S.; Vera, I.; Gandara, R.; Sneddon, S.; Pestell, R.G.; Mercier, I.; Martinez-Outschoorn, U.E.; Whitaker-Menezes, D.; Howell, A.; Sotgia, F.; et al. Warburg meets autophagy: Cancer-associated fibroblasts accelerate tumor growth and metastasis via oxidative stress, mitophagy, and aerobic glycolysis. *Antioxid. Redox Signal.* **2012**, *16*, 1264–1284. [CrossRef]

105. Sebolt-Leopold, J.S.; English, J.M. Mechanisms of drug inhibition of signalling molecules. *Nature* **2006**, *441*, 457–462. [CrossRef] [PubMed]

106. Zhu, Y.-j.; Zheng, B.; Wang, H.-y.; Chen, L. New knowledge of the mechanisms of sorafenib resistance in liver cancer. *Acta Pharmacol. Sin.* **2017**, *38*, 614–622. [CrossRef]

107. Mendez-Blanco, C.; Fondevila, F.; Garcia-Palomo, A.; Gonzalez-Gallego, J.; Mauriz, L.J. Sorafenib resistance in hepatocarcinoma: Role of hypoxia-inducible factors. *Exp. Mol. Med.* **2018**, *50*, 134. [CrossRef]

108. Vanecek, J. Cellular mechanisms of melatonin action. *Physiol. Rev.* **1998**, *78*, 687–721. [CrossRef]

109. Prieto-Domínguez, N.; Ordóñez, R.; Fernández, A.; Méndez-Blanco, C.; Baulies, A.; Garcia-Ruiz, C.; Fernández-Checa, J.C.; Mauriz, J.L.; González-Gallego, J. Melatonin-induced increase in sensitivity of human hepatocellular carcinoma cells to sorafenib is associated with reactive oxygen species production and mitophagy. *J. Pineal Res.* **2016**, *61*, 396–407. [CrossRef]

110. Prieto-Dominguez, N.; Mendez-Blanco, C.; Carbajo-Pescador, S.; Fondevila, F.; Garcia-Palomo, A.; Gonzalez-Gallego, J.; Mauriz, J.L. Melatonin enhances sorafenib actions in human hepatocarcinoma cells by inhibiting mTORC1/p70S6K/HIF-1α and hypoxia-mediated mitophagy. *Oncotarget* **2017**, *8*, 91402–91414. [CrossRef]

111. Herbst, R.S.; Heymach, J.V.; Lippman, S.M. Lung cancer. *N. Engl. J. Med.* **2008**, *359*, 1367–1380. [CrossRef]

112. Chen, Z.; Fillmore, C.M.; Hammerman, P.S.; Kim, C.F.; Wong, K.K. Non-small-cell lung cancers: A heterogeneous set of diseases. *Nat. Rev. Cancer* **2014**, *14*, 535–546. [CrossRef]

113. Nguyen, L.V.; Vanner, R.; Dirks, P.; Eaves, C.J. Cancer stem cells: An evolving concept. *Nat. Rev. Cancer* **2012**, *12*, 133–143. [CrossRef]

114. Tang, D.G. Understanding cancer stem cell heterogeneity and plasticity. *Cell Res.* **2012**, *22*, 457–472. [CrossRef] [PubMed]

115. Bonnet, D.; Dick, J.E. Human acute myeloid leukemia is organized as a hierarchy that originates from a primitive hematopoietic cell. *Nat. Med.* **1997**, *3*, 730–737. [CrossRef]

116. Zhang, W.C.; Yang, H.; Soh, B.S.; Sun, L.L.; Chin, T.M.; Lim, E.H.; Lim, B. Abstract 487: Evidence for tumor initiating stem cells in lung cancer. *Cancer Res.* **2011**, *71*, 487.

117. Tam, W.L.; Lu, H.; Buikhuisen, J.; Soh, B.S.; Lim, E.; Reinhardt, F.; Wu, Z.J.; Krall, J.A.; Bierie, B.; Guo, W.; et al. Protein kinase C α is a central signaling node and therapeutic target for breast cancer stem cells. *Cancer Cell* **2013**, *24*, 347–364. [CrossRef] [PubMed]

118. Singh, S.K.; Hawkins, C.; Clarke, I.D.; Squire, J.A.; Bayani, J.; Hide, T.; Henkelman, R.M.; Cusimano, M.D.; Dirks, P.B. Identification of human brain tumour initiating cells. *Nature* **2004**, *432*, 396–401. [CrossRef]

119. Ricci-Vitiani, L.; Lombardi, D.G.; Pilozzi, E.; Biffoni, M.; Todaro, M.; Peschle, C.; De Maria, R. Identification and expansion of human colon-cancer-initiating cells. *Nature* **2007**, *445*, 111–115. [CrossRef]

120. Hoe, S.L.L.; Tan, L.P.; Jamal, J.; Peh, S.C.; Ng, C.C.; Zhang, W.C.; Ahmad, M.; Khoo, A.S.B. Evaluation of stem-like side population cells in a recurrent nasopharyngeal carcinoma cell line. *Cancer Cell Int.* **2014**, *14*, 101. [CrossRef] [PubMed]

121. Gupta, P.B.; Onder, T.T.; Jiang, G.; Tao, K.; Kuperwasser, C.; Weinberg, R.A.; Lander, E.S. Identification of selective inhibitors of cancer stem cells by high-throughput screening. *Cell* **2009**, *138*, 645–659. [CrossRef] [PubMed]

122. Ito, K.; Hirao, A.; Arai, F.; Matsuoka, S.; Takubo, K.; Hamaguchi, I.; Nomiyama, K.; Hosokawa, K.; Sakurada, K.; Nakagata, N.; et al. Regulation of oxidative stress by ATM is required for self-renewal of haematopoietic stem cells. *Nature* **2004**, *431*, 997–1002. [CrossRef]

123. Diehn, M.; Cho, R.W.; Lobo, N.A.; Kalisky, T.; Dorie, M.J.; Kulp, A.N.; Qian, D.; Lam, J.S.; Ailles, L.E.; Wong, M.; et al. Association of reactive oxygen species levels and radioresistance in cancer stem cells. *Nature* **2009**, *458*, 780–783. [CrossRef]

124. Dando, I.; Cordani, M.; Dalla Pozza, E.; Biondani, G.; Donadelli, M.; Palmieri, M. Antioxidant mechanisms and ROS-related microRNAs in cancer stem cells. *Oxid. Med. Cell Longev.* **2015**, *4257*, 201508. [CrossRef]

125. Bao, S.; Wu, Q.; McLendon, R.E.; Hao, Y.; Shi, Q.; Hjelmeland, A.B.; Dewhirst, M.W.; Bigner, D.D.; Rich, J.N. Glioma stem cells promote radioresistance by preferential activation of the DNA damage response. *Nature* **2006**, *444*, 756–760. [CrossRef]

126. Curtis, S.J.; Sinkevicius, K.W.; Li, D.; Lau, A.N.; Roach, R.R.; Zamponi, R.; Woolfenden, A.E.; Kirsch, D.G.; Wong, K.K.; Kim, C.F. Primary tumor genotype is an important determinant in identification of lung cancer propagating cells. *Cell Stem Cell* **2010**, *7*, 127–133. [CrossRef]

127. Houessinon, A.; Francois, C.; Sauzay, C.; Louandre, C.; Mongelard, G.; Godin, C.; Bodeau, S.; Takahashi, S.; Saidak, Z.; Gutierrez, L.; et al. Metallothionein-1 as a biomarker of altered redox metabolism in hepatocellular carcinoma cells exposed to sorafenib. *Mol. Cancer* **2016**, *15*, 38. [CrossRef]

128. Li, L.; Abdel Fattah, E.; Cao, G.; Ren, C.; Yang, G.; Goltsov, A.A.; Chinault, A.C.; Cai, W.W.; Timme, T.L.; Thompson, T.C. Glioma pathogenesis-related protein 1 exerts tumor suppressor activities through proapoptotic reactive oxygen species-c-Jun-NH$_2$ kinase signaling. *Cancer Res.* **2008**, *68*, 434–443. [CrossRef]

129. Yuan, D.; Xu, J.; Wang, J.; Pan, Y.; Fu, J.; Bai, Y.; Zhang, J.; Shao, C. Extracellular miR-1246 promotes lung cancer cell proliferation and enhances radioresistance by directly targeting DR5. *Oncotarget* **2016**, *7*, 32707–32722. [CrossRef]

130. Golestaneh, A.F.; Atashi, A.; Langroudi, L.; Shafiee, A.; Ghaemi, N.; Soleimani, M. MiRNAs expressed differently in cancer stem cells and cancer cells of human gastric cancer cell line MKN-45. *Cell Biochem. Funct.* **2012**, *30*, 411–418. [CrossRef]

131. Jackman, D.M.; Johnson, B.E. Small-cell lung cancer. *Lancet* **2005**, *366*, 1385–1396. [CrossRef]

132. Oser, M.G.; Niederst, M.J.; Sequist, L.V.; Engelman, J.A. Transformation from non-small-cell lung cancer to small-cell lung cancer: molecular drivers and cells of origin. *Lancet Oncol.* **2015**, *16*, e165–e172. [CrossRef]

133. Sequist, L.V.; Waltman, B.A.; Dias-Santagata, D.; Digumarthy, S.; Turke, A.B.; Fidias, P.; Bergethon, K.; Shaw, A.T.; Gettinger, S.; Cosper, A.K.; et al. Genotypic and histological evolution of lung cancers acquiring resistance to EGFR inhibitors. *Sci. Transl. Med.* **2011**, *3*, 75ra26. [CrossRef]

134. Marcoux, N.; Gettinger, S.N.; O'Kane, G.; Arbour, K.C.; Neal, J.W.; Husain, H.; Evans, T.L.; Brahmer, J.R.; Muzikansky, A.; Bonomi, P.D.; et al. EGFR-mutant adenocarcinomas that transform to small-cell lung cancer and other neuroendocrine carcinomas: clinical outcomes. *J. Clin. Oncol.* **2018**, *37*, 278–285. [CrossRef]

135. Hayashita, Y.; Osada, H.; Tatematsu, Y.; Yamada, H.; Yanagisawa, K.; Tomida, S.; Yatabe, Y.; Kawahara, K.; Sekido, Y.; Takahashi, T. A polycistronic microRNA cluster, miR-17-92, is overexpressed in human lung cancers and enhances cell proliferation. *Cancer Res.* **2005**, *65*, 9628–9632. [CrossRef]

136. Ebi, H.; Sato, T.; Sugito, N.; Hosono, Y.; Yatabe, Y.; Matsuyama, Y.; Yamaguchi, T.; Osada, H.; Suzuki, M.; Takahashi, T. Counterbalance between RB inactivation and miR-17-92 overexpression in reactive oxygen species and DNA damage induction in lung cancers. *Oncogene* **2009**, *28*, 3371–3379. [CrossRef]

137. Kalluri, R.; Neilson, E.G. Epithelial-mesenchymal transition and its implications for fibrosis. *J. Clin. Investig.* **2003**, *112*, 1776–1784. [CrossRef]

138. Iliopoulos, D.; Lindahl-Allen, M.; Polytarchou, C.; Hirsch, H.A.; Tsichlis, P.N.; Struhl, K. Loss of miR-200 inhibition of Suz12 leads to polycomb-mediated repression required for the formation and maintenance of cancer stem cells. *Mol. Cell* **2010**, *39*, 761–772. [CrossRef]

139. Shimono, Y.; Zabala, M.; Cho, R.W.; Lobo, N.; Dalerba, P.; Qian, D.; Diehn, M.; Liu, H.; Panula, S.P.; Chiao, E.; et al. Downregulation of miRNA-200c links breast cancer stem cells with normal stem cells. *Cell* **2009**, *138*, 592–603. [CrossRef]

140. Gregory, P.A.; Bert, A.G.; Paterson, E.L.; Barry, S.C.; Tsykin, A.; Farshid, G.; Vadas, M.A.; Khew-Goodall, Y.; Goodall, G.J. The miR-200 family and miR-205 regulate epithelial to mesenchymal transition by targeting ZEB1 and SIP1. *Nat. Cell Biol.* **2008**, *10*, 593–601. [CrossRef]

141. Mateescu, B.; Batista, L.; Cardon, M.; Gruosso, T.; de Feraudy, Y.; Mariani, O.; Nicolas, A.; Meyniel, J.-P.; Cottu, P.; Sastre-Garau, X.; et al. MiR-141 and miR-200a act on ovarian tumorigenesis by controlling oxidative stress response. *Nat. Med.* **2011**, *17*, 1627–1635. [CrossRef]

142. DeNicola, G.M.; Karreth, F.A.; Humpton, T.J.; Gopinathan, A.; Wei, C.; Frese, K.; Mangal, D.; Yu, K.H.; Yeo, C.J.; Calhoun, E.S.; et al. Oncogene-induced Nrf2 transcription promotes ROS detoxification and tumorigenesis. *Nature* **2011**, *475*, 106–109. [CrossRef] [PubMed]

143. Magenta, A.; Cencioni, C.; Fasanaro, P.; Zaccagnini, G.; Greco, S.; Sarra-Ferraris, G.; Antonini, A.; Martelli, F.; Capogrossi, M.C. MiR-200c is upregulated by oxidative stress and induces endothelial cell apoptosis and senescence via Zeb1 inhibition. *Cell Death Differ.* **2011**, *18*, 1628–1639. [CrossRef] [PubMed]

144. Huang, X.; Ding, L.; Bennewith, K.L.; Tong, R.T.; Welford, S.M.; Ang, K.K.; Story, M.; Le, Q.T.; Giaccia, A.J. Hypoxia-inducible miR-210 regulates normoxic gene expression involved in tumor initiation. *Mol. Cell* **2009**, *35*, 856–867. [CrossRef] [PubMed]

145. Guzy, R.D.; Schumacker, P.T. Oxygen sensing by mitochondria at complex III: The paradox of increased reactive oxygen species during hypoxia. *Exp. Physiol.* **2006**, *91*, 807–819. [CrossRef]

146. Chan, S.Y.; Zhang, Y.Y.; Hemann, C.; Mahoney, C.E.; Zweier, J.L.; Loscalzo, J. MicroRNA-210 controls mitochondrial metabolism during hypoxia by repressing the iron-sulfur cluster assembly proteins ISCU1/2. *Cell Metab.* **2009**, *10*, 273–284. [CrossRef] [PubMed]

147. Favaro, E.; Ramachandran, A.; McCormick, R.; Gee, H.; Blancher, C.; Crosby, M.; Devlin, C.; Blick, C.; Buffa, F.; Li, J.L.; et al. MicroRNA-210 regulates mitochondrial free radical response to hypoxia and krebs cycle in cancer cells by targeting iron sulfur cluster protein ISCU. *PLoS ONE* **2010**, *5*, e10345. [CrossRef]

148. Chen, Z.; Li, Y.; Zhang, H.; Huang, P.; Luthra, R. Hypoxia-regulated microRNA-210 modulates mitochondrial function and decreases ISCU and COX10 expression. *Oncogene* **2010**, *29*, 4362–4368. [CrossRef]

149. Yang, W.; Sun, T.; Cao, J.; Liu, F.; Tian, Y.; Zhu, W. Downregulation of miR-210 expression inhibits proliferation, induces apoptosis and enhances radiosensitivity in hypoxic human hepatoma cells in vitro. *Exp. Cell Res.* **2012**, *318*, 944–954. [CrossRef]

150. Yang, W.; Wei, J.; Guo, T.; Shen, Y.; Liu, F. Knockdown of miR-210 decreases hypoxic glioma stem cells stemness and radioresistance. *Exp. Cell Res.* **2014**, *326*, 22–35. [CrossRef]

151. Ivan, M.; Huang, X. MiR-210: Fine-tuning the hypoxic response. *Adv. Exp. Med. Biol.* **2014**, *772*, 205–227.

152. Calvert, A.E.; Chalastanis, A.; Wu, Y.; Hurley, L.A.; Kouri, F.M.; Bi, Y.; Kachman, M.; May, J.L.; Bartom, E.; Hua, Y.; et al. Cancer-associated IDH1 promotes growth and resistance to targeted therapies in the absence of mutation. *Cell Rep.* **2017**, *19*, 1858–1873. [CrossRef]

153. Keith, B.; Johnson, R.S.; Simon, M.C. HIF1α and HIF2α: Sibling rivalry in hypoxic tumour growth and progression. *Nat. Rev. Cancer* **2011**, *12*, 9–22. [CrossRef] [PubMed]

154. Samanta, D.; Gilkes, D.M.; Chaturvedi, P.; Xiang, L.; Semenza, G.L. Hypoxia-inducible factors are required for chemotherapy resistance of breast cancer stem cells. *Proc. Natl. Acad. Sci. USA* **2014**, *111*, E5429–E5438. [CrossRef]

155. Kaelin, W.G., Jr. The von Hippel-Lindau tumour suppressor protein: O_2 sensing and cancer. *Nat. Rev. Cancer* **2008**, *8*, 865–873. [CrossRef]

156. Guo, J.; Chakraborty, A.A.; Liu, P.; Gan, W.; Zheng, X.; Inuzuka, H.; Wang, B.; Zhang, J.; Zhang, L.; Yuan, M.; et al. pVHL suppresses kinase activity of Akt in a proline-hydroxylation-dependent manner. *Science* **2016**, *353*, 929–932. [CrossRef]

157. Lee, S.B.; Frattini, V.; Bansal, M.; Castano, A.M.; Sherman, D.; Hutchinson, K.; Bruce, J.N.; Califano, A.; Liu, G.; Cardozo, T.; et al. An ID2-dependent mechanism for VHL inactivation in cancer. *Nature* **2016**, *529*, 172–177. [CrossRef]

158. MacKenzie, E.D.; Selak, M.A.; Tennant, D.A.; Payne, L.J.; Crosby, S.; Frederiksen, C.M.; Watson, D.G.; Gottlieb, E. Cell-permeating α-ketoglutarate derivatives alleviate pseudohypoxia in succinate dehydrogenase-deficient cells. *Mol. Cell Biol.* **2007**, *27*, 3282–3289. [CrossRef]

159. Gomes, A.P.; Price, N.L.; Ling, A.J.; Moslehi, J.J.; Montgomery, M.K.; Rajman, L.; White, J.P.; Teodoro, J.S.; Wrann, C.D.; Hubbard, B.P.; et al. Declining NAD^+ induces a pseudohypoxic state disrupting nuclear-mitochondrial communication during aging. *Cell* **2013**, *155*, 1624–1638. [CrossRef]

160. Wilson, T.R.; Fridlyand, J.; Yan, Y.; Penuel, E.; Burton, L.; Chan, E.; Peng, J.; Lin, E.; Wang, Y.; Sosman, J.; et al. Widespread potential for growth-factor-driven resistance to anticancer kinase inhibitors. *Nature* **2012**, *487*, 505–509. [CrossRef]

161. Niederst, M.J.; Engelman, J.A. Bypass mechanisms of resistance to receptor tyrosine kinase inhibition in lung cancer. *Sci. Signal.* **2013**, *6*, re6. [CrossRef]

162. DeBerardinis, R.J.; Chandel, N.S. Fundamentals of cancer metabolism. *Sci. Adv.* **2016**, *2*, e1600200. [CrossRef] [PubMed]

163. Hensley, C.T.; Faubert, B.; Yuan, Q.; Lev-Cohain, N.; Jin, E.; Kim, J.; Jiang, L.; Ko, B.; Skelton, R.; Loudat, L.; et al. Metabolic heterogeneity in human lung tumors. *Cell* **2016**, *164*, 681–694. [CrossRef] [PubMed]

164. DeBerardinis, R.J.; Mancuso, A.; Daikhin, E.; Nissim, I.; Yudkoff, M.; Wehrli, S.; Thompson, C.B. Beyond aerobic glycolysis: transformed cells can engage in glutamine metabolism that exceeds the requirement for protein and nucleotide synthesis. *Proc. Natl. Acad. Sci. USA* **2007**, *104*, 19345–19350. [CrossRef]

165. Yuneva, M.; Zamboni, N.; Oefner, P.; Sachidanandam, R.; Lazebnik, Y. Deficiency in glutamine but not glucose induces MYC-dependent apoptosis in human cells. *J. Cell Biol.* **2007**, *178*, 93–105. [CrossRef]

166. Imielinski, M.; Guo, G.; Meyerson, M. Insertions and deletions target lineage-defining genes in human cancers. *Cell* **2017**, *168*, 460.e14–472.e14. [CrossRef] [PubMed]

167. Li, B.; Ren, S.; Li, X.; Wang, Y.; Garfield, D.; Zhou, S.; Chen, X.; Su, C.; Chen, M.; Kuang, P.; et al. MiR-21 overexpression is associated with acquired resistance of EGFR-TKI in non-small cell lung cancer. *Lung Cancer* **2014**, *83*, 146–153. [CrossRef]

168. Zheng, G.; Li, N.; Jia, X.; Peng, C.; Luo, L.; Deng, Y.; Yin, J.; Song, Y.; Liu, H.; Lu, M.; et al. MYCN-mediated miR-21 overexpression enhances chemo-resistance via targeting CADM1 in tongue cancer. *J. Mol. Med. (Berl.)* **2016**, *94*, 1129–1141. [CrossRef]

169. Nishikura, K. A-to-I editing of coding and non-coding RNAs by ADARs. *Nat. Rev. Mol. Cell Biol.* **2016**, *17*, 83–96. [CrossRef]

170. Yang, W.; Chendrimada, T.P.; Wang, Q.; Higuchi, M.; Seeburg, P.H.; Shiekhattar, R.; Nishikura, K. Modulation of microRNA processing and expression through RNA editing by ADAR deaminases. *Nat. Struct. Mol. Biol.* **2006**, *13*, 13–21. [CrossRef]

171. Lazzari, E.; Crews, L.A.; Wu, C.; Leu, H.; Ali, S.; Chiaramonte, R.; Minden, M.; Costello, C.; Jamieson, C.H.M. Abstract 2414: ADAR1-dependent RNA editing is a mechanism of therapeutic resistance in human plasma cell malignancies. *Cancer Res.* **2016**, *76*, 2414.

172. Zhang, W.C.; Slack, F.J. ADARs edit microRNAs to promote leukemic stem cell activity. *Cell Stem Cell* **2016**, *19*, 141–142. [CrossRef] [PubMed]

173. Zipeto, M.A.; Court, A.C.; Sadarangani, A.; Delos Santos, N.P.; Balaian, L.; Chun, H.J.; Pineda, G.; Morris, S.R.; Mason, C.N.; Geron, I.; et al. ADAR1 activation drives leukemia stem cell self-renewal by impairing Let-7 biogenesis. *Cell Stem Cell* **2016**, *19*, 177–191. [CrossRef]

174. Jiang, Q.; Isquith, J.; Zipeto, M.A.; Diep, R.H.; Pham, J.; Delos Santos, N.; Reynoso, E.; Chau, J.; Leu, H.; Lazzari, E.; et al. Hyper-editing of cell-cycle regulatory and tumor suppressor RNA promotes malignant progenitor propagation. *Cancer Cell* **2019**, *35*, 81.e7–94.e7. [CrossRef] [PubMed]

175. Zaretsky, J.M.; Garcia-Diaz, A.; Shin, D.S.; Escuin-Ordinas, H.; Hugo, W.; Hu-Lieskovan, S.; Torrejon, D.Y.; Abril-Rodriguez, G.; Sandoval, S.; Barthly, L.; et al. Mutations associated with acquired resistance to PD-1 blockade in melanoma. *N. Engl. J. Med.* **2016**, *375*, 819–829. [CrossRef]

176. Sade-Feldman, M.; Jiao, Y.J.; Chen, J.H.; Rooney, M.S.; Barzily-Rokni, M.; Eliane, J.-P.; Bjorgaard, S.L.; Hammond, M.R.; Vitzthum, H.; Blackmon, S.M.; et al. Resistance to checkpoint blockade therapy through inactivation of antigen presentation. *Nat. Commun.* **2017**, *8*, 1136. [CrossRef]

177. Ishizuka, J.J.; Manguso, R.T.; Cheruiyot, C.K.; Bi, K.; Panda, A.; Iracheta-Vellve, A.; Miller, B.C.; Du, P.P.; Yates, K.B.; Dubrot, J.; et al. Loss of ADAR1 in tumours overcomes resistance to immune checkpoint blockade. *Nature* **2019**, *565*, 43–48. [CrossRef]

178. Chen, L.; Rio, D.C.; Haddad, G.G.; Ma, E. Regulatory role of dADAR in ROS metabolism in drosophila CNS. *Brain Res. Mol. Brain Res.* **2004**, *131*, 93–100. [CrossRef]

179. Dominissini, D.; Moshitch-Moshkovitz, S.; Schwartz, S.; Salmon-Divon, M.; Ungar, L.; Osenberg, S.; Cesarkas, K.; Jacob-Hirsch, J.; Amariglio, N.; Kupiec, M.; et al. Topology of the human and mouse m^{6}A RNA methylomes revealed by m^{6}A-seq. *Nature* **2012**, *485*, 201–206. [CrossRef] [PubMed]

180. Geula, S.; Moshitch-Moshkovitz, S.; Dominissini, D.; Mansour, A.A.; Kol, N.; Salmon-Divon, M.; Hershkovitz, V.; Peer, E.; Mor, N.; Manor, Y.S.; et al. M^{6}A mRNA methylation facilitates resolution of naive pluripotency toward differentiation. *Science* **2015**, *347*, 1002–1006. [CrossRef]

181. Lin, S.; Choe, J.; Du, P.; Triboulet, R.; Gregory, R.I. The m^{6}A methyltransferase METTL3 promotes translation in human cancer cells. *Mol. Cell* **2016**, *62*, 335–345. [CrossRef]

182. Choe, J.; Lin, S.; Zhang, W.; Liu, Q.; Wang, L.; Ramirez-Moya, J.; Du, P.; Kim, W.; Tang, S.; Sliz, P.; et al. MRNA circularization by METTL3-eIF3h enhances translation and promotes oncogenesis. *Nature* **2018**, *561*, 556–560. [CrossRef] [PubMed]

183. Lan, Q.; Liu, P.Y.; Haase, J.; Bell, J.L.; Huttelmaier, S.; Liu, T. The critical role of RNA m^{6}A methylation in cancer. *Cancer Res.* **2019**, *79*, 1285–1292. [CrossRef] [PubMed]

184. Shi, H.; Wei, J.; He, C. Where, when, and how: Context-dependent functions of RNA methylation writers, readers, and erasers. *Mol. Cell* **2019**, *74*, 640–650. [CrossRef] [PubMed]

185. Visvanathan, A.; Patil, V.; Arora, A.; Hegde, S.A.; Arivazhagan, A.; Santosh, V.; Somasundaram, K. Essential Role of Mettl3-Mediated M6a Modification in Glioma Stem-Like Cells Maintenance and Radioresistance. *Oncogene* **2018**, *37*, 522–533. [CrossRef]

186. Taketo, K.; Konno, M.; Asai, A.; Koseki, J.; Toratani, M.; Satoh, T.; Doki, Y.; Mori, M.; Ishii, H.; Ogawa, K. The Epitranscriptome M6a Writer Mettl3 Promotes Chemo- and Radioresistance in Pancreatic Cancer Cells. *Int. J. Oncol.* **2018**, *52*, 621–629. [CrossRef]

187. Yan, F.; Al-Kali, A.; Zhang, Z.; Liu, J.; Pang, J.; Zhao, N.; He, C.; Litzow, M.R.; Liu, S. A dynamic N 6-methyladenosine methylome regulates intrinsic and acquired resistance to tyrosine kinase inhibitors. *Cell Res.* **2018**, *28*, 1062–1076. [CrossRef]

188. Wang, J.; Ishfaq, M.; Xu, L.; Xia, C.; Chen, C.; Li, J. METTL3/m^{6}A/miRNA-873–5p attenuated oxidative stress and apoptosis in colistin-induced kidney injury by modulating Keap1/Nrf2 pathway. *Front. Pharmacol.* **2019**. [CrossRef]

189. Topol, E.J. High-performance medicine: The convergence of human and artificial intelligence. *Nat. Med.* **2019**, *25*, 44–56. [CrossRef]

190. Romm, E.L.; Tsigelny, I.F. Artificial intelligence in drug treatment. *Annu. Rev. Pharmacol. Toxicol.* **2020**, *60*. [CrossRef]

191. Mason, D.J.; Eastman, R.T.; Lewis, R.P.I.; Stott, I.P.; Guha, R.; Bender, A. Using machine learning to predict synergistic antimalarial compound combinations with novel structures. *Front. Pharmacol.* **2018**. [CrossRef]

192. Li, X.; Xu, Y.; Cui, H.; Huang, T.; Wang, D.; Lian, B.; Li, W.; Qin, G.; Chen, L.; Xie, L. Prediction of synergistic anti-cancer drug combinations based on drug target network and drug induced gene expression profiles. *Artif. Intell. Med.* **2017**, *83*, 35–43. [CrossRef]

193. Xia, F.; Shukla, M.; Brettin, T.; Garcia-Cardona, C.; Cohn, J.; Allen, J.E.; Maslov, S.; Holbeck, S.L.; Doroshow, J.H.; Evrard, Y.A.; et al. Predicting tumor cell line response to drug pairs with deep learning. *BMC Bioinform.* **2018**, *19*, 486. [CrossRef] [PubMed]

194. McQuade, J.L.; Daniel, C.R.; Helmink, B.A.; Wargo, J.A. Modulating the microbiome to improve therapeutic response in cancer. *Lancet Oncol.* **2019**, *20*, e77–e91. [CrossRef]

195. Peek, R.M., Jr.; Blaser, M.J. Helicobacter pylori and gastrointestinal tract adenocarcinomas. *Nat. Rev. Cancer* **2002**, *2*, 28–37. [CrossRef] [PubMed]

196. Bullman, S.; Pedamallu, C.S.; Sicinska, E.; Clancy, T.E.; Zhang, X.; Cai, D.; Neuberg, D.; Huang, K.; Guevara, F.; Nelson, T.; et al. Analysis of fusobacterium persistence and antibiotic response in colorectal cancer. *Science* **2017**, *358*, 1443–1448. [CrossRef] [PubMed]

197. Ma, C.; Han, M.; Heinrich, B.; Fu, Q.; Zhang, Q.; Sandhu, M.; Agdashian, D.; Terabe, M.; Berzofsky, J.A.; Fako, V.; et al. Gut microbiome-mediated bile acid metabolism regulates liver cancer via NKT cells. *Science* **2018**, *360*, eaan5931. [CrossRef]

198. Pushalkar, S.; Hundeyin, M.; Daley, D.; Zambirinis, C.P.; Kurz, E.; Mishra, A.; Mohan, N.; Aykut, B.; Usyk, M.; Torres, L.E.; et al. The pancreatic cancer microbiome promotes oncogenesis by induction of innate and adaptive immune suppression. *Cancer Discov.* **2018**, *8*, 403–416. [CrossRef]

199. Sam, Q.H.; Chang, M.W.; Chai, L.Y. The fungal mycobiome and its interaction with gut bacteria in the host. *Int. J. Mol. Sci.* **2017**, *18*, 330. [CrossRef]

200. Aykut, B.; Pushalkar, S.; Chen, R.; Li, Q.; Abengozar, R.; Kim, J.I.; Shadaloey, S.A.; Wu, D.; Preiss, P.; Verma, N.; et al. The fungal mycobiome promotes pancreatic oncogenesis via activation of MBL. *Nature* **2019**, *574*, 264–267. [CrossRef]

201. Moore, P.S.; Chang, Y. Why do viruses cause cancer? Highlights of the first century of human tumour virology. *Nat. Rev. Cancer* **2010**, *10*, 878–889. [CrossRef]

202. Geller, L.T.; Barzily-Rokni, M.; Danino, T.; Jonas, O.H.; Shental, N.; Nejman, D.; Gavert, N.; Zwang, Y.; Cooper, Z.A.; Shee, K.; et al. Potential role of intratumor bacteria in mediating tumor resistance to the chemotherapeutic drug gemcitabine. *Science* **2017**, *357*, 1156–1160. [CrossRef]

203. Yu, T.; Guo, F.; Yu, Y.; Sun, T.; Ma, D.; Han, J.; Qian, Y.; Kryczek, I.; Sun, D.; Nagarsheth, N.; et al. Fusobacterium nucleatum promotes chemoresistance to colorectal cancer by modulating autophagy. *Cell* **2017**, *170*, 548.e16–563.e16. [CrossRef] [PubMed]

204. Tsang, W.P.; Kwok, T.T. The miR-18a* microRNA functions as a potential tumor suppressor by targeting on K-Ras. *Carcinogenesis* **2009**, *30*, 953–959. [CrossRef]

205. Lievre, A.; Bachet, J.B.; Le Corre, D.; Boige, V.; Landi, B.; Emile, J.F.; Cote, J.F.; Tomasic, G.; Penna, C.; Ducreux, M.; et al. KRAS mutation status is predictive of response to cetuximab therapy in colorectal cancer. *Cancer Res.* **2006**, *66*, 3992–3995. [CrossRef]

206. Misale, S.; Yaeger, R.; Hobor, S.; Scala, E.; Janakiraman, M.; Liska, D.; Valtorta, E.; Schiavo, R.; Buscarino, M.; Siravegna, G.; et al. Emergence of KRAS mutations and acquired resistance to anti-EGFR therapy in colorectal cancer. *Nature* **2012**, *486*, 532–536. [CrossRef] [PubMed]

207. Haqshenas, G.; Doerig, C. Targeting of host cell receptor tyrosine kinases by intracellular pathogens. *Sci. Signal.* **2019**, *12*, eaau9894. [CrossRef]

208. Schor, S.; Einav, S. Combating intracellular pathogens with repurposed host-targeted drugs. *ACS Infect. Dis.* **2018**, *4*, 88–92. [CrossRef] [PubMed]

209. Stewart, A.C.; Gay, C.M.; Xi, Y.; Fujimoto, J.; Kalhor, N.; Hartsfield, P.M.; Tran, H.; Fernandez, L.; Lu, D.; Wang, Y.; et al. Abstract 2899: Single-cell analyses reveal increasing intratumoral heterogeneity as an essential component of treatment resistance in small cell lung cancer. *Cancer Res.* **2019**, *79*, 2899.

210. Rambow, F.; Rogiers, A.; Marin-Bejar, O.; Aibar, S.; Femel, J.; Dewaele, M.; Karras, P.; Brown, D.; Chang, Y.H.; Debiec-Rychter, M.; et al. Toward minimal residual disease-directed therapy in melanoma. *Cell* **2018**, *174*, 843.e19–855.e19. [CrossRef]

211. Bivona, T.G.; Doebele, R.C. A framework for understanding and targeting residual disease in oncogene-driven solid cancers. *Nat. Med.* **2016**, *22*, 472–478. [CrossRef]

212. Ponz-Sarvise, M.; Corbo, V.; Tiriac, H.; Engle, D.D.; Frese, K.K.; Oni, T.E.; Hwang, C.I.; Ohlund, D.; Chio, I.I.C.; Baker, L.A.; et al. Identification of resistance pathways specific to malignancy using organoid models of pancreatic cancer. *Clin. Cancer Res.* **2019**, *25*, 6742–6755. [CrossRef]

213. Usui, T.; Sakurai, M.; Umata, K.; Elbadawy, M.; Ohama, T.; Yamawaki, H.; Hazama, S.; Takenouchi, H.; Nakajima, M.; Tsunedomi, R.; et al. Hedgehog signals mediate anti-cancer drug resistance in three-dimensional primary colorectal cancer organoid culture. *Int. J. Mol. Sci.* **2018**, *19*, 1098. [CrossRef]

214. Hubert, C.G.; Rivera, M.; Spangler, L.C.; Wu, Q.; Mack, S.C.; Prager, B.C.; Couce, M.; McLendon, R.E.; Sloan, A E.; Rich, N.J. A three-dimensional organoid culture system derived from human glioblastomas recapitulates the hypoxic gradients and cancer stem cell heterogeneity of tumors found in vivo. *Cancer Res.* **2016**, *76*, 2465–2477. [CrossRef] [PubMed]

215. Chakrabarti, J.; Holokai, L.; Syu, L.; Steele, N.; Chang, J.; Dlugosz, A.; Zavros, Y. Mouse-derived gastric organoid and immune cell co-culture for the study of the tumor microenvironment. *Methods Mol. Biol.* **2018**, *1817*, 157–168. [PubMed]

216. Neal, J.T.; Li, X.; Zhu, J.; Giangarra, V.; Grzeskowiak, C.L.; Ju, J.; Liu, I.H.; Chiou, S.H.; Salahudeen, A.A.; Smith, A.R.; et al. Organoid modeling of the tumor immune microenvironment. *Cell* **2018**, *175*, 1972.e16–1988.e16. [CrossRef] [PubMed]

217. Sozzi, G.; Boeri, M.; Rossi, M.; Verri, C.; Suatoni, P.; Bravi, F.; Roz, L.; Conte, D.; Grassi, M.; Sverzellati, N.; et al. Clinical utility of a plasma-based miRNA signature classifier within computed tomography lung cancer screening: A correlative mild trial study. *J. Clin. Oncol.* **2014**, *32*, 768–773. [CrossRef]

218. Siravegna, G.; Bardelli, A. Genotyping cell-free tumor DNA in the blood to detect residual disease and drug resistance. *Genome Biol.* **2014**, *15*, 449. [CrossRef]

International Journal of
Molecular Sciences

Review

Interplay between MicroRNAs and Oxidative Stress in Neurodegenerative Diseases

Julia Konovalova [1], Dmytro Gerasymchuk [1,2], Ilmari Parkkinen [1], Piotr Chmielarz [3] and Andrii Domanskyi [1,*]

[1] Institute of Biotechnology, HiLIFE, University of Helsinki, 00014 Helsinki, Finland;
    julia.konovalova@helsinki.fi (J.K.); dmytro.gerasymchuk@helsinki.fi (D.G.);
    ilmari.parkkinen@helsinki.fi (I.P.)
[2] Institute of Molecular Biology and Genetics, NASU, Kyiv 03143, Ukraine
[3] Department of Brain Biochemistry, Maj Institute of Pharmacology, Polish Academy of Sciences,
    31-343 Krakow, Poland; chmiel@if-pan.krakow.pl
* Correspondence: andrii.domanskyi@helsinki.fi; Tel.: +358-50-448-4545

Received: 28 October 2019; Accepted: 28 November 2019; Published: 30 November 2019

Abstract: MicroRNAs are post-transcriptional regulators of gene expression, crucial for neuronal differentiation, survival, and activity. Age-related dysregulation of microRNA biogenesis increases neuronal vulnerability to cellular stress and may contribute to the development and progression of neurodegenerative diseases. All major neurodegenerative disorders are also associated with oxidative stress, which is widely recognized as a potential target for protective therapies. Albeit often considered separately, microRNA networks and oxidative stress are inextricably entwined in neurodegenerative processes. Oxidative stress affects expression levels of multiple microRNAs and, conversely, microRNAs regulate many genes involved in an oxidative stress response. Both oxidative stress and microRNA regulatory networks also influence other processes linked to neurodegeneration, such as mitochondrial dysfunction, deregulation of proteostasis, and increased neuroinflammation, which ultimately lead to neuronal death. Modulating the levels of a relatively small number of microRNAs may therefore alleviate pathological oxidative damage and have neuroprotective activity. Here, we review the role of individual microRNAs in oxidative stress and related pathways in four neurodegenerative conditions: Alzheimer's (AD), Parkinson's (PD), Huntington's (HD) disease, and amyotrophic lateral sclerosis (ALS). We also discuss the problems associated with the use of oversimplified cellular models and highlight perspectives of studying microRNA regulation and oxidative stress in human stem cell-derived neurons.

Keywords: microRNA; oxidative stress; ROS; translation regulation; neurodegeneration; Alzheimer's disease; Parkinson's disease; Huntington's disease; ALS

---

1. Introduction

Neurodegenerative diseases, such as Alzheimer's (AD), Parkinson's (PD), Huntington's (HD) disease, and Amyotrophic Lateral Sclerosis (ALS), are devastating and currently incurable conditions causing severe cognitive and/or motor impairments predominantly in aged people [1,2]. The incidence of age-related neurodegeneration is expected to increase due to aging population and increased life expectancy in the developed countries. Alzheimer's disease (AD) and other dementias are estimated to affect up to 50 million people worldwide [3]. Another 10 million patients are suffering from Parkinson's disease (PD), which occurs in ≈2% of people over 70 years of age [4]. To develop curative therapies for neurodegenerative diseases, it is crucial to elucidate molecular mechanisms regulating neuron survival and degeneration.

Oxidative stress has been implicated in predisposing neurons to death either directly or indirectly as a consequence of mitochondrial dysfunction, pathological protein aggregation, specific neurotransmitter (dopamine) metabolism, inflammation, or deregulation of antioxidant pathways [5–10]. The brain is particularly susceptible to oxidative stress due to high oxygen consumption (reflecting high ATP demand) and the reliance on mitochondrial activity, intracellular calcium, and a relatively weak endogenous antioxidant defense, among other reasons [11,12]. Reactive oxygen species (ROS) cause oxidative damage to proteins, lipids, and nucleic acids, compromising critical cellular functions and activating cell death pathways [13]. Oxidative stress and oxidative damage are commonly observed in different neurodegenerative diseases and, therefore, therapies aiming to reduce cellular ROS levels may offer neuroprotective treatments for multiple neurodegenerative conditions. However, attempts to treat neurodegenerative diseases with antioxidant drugs have mostly been unsuccessful, in part, due to insufficient blood–brain barrier penetration, short treatment duration, or incorrect timing of therapy application [13–15]. Alternative therapeutic interventions may aim to counteract oxidative damage by stimulating endogenous neuronal antioxidant defense pathways [16]. In this review, we explore the concept of targeting specific microRNAs regulating or regulated by these pathways as a strategy to protect neurons in neurodegenerative diseases.

MicroRNAs are short regulatory RNA molecules which affect translation and stability of their mRNA targets by guiding RNA-induced silencing complex (RISC) predominantly to 3′ untranslated region (UTR) [17,18]. MicroRNAs are predicted to regulate the activity of about a half of all protein coding genes, reducing fluctuations in protein expression [19,20].

MicroRNAs are expressed as precursor hairpins which undergo sequential processing in the nucleus and cytoplasm by specific protein complexes containing ribonucleases Drosha and Dicer; mature functional microRNAs are then loaded to Argonaute family protein Ago2, a central component of the RISC complex [18]. MicroRNAs are critical for neuronal functions both during development and in the adult brain [21,22]. Loss of mature microRNA functions by genetic deletion of Dicer or Ago2 is embryonic lethal [23,24], whereas deletion of Dicer during embryogenesis severely impairs neuronal development [21,25–29]. Inducible deletion of Dicer in postnatal Purkinje cells and in adult forebrain and dopamine neurons causes their progressive loss and severe behavioral phenotypes [30–34]. While some neuronal populations survive Dicer deletion, their functions are clearly affected [35–38]. Similarly, loss of Ago2 in adult neurons is dispensable for their survival, but it nevertheless affects neuronal functions resulting in a behavioral phenotype [39]. MicroRNAs have been implicated in modulation of neuronal signaling by regulating neuronal excitability, dendritogenesis, local translation in dendritic spines, and neurotransmitter release [21,40–42]. Age and disease-related downregulation of the microRNA biogenesis pathway in adult neurons can lead to changes in their survival, functions, and connectivity. Inhibition of Dicer activity and resulting changes in microRNA expression levels have been observed in aging and in neurological and neurodegenerative diseases [30,42–50]. Deregulation of microRNA biogenesis is causing cellular stress and, vice versa, increased stress causes deregulation of microRNA biogenesis, creating a vicious cycle leading to eventual cell death [51–53]. In line with this hypothesis, stimulation of microRNA biogenesis is neuroprotective in mouse models of ALS and PD [30,52,54].

A small number of microRNAs can regulate hundreds of transcripts and may enable a crosstalk between different cellular pathways [55,56]. For example, several microRNAs can be targeting many genes involved in antioxidant defense pathways [53]. Modulating the levels of a relatively small number of microRNAs which regulate the oxidative stress response in neurons may therefore alleviate pathological oxidative damage and have neuroprotective activity. However, it is not trivial to identify such microRNAs and their target genes in aged and degenerating neuronal populations. Below, we review the current literature addressing the interplay between oxidative stress and microRNAs in major neurodegenerative diseases.

2. Alzheimer's Disease

Dementia is estimated to affect more than 50 million people worldwide with the prognosis of doubling in the next 20 years [57]. AD, which is an irreversible neurodegenerative disorder affecting both cognition and emotional behavior of affected persons (usually at the age of 65 and older) [58], is considered to be the cause of 50–75% of all dementia cases with no effective treatment to stop or slow down the disease [59,60].

Despite intensive studies, the real causes of AD development are still not clear. Extracellular accumulation of amyloid-β (Aβ) peptides and hyperphosphorylation of the microtubule-associated Tau protein are the main hallmarks of AD development at the molecular and cellular level leading to the accumulation of senile plagues and neurofibrillary tangles, respectively. Mutations in *PSEN1*, *PSEN2*, *APP* genes, variants of *APOE* gene, and posttranscriptional modifications of AD-associated proteins can also contribute to the development of this neurodegenerative disease. Taken together, these changes result in synaptic loss, neuronal cell death, and cognitive impairment reviewed in [61,62].

According to numerous studies, microRNA contribute to the development of AD regulating accumulation of Aβ peptides and Tau phosphorylation [63–68]. However, accumulation of insoluble protein aggregates is not the only, and, possibly, not the main pathological process driving AD progression. Oxidative stress is of particular importance for AD development as it causes chronic inflammation at the early stages of neurodegeneration, which leads to mitochondrial dysfunction, oxidative damage of nucleic acids, changes in genes expression, and abnormal modifications of lipids and proteins [69]. Oxidative stress causes both up- and downregulation of different microRNAs and, conversely, many microRNAs can regulate oxidative stress response [70] (Figure 1).

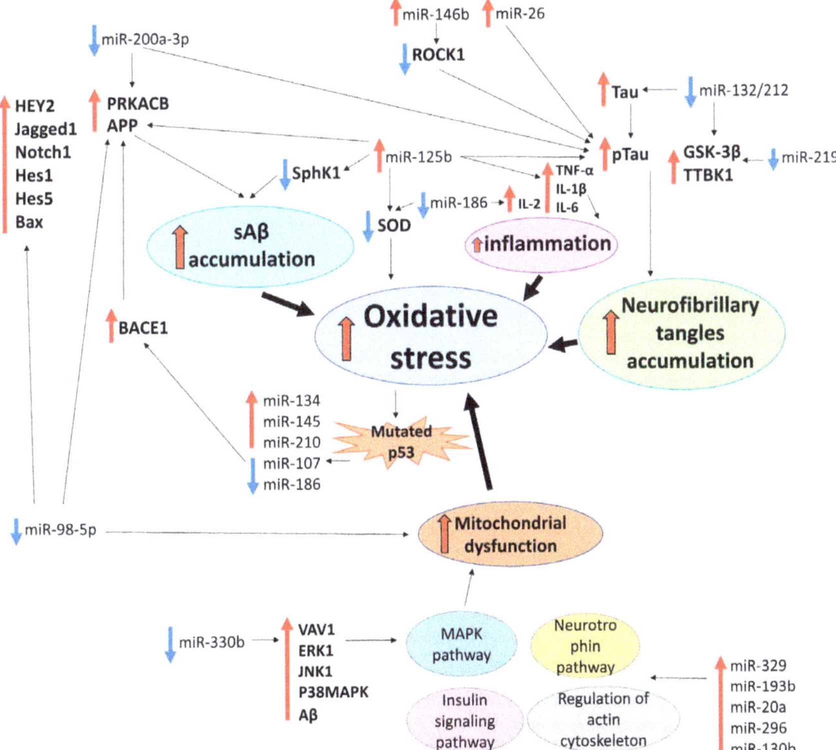

Figure 1. MicroRNAs implicated in oxidative stress-related cellular pathways in Alzheimer's disease.

Li et al. demonstrated that soluble Aβ peptides (sAβ) known to generate ROS [71] reliably induced expression of miR-134, miR-145 and miR-210. In the same study, expression of miR-107 was markedly reduced, supporting a bilateral effect of sAβ-induced ROS on microRNA expression [72]. Decreased levels of miR-107 is associated with early stages of AD progression. This microRNA directly targets BACE1 mRNA encoding β-secretase enzyme that processes APP to Aβ peptides [73]. In AD patients with the APOE4 genotype, decreased levels of miR-107 have been demonstrated along with the increased production of Aβ peptides. Accumulation of Aβ induced oxidative stress in APOE4 leads to the deregulation of the *TP53* gene. In addition to its role in cancer, p53 protein (encoded by a *TP53* gene) can be involved in cell death in AD patients with upregulation at the early stages of the disease and downregulation during neurodegeneration [74]. Previously, p53 mutations that may be associated with oxidative stress were observed in AD patients and AD animal models [75,76]. Since miR-107 is downregulated in cell lines with mutated p53 [77], p53 mutations and accumulation of Aβ may result in the decrease of miR-107 levels in AD patients. Moreover, 8-oxo-2′deoxyguanosine RNA modifications caused by oxidative stress can serve as an additional factor of decreasing miR-107 levels [78]. Levels of another microRNA, miR-186, are decreased through aging. This microRNA targets 3′UTR of BACE1 and is implicated in the mitigation of the oxidative stress effects in AD pathogenesis [79].

Another study revealed that the upregulation of miR-342-5p is important for neurogenesis and neuroprotection in an AD mouse model. Downregulation of Ankylin G, a direct target of miR-342-5p, results in AD axonopathy [80]. Liang et al. showed a decrease of miR-153 expression following sAβ treatment of M17 human neuroblastoma cells in combination with H_2O_2. APP and APLP2, an APP homologue, are confirmed as direct targets of miR-153, providing additional evidence of microRNA-based regulation of the essential stage of AD progression and the role of oxidative stress in this process [81].

Phosphorylation of Tau protein followed by the accumulation of neurofibrillary tangles is affected by the formation of ROS. Numerous studies confirmed the role of oxidative stress on Tau acetylation and subsequent phosphorylation by GSK-3 kinase or other pathways [82–84]. Several microRNAs also contribute to the regulation of Tau phosphorylation. MiR-200a-3p targets BACE1 and PRKACB (catalytic subunit of PKA), reducing Aβ accumulation and Tau hyperphosphorylation, respectively [85]. Li et al. identified overexpressed miR-219 in brains of AD patients. In the SH-SY5Y cell line, miR-219 downregulated Tau phosphorylation by targeting TTBK1 and GSK-3β [86]. GSK-3β alongside with Rbfox1, EP300, and Calpain 2 are directly targeted by miR-132/212, which are among the most downregulated microRNAs in AD [87]. Moreover, Tau mRNA is directly targeted by miR-132/212 [88]. In contrast to the abovementioned cases, overexpression of miR-146b in the AD brain induced abnormal Tau phosphorylation by targeting ROCK1 kinase [89]. Absalon et al. demonstrated a neuroprotective effect of sequence-specific inhibition of miR-26 in primary cortical neurons treated with H_2O_2. miR-26 is known to be upregulated in AD patients and contributes to Tau hyperphosphorylation and Aβ accumulation [90].

Screening of AD-associated microRNAs in H_2O_2-treated primary hippocampal neurons and a senescent mouse model demonstrated strong upregulation of miR-329, miR-193b, miR-20a, miR-296, and miR-130b. Expression of miR-329 played a critical role in the activity-dependent dendritic outgrowth of hippocampal neurons, whereas miR-130b expressed in the hippocampus was related to chronic stress-induced depression. miR-20a targeted neuronal differentiation markers BCL2, MEF2D and MAP3K12 (ZPK/MUK/DLK), suggesting its key role in the regulation of gene expression during brain development. According to KEGG analysis, upregulated microRNAs participated in the cellular processes closely connected to the occurrence and development of AD, in particular, neurotrophin signaling pathway, MAPK pathway, insulin signaling pathway, and regulation of actin cytoskeleton. This can indicate the importance of abovementioned microRNAs for the development of AD. [91]. miR-330a has also been reported to contribute to alleviation of oxidative stress and mitochondria dysfunction in AD by targeting mRNAs of VAV1, ERK1, JNK1, P38MAPK, and Aβ, which are all upregulated in AD mice, indicating the involvement of the MAPK pathway in AD [92].

The Notch pathway is among important cellular processes that can be associated with oxidative stress [93–95]. The Notch-HEY2 pathway in the hippocampal neurons of AD mice was activated with the downregulation of miR-98-5p compared to normal hippocampal neurons. *APP* correlated with levels of miR-98-5p, alongside *HEY2*, *Jagged1*, *Notch1*, *Hes1*, *Hes5*, and *Bax* genes of the Notch pathway, indicating the inhibitory effect of miR-98-5p on these genes and, thus, AD progression. Furthermore, miR-98-5p promoted the growth of hippocampal neurons, inhibited neuronal apoptosis, and improved oxidative stress and mitochondrial dysfunction of AD mice, whereas HEY2 was reported to have opposite effects. These results contradict previous data about promotion of Aβ production by miR-98-5p and upregulation of this microRNA in AD mouse models. Further studies would possibly clarify a more precise role of miR-98 in AD [96].

Despite being affected, microRNAs themselves can trigger oxidative stress in neurons promoting neurodegeneration. miR-125b is known as an important factor of AD progression, promoting APP, BACE1, and Tau overexpression and hyperphosphorylation [97]. In mouse neuroblastoma Neuro2a APPSwe/Δ9 cells, overexpression of miR-125b enhanced oxidative stress by decreasing levels of superoxide dismutase (SOD) together with the stimulation of apoptosis. Additionally, this microRNA stimulates overexpression of TNF-α, IL-1β, and IL-6 inflammatory cytokines, further supporting the connection between inflammation and oxidative stress in degeneration of neurons. Moreover, miR-125b significantly decreased expression of SphK1, which improves memory, learning, and suppresses formation of Aβ peptides [98]. The biological activities of IL2, another inflammatory cytokine, correlate with the JAK/STAT pathway involved in AD development by inducing astrocyte reactivity. A recent study by Wu et al. demonstrated the role of miR-186, a tumor suppressor microRNA, in the downregulation of IL2. Rat brains with decreased expression of miR-186 had been characterized by the elevated levels of IL2, JAK/STAT, Bax, and Cleaved-caspase 3 genes and ROS, whereas BCL2 and SOD activity were downregulated [99].

3. Parkinson's Disease

PD is a common progressive neurodegenerative disorder. PD is primarily characterized by degeneration of dopamine neurons in the substantia nigra pars compacta (SNpc) and their projections to the corpus striatum. Dopamine neuron loss leads to manifestation of PD motor symptoms, such as bradykinesia, resting tremor, postural instability, and rigidity [6]. Additionally, PD patients exhibit a broad range of non-motor symptoms, such as depression, sleep disorders, and dementia. Many of them precede the appearance of motor symptoms and worsen with progression of PD [100].

PD is an age-related disorder, affecting approximately 1% of the population over 60 years old and this number reaches 4-5% in the population over 85-years old. Despite many years of research, mechanisms underlying the pathology of PD are still not well understood. Several genetic mutations associated with PD have been identified and account for at least 5–10% of PD cases; however, in most of the cases, the etiology of PD is unknown [101].

Many different mechanisms have been proposed to drive neuronal death in PD, including oxidative stress. Major sources of oxidative stress in dopamine neurons include dopamine metabolism, mitochondrial dysfunction, impairment of the endogenous antioxidant system, aggregation of the α-synuclein protein, and neuroinflammation (Figure 2) [102].

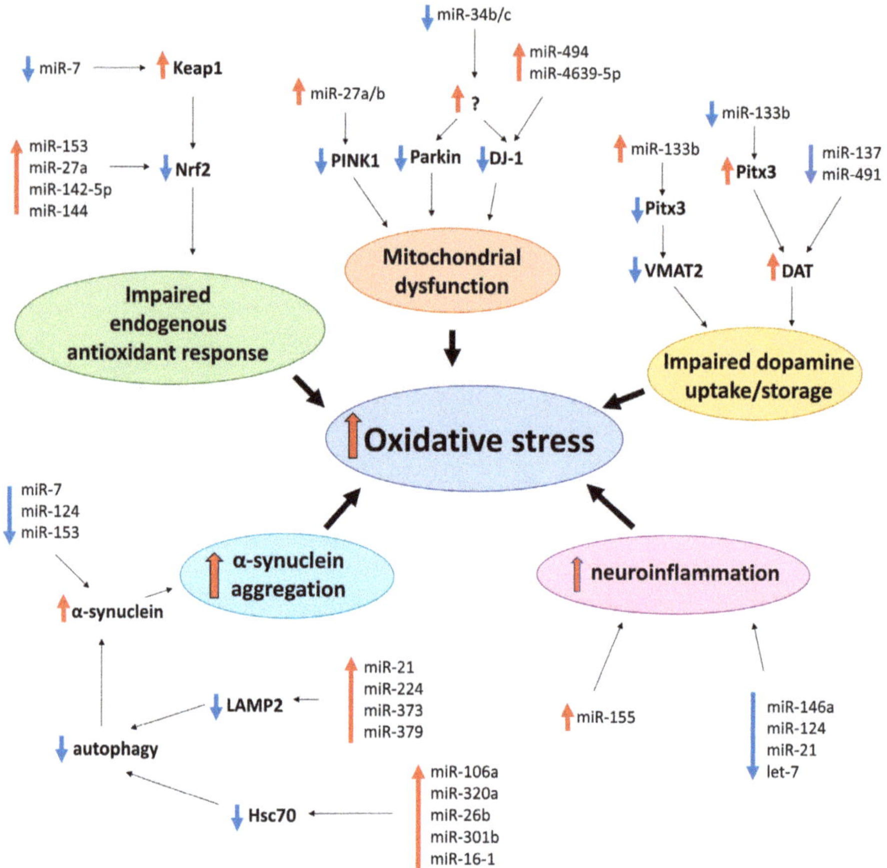

Figure 2. MicroRNAs implicated in oxidative stress-related cellular pathways in Parkinson's disease.

Selective vulnerability of dopamine neurons suggests a role of dopamine itself in pathogenesis of PD. Normally, dopamine that is newly synthesized or uptaken from the synaptic cleft is removed from the cytosol and stored in synaptic vesicles by vesicular monoamine transporter 2 (VMAT2). Excess of cytosolic dopamine readily oxidizes and forms ROS [6]. MiR-133b indirectly inhibits expression of VMAT2 via downregulation of Pitx3 [103,104]. Therefore, its upregulation may contribute to PD pathology, since dopamine neurons with reduced VMAT2 expression showed increased sensitivity to dopamine-mediated toxicity [105]. Additionally, increased dopamine transporter (DAT)-mediated dopamine uptake may result in oxidative damage and neuronal degeneration [106]. Interestingly, miR-133b can also alter expression of DAT via the same route as VMAT2 [103]. Therefore, decreased levels of miR-133b may result in elevated levels of DAT, contributing to oxidative stress. This suggestion is particularly interesting in the light of findings that miR-133b is downregulated in the midbrain of PD patients [26]. MiR-137 and miR-491 negatively regulate DAT expression and uptake of dopamine by DAT in vitro [107], and decreased expression of these microRNAs may also implicate them in oxidative stress in PD.

Dysfunctional mitochondria is one of the main sources of ROS. Several mutations in genes encoding proteins PINK1, Parkin, and DJ-1 can affect mitochondrial function, increase oxidative stress, and cause autosomal recessive PD in humans [6]. PINK1, together with Parkin, are mitochondrial quality control regulators: they induce disposal of dysfunctional mitochondria reviewed in [108].

PINK1 exhibits a neuroprotective effect in dopamine neurons by inhibiting ROS production [109], while PINK1 knockout in human and mouse dopamine neurons causes increased ROS generation [110]. MiR-27a and miR-27b suppress expression of PINK1 [111], which potentially can induce oxidative stress. Additionally, miR-27a may be implicated in downregulation of mitochondrial complex I subunit NDUFS4 and, together with miR-155, mitochondrial complex V subunit ATP5G3 [112].

DJ-1 is a multifunctional protein and, amongst various roles, it is a regulator of mitochondrial activity and an important player in mediating the oxidative stress response [113,114]. In addition to its role in familial cases of PD, damaged by irreversible oxidation DJ-1 was also reported in the brains of sporadic PD patients [115]. Increased levels of miR-494 downregulate DJ-1 levels and increase cell vulnerability to oxidative stress both in vitro and in vivo [116]. Upregulation of mir-4639-5p, also targeting DJ-1 expression, increases oxidative stress and causes cell death in SH-SY5Y cells, a frequently used dopamine neuron-like model, and its increased expression was reported in PD patients [117]. In addition, miR-34b and miR-34c are downregulated in PD patients (particularly in the SNpc), and their depletion was correlated with mitochondrial dysfunction, increased oxidative stress, and a moderate decrease of SH-SY5Y cell viability. Decreased expression of miR-34b/c was coupled with downregulated expression of Parkin and DJ-1, although mechanism of their action is unclear [118].

The Nrf2-antioxidant response element (ARE) pathway is an endogenous antioxidant system, shown to be downregulated in neurodegenerative diseases. Nrf2 is regulated by Keap1, which facilitates its degradation. Oxidative stress induces translocation of Nrf2 to the nucleus, activating expression of genes, which encode proteins involved in the oxidative stress response, such as SOD1 and GSH (for more details, see [119,120]). miR-7 is capable of repressing Keap1 [121]; what is particularly interesting in the light of this report is that miR-7 is downregulated in the SNpc of PD patients, and its downregulation results in a loss of dopamine neurons in vivo [122]. MiR-153, miR-27a, miR-142-5p, and miR-144 can directly downregulate Nrf2 expression in SH-SY5Y cells [123], potentially contributing to an impaired oxidative stress response.

Histopathologically, PD is characterized by formation of inclusions in neuronal soma (Lewy bodies) or processes (Lewy neurites) with the protein α-synuclein as a major component [124]. Mutations in encoding α-synuclein gene, *SNCA*, and its duplication and triplication were reported to cause familial cases of PD [125]. α-synuclein is capable of inducing oxidative stress and increased levels of ROS, although the exact mechanism is still unclear [126–130]. Multiple microRNAs were reported to control α-synuclein expression, including miR-7, miR-214, miR-153, and miR-34b/c, and their downregulation may contribute to α-synuclein-mediated neurotoxicity in PD [131–134]. α-synuclein aggregation can also be mediated through its impaired removal by chaperon-mediated autophagy. For example, miR-21, miR-224, miR-373, and miR-379 were demonstrated to downregulate LAMP2 expression, and miR-26b, miR-106a, miR-301b, miR-320a, and miR-16-1 were shown to suppress expression of Hsc70 [135–137]. Upregulation of some of these microRNAs were detected in PD patients [120]. MicroRNA regulation of α-synuclein expression has recently been systematically reviewed elsewhere [138]. Altogether, the literature describes multiple mechanisms for microRNAs to contribute to α-synuclein accumulation, which consequently could lead to oxidative stress.

Neuroinflammation, mediated by microglia and to a lesser extent by astrocytes and oligodendrocytes, was shown to play an important role in PD pathophysiology. Particularly, activated microglia can produce numerous cytotoxic substances, including superoxide, and therefore contribute to oxidative stress in the brain (for more details, see [139,140]). Some microRNAs were reported to be implicated in neuroinflammation, such as miR-155 (pro-inflammatory), and miR-146a and miR-124 (anti-inflammatory) [124]. Interestingly, miR-155 was found to be upregulated in an α-synuclein in vivo model of PD and was proposed to mediate α-synuclein-induced inflammation [141,142]. Additionally, increased levels of miR-155 were reported in PD patients. In the same study, downregulation of miR-146a was also demonstrated [143]. MiR-124 attenuates microglia activation and improves survival of dopamine neurons in the MPTP model of PD [144]. PD-associated proteins, including Parkin, DJ-1, and α-synuclein, can induce neuroinflammation by activating microglia [145].

4. Amyotrophic Lateral Sclerosis

Amyotrophic lateral sclerosis (ALS) is a neurodegenerative disease characterized by a loss of upper and lower motor neurons in the brain and spinal cord [146], which leads to loss of voluntary control over muscles and subsequent muscle atrophy. Patients gradually experience worsening symptoms of muscle weakness, problems with speaking, chewing and swallowing, and eventually breathing difficulties most often leading to death due to respiratory failure. About one out of 300–500 humans is affected by ALS, with the incidence being higher in men. The risk increases with age and survival is estimated at 3-4 years after onset. ALS presents either in a sporadic or a familial form. There are many genes associated with the familial form and a few mutations which are known to be the cause, the most common ones being on RNA binding protein FUS (*FUS*), TAR DNA-binding protein 43 (*TARDBP*), chromosome 9 open reading frame 72 (*C9orf72*) and Cu^{2+}/Zn^{2+} superoxide dismutase (*SOD1*) [147]. Notably, *TARDBP* and *FUS* are involved in RNA biology, including microRNA processing [148]. Non-genetic factors are also implicated in ALS. For instance, environmental insults can cause oxidative stress through the release of free radicals, mainly ROS and reactive nitrogen species, which may lead to epigenetic modifications and changes in gene expression relevant for ALS [149].

Supporting evidence for the role of oxidative stress in ALS was demonstrated by a recent meta-analysis which showed that malondialdehyde, 8-hydroxyguanosine, and Advanced Oxidation Protein Product were significantly elevated in the peripheral blood of ALS patients when compared to controls, as opposed to levels of antioxidant glutathione and uric acid which were downregulated [150]. Other oxidative stress markers such as Cu, SOD, glutathione peroxidase, Co-Q10, and transferrin did not have a link to ALS.

The progressive loss of motor neurons happens relatively fast compared to other neurodegenerative diseases and causes a wide variety of clinical symptoms related to motor deficits, making early diagnosis of ALS challenging. Thus, there is an active search for biomarkers of the disease and microRNAs could represent one option as their expression signatures have been studied in patients. Many studies have identified differential expression of small RNAs, including microRNAs, in the muscle, cerebrospinal fluid, motor neuron progenitors, and blood as well as in *post mortem* tissue samples (spinal cord, brain stem, and the brain) of both sporadic and familial ALS patients compared to healthy controls [151–155]. Besides being valuable as biomarkers, many microRNAs are also studied from a therapeutic point of view as regulating them may provide an option to treat ALS. For example, using an AAV-mediated artificial microRNA targeting *SOD1*, which is involved in reducing ROS and one of the causal genes of ALS, has shown efficient silencing of the gene in macaques [156].

A number of differentially expressed miRNAs in ALS patients versus controls regulate genes involved in oxidative stress, e.g., reducing or counteracting ROS/reactive nitrogen species and may be useful as biomarkers and/or therapeutics. For example, miR-27a, miR-34a, miR-155, miR-142-5p, and miR-338-3p have been studied as biomarkers and potential therapeutic targets in relation to ALS and are involved in oxidative stress directly or indirectly [153,155,157,158] (Figure 3).

miR-34a regulates an X-linked inhibitor of apoptosis (XIAP) that is linked to oxidative stress-induced senescence and Sirtuin 1 (SIRT1), which is protective against oxidative stress-induced apoptosis [154,159]. Of interest, SIRT1 is downregulated in PD [160]. Moreover, ALS patient-derived cell lines have a reduction of miR-34a, which is rescued by treatment with enoxacin, a small-molecule drug stimulating microRNA biogenesis [154]. Thus, enoxacin and other microRNA biogenesis stimulating drugs can potentially be used as ALS therapy [52].

The Nrf2-ARE pathway regulates many genes involved in redox reactions and has been linked to ALS [161]. It is regulated by several microRNAs, directly by e.g., aforementioned miR-27a and miR-34a and indirectly by e.g., miR-7 and miR-494, which regulate Nrf2 modulating proteins [116,121,123,162]. Furthermore, inhibiting miR-142-5p reduces oxidative stress via upregulation of the Nrf2-ARE signaling pathway, and it is downregulated in the CSF of sporadic ALS patients [155,163].

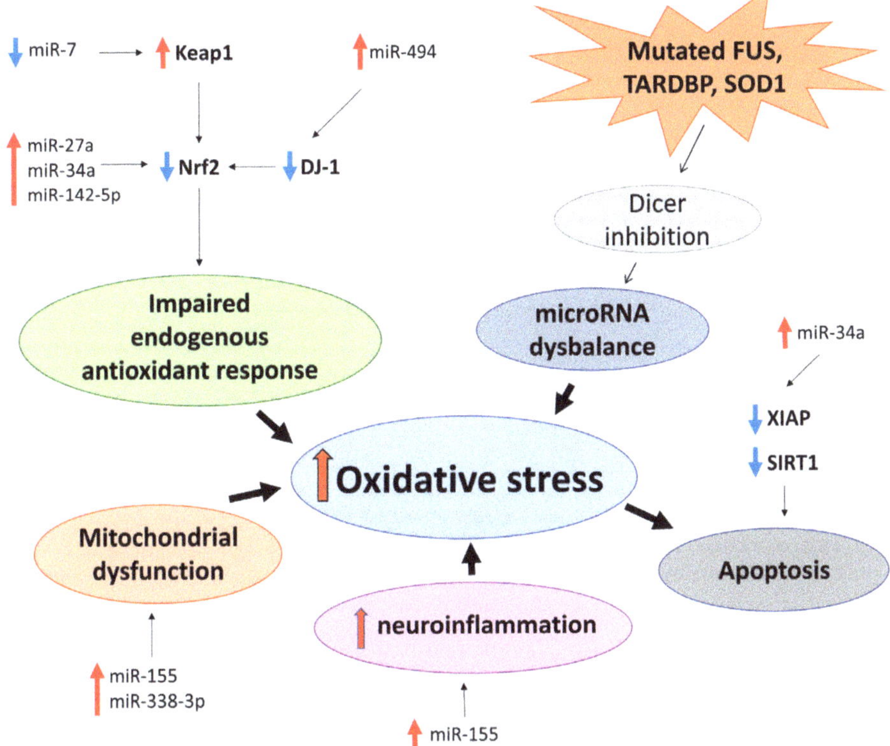

Figure 3. MicroRNAs implicated in oxidative stress-related cellular pathways in Amyotrophic Lateral Sclerosis (AMS).

MiR-155 has been shown to be upregulated in both sporadic and familial ALS patients, and inhibiting it in the brains of SOD1G93A mice increases both survival and disease duration [157]. Additionally, miR-338-3p regulates certain subunits of mitochondrial OXPHOS complexes [164] and is also implicated in ALS in human patients and mouse models [158,165]. A broader microRNA dysregulation has also been observed in human ALS patient motor neurons and overexpression of ALS-causing genes *FUS*, *TARDBP*, and *SOD1* seem to inhibit pre-miRNA processing by Dicer. Enhancing Dicer with enoxacin improves neuromuscular function in two separate ALS mouse models [52]. Therefore, a treatment strategy not only taking into account oxidative stress, but also microRNA dysregulation could prove to be useful for ALS patients. However, this and the relationship of microRNAs and oxidative stress should be studied much more carefully before engagement of clinical trials.

5. Huntington's Disease

HD is a relatively rare hereditary disorder with the highest prevalence in the white Caucasian population (about 1:10,000 to 1:20,000) reviewed in [166]. HD is caused by abnormal expansion of a repeated trinucleotide (CAG) sequence in the huntingtin (*HTT*) gene, translated to a long polyglutamine stretch in mutant huntingtin (mHTT) protein or, via repeat associated non-ATG (RAN) translation, to homopolymeric proteins prone to aggregation (for detailed review, see [166–168]). Longer CAG repeats correlate with an earlier age of onset of disease symptoms, which include severe motor (chorea, bradykinesia, and dystonia), cognitive (executive function, memory, attention and visuospatial functions) and psychiatric (anxiety, aggression, apathy and depression) disturbances, combined with sleep and circadian disorders, weight loss, skeletal muscle wasting, testicular atrophy, and peripheral immune system alterations [166,168]. The majority of these symptoms are caused by degeneration of striatal GABAergic medium spiny neurons and the cortical neurons projecting to them, accompanied by astrogliosis and microglia activation. Glutamate excitotoxicity, caused by reduced astrocyte glutamate uptake, further exacerbates neurodegeneration. Progressive atrophy of the striatum and cerebral cortex leads to patient death at 15–20 years from the disease onset [166,168].

On a molecular level, HD is characterized by the presence of nuclear inclusions and cytoplasmic aggregates containing mHTT and RAN translation proteins, transcription dysregulation (including large changes in microRNAs), inhibition of proteasome activity and autophagy, defects in synaptic neurotransmission, mitochondrial dysfunction, and oxidative stress [8,166,168–170]. A direct link between microRNA dysregulation in HD and oxidative stress has not been evidently described in the literature; however, both are highly relevant for HD as summarized below. Moreover, drawing from research on other neurodegenerative disorders (particularly ALS and PD), it seems plausible that global dysregulation of microRNAs in HD and oxidative stress might form a vicious cycle exacerbating each other and potentially worsening disease progression [51].

Dysregulation of transcription caused by interaction of mHTT with Repressor Element 1 Silencing Transcription Factor (REST) affected, among other targets, the expression of several REST-regulated microRNAs in mouse HD models and, importantly, in *post mortem* cortex samples of HD patients, where upregulation of miR-29a and miR-330 and downregulation of miR-132 was observed [171]. Similarly, analysis of cortical microRNA expression in the brains of patients at different HD stages identified progressive downregulation of miR-9, miR-9*, miR-29b, and miR-124a, whereas, in contrast to the study of Johnson et al. [171], no changes of miR-29a and significant upregulation of miR-132 at late disease stages were observed [172]. Interestingly, both wild-type and mHTT interact with Ago2 and localize to P bodies, suggesting that mHTT can affect Ago2 and, consequently, RISC complex activity in HD [173]. Importantly, recent results confirmed the effect of mHTT on Ago2, demonstrating that aggregation of mHTT, through autophagy impairment, can lead to Ago2 accumulation in a mouse HD model and HD patients and, consequently, to global dysregulation of microRNA levels and activity [174]. mHTT mRNA can also lead to generation of small CAG-repeated RNAs, whose generation and neurotoxic activity depend on Dicer and Ago2, potentially affecting microRNA biogenesis [175]. Thus, both transcription and processing of microRNAs appear to be dysregulated in HD. Indeed, analysis of HD mouse models identified common downregulation of miR-22, miR-29c, miR-128, miR-132, miR-138, miR-218, miR-222, miR-344, and miR-674*, as well as reduced levels of Drosha and Dicer mRNA [176]. In line with these results, microRNA sequencing and differential expression analysis demonstrated deregulation of multiple microRNAs in the frontal cortex and striatum of HD patients [177]. Moreover, because microRNA silencing machinery may be impeded in HD due to Ago2 translocation to stress granules [173,174], observed changes in specific microRNAs should be interpreted with caution as they might not reflect a functional outcome on target mRNA regulation (Figure 4).

The unequivocal cause for HD is the CAG expanse in the *HTT* gene and a higher number of CAG repeats leads to an earlier manifestation of the disease. However, large variations in age of disease onset among individuals with moderate (<55) CAG repeat numbers, together with variations in disease progression, strongly imply genetic and environmental modifiers of the disease which could exacerbate detrimental effects of *mHTT* explaining observed variability [178,179]. Oxidative stress or, conversely, capacity of antioxidant defense systems seem highly plausible as modifiers of HD [8,180]. Markers of oxidative stress rise with transition from the asymptomatic to symptomatic phase in HD patients [181], and oxidative stress is widely described as the main contributor to cell death in HD [8]. While there are no studies specifically addressing the link between microRNAs and oxidative stress in HD, some of the above-mentioned microRNAs, such as miR-9, miR-29, miR-124, and miR-128, changed in HD models and patients, have also been predicted to target genes involved in the oxidative stress response [53]. Similarly, general dysregulation of the microRNA network observed in HD will affect neuronal susceptibility to stress, including oxidative stress [182,183], which putatively could affect pace of disease progression. Conversely, it is tempting to speculate that strategies based on boosting microRNAs processing machinery could slow down demise of neurons in HD similarly to what we and others have shown in models of ALS [52] and PD [30].

Overall, while both oxidative stress and microRNA dysregulation are established features in HD, their interaction remains largely unexplored, yet an intriguing and promising topic for further studies.

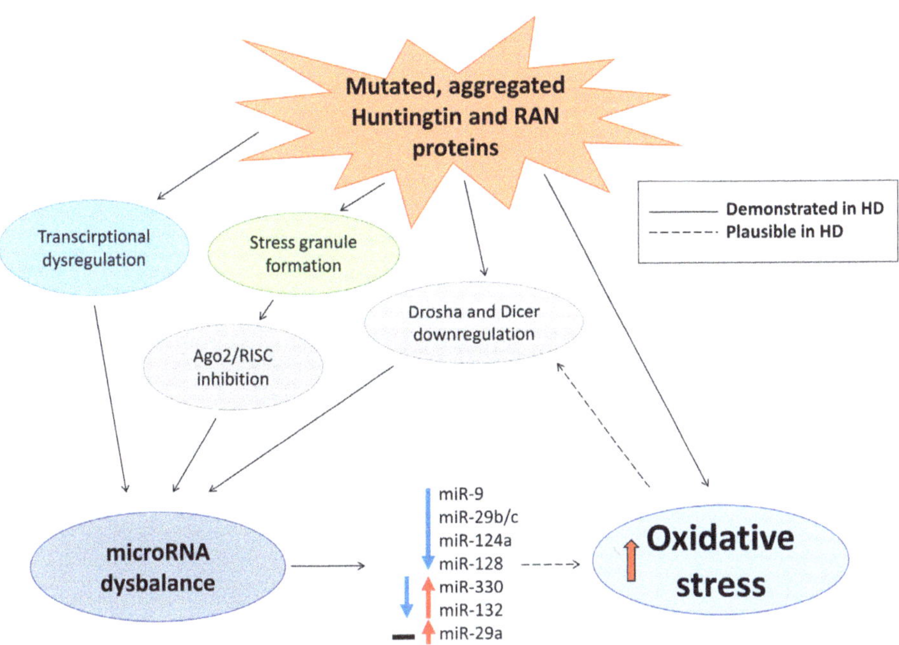

Figure 4. MicroRNAs implicated in oxidative stress-related cellular pathways in Huntington's Disease.

6. Common and Unique MicroRNAs Affecting Oxidative Stress in Neurodegenerative Diseases

As discussed above, neurodegenerative diseases share many similarities, including mitochondrial dysfunction, formation, and spread of insoluble protein inclusions and, as reviewed here, oxidative stress and deregulation of microRNA networks. Among multiple microRNAs associated with neurodegenerative diseases, we have focused on those implicated in the oxidative stress response (Figures 1–4 and Table 1). For many microRNAs, association with oxidative stress was not reported in the original publication, which frequently only demonstrated the change in its level in a selected neurodegenerative condition. In such cases, we consulted other studies, like [53], to identify if a particular microRNA can be involved in the oxidative stress response. Comparison of microRNAs associated with each of the four neurodegenerative diseases reviewed here identified only a small set of common microRNAs affecting the oxidative stress response in different neurodegenerative conditions (Table 1), and no single common oxidative stress-implicated microRNA was reported to be associated with three or four diseases. However, compared to AD and PD, relatively few studies have addressed changes in microRNA levels in ALS and HD, and, therefore, it is reasonable to expect that many more microRNAs associated with these diseases are awaiting their discovery. Nevertheless, many microRNAs are common at least between two neurodegenerative conditions (Table 1); among them are miR-34, miR-124, miR-132, miR-26, mir-7 which are highly expressed in the brain [184,185] and regulate multiple oxidative stress-related pathways (Figures 1–4). Such microRNAs are particularly attractive as potential therapeutic targets for the treatment of neurodegeneration. However, the unique microRNAs also deserve attention, as they may be reflecting fundamental differences in the development and progression of particular neurodegenerative disease and serve as specific biomarkers, facilitating and accelerating disease diagnosis.

Table 1. MicroRNAs associated with neurodegenerative diseases (AD, PD, ALS, and HD) and implicated in regulation of oxidative stress and related cellular pathways. Bolded are microRNAs associated with more than one neurodegenerative disease.

Disease	Associated microRNAs Involved in Oxidative Stress Regulation	References
Alzheimer's disease	miR-107	[72,73,78]
	miR-125b	[97]
	miR-130b	[91]
	miR-132/212	[87,88]
	miR-134	[72]
	miR-145	[72]
	miR-146b	[89]
	miR-153	[81]
	miR-186	[79,99]
	miR-193b	[91]
	miR-200a-3p	[85]
	miR-20a	[91]
	miR-210	[72]
	miR-219	[86]
	miR-26	[90]
	miR-296	[91]
	miR-329	[91]
	miR-330a	[92]
	miR-342-5p	[80]
	miR-98-5p	[96]
Parkinson's disease	miR-106a	[135]
	miR-124	[124,144]
	miR-133b	[103]
	miR-137	[106]
	miR-142-5p	[123]
	miR-144	[123]
	miR-146a	[124]
	miR-153	[123,133]
	miR-155	[124,141]
	miR-16-1	[137]
	miR-214	[132]
	miR-224	[135]
	miR-26b	[135]
	miR-27a/b	[110,111,123]
	miR-301b	[135]
	miR-320	[136]
	miR-34b/c	[118,134]
	miR-373	[135]
	miR-379	[135]
	mir-4639-5p	[117]
	miR-491	[106]
	miR-494	[116]
	miR-7	[121,122,131,133]
ALS	miR-142-5p	[155,163]
	miR-155	[157]
	miR-27a	[158,165]
	miR-338-3p	[162]
	miR-34a	[116,154]
Huntington's disease	**miR-124a**	[172]
	miR-128	[176]
	miR-132	[171,172,176]
	miR-29a/b/c	[171,172,176]
	miR-330	[171]
	miR-9	[172]

7. Challenges and Perspectives

The above reviewed results clearly demonstrate the intrinsic link between oxidative stress and microRNAs in ageing and disease. However, there are many questions, experimental details, and technical difficulties that need to be solved to bring microRNA-based therapies to clinical use. We undoubtedly have learned a lot about microRNAs and oxidative stress from experiments in cultured cells and extrapolating results from cancer research, but we should exercise caution in translating the findings obtained in cell culture to human neurons. Despite continuous improvement of computational algorithms, prediction and validation of microRNA-mRNA regulation remains challenging [186,187]. Many reported results on microRNA-mRNA regulation are obtained using luciferase reporter assays and transient transfection of microRNA mimics, which are known to cause unspecific general effects on the microRNA biogenesis pathway [188]. The use of proper controls (scrambled microRNAs and reporters with mutated putative binding sites) in such studies is, therefore, absolutely crucial for their validity. Additional caution in interpretation of microRNA overexpression studies should be taken since achieved and functionally effective overexpression levels might be orders of magnitude higher than normally observed.

MicroRNA expression profiles in neurons and glia in vivo are cell type-specific and different from cultured immortalized cells, as are 3′UTR isoforms [189,190], and, moreover, they change with age and the stage of the disease. Furthermore, expression patterns of microRNAs and their putative targets are distinct in different neuronal populations [191]. Thus, ideally, we should address regulation of endogenous mRNA by endogenous microRNAs, for example, by utilizing target protectors introduced to post-mitotic neurons at the lowest possible concentrations, using proper controls [192,193]. Development of new genetic methods, such as CRISPR/Cas9-mediated gene knockout [194–196], greatly facilitated loss-of-function genetic studies, enabling relatively easy deletion of both individual microRNAs and whole microRNA families in cultured cells and in vivo [197–200]. Both knockout and base editing using CRISPR/Cas9 [201–204] can be further utilized to selectively mutate or create microRNA binding site(s) on 3′UTR of a particular gene, allowing for precisely addressing the consequences of modulation of individual microRNA-mRNA binding. Identified neuroprotective microRNAs can be introduced to the brain using gene therapy vectors, similar to the ones used in clinical trials for neurotrophic factor expression in neurodegenerative disorders [205].

Translation of the results from animal to human settings has long been an issue in neurodegeneration research, with many neuroprotective treatments successfully working in rodent and even primate models, but not in human patients, failing at the stage of double-blinded randomized clinical trials [206]. Neither genetic nor toxin-based rodent models fully recapitulate features of neurodegenerative diseases. While many AD and PD models focus on protein aggregation, other factors contributing to neurodegeneration clearly exist. Mouse and human midbrain progenitors and dopamine neurons have distinct RNA expression profiles and species-specific differences, for example, the presence of neuromelanin and differences in dopamine oxidation [10,207]. Genetic mutations, which lead to early onset familial PD in humans, do not recapitulate the disease when introduced to rodents [208–210]. The lack of appropriate neurodegenerative disease models greatly impairs studies of the disease-related microRNAs. While the majority of microRNAs are conserved between rodents and humans, a number of primate- and human-specific microRNAs have been identified [211,212]. Furthermore, existing data demonstrate that some genes may exhibit human-specific regulation by microRNAs [213,214]. These questions have been partly addressed by the analysis of microRNA–mRNA interactions in neurons derived from patients at different stages of disease progression; however, obtaining high quality RNA in sufficient amounts from specific neuronal populations in *post mortem* brain samples is technically very challenging. Development of more sensitive methods, such as single cell microRNA–mRNA co-sequencing [215] would greatly improve the analysis of patient samples. Studies of post mortem tissue samples are also limited in that they only provide a snapshot of microRNAs changed at a particular disease stage, whereas longitudinal studies would have been much more informative.

Fortunately, current advances in differentiation of patient-derived induced pluripotent stem cells towards specific neuronal populations have finally allowed studying neurodegeneration and, particularly, microRNA alterations, in human disease models [216,217]. However, the protocols for human stem cell reprogramming and differentiation are still challenging, and the obtained neurons have embryonic or early postnatal phenotype, rather than adult neurons affected by neurodegeneration in patients. Culturing cells in artificial in vitro environments can affect their mRNA and microRNA expression patterns and oxidative stress levels (for a review of the current state of the field and challenges, see [216]). We are still lacking the methods to reliably detect and monitor levels of oxidative damage in live cells [218]. Development of such experimental techniques and models would also enable longitudinal studies to address the question on whether oxidative stress is a cause or consequence of other processes affecting neuronal survival, such as mitochondrial dysfunction, protein aggregation, or microRNA biogenesis disruption.

Focusing exclusively on neurons will not be sufficient to understand neurodegeneration—astrocytes, oligodendrocytes, and microglia are important players which may also be involved in modulating oxidative stress effects, for example, by regulating neuroinflammation. Therefore, to uncover molecular mechanisms behind human neurodegenerative diseases, we need to study human models representing and recapitulating the interaction between several neural cell types. Three-dimensional human brain organoids offer a great hope for neurodegeneration modeling [219], though it remains to be seen whether such organoids could be developed to the stage mature enough to model properties of the aged or even the adult brain. In this respect, one very promising direction would be to establish humanized animal models based on transplantation of human neural cell precursors to the rodent brain. A similar strategy has already been successfully implemented to obtain humanized mice with brains chimeric for human glia [220]. For example, it has recently been shown that, after transplantation to the rat midbrain, a proportion of human embryonic stem cell-derived neuron precursors will differentiate to nigral dopamine neurons, integrate into appropriate neuronal circuits, and regrow axons to innervate their natural targets [221–223]. It is therefore possible in principle to obtain rodents containing human glia, microglia, and neurons correctly differentiated and integrated into host neuronal circuits and use these humanized animals to model degeneration of human neurons in a human-specific cell environment. Clearly, more work is needed to overcome technical and ethical hurdles; however, recent progress in the development of molecular tools and cellular models gives a strong hope that successful treatments to cure neurodegenerative diseases may finally be available.

Author Contributions: Conceptualization: J.K., D.G., I.P., P.C. and A.D.; writing—original draft preparation, J.K., D.G., I.P., P.C. and A.D.; writing—review and editing, J.K., D.G., I.P., P.C. and A.D.; supervision, A.D.; funding acquisition, D.G., I.P. and A.D.

Funding: This research was funded by the Academy of Finland grants #293392, #319195, Instrumentarium Science Foundation, the Doctoral Program for Drug Research, and the Finnish National Agency for Education (EDUFI) and statutory funds of the Maj Institute of Pharmacology, PAS, Poland. Open access funding provided by the University of Helsinki.

Acknowledgments: We thank Katrina Albert for careful proofreading and critical comments on the manuscript.

Conflicts of Interest: The authors declare no conflict of interest.

Abbreviations

Aβ	Amyloid-β
AD	Alzheimer's disease
ALS	Amyotrophic Lateral Sclerosis
APP	Amyloid precursor protein
ARE	Antioxidant response element
ATP	Adenosine triphosphate
BACE1	Beta-secretase 1
CRISPR	Clustered regularly interspaced short palindromic repeats
DAT	Dopamine transporter
GSH	Glutathione
HD	Huntington's disease
mHTT	mutant Huntingtin
PD	Parkinson's disease
PSEN	Presenilin
RAN	Repeat associated non-ATG
REST	Repressor Element 1 Silencing Transcription Factor
RISC	RNA-induced silencing complex
ROS	Reactive oxygen species
sAβ	soluble amyloid-β
SNpc	Substantia nigra pars compacta
SOD	Superoxide dismutase
VMAT2	Vesicular monoamine transporter 2
UTR	Untranslated region

References

1. Erkkinen, M.G.; Kim, M.O.; Geschwind, M.D. Clinical Neurology and Epidemiology of the Major Neurodegenerative Diseases. *Cold Spring Harb Perspect. Biol.* **2018**, *10*. [CrossRef]
2. Brown, R.H.; Al-Chalabi, A. Amyotrophic Lateral Sclerosis. *N. Engl. J. Med.* **2017**, *377*, 162–172. [CrossRef]
3. Collaborators, G.B.D.N. Global, regional, and national burden of neurological disorders, 1990-2016: A systematic analysis for the Global Burden of Disease Study 2016. *Lancet Neurol.* **2019**, *18*, 459–480. [CrossRef]
4. Pringsheim, T.; Jette, N.; Frolkis, A.; Steeves, T.D. The prevalence of Parkinson's disease: A systematic review and meta-analysis. *Mov. Disord.* **2014**, *29*, 1583–1590. [CrossRef] [PubMed]
5. Lin, M.T.; Beal, M.F. Mitochondrial dysfunction and oxidative stress in neurodegenerative diseases. *Nature* **2006**, *443*, 787–795. [CrossRef] [PubMed]
6. Blesa, J.; Trigo-Damas, I.; Quiroga-Varela, A.; Jackson-Lewis, V.R. Oxidative stress and Parkinson's disease. *Front. Neuroanat.* **2015**, *9*, 91. [CrossRef] [PubMed]
7. Tonnies, E.; Trushina, E. Oxidative Stress, Synaptic Dysfunction, and Alzheimer's Disease. *J. Alzheimers Dis.* **2017**, *57*, 1105–1121. [CrossRef]
8. Kumar, A.; Ratan, R.R. Oxidative stress and Huntington's disease: The good, the bad, and the ugly. *J. Huntingtons Dis.* **2016**, *5*, 217–237. [CrossRef]
9. Niedzielska, E.; Smaga, I.; Gawlik, M.; Moniczewski, A.; Stankowicz, P.; Pera, J.; Filip, M. Oxidative stress in neurodegenerative diseases. *Mol. Neurobiol.* **2016**, *53*, 4094–4125. [CrossRef]
10. Burbulla, L.F.; Song, P.; Mazzulli, J.R.; Zampese, E.; Wong, Y.C.; Jeon, S.; Santos, D.P.; Blanz, J.; Obermaier, C.D.; Strojny, C.; et al. Dopamine oxidation mediates mitochondrial and lysosomal dysfunction in Parkinson's disease. *Science* **2017**, *357*, 1255–1261. [CrossRef]
11. Cobley, J.N.; Fiorello, M.L.; Bailey, D.M. 13 reasons why the brain is susceptible to oxidative stress. *Redox Biol.* **2018**, *15*, 490–503. [CrossRef] [PubMed]
12. Pacelli, C.; Giguere, N.; Bourque, M.J.; Levesque, M.; Slack, R.S.; Trudeau, L.E. Elevated mitochondrial bioenergetics and axonal arborization size are key contributors to the vulnerability of dopamine neurons. *Curr. Biol.* **2015**, *25*, 2349–2360. [CrossRef] [PubMed]

13. Liu, Z.; Zhou, T.; Ziegler, A.C.; Dimitrion, P.; Zuo, L. Oxidative stress in neurodegenerative diseases: From molecular mechanisms to clinical applications. *Oxid. Med. Cell Longev.* **2017**, *2017*, 2525967. [CrossRef] [PubMed]

14. Kim, G.H.; Kim, J.E.; Rhie, S.J.; Yoon, S. The role of oxidative stress in neurodegenerative diseases. *Exp. Neurobiol.* **2015**, *24*, 325–340. [CrossRef] [PubMed]

15. Murphy, M.P. Antioxidants as therapies: Can we improve on nature? *Free Radic. Biol. Med.* **2014**, *66*, 20–23. [CrossRef]

16. Vasconcelos, A.R.; Dos Santos, N.B.; Scavone, C.; Munhoz, C.D. Nrf2/ARE pathway modulation by dietary energy regulation in neurological disorders. *Front. Pharmacol.* **2019**, *10*, 33. [CrossRef]

17. Bartel, D.P. MicroRNAs: Genomics, biogenesis, mechanism, and function. *Cell* **2004**, *116*, 281–297. [CrossRef]

18. Bartel, D.P. Metazoan MicroRNAs. *Cell* **2018**, *173*, 20–51. [CrossRef]

19. Krol, J.; Loedige, I.; Filipowicz, W. The widespread regulation of microRNA biogenesis, function and decay. *Nat. Rev. Genet.* **2010**, *11*, 597–610. [CrossRef]

20. Schmiedel, J.M.; Klemm, S.L.; Zheng, Y.; Sahay, A.; Bluthgen, N.; Marks, D.S.; van Oudenaarden, A. Gene expression. MicroRNA control of protein expression noise. *Science* **2015**, *348*, 128–132. [CrossRef]

21. Rajman, M.; Schratt, G. MicroRNAs in neural development: From master regulators to fine-tuners. *Development* **2017**, *144*, 2310–2322. [CrossRef] [PubMed]

22. Im, H.I.; Kenny, P.J. MicroRNAs in neuronal function and dysfunction. *Trends Neurosci.* **2012**, *35*, 325–334. [CrossRef] [PubMed]

23. Bernstein, E.; Kim, S.Y.; Carmell, M.A.; Murchison, E.P.; Alcorn, H.; Li, M.Z.; Mills, A.A.; Elledge, S.J.; Anderson, K.V.; Hannon, G.J. Dicer is essential for mouse development. *Nat. Genet.* **2003**, *35*, 215–217. [CrossRef] [PubMed]

24. Morita, S.; Horii, T.; Kimura, M.; Goto, Y.; Ochiya, T.; Hatada, I. One Argonaute family member, Eif2c2 (Ago2), is essential for development and appears not to be involved in DNA methylation. *Genomics* **2007**, *89*, 687–696. [CrossRef]

25. Kawase-Koga, Y.; Otaegi, G.; Sun, T. Different timings of Dicer deletion affect neurogenesis and gliogenesis in the developing mouse central nervous system. *Dev. Dyn.* **2009**, *238*, 2800–2812. [CrossRef]

26. Kim, J.; Inoue, K.; Ishii, J.; Vanti, W.B.; Voronov, S.V.; Murchison, E.; Hannon, G.; Abeliovich, A. A MicroRNA feedback circuit in midbrain dopamine neurons. *Science* **2007**, *317*, 1220–1224. [CrossRef]

27. Huang, T.; Liu, Y.; Huang, M.; Zhao, X.; Cheng, L. Wnt1-cre-mediated conditional loss of Dicer results in malformation of the midbrain and cerebellum and failure of neural crest and dopaminergic differentiation in mice. *J. Mol. Cell Biol.* **2010**, *2*, 152–163. [CrossRef]

28. Davis, T.H.; Cuellar, T.L.; Koch, S.M.; Barker, A.J.; Harfe, B.D.; McManus, M.T.; Ullian, E.M. Conditional loss of Dicer disrupts cellular and tissue morphogenesis in the cortex and hippocampus. *J. Neurosci.* **2008**, *28*, 4322–4330. [CrossRef]

29. De Pietri Tonelli, D.; Pulvers, J.N.; Haffner, C.; Murchison, E.P.; Hannon, G.J.; Huttner, W.B. miRNAs are essential for survival and differentiation of newborn neurons but not for expansion of neural progenitors during early neurogenesis in the mouse embryonic neocortex. *Development* **2008**, *135*, 3911–3921. [CrossRef]

30. Chmielarz, P.; Konovalova, J.; Najam, S.S.; Alter, H.; Piepponen, T.P.; Erfle, H.; Sonntag, K.C.; Schutz, G.; Vinnikov, I.A.; Domanskyi, A. Dicer and microRNAs protect adult dopamine neurons. *Cell Death Dis.* **2017**, *8*, e2813. [CrossRef]

31. Kaneko, H.; Dridi, S.; Tarallo, V.; Gelfand, B.D.; Fowler, B.J.; Cho, W.G.; Kleinman, M.E.; Ponicsan, S.L.; Hauswirth, W.W.; Chiodo, V.A.; et al. DICER1 deficit induces Alu RNA toxicity in age-related macular degeneration. *Nature* **2011**, *471*, 325–330. [CrossRef] [PubMed]

32. Pang, X.; Hogan, E.M.; Casserly, A.; Gao, G.; Gardner, P.D.; Tapper, A.R. Dicer expression is essential for adult midbrain dopaminergic neuron maintenance and survival. *Mol. Cell Neurosci.* **2014**, *58*, 22–28. [CrossRef] [PubMed]

33. Hebert, S.S.; Papadopoulou, A.S.; Smith, P.; Galas, M.C.; Planel, E.; Silahtaroglu, A.N.; Sergeant, N.; Buee, L.; De Strooper, B. Genetic ablation of Dicer in adult forebrain neurons results in abnormal tau hyperphosphorylation and neurodegeneration. *Hum. Mol. Genet.* **2010**, *19*, 3959–3969. [CrossRef] [PubMed]

34. Schaefer, A.; O'Carroll, D.; Tan, C.L.; Hillman, D.; Sugimori, M.; Llinas, R.; Greengard, P. Cerebellar neurodegeneration in the absence of microRNAs. *J. Exp. Med.* **2007**, *204*, 1553–1558. [CrossRef]

35. Konopka, W.; Kiryk, A.; Novak, M.; Herwerth, M.; Parkitna, J.R.; Wawrzyniak, M.; Kowarsch, A.; Michaluk, P.; Dzwonek, J.; Arnsperger, T.; et al. MicroRNA loss enhances learning and memory in mice. *J. Neurosci.* **2010**, *30*, 14835–14842. [CrossRef]

36. Vinnikov, I.A.; Hajdukiewicz, K.; Reymann, J.; Beneke, J.; Czajkowski, R.; Roth, L.C.; Novak, M.; Roller, A.; Dörner, N.; Starkuviene, V.; et al. Hypothalamic miR-103 protects from hyperphagic obesity in mice. *J. Neurosci.* **2014**, *34*, 10659–10674. [CrossRef]

37. Cuellar, T.L.; Davis, T.H.; Nelson, P.T.; Loeb, G.B.; Harfe, B.D.; Ullian, E.; McManus, M.T. Dicer loss in striatal neurons produces behavioral and neuroanatomical phenotypes in the absence of neurodegeneration. *Proc. Natl. Acad. Sci. USA* **2008**, *105*, 5614–5619. [CrossRef]

38. Mang, G.M.; Pradervand, S.; Du, N.H.; Arpat, A.B.; Preitner, F.; Wigger, L.; Gatfield, D.; Franken, P. A neuron-specific deletion of the microRNA-processing enzyme DICER induces severe but transient obesity in mice. *PLoS ONE* **2015**, *10*, e0116760. [CrossRef]

39. Schaefer, A.; Im, H.I.; Veno, M.T.; Fowler, C.D.; Min, A.; Intrator, A.; Kjems, J.; Kenny, P.J.; O'Carroll, D.; Greengard, P. Argonaute 2 in dopamine 2 receptor-expressing neurons regulates cocaine addiction. *J. Exp. Med.* **2010**, *207*, 1843–1851. [CrossRef]

40. Schratt, G. microRNAs at the synapse. *Nat. Rev. Neurosci* **2009**, *10*, 842–849. [CrossRef]

41. Antoniou, A.; Khudayberdiev, S.; Idziak, A.; Bicker, S.; Jacob, R.; Schratt, G. The dynamic recruitment of TRBP to neuronal membranes mediates dendritogenesis during development. *EMBO Rep.* **2018**, *19*. [CrossRef] [PubMed]

42. Thomas, K.T.; Gross, C.; Bassell, G.J. microRNAs Sculpt Neuronal Communication in a Tight Balance That Is Lost in Neurological Disease. *Front. Mol. Neurosci.* **2018**, *11*, 455. [CrossRef] [PubMed]

43. Dimmeler, S.; Nicotera, P. MicroRNAs in age-related diseases. *EMBO Mol. Med.* **2013**, *5*, 180–190. [CrossRef] [PubMed]

44. Eacker, S.M.; Dawson, T.M.; Dawson, V.L. Understanding microRNAs in neurodegeneration. *Nat. Rev. Neurosci.* **2009**, *10*, 837–841. [CrossRef] [PubMed]

45. Eacker, S.M.; Dawson, T.M.; Dawson, V.L. The interplay of microRNA and neuronal activity in health and disease. *Front. Cell Neurosci.* **2013**, *7*, 136. [CrossRef]

46. Heman-Ackah, S.M.; Hallegger, M.; Rao, M.S.; Wood, M.J. RISC in PD: The impact of microRNAs in Parkinson's disease cellular and molecular pathogenesis. *Front. Mol. Neurosci.* **2013**, *6*, 40. [CrossRef]

47. Mouradian, M.M. MicroRNAs in Parkinson's disease. *Neurobiol. Dis.* **2012**, *46*, 279–284. [CrossRef]

48. Sonntag, K.C. MicroRNAs and deregulated gene expression networks in neurodegeneration. *Brain Res.* **2010**, *1338*, 48–57. [CrossRef]

49. Kurzynska-Kokorniak, A.; Koralewska, N.; Pokornowska, M.; Urbanowicz, A.; Tworak, A.; Mickiewicz, A.; Figlerowicz, M. The many faces of Dicer: The complexity of the mechanisms regulating Dicer gene expression and enzyme activities. *Nucleic Acids Res.* **2015**, *43*, 4365–4380. [CrossRef]

50. Leggio, L.; Vivarelli, S.; L'Episcopo, F.; Tirolo, C.; Caniglia, S.; Testa, N.; Marchetti, B.; Iraci, N. microRNAs in Parkinson's disease: From pathogenesis to novel diagnostic and therapeutic approaches. *Int. J. Mol. Sci.* **2017**, *18*, 2698. [CrossRef]

51. Emde, A.; Hornstein, E. miRNAs at the interface of cellular stress and disease. *EMBO J.* **2014**, *33*, 1428–1437. [CrossRef] [PubMed]

52. Emde, A.; Eitan, C.; Liou, L.L.; Libby, R.T.; Rivkin, N.; Magen, I.; Reichenstein, I.; Oppenheim, H.; Eilam, R.; Silvestroni, A.; et al. Dysregulated miRNA biogenesis downstream of cellular stress and ALS-causing mutations: A new mechanism for ALS. *EMBO J.* **2015**, *34*, 2633–2651. [CrossRef] [PubMed]

53. Engedal, N.; Zerovnik, E.; Rudov, A.; Galli, F.; Olivieri, F.; Procopio, A.D.; Rippo, M.R.; Monsurro, V.; Betti, M.; Albertini, M.C. From oxidative stress damage to pathways, networks, and autophagy via MicroRNAs. *Oxid. Med. Cell Longev.* **2018**, *2018*, 4968321. [CrossRef] [PubMed]

54. Vinnikov, I.A.; Domanskyi, A. Can we treat neurodegenerative diseases by preventing an age-related decline in microRNA expression? *Neural Regen. Res.* **2017**, *12*, 1602–1604. [CrossRef]

55. Salmena, L.; Poliseno, L.; Tay, Y.; Kats, L.; Pandolfi, P.P. A ceRNA hypothesis: The Rosetta Stone of a hidden RNA language? *Cell* **2011**, *146*, 353–358. [CrossRef]

56. Sumazin, P.; Yang, X.; Chiu, H.S.; Chung, W.J.; Iyer, A.; Llobet-Navas, D.; Rajbhandari, P.; Bansal, M.; Guarnieri, P.; Silva, J.; et al. An extensive microRNA-mediated network of RNA-RNA interactions regulates established oncogenic pathways in glioblastoma. *Cell* **2011**, *147*, 370–381. [CrossRef]

57. Prince, M.; Ali, G.-C.; Guerchet, M.; Prina, A.M.; Albanese, E.; Wu, Y.-T. Recent global trends in the prevalence and incidence of dementia, and survival with dementia. *Alzheimer's Res. Ther.* **2016**, *8*, 23. [CrossRef]

58. Tramutola, A.; Lanzillotta, C.; Perluigi, M.; Butterfield, D.A. Oxidative stress, protein modification and Alzheimer disease. *Brain Res. Bull.* **2017**, *133*, 88–96. [CrossRef]

59. Van Cauwenberghe, C.; Van Broeckhoven, C.; Sleegers, K. The genetic landscape of Alzheimer disease: Clinical implications and perspectives. *Genetics Med.* **2015**, *18*, 421. [CrossRef]

60. Qiu, C.; De Ronchi, D.; Fratiglioni, L. The epidemiology of the dementias: An update. *Curr. Opin. Psychiatry* **2007**, *20*, 380–385. [CrossRef]

61. Prasad, K.N. Oxidative stress and pro-inflammatory cytokines may act as one of the signals for regulating microRNAs expression in Alzheimer's disease. *Mech. Ageing Dev.* **2017**, *162*, 63–71. [CrossRef] [PubMed]

62. Lanoiselée, H.-M.; Nicolas, G.; Wallon, D.; Rovelet-Lecrux, A.; Lacour, M.; Rousseau, S.; Richard, A.-C.; Pasquier, F.; Rollin-Sillaire, A.; Martinaud, O.; et al. APP, PSEN1, and PSEN2 mutations in early-onset Alzheimer disease: A genetic screening study of familial and sporadic cases. *PLoS Med.* **2017**, *14*, e1002270. [CrossRef] [PubMed]

63. Patel, N.; Hoang, D.; Miller, N.; Ansaloni, S.; Huang, Q.; Rogers, J.T.; Lee, J.C.; Saunders, A.J. MicroRNAs can regulate human APP levels. *Mol. Neurodegener.* **2008**, *3*, 10. [CrossRef] [PubMed]

64. Hébert, S.S.; Horré, K.; Nicolaï, L.; Bergmans, B.; Papadopoulou, A.S.; Delacourte, A.; De Strooper, B. MicroRNA regulation of Alzheimer's Amyloid precursor protein expression. *Neurobiol. Dis.* **2009**, *33*, 422–428. [CrossRef] [PubMed]

65. Liu, C.G.; Wang, J.L.; Li, L.; Xue, L.X.; Zhang, Y.Q.; Wang, P.C. MicroRNA-135a and -200b, potential Biomarkers for Alzheimer's disease, regulate β secretase and amyloid precursor protein. *Brain Res.* **2014**, *1583*, 55–64. [CrossRef]

66. Kang, Q.; Xiang, Y.; Li, D.; Liang, J.; Zhang, X.; Zhou, F.; Qiao, M.; Nie, Y.; He, Y.; Cheng, J.; et al. MiR-124-3p attenuates hyperphosphorylation of Tau protein-induced apoptosis via caveolin-1-PI3K/Akt/GSK3β pathway in N2a/APP695swe cells. *Oncotarget* **2017**, *8*, 24314–24326. [CrossRef]

67. Long, J.M.; Ray, B.; Lahiri, D.K. MicroRNA-153 physiologically inhibits expression of amyloid-β precursor protein in cultured human fetal brain cells and is dysregulated in a subset of Alzheimer disease patients. *J. Biol. Chem.* **2012**, *287*, 31298–31310. [CrossRef]

68. Wang, X.; Tan, L.; Lu, Y.; Peng, J.; Zhu, Y.; Zhang, Y.; Sun, Z. MicroRNA-138 promotes tau phosphorylation by targeting retinoic acid receptor alpha. *FEBS Lett.* **2015**, *589*, 726–729. [CrossRef]

69. Prasad, K.N. Simultaneous activation of Nrf2 and elevation of antioxidant compounds for reducing oxidative stress and chronic inflammation in human Alzheimer's disease. *Mech. Ageing Dev.* **2016**, *153*, 41–47. [CrossRef]

70. Amakiri, N.; Kubosumi, A.; Tran, J.; Reddy, P.H. Amyloid beta and microRNAs in Alzheimer's disease. *Front. Neurosci.* **2019**, *13*. [CrossRef]

71. Varadarajan, S.; Yatin, S.; Aksenova, M.; Butterfield, D.A. Review: Alzheimer's amyloid β-peptide-associated free radical oxidative stress and neurotoxicity. *J. Struct. Biol.* **2000**, *130*, 184–208. [CrossRef] [PubMed]

72. Li, J.J.; Dolios, G.; Wang, R.; Liao, F.-F. Soluble beta-amyloid peptides, but not insoluble fibrils, have specific effect on neuronal icroRNA expression. *PLoS ONE* **2014**, *9*, e90770. [CrossRef]

73. Wang, W.-X.; Rajeev, B.W.; Stromberg, A.J.; Ren, N.; Tang, G.; Huang, Q.; Rigoutsos, I.; Nelson, P.T. The expression of microRNA miR-107 decreases early in Alzheimer's disease and may accelerate disease progression through regulation of β-site amyloid precursor protein-cleaving enzyme 1. *J. Neurosci.* **2008**, *28*, 1213–1223. [CrossRef] [PubMed]

74. Monte, S.; Wands, J. Molecular indices of oxidative stress and mitochondrial dysfunction occur early and often progress with severity of Alzheimer's disease. *J. Alzheimer's Dis.* **2006**, *9*, 167–181. [CrossRef] [PubMed]

75. Dorszewska, J.; Oczkowska, A.; Suwalska, M.; Rozycka, A.; Florczak-Wyspianska, J.; Dezor, M.; Lianeri, M.; Jagodzinski, P.P.; Kowalczyk, M.J.; Prendecki, M.; et al. Mutations in the exon 7 of Trp53 gene and the level of p53 protein in double transgenic mouse model of Alzheimer's disease. *Folia Neuropathol.* **2014**, *52*, 30–40. [CrossRef]

76. Dorszewska, J.; Oczkowska, A.; Suwalska, M.; Rozycka, A.; Florczak, J.; Dezor, M.; Lianeri, M.; Jagodzinski, P.; Kowalczyk, M.; Prendecki, M.; et al. Mutations of TP53 gene and oxidative stress in Alzheimer's disease patients. *Adv. Alzheimer's Dis.* **2014**, *3*, 24–32. [CrossRef]

77. Chen, L.; Zhang, R.; Li, P.; Liu, Y.; Qin, K.; Fa, Z.Q.; Liu, Y.J.; Ke, Y.Q.; Jiang, X.D. P53-induced microRNA-107 inhibits proliferation of glioma cells and down-regulates the expression of CDK6 and Notch-2. *Neurosci. Lett.* **2013**, *534*, 327–332. [CrossRef]

78. Prendecki, M.; Florczak-Wyspianska, J.; Kowalska, M.; Ilkowski, J.; Grzelak, T.; Bialas, K.; Kozubski, W.; Dorszewska, J. APOE genetic variants and apoE, miR-107 and miR-650 levels in Alzheimer's disease. *Folia Neuropathol.* **2019**, *57*, 106–116. [CrossRef]

79. Kim, J.; Yoon, H.; Chung, D.E.; Brown, J.L.; Belmonte, K.C.; Kim, J. miR-186 is decreased in aged brain and suppresses BACE1 expression. *J. Neurochem.* **2016**, *137*, 436–445. [CrossRef]

80. Sun, X.; Wu, Y.; Gu, M.; Zhang, Y. MiR-342-5p decreases ankyrin G levels in Alzheimer's disease transgenic mouse models. *Cell Rep.* **2014**, *6*, 264–270. [CrossRef]

81. Liang, C.; Zhu, H.; Xu, Y.; Huang, L.; Ma, C.; Deng, W.; Liu, Y.; Qin, C. MicroRNA-153 negatively regulates the expression of amyloid precursor protein and amyloid precursor-like protein 2. *Brain Res.* **2012**, *1455*, 103–113. [CrossRef] [PubMed]

82. Guo, X.D.; Sun, G.L.; Zhou, T.T.; Wang, Y.Y.; Xu, X.; Shi, X.F.; Zhu, Z.Y.; Rukachaisirikul, V.; Hu, L.H.; Shen, X. LX2343 alleviates cognitive impairments in AD model rats by inhibiting oxidative stress-induced neuronal apoptosis and tauopathy. *Acta Pharmacol. Sin.* **2017**, *38*, 1104–1119. [CrossRef] [PubMed]

83. Salminen, A.; Kaarniranta, K.; Kauppinen, A. Crosstalk between oxidative stress and SIRT1: Impact on the aging process. *Int. J. Mol. Sci.* **2013**, *14*, 3834–3859. [CrossRef] [PubMed]

84. Dias-Santagata, D.; Fulga, T.A.; Duttaroy, A.; Feany, M.B. Oxidative stress mediates tau-induced neurodegeneration in Drosophila. *J. Clin. Investig.* **2007**, *117*, 236–245. [CrossRef] [PubMed]

85. Wang, L.; Liu, J.; Wang, Q.; Jiang, H.; Zeng, L.; Li, Z.; Liu, R. MicroRNA-200a-3p mediates neuroprotection in Alzheimer-related deficits and attenuates amyloid-beta overproduction and tau hyperphosphorylation via coregulating BACE1 and PRKACB. *Front. Pharmacol.* **2019**, *10*. [CrossRef]

86. Li, J.; Chen, W.; Yi, Y.; Tong, Q. miR-219-5p inhibits tau phosphorylation by targeting TTBK1 and GSK-3β in Alzheimer's disease. *J. Cell. Biochem.* **2019**, *120*, 9936–9946. [CrossRef]

87. El Fatimy, R.; Li, S.; Chen, Z.; Mushannen, T.; Gongala, S.; Wei, Z.; Balu, D.T.; Rabinovsky, R.; Cantlon, A.; Elkhal, A.; et al. MicroRNA-132 provides neuroprotection for tauopathies via multiple signaling pathways. *Acta Neuropathol.* **2018**, *136*, 537–555. [CrossRef]

88. Smith, P.Y.; Hernandez-Rapp, J.; Jolivette, F.; Lecours, C.; Bisht, K.; Goupil, C.; Dorval, V.; Parsi, S.; Morin, F.; Planel, E.; et al. MiR-132/212 deficiency impairs tau metabolism and promotes pathological aggregation in vivo. *Hum. Mol. Gen.* **2015**, *24*, 6721–6735. [CrossRef]

89. Wang, G.; Huang, Y.; Wang, L.-L.; Zhang, Y.-F.; Xu, J.; Zhou, Y.; Lourenco, G.F.; Zhang, B.; Wang, Y.; Ren, R.-J.; et al. MicroRNA-146a suppresses ROCK1 allowing hyperphosphorylation of tau in Alzheimer's disease. *Sci. Rep.* **2016**, *6*, 26697. [CrossRef]

90. Absalon, S.; Kochanek, D.M.; Raghavan, V.; Krichevsky, A.M. MiR-26b, upregulated in Alzheimer's disease, activates cell cycle entry, tau-phosphorylation, and apoptosis in postmitotic neurons. *J. Neurosci.* **2013**, *33*, 14645–14659. [CrossRef]

91. Zhang, R.; Zhang, Q.; Niu, J.; Lu, K.; Xie, B.; Cui, D.; Xu, S. Screening of microRNAs associated with Alzheimer's disease using oxidative stress cell model and different strains of senescence accelerated mice. *J. Neurol. Sci.* **2014**, *338*, 57–64. [CrossRef] [PubMed]

92. Zhou, Y.; Wang, Z.-F.; Li, W.; Hong, H.; Chen, J.; Tian, Y.; Liu, Z.-Y. Protective effects of microRNA-330 on amyloid β-protein production, oxidative stress, and mitochondrial dysfunction in Alzheimer's disease by targeting VAV1 via the MAPK signaling pathway. *J. Cell. Biochem.* **2018**, *119*, 5437–5448. [CrossRef] [PubMed]

93. Gonzalez-Foruria, I.; Santulli, P.; Chouzenoux, S.; Carmona, F.; Chapron, C.; Batteux, F. Dysregulation of the ADAM17/Notch signalling pathways in endometriosis: From oxidative stress to fibrosis. *Mol. Hum. Reprod.* **2017**, *23*, 488–499. [CrossRef] [PubMed]

94. Jiao, W.E.; Wei, J.F.; Kong, Y.G.; Xu, Y.; Tao, Z.Z.; Chen, S.M. Notch signaling promotes development of allergic rhinitis by suppressing Foxp3 expression and treg cell differentiation. *Int. Arch. Allergy Immunol.* **2019**, *178*, 33–44. [CrossRef]

95. Zhu, P.; Yang, M.; He, H.; Kuang, Z.; Liang, M.; Lin, A.; Liang, S.; Wen, Q.; Cheng, Z.; Sun, C. Curcumin attenuates hypoxia/reoxygenationinduced cardiomyocyte injury by downregulating Notch signaling. *Mol. Med. Rep.* **2019**, *20*, 1541–1550. [CrossRef]

96. Chen, F.-Z.; Zhao, Y.; Chen, H.-Z. MicroRNA-98 reduces amyloid β-protein production and improves oxidative stress and mitochondrial dysfunction through the Notch signaling pathway via HEY2 in Alzheimer's disease mice. *Int. J. Mol. Med.* **2019**, *43*, 91–102. [CrossRef]

97. Zhang, L.; Dong, H.; Si, Y.; Wu, N.; Cao, H.; Mei, B.; Meng, B. miR-125b promotes tau phosphorylation by targeting the neural cell adhesion molecule in neuropathological progression. *Neurobiol. Aging* **2019**, *73*, 41–49. [CrossRef]

98. Jin, Y.; Tu, Q.; Liu, M. MicroRNA125b regulates Alzheimer's disease through SphK1 regulation. *Mol. Med. Rep.* **2018**, *18*, 2373–2380. [CrossRef]

99. Wu, D.-M.; Wen, X.; Wang, Y.-J.; Han, X.-R.; Wang, S.; Shen, M.; Fan, S.-H.; Zhuang, J.; Zhang, Z.-F.; Shan, Q.; et al. Effect of microRNA-186 on oxidative stress injury of neuron by targeting interleukin 2 through the janus kinase-signal transducer and activator of transcription pathway in a rat model of Alzheime's disease. *J. Cell. Physiol.* **2018**, *233*, 9488–9502. [CrossRef]

100. Jellinger, K.A. Neuropathobiology of non-motor symptoms in Parkinson disease. *J. Neural Transm.* **2015**, *122*, 1429–1440. [CrossRef]

101. Tysnes, O.B.; Storstein, A. Epidemiology of Parkinson's disease. *J. Neural Transm.* **2017**, *124*, 901–905. [CrossRef] [PubMed]

102. Wei, Z.; Li, X.; Li, X.; Liu, Q.; Cheng, Y. Oxidative stress in Parkinson's disease: A systematic review and meta-analysis. *Front. Mol. Neurosci.* **2018**, *11*, 236. [CrossRef] [PubMed]

103. Hwang, D.Y.; Hong, S.; Jeong, J.W.; Choi, S.; Kim, H.; Kim, J.; Kim, K.S. Vesicular monoamine transporter 2 and dopamine transporter are molecular targets of Pitx3 in the ventral midbrain dopamine neurons. *J. Neurochem.* **2009**, *111*, 1202–1212. [CrossRef] [PubMed]

104. Li, Y.; Li, C.; Chen, Z.; He, J.; Tao, Z.; Yin, Z.Q. A microRNA, mir133b, suppresses melanopsin expression mediated by failure dopaminergic amacrine cells in RCS rats. *Cell Signal.* **2012**, *24*, 685–698. [CrossRef]

105. Caudle, W.M.; Richardson, J.R.; Wang, M.Z.; Taylor, T.N.; Guillot, T.S.; McCormack, A.L.; Colebrooke, R.E.; Di Monte, D.A.; Emson, P.C.; Miller, G.W. Reduced vesicular storage of dopamine causes progressive nigrostriatal neurodegeneration. *J. Neurosci.* **2007**, *27*, 8138–8148. [CrossRef]

106. Masoud, S.T.; Vecchio, L.M.; Bergeron, Y.; Hossain, M.M.; Nguyen, L.T.; Bermejo, M.K.; Kile, B.; Sotnikova, T.D.; Siesser, W.B.; Gainetdinov, R.R.; et al. Increased expression of the dopamine transporter leads to loss of dopamine neurons, oxidative stress and l-DOPA reversible motor deficits. *Neurobiol. Dis.* **2015**, *74*, 66–75. [CrossRef]

107. Jia, X.; Wang, F.; Han, Y.; Geng, X.; Li, M.; Shi, Y.; Lu, L.; Chen, Y. MiR-137 and miR-491 negatively regulate dopamine transporter expression and function in neural cells. *Neurosci. Bull.* **2016**, *32*, 512–522. [CrossRef]

108. Barodia, S.K.; Creed, R.B.; Goldberg, M.S. Parkin and PINK1 functions in oxidative stress and neurodegeneration. *Brain Res. Bull.* **2017**, *133*, 51–59. [CrossRef]

109. Wang, H.L.; Chou, A.H.; Wu, A.S.; Chen, S.Y.; Weng, Y.H.; Kao, Y.C.; Yeh, T.H.; Chu, P.J.; Lu, C.S. PARK6 PINK1 mutants are defective in maintaining mitochondrial membrane potential and inhibiting ROS formation of substantia nigra dopaminergic neurons. *Biochim. Biophys. Acta* **2011**, *1812*, 674–684. [CrossRef]

110. Wood-Kaczmar, A.; Gandhi, S.; Yao, Z.; Abramov, A.Y.; Miljan, E.A.; Keen, G.; Stanyer, L.; Hargreaves, I.; Klupsch, K.; Deas, E.; et al. PINK1 is necessary for long term survival and mitochondrial function in human dopaminergic neurons. *PLoS ONE* **2008**, *3*, e2455. [CrossRef]

111. Kim, J.; Fiesel, F.C.; Belmonte, K.C.; Hudec, R.; Wang, W.X.; Kim, C.; Nelson, P.T.; Springer, W.; Kim, J. miR-27a and miR-27b regulate autophagic clearance of damaged mitochondria by targeting PTEN-induced putative kinase 1 (PINK1). *Mol. Neurodegener.* **2016**, *11*, 55. [CrossRef] [PubMed]

112. Prajapati, P.; Sripada, L.; Singh, K.; Bhatelia, K.; Singh, R.; Singh, R. TNF-alpha regulates miRNA targeting mitochondrial complex-I and induces cell death in dopaminergic cells. *Biochim. Biophys. Acta* **2015**, *1852*, 451–461. [CrossRef] [PubMed]

113. Hayashi, T.; Ishimori, C.; Takahashi-Niki, K.; Taira, T.; Kim, Y.C.; Maita, H.; Maita, C.; Ariga, H.; Iguchi-Ariga, S.M. DJ-1 binds to mitochondrial complex I and maintains its activity. *Biochem. Biophys. Res. Commun.* **2009**, *390*, 667–672. [CrossRef] [PubMed]

114. Ariga, H.; Takahashi-Niki, K.; Kato, I.; Maita, H.; Niki, T.; Iguchi-Ariga, S.M. Neuroprotective function of DJ-1 in Parkinson's disease. *Oxid. Med. Cell Longev.* **2013**, *2013*, 683920. [CrossRef]

115. Choi, J.; Sullards, M.C.; Olzmann, J.A.; Rees, H.D.; Weintraub, S.T.; Bostwick, D.E.; Gearing, M.; Levey, A.I.; Chin, L.S.; Li, L. Oxidative damage of DJ-1 is linked to sporadic Parkinson and Alzheimer diseases. *J. Biol. Chem.* **2006**, *281*, 10816–10824. [CrossRef]

116. Xiong, R.; Wang, Z.; Zhao, Z.; Li, H.; Chen, W.; Zhang, B.; Wang, L.; Wu, L.; Li, W.; Ding, J.; et al. MicroRNA-494 reduces DJ-1 expression and exacerbates neurodegeneration. *Neurobiol. Aging* **2014**, *35*, 705–714. [CrossRef]

117. Chen, Y.; Gao, C.; Sun, Q.; Pan, H.; Huang, P.; Ding, J.; Chen, S. MicroRNA-4639 is a regulator of DJ-1 expression and a potential early diagnostic marker for Parkinson's disease. *Front. Aging Neurosci.* **2017**, *9*, 232. [CrossRef]

118. Minones-Moyano, E.; Porta, S.; Escaramis, G.; Rabionet, R.; Iraola, S.; Kagerbauer, B.; Espinosa-Parrilla, Y.; Ferrer, I.; Estivill, X.; Marti, E. MicroRNA profiling of Parkinson's disease brains identifies early downregulation of miR-34b/c which modulate mitochondrial function. *Hum. Mol. Genet.* **2011**, *20*, 3067–3078. [CrossRef]

119. Ramsey, C.P.; Glass, C.A.; Montgomery, M.B.; Lindl, K.A.; Ritson, G.P.; Chia, L.A.; Hamilton, R.L.; Chu, C.T.; Jordan-Sciutto, K.L. Expression of Nrf2 in neurodegenerative diseases. *J. Neuropathol. Exp. Neurol.* **2007**, *66*, 75–85. [CrossRef]

120. Xie, Y.; Chen, Y. MicroRNAs: Emerging targets regulating oxidative stress in the models of Parkinson's disease. *Front. Neurosci.* **2016**, *10*, 298. [CrossRef]

121. Kabaria, S.; Choi, D.C.; Chaudhuri, A.D.; Jain, M.R.; Li, H.; Junn, E. MicroRNA-7 activates Nrf2 pathway by targeting Keap1 expression. *Free Radic. Biol. Med.* **2015**, *89*, 548–556. [CrossRef] [PubMed]

122. McMillan, K.J.; Murray, T.K.; Bengoa-Vergniory, N.; Cordero-Llana, O.; Cooper, J.; Buckley, A.; Wade-Martins, R.; Uney, J.B.; O'Neill, M.J.; Wong, L.F.; et al. Loss of MicroRNA-7 regulation leads to alpha-synuclein accumulation and dopaminergic neuronal loss in vivo. *Mol. Ther.* **2017**, *25*, 2404–2414. [CrossRef] [PubMed]

123. Narasimhan, M.; Patel, D.; Vedpathak, D.; Rathinam, M.; Henderson, G.; Mahimainathan, L. Identification of novel microRNAs in post-transcriptional control of Nrf2 expression and redox homeostasis in neuronal, SH-SY5Y cells. *PLoS ONE* **2012**, *7*, e51111. [CrossRef] [PubMed]

124. Slota, J.A.; Booth, S.A. MicroRNAs in neuroinflammation: Implications in disease pathogenesis, biomarker discovery and therapeutic applications. *Noncoding RNA* **2019**, *5*, 35. [CrossRef]

125. Farrer, M.J. Genetics of Parkinson disease: Paradigm shifts and future prospects. *Nat. Rev. Genet.* **2006**, *7*, 306–318. [CrossRef]

126. Hsu, L.J.; Sagara, Y.; Arroyo, A.; Rockenstein, E.; Sisk, A.; Mallory, M.; Wong, J.; Takenouchi, T.; Hashimoto, M.; Masliah, E. alpha-synuclein promotes mitochondrial deficit and oxidative stress. *Am. J. Pathol.* **2000**, *157*, 401–410. [CrossRef]

127. Martin, L.J.; Pan, Y.; Price, A.C.; Sterling, W.; Copeland, N.G.; Jenkins, N.A.; Price, D.L.; Lee, M.K. Parkinson's disease alpha-synuclein transgenic mice develop neuronal mitochondrial degeneration and cell death. *J. Neurosci.* **2006**, *26*, 41–50. [CrossRef]

128. Perfeito, R.; Ribeiro, M.; Rego, A.C. Alpha-synuclein-induced oxidative stress correlates with altered superoxide dismutase and glutathione synthesis in human neuroblastoma SH-SY5Y cells. *Arch. Toxicol.* **2017**, *91*, 1245–1259. [CrossRef]

129. Deas, E.; Cremades, N.; Angelova, P.R.; Ludtmann, M.H.; Yao, Z.; Chen, S.; Horrocks, M.H.; Banushi, B.; Little, D.; Devine, M.J.; et al. Alpha-synuclein oligomers interact with metal ions to induce oxidative stress and neuronal death in Parkinson's disease. *Antioxid. Redox Signal.* **2016**, *24*, 376–391. [CrossRef]

130. Di Maio, R.; Barrett, P.J.; Hoffman, E.K.; Barrett, C.W.; Zharikov, A.; Borah, A.; Hu, X.; McCoy, J.; Chu, C.T.; Burton, E.A.; et al. alpha-Synuclein binds to TOM20 and inhibits mitochondrial protein import in Parkinson's disease. *Sci. Transl. Med.* **2016**, *8*, 342ra378. [CrossRef]

131. Junn, E.; Lee, K.W.; Jeong, B.S.; Chan, T.W.; Im, J.Y.; Mouradian, M.M. Repression of alpha-synuclein expression and toxicity by microRNA-7. *Proc. Natl. Acad. Sci. USA* **2009**, *106*, 13052–13057. [CrossRef] [PubMed]

132. Wang, Z.H.; Zhang, J.L.; Duan, Y.L.; Zhang, Q.S.; Li, G.F.; Zheng, D.L. MicroRNA-214 participates in the neuroprotective effect of Resveratrol via inhibiting alpha-synuclein expression in MPTP-induced Parkinson's disease mouse. *Biomed. Pharmacother.* **2015**, *74*, 252–256. [CrossRef] [PubMed]

133. Doxakis, E. Post-transcriptional regulation of alpha-synuclein expression by mir-7 and mir-153. *J. Biol. Chem.* **2010**, *285*, 12726–12734. [CrossRef] [PubMed]

134. Kabaria, S.; Choi, D.C.; Chaudhuri, A.D.; Mouradian, M.M.; Junn, E. Inhibition of miR-34b and miR-34c enhances alpha-synuclein expression in Parkinson's disease. *FEBS Lett.* **2015**, *589*, 319–325. [CrossRef] [PubMed]

135. Alvarez-Erviti, L.; Seow, Y.; Schapira, A.H.; Rodriguez-Oroz, M.C.; Obeso, J.A.; Cooper, J.M. Influence of microRNA deregulation on chaperone-mediated autophagy and alpha-synuclein pathology in Parkinson's disease. *Cell Death Dis.* **2013**, *4*, e545. [CrossRef] [PubMed]

136. Li, G.; Yang, H.; Zhu, D.; Huang, H.; Liu, G.; Lun, P. Targeted suppression of chaperone-mediated autophagy by miR-320a promotes alpha-synuclein aggregation. *Int. J. Mol. Sci.* **2014**, *15*, 15845–15857. [CrossRef] [PubMed]

137. Zhang, Z.; Cheng, Y. miR-16-1 promotes the aberrant alpha-synuclein accumulation in parkinson disease via targeting heat shock protein 70. *Sci. World J.* **2014**, *2014*, 938348. [CrossRef]

138. Recasens, A.; Perier, C.; Sue, C.M. Role of microRNAs in the Regulation of alpha-Synuclein Expression: A Systematic Review. *Front. Mol. Neurosci.* **2016**, *9*, 128. [CrossRef]

139. Dias, V.; Junn, E.; Mouradian, M.M. The role of oxidative stress in Parkinson's disease. *J. Parkinsons Dis.* **2013**, *3*, 461–491. [CrossRef]

140. Block, M.L.; Zecca, L.; Hong, J.S. Microglia-mediated neurotoxicity: Uncovering the molecular mechanisms. *Nat. Rev. Neurosci.* **2007**, *8*, 57–69. [CrossRef]

141. Thome, A.D.; Harms, A.S.; Volpicelli-Daley, L.A.; Standaert, D.G. microRNA-155 regulates alpha-synuclein-induced inflammatory responses in models of Parkinson disease. *J. Neurosci.* **2016**, *36*, 2383–2390. [CrossRef] [PubMed]

142. Zhou, Y.; Lu, M.; Du, R.H.; Qiao, C.; Jiang, C.Y.; Zhang, K.Z.; Ding, J.H.; Hu, G. MicroRNA-7 targets Nod-like receptor protein 3 inflammasome to modulate neuroinflammation in the pathogenesis of Parkinson's disease. *Mol. Neurodegener.* **2016**, *16*, 11–28. [CrossRef] [PubMed]

143. Caggiu, E.; Paulus, K.; Mameli, G.; Arru, G.; Sechi, G.P.; Sechi, L.A. Differential expression of miRNA 155 and miRNA 146a in Parkinson's disease patients. *eNeurologicalSci* **2018**, *13*, 1–4. [CrossRef]

144. Yao, L.; Ye, Y.; Mao, H.; Lu, F.; He, X.; Lu, G.; Zhang, S. MicroRNA-124 regulates the expression of MEKK3 in the inflammatory pathogenesis of Parkinson's disease. *J. Neuroinflamm.* **2018**, *15*, 13. [CrossRef] [PubMed]

145. Wilhelmus, M.M.; Nijland, P.G.; Drukarch, B.; de Vries, H.E.; van Horssen, J. Involvement and interplay of Parkin, PINK1, and DJ1 in neurodegenerative and neuroinflammatory disorders. *Free Radic. Biol. Med.* **2012**, *53*, 983–992. [CrossRef]

146. Van Es, M.A.; Hardiman, O.; Chio, A.; Al-Chalabi, A.; Pasterkamp, R.J.; Veldink, J.H.; van den Berg, L.H. Amyotrophic lateral sclerosis. *Lancet* **2017**, *390*, 2084–2098. [CrossRef]

147. Volk, A.E.; Weishaupt, J.H.; Andersen, P.M.; Ludolph, A.C.; Kubisch, C. Current knowledge and recent insights into the genetic basis of amyotrophic lateral sclerosis. *Med. Genet.* **2018**, *30*, 252–258. [CrossRef]

148. Guerrero, E.N.; Wang, H.; Mitra, J.; Hegde, P.M.; Stowell, S.E.; Liachko, N.F.; Kraemer, B.C.; Garruto, R.M.; Rao, K.S.; Hegde, M.L. TDP-43/FUS in motor neuron disease: Complexity and challenges. *Prog. Neurobiol.* **2016**, *145–146*, 78–97. [CrossRef]

149. Belzil, V.V.; Katzman, R.B.; Petrucelli, L. ALS and FTD: An epigenetic perspective. *Acta Neuropathol.* **2016**, *132*, 487–502. [CrossRef]

150. Wang, Z.; Bai, Z.; Qin, X.; Cheng, Y. Aberrations in oxidative stress markers in amyotrophic lateral sclerosis: A systematic review and meta-analysis. *Oxid. Med. Cell Longev.* **2019**, *2019*, 1712323. [CrossRef]

151. Joilin, G.; Leigh, P.N.; Newbury, S.F.; Hafezparast, M. An overview of MicroRNAs as biomarkers of ALS. *Front. Neurol.* **2019**, *10*, 186. [CrossRef] [PubMed]

152. Kovanda, A.; Leonardis, L.; Zidar, J.; Koritnik, B.; Dolenc-Groselj, L.; Ristic Kovacic, S.; Curk, T.; Rogelj, B. Differential expression of microRNAs and other small RNAs in muscle tissue of patients with ALS and healthy age-matched controls. *Sci. Rep.* **2018**, *8*, 5609. [CrossRef] [PubMed]

153. Ricci, C.; Marzocchi, C.; Battistini, S. MicroRNAs as biomarkers in amyotrophic lateral sclerosis. *Cells* **2018**, *7*, 219. [CrossRef] [PubMed]

154. Rizzuti, M.; Filosa, G.; Melzi, V.; Calandriello, L.; Dioni, L.; Bollati, V.; Bresolin, N.; Comi, G.P.; Barabino, S.; Nizzardo, M.; et al. MicroRNA expression analysis identifies a subset of downregulated miRNAs in ALS motor neuron progenitors. *Sci. Rep.* **2018**, *8*, 10105. [CrossRef]

155. Waller, R.; Wyles, M.; Heath, P.R.; Kazoka, M.; Wollff, H.; Shaw, P.J.; Kirby, J. Small RNA sequencing of sporadic amyotrophic lateral sclerosis cerebrospinal fluid reveals differentially expressed miRNAs related to neural and glial activity. *Front. Neurosci.* **2017**, *11*, 731. [CrossRef]

156. Borel, F.; Gernoux, G.; Sun, H.; Stock, R.; Blackwood, M.; Brown, R.H., Jr.; Mueller, C. Safe and effective superoxide dismutase 1 silencing using artificial microRNA in macaques. *Sci. Transl. Med.* **2018**, *10*. [CrossRef]

157. Koval, E.D.; Shaner, C.; Zhang, P.; du Maine, X.; Fischer, K.; Tay, J.; Chau, B.N.; Wu, G.F.; Miller, T.M. Method for widespread microRNA-155 inhibition prolongs survival in ALS-model mice. *Hum. Mol. Genet.* **2013**, *22*, 4127–4135. [CrossRef]

158. Li, C.; Wei, Q.; Gu, X.; Chen, Y.; Chen, X.; Cao, B.; Ou, R.; Shang, H. Decreased glycogenolysis by miR-338-3p promotes regional glycogen accumulation within the spinal cord of amyotrophic lateral sclerosis mice. *Front. Mol. Neurosci.* **2019**, *12*, 114. [CrossRef]

159. Yamakuchi, M.; Ferlito, M.; Lowenstein, C.J. miR-34a repression of SIRT1 regulates apoptosis. *Proc. Natl. Acad. Sci. USA* **2008**, *105*, 13421–13426. [CrossRef]

160. Singh, P.; Hanson, P.S.; Morris, C.M. SIRT1 ameliorates oxidative stress induced neural cell death and is down-regulated in Parkinson's disease. *BMC Neurosci.* **2017**, *18*, 46. [CrossRef]

161. Paladino, S.; Conte, A.; Caggiano, R.; Pierantoni, G.M.; Faraonio, R. Nrf2 pathway in age-related neurological disorders: Insights into MicroRNAs. *Cell Physiol. Biochem.* **2018**, *47*, 1951–1976. [CrossRef] [PubMed]

162. Ba, Q.; Cui, C.; Wen, L.; Feng, S.; Zhou, J.; Yang, K. Schisandrin B shows neuroprotective effect in 6-OHDA-induced Parkinson's disease via inhibiting the negative modulation of miR-34a on Nrf2 pathway. *Biomed. Pharmacother.* **2015**, *75*, 165–172. [CrossRef] [PubMed]

163. Wang, N.; Zhang, L.; Lu, Y.; Zhang, M.; Zhang, Z.; Wang, K.; Lv, J. Down-regulation of microRNA-142-5p attenuates oxygen-glucose deprivation and reoxygenation-induced neuron injury through up-regulating Nrf2/ARE signaling pathway. *Biomed. Pharmacother.* **2017**, *89*, 1187–1195. [CrossRef] [PubMed]

164. Aschrafi, A.; Kar, A.N.; Natera-Naranjo, O.; MacGibeny, M.A.; Gioio, A.E.; Kaplan, B.B. MicroRNA-338 regulates the axonal expression of multiple nuclear-encoded mitochondrial mRNAs encoding subunits of the oxidative phosphorylation machinery. *Cell Mol. Life Sci.* **2012**, *69*, 4017–4027. [CrossRef]

165. Shioya, M.; Obayashi, S.; Tabunoki, H.; Arima, K.; Saito, Y.; Ishida, T.; Satoh, J. Aberrant microRNA expression in the brains of neurodegenerative diseases: miR-29a decreased in Alzheimer disease brains targets neurone navigator 3. *Neuropathol. Appl. Neurobiol.* **2010**, *36*, 320–330. [CrossRef]

166. McColgan, P.; Tabrizi, S.J. Huntington's disease: A clinical review. *Eur. J. Neurol.* **2018**, *25*, 24–34. [CrossRef]

167. Banez-Coronel, M.; Ayhan, F.; Tarabochia, A.D.; Zu, T.; Perez, B.A.; Tusi, S.K.; Pletnikova, O.; Borchelt, D.R.; Ross, C.A.; Margolis, R.L.; et al. RAN translation in huntington disease. *Neuron* **2015**, *88*, 667–677. [CrossRef]

168. Ross, C.A.; Tabrizi, S.J. Huntington's disease: From molecular pathogenesis to clinical treatment. *Lancet Neurol.* **2011**, *10*, 83–98. [CrossRef]

169. Moumne, L.; Betuing, S.; Caboche, J. Multiple aspects of gene dysregulation in Huntington's disease. *Front. Neurol.* **2013**, *4*, 127. [CrossRef]

170. Zheng, J.; Winderickx, J.; Franssens, V.; Liu, B. A Mitochondria-associated oxidative stress perspective on Huntington's disease. *Front. Mol. Neurosci.* **2018**, *11*, 329. [CrossRef]

171. Johnson, R.; Zuccato, C.; Belyaev, N.D.; Guest, D.J.; Cattaneo, E.; Buckley, N.J. A microRNA-based gene dysregulation pathway in Huntington's disease. *Neurobiol. Dis.* **2008**, *29*, 438–445. [CrossRef] [PubMed]

172. Packer, A.N.; Xing, Y.; Harper, S.Q.; Jones, L.; Davidson, B.L. The bifunctional microRNA miR-9/miR-9* regulates REST and CoREST and is downregulated in Huntington's disease. *J. Neurosci.* **2008**, *28*, 14341–14346. [CrossRef] [PubMed]

173. Savas, J.N.; Makusky, A.; Ottosen, S.; Baillat, D.; Then, F.; Krainc, D.; Shiekhattar, R.; Markey, S.P.; Tanese, N. Huntington's disease protein contributes to RNA-mediated gene silencing through association with Argonaute and P bodies. *Proc. Natl. Acad. Sci. USA* **2008**, *105*, 10820–10825. [CrossRef] [PubMed]

174. Pircs, K.; Petri, R.; Madsen, S.; Brattas, P.L.; Vuono, R.; Ottosson, D.R.; St-Amour, I.; Hersbach, B.A.; Matusiak-Bruckner, M.; Lundh, S.H.; et al. Huntingtin aggregation impairs autophagy, leading to argonaute-2 accumulation and global MicroRNA dysregulation. *Cell Rep.* **2018**, *24*, 1397–1406. [CrossRef]

175. Banez-Coronel, M.; Porta, S.; Kagerbauer, B.; Mateu-Huertas, E.; Pantano, L.; Ferrer, I.; Guzman, M.; Estivill, X.; Marti, E. A pathogenic mechanism in Huntington's disease involves small CAG-repeated RNAs with neurotoxic activity. *PLoS Genet.* **2012**, *8*, e1002481. [CrossRef]

176. Lee, S.T.; Chu, K.; Im, W.S.; Yoon, H.J.; Im, J.Y.; Park, J.E.; Park, K.H.; Jung, K.H.; Lee, S.K.; Kim, M.; et al. Altered microRNA regulation in Huntington's disease models. *Exp. Neurol.* **2011**, *227*, 172–179. [CrossRef]

177. Marti, E.; Pantano, L.; Banez-Coronel, M.; Llorens, F.; Minones-Moyano, E.; Porta, S.; Sumoy, L.; Ferrer, I.; Estivill, X. A myriad of miRNA variants in control and Huntington's disease brain regions detected by massively parallel sequencing. *Nucleic Acids Res.* **2010**, *38*, 7219–7235. [CrossRef]

178. Wexler, N.S.; Lorimer, J.; Porter, J.; Gomez, F.; Moskowitz, C.; Shackell, E.; Marder, K.; Penchaszadeh, G.; Roberts, S.A.; Gayan, J.; et al. Venezuelan kindreds reveal that genetic and environmental factors modulate Huntington's disease age of onset. *Proc. Natl. Acad. Sci. USA* **2004**, *101*, 3498–3503. [CrossRef]

179. Rosenblatt, A.; Liang, K.Y.; Zhou, H.; Abbott, M.H.; Gourley, L.M.; Margolis, R.L.; Brandt, J.; Ross, C.A. The association of CAG repeat length with clinical progression in Huntington disease. *Neurology* **2006**, *66*, 1016–1020. [CrossRef]

180. Johri, A.; Beal, M.F. Antioxidants in Huntington's disease. *Biochim. Biophys. Acta* **2012**, *1822*, 664–674. [CrossRef]

181. Duran, R.; Barrero, F.J.; Morales, B.; Luna, J.D.; Ramirez, M.; Vives, F. Oxidative stress and plasma aminopeptidase activity in Huntington's disease. *J. Neural. Transm.* **2010**, *117*, 325–332. [CrossRef] [PubMed]

182. Olejniczak, M.; Kotowska-Zimmer, A.; Krzyzosiak, W. Stress-induced changes in miRNA biogenesis and functioning. *Cell Mol. Life Sci.* **2018**, *75*, 177–191. [CrossRef] [PubMed]

183. Asada, S.; Takahashi, T.; Isodono, K.; Adachi, A.; Imoto, H.; Ogata, T.; Ueyama, T.; Matsubara, H.; Oh, H. Downregulation of Dicer expression by serum withdrawal sensitizes human endothelial cells to apoptosis. *Am. J. Physiol. Heart Circ. Physiol.* **2008**, *295*, H2512–H2521. [CrossRef] [PubMed]

184. Inukai, S.; de Lencastre, A.; Turner, M.; Slack, F. Novel microRNAs differentially expressed during aging in the mouse brain. *PLoS ONE* **2012**, *7*, e40028. [CrossRef]

185. Ludwig, N.; Leidinger, P.; Becker, K.; Backes, C.; Fehlmann, T.; Pallasch, C.; Rheinheimer, S.; Meder, B.; Stahler, C.; Meese, E.; et al. Distribution of miRNA expression across human tissues. *Nucleic Acids Res.* **2016**, *44*, 3865–3877. [CrossRef]

186. Agarwal, V.; Bell, G.W.; Nam, J.W.; Bartel, D.P. Predicting effective microRNA target sites in mammalian mRNAs. *Elife* **2015**, *4*. [CrossRef]

187. Witkos, T.M.; Koscianska, E.; Krzyzosiak, W.J. Practical aspects of microRNA target prediction. *Curr. Mol. Med.* **2011**, *11*, 93–109. [CrossRef]

188. Jin, H.Y.; Gonzalez-Martin, A.; Miletic, A.V.; Lai, M.; Knight, S.; Sabouri-Ghomi, M.; Head, S.R.; Macauley, M.S.; Rickert, R.C.; Xiao, C. Transfection of microRNA mimics should be used with caution. *Front. Genet.* **2015**, *6*, 340. [CrossRef]

189. Jovicic, A.; Roshan, R.; Moisoi, N.; Pradervand, S.; Moser, R.; Pillai, B.; Luthi-Carter, R. Comprehensive expression analyses of neural cell-type-specific miRNAs identify new determinants of the specification and maintenance of neuronal phenotypes. *J. Neurosci.* **2013**, *33*, 5127–5137. [CrossRef]

190. Nam, J.W.; Rissland, O.S.; Koppstein, D.; Abreu-Goodger, C.; Jan, C.H.; Agarwal, V.; Yildirim, M.A.; Rodriguez, A.; Bartel, D.P. Global analyses of the effect of different cellular contexts on microRNA targeting. *Mol. Cell* **2014**, *53*, 1031–1043. [CrossRef]

191. He, M.; Liu, Y.; Wang, X.; Zhang, M.Q.; Hannon, G.J.; Huang, Z.J. Cell-type-based analysis of microRNA profiles in the mouse brain. *Neuron* **2012**, *73*, 35–48. [CrossRef] [PubMed]

192. Svoboda, P. A toolbox for miRNA analysis. *FEBS Lett.* **2015**, *589*, 1694–1701. [CrossRef] [PubMed]

193. Knauss, J.L.; Bian, S.; Sun, T. Plasmid-based target protectors allow specific blockade of miRNA silencing activity in mammalian developmental systems. *Front. Cell Neurosci.* **2013**, *7*, 163. [CrossRef] [PubMed]

194. Ran, F.A.; Hsu, P.D.; Wright, J.; Agarwala, V.; Scott, D.A.; Zhang, F. Genome engineering using the CRISPR-Cas9 system. *Nat. Protoc.* **2013**, *8*, 2281–2308. [CrossRef]

195. Jinek, M.; Chylinski, K.; Fonfara, I.; Hauer, M.; Doudna, J.A.; Charpentier, E. A programmable dual-RNA-guided DNA endonuclease in adaptive bacterial immunity. *Science* **2012**, *337*, 816–821. [CrossRef]

196. Gasiunas, G.; Barrangou, R.; Horvath, P.; Siksnys, V. Cas9-crRNA ribonucleoprotein complex mediates specific DNA cleavage for adaptive immunity in bacteria. *Proc. Natl. Acad. Sci. USA* **2012**, *109*, E2579–E2586. [CrossRef]

197. Amin, N.D.; Bai, G.; Klug, J.R.; Bonanomi, D.; Pankratz, M.T.; Gifford, W.D.; Hinckley, C.A.; Sternfeld, M.J.; Driscoll, S.P.; Dominguez, B.; et al. Loss of motoneuron-specific microRNA-218 causes systemic neuromuscular failure. *Science* **2015**, *350*, 1525–1529. [CrossRef]

198. Chang, H.; Yi, B.; Ma, R.; Zhang, X.; Zhao, H.; Xi, Y. CRISPR/cas9, a novel genomic tool to knock down microRNA in vitro and in vivo. *Sci. Rep.* **2016**, *6*, 22312. [CrossRef]

199. Kurata, J.S.; Lin, R.J. MicroRNA-focused CRISPR-Cas9 library screen reveals fitness-associated miRNAs. *RNA* **2018**, *24*, 966–981. [CrossRef]

200. Back, S.; Necarsulmer, J.; Whitaker, L.R.; Coke, L.M.; Koivula, P.; Heathward, E.J.; Fortuno, L.V.; Zhang, Y.; Yeh, C.G.; Baldwin, H.A.; et al. Neuron-specific genome modification in the adult rat brain using CRISPR-Cas9 transgenic rats. *Neuron* **2019**, *102*, 105–119. [CrossRef]

201. Hess, G.T.; Tycko, J.; Yao, D.; Bassik, M.C. Methods and applications of CRISPR-mediated base editing in eukaryotic genomes. *Mol. Cell* **2017**, *68*, 26–43. [CrossRef] [PubMed]

202. Zafra, M.P.; Schatoff, E.M.; Katti, A.; Foronda, M.; Breinig, M.; Schweitzer, A.Y.; Simon, A.; Han, T.; Goswami, S.; Montgomery, E.; et al. Optimized base editors enable efficient editing in cells, organoids and mice. *Nat. Biotechnol.* **2018**, *36*, 888–893. [CrossRef] [PubMed]

203. Hu, J.H.; Miller, S.M.; Geurts, M.H.; Tang, W.; Chen, L.; Sun, N.; Zeina, C.M.; Gao, X.; Rees, H.A.; Lin, Z.; et al. Evolved Cas9 variants with broad PAM compatibility and high DNA specificity. *Nature* **2018**, *556*, 57–63. [CrossRef] [PubMed]

204. Anzalone, A.V.; Randolph, P.B.; Davis, J.R.; Sousa, A.A.; Koblan, L.W.; Levy, J.M.; Chen, P.J.; Wilson, C.; Newby, G.A.; Raguram, A.; et al. Search-and-replace genome editing without double-strand breaks or donor DNA. *Nature* **2019**. [CrossRef] [PubMed]

205. Domanskyi, A.; Saarma, M.; Airavaara, M. Prospects of neurotrophic factors for Parkinson's disease: Comparison of protein and gene therapy. *Hum. Gene Ther.* **2015**, *26*, 550–559. [CrossRef]

206. Espay, A.J.; Brundin, P.; Lang, A.E. Precision medicine for disease modification in Parkinson disease. *Nat. Rev. Neurol.* **2017**, *13*, 119–126. [CrossRef]

207. La Manno, G.; Gyllborg, D.; Codeluppi, S.; Nishimura, K.; Salto, C.; Zeisel, A.; Borm, L.E.; Stott, S.R.W.; Toledo, E.M.; Villaescusa, J.C.; et al. Molecular diversity of midbrain development in mouse, human, and stem cells. *Cell* **2016**, *167*, 566–580. [CrossRef]

208. Creed, R.B.; Goldberg, M.S. New developments in genetic rat models of Parkinson's disease. *Mov. Disord.* **2018**, *33*, 717–729. [CrossRef]

209. Chesselet, M.F.; Richter, F. Modelling of Parkinson's disease in mice. *Lancet Neurol.* **2011**, *10*, 1108–1118. [CrossRef]

210. Blesa, J.; Przedborski, S. Parkinson's disease: Animal models and dopaminergic cell vulnerability. *Front. Neuroanat.* **2014**, *8*, 155. [CrossRef]

211. Berezikov, E.; Thuemmler, F.; van Laake, L.W.; Kondova, I.; Bontrop, R.; Cuppen, E.; Plasterk, R.H. Diversity of microRNAs in human and chimpanzee brain. *Nat. Genet.* **2006**, *38*, 1375–1377. [CrossRef] [PubMed]

212. Londin, E.; Loher, P.; Telonis, A.G.; Quann, K.; Clark, P.; Jing, Y.; Hatzimichael, E.; Kirino, Y.; Honda, S.; Lally, M.; et al. Analysis of 13 cell types reveals evidence for the expression of numerous novel primate- and tissue-specific microRNAs. *Proc. Natl. Acad. Sci. USA* **2015**, *112*, E1106–E1115. [CrossRef] [PubMed]

213. McLoughlin, H.S.; Wan, J.; Spengler, R.M.; Xing, Y.; Davidson, B.L. Human-specific microRNA regulation of FOXO1: Implications for microRNA recognition element evolution. *Hum. Mol. Genet.* **2014**, *23*, 2593–2603. [CrossRef]

214. Franca, G.S.; Hinske, L.C.; Galante, P.A.; Vibranovski, M.D. Unveiling the impact of the genomic architecture on the evolution of vertebrate microRNAs. *Front. Genet.* **2017**, *8*, 34. [CrossRef] [PubMed]

215. Wang, N.; Zheng, J.; Chen, Z.; Liu, Y.; Dura, B.; Kwak, M.; Xavier-Ferrucio, J.; Lu, Y.C.; Zhang, M.; Roden, C.; et al. Single-cell microRNA-mRNA co-sequencing reveals non-genetic heterogeneity and mechanisms of microRNA regulation. *Nat. Commun.* **2019**, *10*, 95. [CrossRef] [PubMed]

216. Engle, S.J.; Blaha, L.; Kleiman, R.J. Best practices for translational disease modeling using human iPSC-derived neurons. *Neuron* **2018**, *100*, 783–797. [CrossRef]

217. Tolosa, E.; Botta-Orfila, T.; Morato, X.; Calatayud, C.; Ferrer-Lorente, R.; Marti, M.J.; Fernandez, M.; Gaig, C.; Raya, A.; Consiglio, A.; et al. MicroRNA alterations in iPSC-derived dopaminergic neurons from Parkinson disease patients. *Neurobiol. Aging* **2018**, *69*, 283–291. [CrossRef]

218. Katerji, M.; Filippova, M.; Duerksen-Hughes, P. Approaches and methods to measure oxidative stress in clinical samples: Research applications in the cancer field. *Oxid. Med. Cell Longev.* **2019**, *2019*, 1279250. [CrossRef]

219. Grenier, K.; Kao, J.; Diamandis, P. Three-dimensional modeling of human neurodegeneration: Brain organoids coming of age. *Mol. Psychiatry* **2019**. [CrossRef]

220. Windrem, M.S.; Schanz, S.J.; Morrow, C.; Munir, J.; Chandler-Militello, D.; Wang, S.; Goldman, S.A. A competitive advantage by neonatally engrafted human glial progenitors yields mice whose brains are chimeric for human glia. *J. Neurosci.* **2014**, *34*, 16153–16161. [CrossRef]

221. Cardoso, T.; Adler, A.F.; Mattsson, B.; Hoban, D.B.; Nolbrant, S.; Wahlestedt, J.N.; Kirkeby, A.; Grealish, S.; Bjorklund, A.; Parmar, M. Target-specific forebrain projections and appropriate synaptic inputs of hESC-derived dopamine neurons grafted to the midbrain of parkinsonian rats. *J. Comp. Neurol.* **2018**, *526*, 2133–2146. [CrossRef] [PubMed]

222. Grealish, S.; Diguet, E.; Kirkeby, A.; Mattsson, B.; Heuer, A.; Bramoulle, Y.; Van Camp, N.; Perrier, A.L.; Hantraye, P.; Bjorklund, A.; et al. Human ESC-derived dopamine neurons show similar preclinical efficacy and potency to fetal neurons when grafted in a rat model of Parkinson's disease. *Cell Stem Cell* **2014**, *15*, 653–665. [CrossRef] [PubMed]

223. Adler, A.F.; Cardoso, T.; Nolbrant, S.; Mattsson, B.; Hoban, D.B.; Jarl, U.; Wahlestedt, J.N.; Grealish, S.; Bjorklund, A.; Parmar, M. HESC-derived dopaminergic transplants integrate into basal ganglia circuitry in a preclinical model of Parkinson's disease. *Cell Rep.* **2019**, *28*, 3462–3473. [CrossRef] [PubMed]

International Journal of
Molecular Sciences

Review

The Role of MicroRNAs in Diabetes-Related Oxidative Stress

Mirza Muhammad Fahd Qadir [1,2,†], Dagmar Klein [1,†], Silvia Álvarez-Cubela [1], Juan Domínguez-Bendala [1,2,3,*] and Ricardo Luis Pastori [1,4,*]

[1] Diabetes Research Institute, University of Miami Miller School of Medicine, Miami, FL 33136, USA; fahd.qadir@med.miami.edu (M.M.F.Q.); dklein@med.miami.edu (D.K.); salvarez@med.miami.edu (S.Á.-C.)
[2] Department of Cell Biology and Anatomy, University of Miami Miller School of Medicine, Miami, FL 33136, USA
[3] Department of Surgery, University of Miami Miller School of Medicine, Miami, FL 33136, USA
[4] Department of Medicine, Division of Metabolism, Endocrinology and Diabetes, University of Miami Miller School of Medicine, Miami, FL 33136, USA
*   Correspondence: jdominguez2@med.miami.edu (J.D.-B.); rpastori@med.miami.edu (R.L.P.)
†   These authors contributed equally to this work.

Received: 23 October 2019; Accepted: 28 October 2019; Published: 31 October 2019

Abstract: Cellular stress, combined with dysfunctional, inadequate mitochondrial phosphorylation, produces an excessive amount of reactive oxygen species (ROS) and an increased level of ROS in cells, which leads to oxidation and subsequent cellular damage. Because of its cell damaging action, an association between anomalous ROS production and disease such as Type 1 (T1D) and Type 2 (T2D) diabetes, as well as their complications, has been well established. However, there is a lack of understanding about genome-driven responses to ROS-mediated cellular stress. Over the last decade, multiple studies have suggested a link between oxidative stress and microRNAs (miRNAs). The miRNAs are small non-coding RNAs that mostly suppress expression of the target gene by interaction with its 3'untranslated region (3'UTR). In this paper, we review the recent progress in the field, focusing on the association between miRNAs and oxidative stress during the progression of diabetes.

Keywords: diabetes; beta cells; oxidative stress; microRNAs

---

1. Introduction

Diabetes, which affects approximately 422 million people worldwide, is a disease characterized by the loss of glycemic control, which causes side effects such as polyuria, glycosuria, weight loss, neuropathies, retinopathy, and renal plus vascular diseases. Because diabetes results in the loss of glucose homeostasis, it is associated with high morbidity and mortality [1]. The most prevalent forms of this disease are Type 1 (T1D) and Type 2 diabetes (T2D). Both types are characterized by hyperglycemia due to either insufficient insulin production (T1D) or loss of cellular sensitivity to insulin, known as insulin resistance (T2D). Insulin-producing beta cells reside in the pancreas within clusters of endocrine cells called "Islets of Langerhans". Islets are dispersed throughout the pancreas, representing around 2% of the overall pancreatic tissue [2]. Beta cells are essential for blood glucose homeostasis. Their dysregulation is linked to both forms of diabetes. In T1D, the primary targets of autoimmunity are beta cells [3]. In T2D, insulin resistance (i.e., the inability of cells to respond to insulin to take up glucose) leads to excessive insulin production by beta cells, resulting in their exhaustion and eventual death [4]. Strong evidence indicates that T2D is associated with a deficit in beta cell mass [5], which leads to long lasting inefficient glycemic control leading to toxic amount of glucose.

Hyperglycemia is responsible for the development of severe complications such as microvascular, neuropathic, and macrovascular problems, which affect the quality and expectancy of life [6,7].

Since beta cells have notoriously low proliferating rates in adults, replenishing beta cell mass remains one of the greatest challenges of modern biology [8,9]. Even a partial restoration of insulin production in the pancreas could be therapeutically sufficient, judging by the fact that even after 80% loss of beta cell mass, T1D patients remain asymptomatic [10]. Although each of the two diabetes types has a different etiology, they are both greatly affected by cellular oxidative stress. On the one hand, oxidative stress in T1D originates from T cell-mediated autoimmunity targeting beta cells through the generation of proinflammatory cytokines. In addition, low tissue expression of antioxidative enzymes and antioxidative agents make affected individuals vulnerable to damage induced by reactive oxygen species (ROS) and reactive nitrogen species (RNS) originating from hypoxia or cytokine-mediated oxidative stress. A well-balanced equilibrium between oxidative molecules and antioxidative defenses is critical for physiological cell functions. On the other hand, type 2 diabetes is a metabolic syndrome where a group of conditions such as hypertension, glucose intolerance, insulin resistance, obesity, and dyslipidemia result in cellular oxidative stress across tissues [11,12]. Specifically, abdominal obesity has been shown to be a source of proinflammatory cytokines and, consequently, leads to insulin resistance.

Numerous studies have recently reported a strong link between oxidative stress and microRNAs (miRNAs). MiRNAs are post-transcriptional regulators, approximately 18 to 23 nucleotides long, that suppress gene expression by specific interaction with target genes [13]. The miRNAs have a role in controlling cellular redox homeostasis between highly reactive oxidative and antioxidative species. Current reports show that changes in miRNA levels contribute to persistent cellular oxidative stress, eventually leading to the development of diseases. Publications over the last few years increasingly support the link between miRNAs and oxidative stress in diabetes. A better understanding of the molecular mechanisms influencing the relationship between miRNAs and oxidative stress in diabetes could be useful to the development of therapeutic approaches that improve beta cell survival under metabolic stress. In this paper, we review the progress made in this field, describing mechanistic miRNA-driven gene regulation during oxidative stress and diabetes progression.

2. Overview of MicroRNA Biology: MiRNA Regulation and Their Role in Islets and Diabetes

The discovery of microRNA (miRNA) over twenty-five years ago revolutionized the field of cell biology and molecular biology. The first well-characterized small RNAs were lin-4 and let-7 [14–16], both of which have been found to be involved in control of early development, while let-7 has been found highly conserved across animal species [17]. According to a conservative analysis from ENCODE (Encyclopedia of DNA Elements) [18], an international consortium funded by the National Human Genome Research Institute (NHGRI) to study the human genome, 62% of the genome bases are transcribed into RNA of more than 200 bases long, of which only 5% corresponds to exons. Therefore, most of the transcribed RNA does not code for proteins and is designated as non-protein coding RNA (ncRNA). MiRNAs, a subset of ncRNAs, are small single stranded gene products of 18 to 23 nts, with an important role in post-transcriptional regulation of gene expression [13,19]. Almost half of the human miRNA genes are located in intergenic regions of the genome. Most of the other half are located in intronic regions of protein-coding genes, whereas some are found within exons [20]. The most common miRNA biogenesis pathway is known as the canonical pathway, although some miRNAs take alternative biogenesis routes [21,22]. In the canonical pathway, miRNA genes are transcribed by RNA polymerase II (Pol-II) to primary miRNAs (pri-miRNAs), which are processed in the nucleus by a microprocessor complex composed of human ribonuclease III (Drosha) and the DGCR8 (DiGeorge syndrome critical region 8) to a pre-miR stem loop precursor of approximately 60 to 70 nt [13,23]. The pre-miRNA stem loop is actively transported to cytoplasm by exportin 5, where it is cleaved by Dicer, another member of the ribonuclease III protein family, into approximately 18 to 23 nucleotide double-stranded mature miRNA [13]. One strand arises from the 5′ end of the stem-loop and the other strand from the 3′ end, termed -5p and -3p, respectively. The miRNA is then incorporated into a ribonucleoprotein complex known as RISC (RNA-induced silencing complex) containing the essential silencing protein Argonaute 2 (Argo2) [24]. Argonautes belong to a highly conserved protein family. Together with small RNAs, such as miRNAs, they form ribonucleoprotein complexes (RNPs) that regulate post-transcriptional gene pathways. If the complementarity with the target mRNA is extensive, as is the case for the homeobox HOXB8 mRNA and miR-196, the Argonaute protein cleaves the mRNA [25]. However, in eukaryotes, the most frequent forms of silencing are by inhibition of translation or mRNA destabilization by polyA shortening [26].

Only the active mature RNA strand, known as a guide strand, is preserved and loaded on RISC, while the other complementary strand, designated as * strand, and known as a passenger strand, is degraded [24]. Many miRNAs retain both 5′ and 3′ strands, which are then incorporated into RISC complexes, generating miR-5p, as well as miR-3p. The choice of miR-5p or -3p as active mature miRNAs depends mostly on cell type [27]. It appears that the decision to select the guide strand from the miRNA duplex generated by Dicer is partly due to thermodynamics considerations. The strand with the weakest binding at its 5′ end is more likely to become the guide strand. In many human miRNAs, the guide strand is U-biased at the 5′ end with an excess of purines, while the passenger strand is C-biased with an excess of pyrimidines. Proteins such as Dicer, Argo2, and others participate in this decision as well. However, the mechanism is basically unknown [28]. The miRNA leads the RISC to a target mRNA. The single strand miRNA-RISC-Argo2 complex principally functions to inhibit target gene expression through recognition of partially complementary sequences in messenger RNA (mRNA), thus regulating mRNA translation by inhibiting gene expression and protein translation. The recognition sequence on the target mRNA is usually found at the 3′ UTR and is recognized by the "seed" sequence, two to eight nucleotides long, located at the 5′ domain of the miRNA. The MiRNAs target specific genes, which in turn may be targeted by many different miRNAs, hence regulating entire critical cellular expression networks (Figure 1).

It has been estimated that over 60% of human protein-coding genes are targets of miRNAs [29].

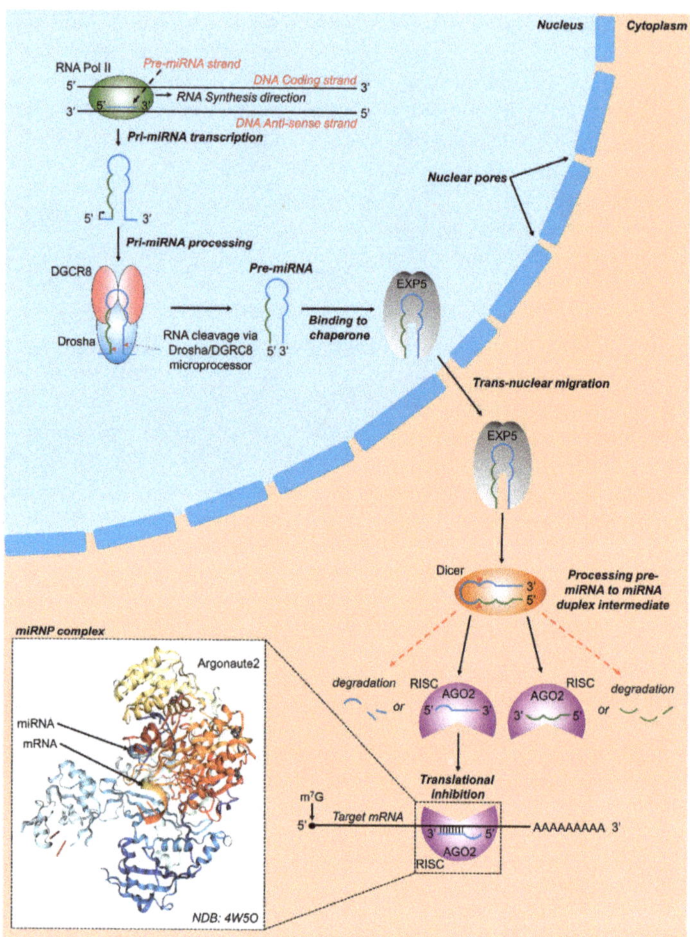

Figure 1. Canonical microRNA biogenesis and RNA targeting. In vertebrates, RNA polymerase-II transcribes primary miRNA genes (pri-miRNAs), which contain a hairpin-loop along with 5′ and 3′ flanking regions. DGCR8 (DiGeorge critical region 8) and a Drosha molecule combine to form the microprocessor complex which binds with pri-miRNA and cleaves it at specific sites (red arrowheads). The resulting precursor miRNA (pre-miRNA) contains a phosphate on its 5′ end and a hydroxyl group on its 3′ end along with a 2 to 3 nucleotide over-hang. Subsequently, the nuclear chaperone Exportin 5 (EXP5) binds to pre-miRNA molecules and transports pre-miRNA molecules to the cytoplasm via transnuclear migration, where Dicer, another RNAse III enzyme, binds to pre-miRNA molecules, cleaves them at specific regions, and releases a miRNA duplex intermediate. Argonaute 2 (AGO2) and other proteins assemble with miRNA molecules released from the miRNA duplex intermediate, together forming the RNA induced silencing complex (RISC). The 3′ or 5′ miRNA containing RISCs may bind to target regions and either result in translational repression, mRNA degradation, or in some cases translational activation. Inset shows a crystal structure of human Argonaute 2 bound to a guide and target RNA [30].

To date, the human genome contains 1917 annotated hairpin precursors, and 2654 mature sequences which are annotated in the Wellcome Trust Sanger Institute miRNA database [31] (http://www.mirbase.org/cgi-bin/mirna_summary.pl?org=hsa). miRNAs play a fundamental role in regulation of gene expression in key biological events such as cell proliferation, differentiation, death, and malignant transformation [13,32–35]. Consequently, impairment of miRNA expression is the underlying cause of many diseases. The miRNAs are mostly intracellular, but they are also found circulating in the body fluids, such as plasma or urine. They are extremely stable in human fluids, and therefore are well suited as clinical biomarkers [36]. They are protected from nucleases either by forming ribonucleoprotein particles (RNPs) with RNA-interacting proteins such as the RISC protein Ago2 or enclosed in extracellular vehicles (EVs) such as exosomes, present in and released by the majority of cell types [37]. The exosome-mediated transfer of mRNAs and miRNAs is a mechanism of cellular communication and genetic exchange among cells. The biogenesis, mode of action and suitability of circulating miRNAs as biomarkers for several diseases, is a hot research topic in biomedicine. Numerous studies suggest that miRNAs have an active role in pancreas organogenesis and in islet function [38–42]. An important study regarding miRNAs and their role in islet development is a report on the deletion of *Dicer1* in pancreatic progenitors. *Dicer1* is an enzyme involved in miRNA maturation, and its loss results in a marked reduction of endocrine cells [40]. Likewise, deletion of *Dicer1* in embryonic beta cells results in fewer beta cells, and impaired glucose tolerance [43,44]. There is evidence that miRNAs are involved in the pathogenesis of diabetes. Comprehensive reviews describing miRNAs in the context of T1D, T2D, and other diabetes models have recently become available. Furthermore, the role of miRNAs in tissues targeted by insulin, and in healthy or stressed islets, have been reported [45–48]. We have previously identified a subset of miRNAs differentially expressed in developing human islets, in human developing pancreas, and in alpha and beta cells of adult human islets [49–52]. These observations set the stage for studies to specifically assess the role of miRNAs and their target molecules in endocrine differentiation. In fact, many studies, including ours, identified individual miRNAs enriching endocrine tissue such as, miR-375 and miR-7, with the role in beta cell differentiation and function [53–57]. The same miRNAs have an important role in in vitro human stem cell differentiation into beta cells [58–61]. On the basis of the information presented above, it can be implied that oxidative stress affecting deregulation of miRNA networks, which is important for acquisition and maintenance of beta cell identity or proper cellular function and metabolism, contributes to the development of diabetes [62].

3. Overview of Oxidative Stress in Glucose Metabolism

The term oxidative stress refers to an imbalance between cellular oxidants and antioxidants [63,64]. Oxidative stress can be classified into the following two major groups: Endogenous (mitochondrial, peroxisomes, lipoxygenases, NADPH oxidase (NOX), and cytochrome P450) and exogenous (UV and ionizing radiation, chemotherapeutics, inflammatory cytokines, and environmental toxins). Oxidative stress is an accumulation of reactive oxygen species (ROS) above physiological levels, where ROS molecules oxidize cellular components stochastically, leading to progressive cellular damage. Under physiological conditions, the utmost ROS generation occurs in mitochondria, accounting for the transformation of 1% to 2% of oxygen molecules into superoxide anions [65]. Adenosine 5′-triphosphate (ATP) molecules are the major cellular energy currency. Generation of ATP in mitochondria, results in the production of ROS which occurs on two occasions with electron transport chain, at complex-I (NADH dehydrogenase) and at complex-III (ubiquinone-cytochrome c reductase). ATPs are first generated in the breakdown of glucose molecules during glycolysis. Glycolysis of one glucose molecule yields two pyruvate molecules with a net gain of only two ATP molecules. The greatest contributor to ATP production is the subsequent metabolism of pyruvate in the mitochondria through the tricarboxylic acid cycle, followed by oxidation of its energy mediators, NADH and FADH2, in the electron transport chain. In this process, known as oxidative phosphorylation, electrons are transferred from electron donors to electron acceptors via redox reactions. Oxidative phosphorylation, hypothetically, generates a maximum of 36 ATP molecules per glucose molecule. Oxygen is the final electron acceptor, generating H_2O. Incomplete transfer of electrons to oxygen results in the production of reactive oxygen species (ROS) such as superoxide or peroxide anions. Superoxide is rapidly converted [66] into peroxide (H_2O_2) by the enzyme superoxide dismutase (SOD). Hydrogen peroxide, in turn, is either neutralized to H_2O and O_2 by glutathione peroxidase (Gpx, in the mitochondria), or detoxified by catalase in peroxisomes. Increased levels of Cu (copper) and Fe (iron) and significantly decreased levels of Zn (zinc) in the serum of T2D patients and their first degree relatives (FDR) could be either triggering factors for the development of diabetes or a consequence of the illness [67]. H_2O_2 can be converted into highly reactive radical hydroxyl (HO·), the neutral form of the hydroxide ion, via the Fenton reaction. Hydroxyl radicals target the DNA base deoxyguanosine with great efficiency [65,68].

A discrete amount of ROS is necessary for efficient cellular physiological function. For example, ROS are one of the metabolic signals for insulin secretion [69] and play an essential role as promoter of natural defenses [70,71]. If the production of ROS during mitochondrial oxidative phosphorylation is not well balanced by antioxidative activity, ROS become toxic [66]. Even though oxidative phosphorylation is a significant contributor to the formation of ROS, recent studies have identified other cellular sources of ROS, such as peroxisomes, endoplasmic reticulum, and plasma membrane, which could contribute to tissue oxidative damage [72]. ROS are free radicals and, because they have unpaired valence electrons, they are extremely reactive with many electron donor molecules such as membrane lipids, proteins, and DNA, leading to potential toxicity. Overproduction of ROS causes oxidative stress associated with numerous diseases and aging.

The interaction of ROS with the cell membrane's polyunsaturated fatty acids generates a lipid peroxidation chain reaction with the production of toxic and highly reactive aldehyde metabolites such as malondialdehyde (MDA) [73,74]. MDA causes a reduction of cell membrane fluidity and function [75]. ROS cause oxidative damage of proteins by direct interaction either on amino acid residues or cofactors or by indirect oxidation via lipid peroxidation end products [76,77]. Likewise, ROS target pyrimidine and purine bases, as well as the deoxyribose moiety of genomic and mitochondrial DNA, causing cellular damage such as strand breakage, nucleotide removal, and DNA-protein binding. Extensive damage that cannot be corrected by cellular DNA repair could result in permanent impairment followed by apoptosis [78].

As far as islet beta cells are concerned, they are highly susceptible to ROS-mediated damage because of insufficient amounts of antioxidative compounds such as glutathione, and the naturally low expression of antioxidative enzymes such as the mitochondrial SOD (Mn-SOD), cytoplasmic Cu/Zn SOD, glutathione peroxidase (GPx), and catalase [79]. Several examples also illustrate the critical role of antioxidative defenses in the vascular system in diabetes. For example, cardiomyocytes in diabetes overexpress SOD or catalase, protecting cardiac mitochondria from extensive oxidative damage. SOD also prevents morphological abnormalities in diabetic hearts, correcting the aberrant contractility [80,81]. Two emerging crucial regulators of antioxidative stress responses are the uncoupling protein 2 (UCP2) and the transcription factor NRF2 (NFE2L2). UCP2, originally thought to function in adaptive thermogenesis similar to UCP1, is now considered to be primarily a regulator of ROS generation in mitochondria. UCP2 is a proton channel protein localized on the inner mitochondrial membrane that reduces the electrochemical gradient on both sides of the membrane, decreases ROS production, and protects against oxidative damage in mitochondria [82]. UCP2 has a critical role in the regulation of glucose homeostasis and in oxidative stress-mediated vascular diseases [83,84]. As for NRF2, it controls the transcription of key components of many antioxidative responses by binding to antioxidant response (ARE) elements in the promoter regions of target genes such as members of the glutathione and thioredoxin antioxidant systems and NAPDH (nicotinamide adenine dinucleotide phosphate) regeneration [85]. NRF2-mediated antioxidative responses are dysfunctional in diabetes [86] and dysregulation of the NRF2 redox pathway affects healing of diabetic wounds [87].

4. Oxidative Stress Generated by T Cell-Mediated Recognition of Beta Cells

T1D is an autoimmune disease characterized by T cell-mediated recognition and destruction of insulin-producing beta cells [88]. The beta cells are destroyed during the inflammatory phase known as insulitis. Insulitis is a significant component of T1D pathology and is characterized by infiltration of islets by immune and inflammatory cells. The leucocytic infiltration in insulitis is relatively subtle and transient, and therefore is detected mostly in cases with recent onset of the disease (less than one year [89]. There is limited knowledge about autoreactive T cells and autoantigens involved in the development of T1D. A primary autoantigen that activates autoreactive T cells is insulin [90]. Current views on T1D onset suggest that autoimmune destruction by insulitis is secondary to primary invasion of macrophages and dendritic cells activated by intercellular ROS from resident pancreatic phagocytes. Stimulated macrophages and dendritic cells will induce inflammatory genes and carry beta cell antigens specifically to lymph nodes, where T cells are activated. The activated T cells will specifically destroy beta cells through proinflammatory cytokine insults and more intracellular ROS formation [91]. So far, there is no cure for autoimmune T1D. Treatment is mostly focused on intensive insulin therapy aiming at tight glycemic control, which can significantly reduce debilitating long-term complications. There is a genetic predisposition for T1D. The strongest associations point at HLA class II, specifically haplotypes DRB1and DQB1 [92]. Although the autoreactive antigens and self-reactive T cells involved in autoimmune attack in T1D are well documented, the mechanism is not yet completely understood, however, the contribution of ROS and proinflammatory cytokines in beta cell death is fully substantiated [93]. The immune-mediated recognition of beta cells by autoreactive T cells and cytotoxic CD8T cells generates ROS and proinflammatory cytokines, inducing beta cell destruction and enhancing the effector response of islet-specific self-reactive CD4 T cells and cytotoxic CD8 T cells [94]. The proinflammatory milieu includes cytokines such as INFg, TNFa, IL-6, IL-12p70 and IL-1b, and ROS [95]. The destructive effect of ROS is amplified by the generation of reactive nitrogen species (RNS), which are extremely toxic free radicals such as free radical nitric oxide (NO) produced by IL-1b in beta cells. The IL-1b activates the enzyme nitric oxide synthase (iNOS), catalyzing production of nitric oxide and ultimately the superoxide ROS [96],. NO interacts with superoxide to generate the highly destructive molecule peroxynitrite. Both NO derived RNS and ROS cause beta cell damage using different pathways [97]. It is important to emphasize that an unbalanced ratio of oxidative to antioxidative events is what causes free radical toxicity. This has been illustrated by a recent study showing the dual role, protective or toxic, of NO in beta cells [98]. As stated above, insulitis and beta cell destruction are the crucial components of T1D pathology, but these are observed only in a limited proportion of islets at any given time, even at the time of diagnosis. Other factors, such as intercellular oxidative stress, precede insulitis [99]. This raises the possibility that in addition to the immune-mediated damaging effect of insulitis, a high level of dysfunction of beta cell contributes to T1D pathology as well. Interestingly, the lipid peroxidation, and oxidative stress detected by the presence of malondialdehyde in plasma of nondiabetic first degree relatives of the patients with T1D [100] supports the observation that oxidative stress can be clinically detected before the onset of diabetes.

5. Oxidative Stress and Metabolic Syndrome and Insulin Resistance in T2 Diabetes

T2D is currently considered a metabolic and inflammatory disease closely associated with metabolic syndrome, a group of conditions such as high blood pressure, glucose intolerance, insulin resistance, obesity, and dyslipidemia [101]. In many cases, a pre-T2D condition known as pre-diabetes is the prelude to the development of the disease. Pre-diabetes is characterized by impaired glucose tolerance and a state of mild hyperglycemia, not high enough to be diagnosed as diabetes, but leading to glucose intolerance. In addition, the main features of pre-diabetes are metabolic abnormalities similar to T2D, with essential roles of proinflammatory cytokines and free fatty acids (FFA), which are elevated in obesity and T2D as well. These factors initiate oxidative stress-mediated pathways, eventually resulting in beta cell dysfunction, impaired insulin secretion, and insulin resistance of peripheral tissue. Many studies indicate that oxidative stress originates before hyperglycemia, which in turn significantly contributes to the later complications of T2D (similar to those of T1D), such as vascular damage, retinopathy, nephropathy, and neuropathy [102]. In vitro and in vivo studies have indicated that the major oxidative stress-mediated pathways activated by hyperglycemia and ROS are JNK/SAPK, p38 MAPK, NF-kB, and the hexosamine biosynthetic pathway [103]. The first two, JNK/SAPK and p38

MAPK, contribute to the development of insulin resistance via direct and indirect phosphorylation of serine and threonine residues of insulin receptors [104,105]. Numerous studies link transcription factor NF-kB with regulation of gene-associated complications of diabetes [106]. In addition, hyperglycemia and oxidative stress mediate their actions through other signaling pathways such as advanced glycation end products (AGEs). AGEs refer to a group of heterogeneous compounds formed by the Maillard reaction process that involves the non-enzymatic glycation of proteins, lipids, and nucleic acids by reducing sugars and aldehydes. AGEs function through the multiligand immunoglobulin superfamily receptor for advanced glycation end products (RAGEs). The AGE compounds directly affect proteins of the mitochondrial respiratory chain to generate reactive oxygen species (ROS) [107]. AGE and RAGE are involved in diabetes vascular pathologies as well [108]. They also activate production of the second messenger signaling lipid diacylglycerol leading to activation of several isoforms of the protein kinase C (PKC). Isoforms of PKC are implicated in generating insulin resistance [109–111]. Last, but not least, AGE increases utilization of the polyol pathway that will decrease the cofactor NAPDH, and therefore directly affects the production of antioxidative glutathione [112,113]. As described above, multiple signaling pathways contribute to oxidative stress-mediated damage leading to T2D. Therefore, dysregulation of miRNAs controlling these pathways can certainly contribute to development and persistence of diabetes.

6. MicroRNAs in Diabetic Oxidative Stress

We reviewed research articles in PubMed, primarily focusing on studies describing changes in the expression of miRNAs due to oxidative stress in the context of diabetes and their target components controlling mechanism of oxidative stress homeostasis.

This review does not include studies dealing with miRNAs induced by proinflammatory cytokines generated by T1D autoimmune attack on beta cells. Thorough reviews have been written on this topic [46,114–116]. Table 1 lists the miRNAs reported as having an effect on oxidative stress in diabetes, the source of oxidative stress and the observed effect, target tissue or organ, and target genes. A few miRNAs, with known target tissue but unknown gene targets are included as well. Ten miRNAs identified in Table 1, overlap with a previous in silico analysis of miRNAs in human cells regulated in vitro by oxidative stress [117]. These are let-7f, miR-9, miR-16, miR-21, miR-22, miR-29b, miR-99a, miR-141, miR-144, and miR-200c. In order to make this overview of miRNAs and their targets in oxidative stress and diabetes easy to follow, we organized the miRNAs by their function in the affected tissues and organs.

Table 1. Selected PubMed articles describing miRNAs in diabetic oxidative stress.

Source of Oxidative Stress	Differentially Expressed miRNAs	Target Tissue/Organ	Target Gene	Reference
T2D	miR-203↓	Cardiac tissue	PIK3CA	[118]
T2D	miR-30e-5p↓	Kidney and vasculature	UCP2, MUC17, UBE2I	[119]
Diabetic retinopathy, hyperglycemia	miR-455-5p↓	Retinal epithelial cells	SOCS3	[120]
Diabetic nephropathy, hyperglycemia	miR-214↓	Kidney tissue	-	[121]
Insulin synthesis	miR-15a↑	Beta cells	UCP2	[122]
Kidney fibrosis	miR-30e↓	Tubular epithelial cells	UCP2	[123]
DCM	miR-30c↓	Cardiac tissue	PGC-1β	[124]
T2D	miR-233↓	Hepatic tissue	KEAP1	[125]
T1D, Diabetic nephropathy	miR-146a↓	Neural tissue, kidney tissue	-	[126,127]
DCM	miR-503↑	Cardiac tissue	NRF2	[128,129]
Diabetic Retinopathy	miR-365↓	Retinal tissue	TIMP3	[130]
Gestational Diabetes	miR-129-2↑	Murine neural tube	PGC-1α	[131]
Hyperglycemia	miR-106b↑	Pancreatic islets	SIRT1	[132]
Diabetic nephropathy	miR-106a↓	Murine neural tissue	ALOX15	[133]
Diabetic retinopathy	miR-7-5p↑	Retinal tissue	EPAC1	[134]
Diabetic neurotoxicity	miR-302↓	Neural tissue	PTEN	[135]
T2D	miR-17↓	Skeletal muscle	GLUT4	[136]

Table 1. *Cont.*

Source of Oxidative Stress	Differentially Expressed miRNAs	Target Tissue/Organ	Target Gene	Reference
Diabetic retinopathy, hyperglycemia	miR-145↓	Retinal epithelial cells	TLR4	[137]
Diabetic nephropathy, hyperglycemia	miR-25↓	Neural tissue, kidney tissue	PTEN, CDC42	[138–140]
TXNIP overexpression	miR-200b↑	Beta cells	ZEB1	[141]
Diabetic mice	miR-200c↑	Vasculature	ZEB1	[142]
Diabetic Mice	miR-200a/b↓	Vasculature	OGT	[143]
DCM	miR-92a↑	Vasculature	HMOX1	[144,145]
T2D	miR-200b/c↑ and miR-429↑	Vasculature	ZEB1	[146]
T2D, T1D	miR-200c↑	Murine arteries	SIRT1, FOXO1, eNOS	[147]
Long-term diabetes	miR-126↑	Vasculature, skeletal muscles	SIRT1, SOD	[148]
T2D	miR-133a↓	Murine gastric smooth muscle cells	RhoA/Rho kinase	[149]
Hyperglycemia, T2D, T1D	miR-21↑	Vasculature, β-cells, Cardiac tissue	KIRT1, FOXO1, NRF2, SOD2, PPARA	[150–152]
T1D model	miR-200b↑	Murine retinal cells	OXR1	[153]
T2D	miR-15a↑	Plasma	AKT3	[154]
Diabetic embryopathy	miR-27a↑	Murine embryos, kidney tissue	NRF2	[129,155]
STZ-diabetic mice	miR-34a↑	β-cells, vasculature	SIRT1	[156]
Endothelial cells, vascular stress	miR-204↑	Vascular wall /endothelium in vivo	SIRT1	[157]
Cardiomyocytes apoptosis	miR-675↓	Vasculature	VDAC1	[158]
T1D, Diabetic retinopathy	miR-195↑	Cardiac tissue, β-cells	CASP3, MFN2	[159,160]
Gestational diabetes, hyperglycemia	miR-322↓	Murine Embryos, Neurons	TRAF3	[161]
T2D	miR-126↓	Vasculature	VEGFR2	[162]
T2D	miR-27b↓	Vasculature, wounds	SHC1, SEMA6A, TSP-1, TSP-2	[163]
Hyperglycemia, Polyol pathway	miR-200a-3p↑, miR-141-3p↑	Kidney tissue	KEAP1, TGFβ1/2	[164]
STZ mice	miR-1↓, miR-499↓, miR-133a/b↓ and miR-21↑	Cardiac tissue	ASPH	[165]
Persistent UPR IRE1α deficiency	miR-200↑, miR-466h-5p↑	Vasculature, wounds	ANGPT1	[166]
T2D, DCM	miR-9-5p↑	Retinal tissue	ELAVL1	[167]
T2D	miR-99a↑	Vasculature	IGF1R, MTOR	[168]
Hyperlipidemia	miR-155-5p↑	β-cells	MAFB	[169]
T1D NOD islets	miR-29c↑	β-cells	MCL1	[170]
T2D, glucose and lipid oxidation	miR-29↑	Skeletal muscle	-	[171]
Diabetic nephropathy	miR-29↑	Regulation of inflammatory cytokines	TTP	[172]
Diabetic heart T2D	miR-29↑	Cardio-metabolic disorders	Lypla 1	[173]
Gestational diabetes	Circular RNAs: circ-5824↓, circ-3636↓, circ-0395↓	Human placenta	(In silico analysis) AGE- and RAGE-related genes	[174]

6.1. Vascular Endothelial Cells, Diabetic Cardiomyopathy, and Muscle

MiR-21 is a miRNA related to diabetes. The expression of miR-21 is increased in the plasma of patients with impaired glucose tolerance and with T2D [150]. It has been proposed that circulating extracellular vesicles carrying miR-21 could be used as a marker of developing type 1 diabetes [175]. It has been found that miR-21 increases susceptibility to oxidative stress induced by fluctuating glucose levels in primary pooled human umbilical vein endothelial cells (HUVECs), by targeting genes regulating homeostasis of intracellular ROS, such as KRIT1, NRF2, and SOD2 [151]. A reduced expression of miR-21 protects against cardiac remodeling in diabetic cardiomyopathy (DCM). An in vivo experiment in mice confirmed, that suppression of miR-21 stimulates the nuclear hormone receptor PPAR (peroxisome proliferator activated receptor), known to regulate homeostasis in response to glucose and lipid levels. The PPAR initiates nuclear translocation of NRF2, and thus the antioxidative response of NRF2 protects from DCM [152]. MiR-21 also regulates the signaling pathway of the intracellular AGE–RAGE interaction and targets TIMP3, an inhibitor of extracellular matrix degradation in diabetic neuropathy [176].

Similarly, in a rat model of DCM, the expression of miR-503 is increased in myocardial cells and has a deleterious role by targeting NRF2 and antioxidant response element (ARE) signaling pathway as well [128]. The cluster of miR-200 is an important player in oxidative response in diabetes [177]. It is formed by the following five evolutionary conserved miRNAs: miR-200a, miR-200b, miR-200c, miR-141, and miR-429. These miRNAs can be grouped according to their seed sequences into subgroup I, miR-200a and miR-141 (AACACUG), and subgroup II composed of miR-200b, miR-200c, and miR-429 (AAUACUG), suggesting that miRNAs in each subgroup will target different genes. Several reports indicate that the miR-200 family has a role in the development of endothelial inflammation present in diabetic vascular complications and cardiovascular diseases. In many instances, the action of miR-200 is via targeting the (zinc finger E-box-binding homeobox) ZEB1. ZEB1 has a role in epithelial–mesenchymal transition (EMT) [141] and is associated with the inhibition of apoptosis. The thioredoxin-interacting protein, TXNIP, is induced in vivo by hyperglycemia and it inhibits the antioxidative function of thioredoxin resulting in accumulation of reactive oxygen species, cellular stress, and induction of the miR-200 family which induces apoptosis through inhibition of ZEB1. Likewise, inhibition of miR-200c restores endothelial function in diabetic mice through upregulation of ZEB1 [177], and in HUVEC under oxidative conditions miR-200 expression is increased which suppress ZEB1 causing apoptosis. Overexpression of ZEB1 in the cells reversed the effect [178]. Downregulation of ZEB1, by miR-200a/b/c and miR-429, contributes to activation of proinflammatory genes in vascular smooth muscle cells of diabetic mice [146]. Furthermore, the miR-200 family negatively regulates beta cell survival in type 2 diabetes in vivo. Overexpression of miR-200, in mice, causes beta cell death and is sufficient to render T2D lethal [179].

In addition, the family of miRNA-200 has been reported to exhibit a protective effect in diabetic oxidative stress by targeting high glucose-induced O-linked N-acetylglucosamine transferase (OGT), whose enzymatic activity is associated with diabetic complications, and endothelial inflammation in mice with diabetes. Experiments with human aortic endothelial cells (HAEC) confirmed miR-200 silencing OGT by direct binding to 3′UTR of mRNA [143].

Another important antioxidative gene that is regulated by the family of miR-200 is Sirtuin 1 (SIRT1) [177]. SIRT1 is NAD+-dependent deacetylase that controls histone chromatin proteins as well as non-histone proteins, many of them are transcription factors such as fork-head box O1 (FOXO)1. To date, seven sirtuins have been identified. They are associated with several cellular processes, such as energy balance, stress resistance, and insulin resistance. Some are located in the cytoplasm and others are located in the nucleus or mitochondria [180]. SIRT1, -2, -3, and -6 have a function in oxidative stress. By targeting SIRT1, endothelial nitric oxide synthase (eNOS) and FOXO1 miR-200 impairs their regulatory circuit and promotes ROS production and endothelial dysfunction [147]. It has been shown that miR-200 targets these three genes in vitro in HUVEC cells. The in vitro results were validated in three in vivo models of oxidative stress, human skin fibroblasts from old donors, femoral arteries from old mice, and a murine model of hindlimb ischemia [147].

In endothelial cells, SIRT1 is targeted by other miRNAs, increasing diabetes-related oxidative stress. Examples include the following: miR-34 induces endothelial inflammation by downregulating SIRT1 [156] and targeting SIRT1; miR-204 promotes vascular endoplasmic reticulum (ER) stress, inflammation, and dysfunction in mice; downregulation of miR-204 activates protection against ER stress through an increase of SIRT1 expression [157]; miR-106b targets SIRT1 in mouse insulinoma cell line NIT-1, rendering them vulnerable to hyperglycemia induced by 30mM glucose; and in vivo suppression of miR-106b increases expression of SIRT1 and reduces cardiovascular damage in diabetic mice [132].

Furthermore, it has been shown, in a mouse model of peripheral arterial disease, that the more abundant circulating form of unacylated ghrelin (UnAG) exerts its protective effect from ROS imbalance in endothelial cells via induction of miR-126, a known endothelial miRNA. By targeting vascular cell adhesion molecule 1 (VCAM1), miR-126 indirectly activates SIRT1 and SOD to induce resistance to oxidative stress [148].

MiR-9 plays a positive role in oxidative stress-mediated cardiomyopathy in T2D. In vitro experiments with immortalized cardiomyocyte culture and samples of failing heart tissue collected at the time of transplantation confirmed that downregulation of miR-9 in human cardiomyocytes results in higher expression of its target ELAV-like protein 1 (ELAVL1), a ubiquitously expressed RNA binding protein that stabilizes inflammatory mRNAs by binding to ARE domains and thus leading to cardiomyocyte death [167]. Another miRNA with a protective role in diabetic cardiomyopathy is miR-30c. MiR-30c targets PGC-1β, one of important coactivators of PPAR alpha and mitochondrial key regulator. Knockdown of PGC1 beta reduces excessive ROS and myocardial lipid accumulation which decreases cardiac dysfunction in diabetes [124].

Numerous studies report miR-29 family participation in oxidative stress-mediated inflammatory response in diabetes. The miR-29 family consists of three members divided into two clusters that are transcribed polycistronically; the miR-29a/b-1 cluster is localized on human chromosome 7 and the miR-29c/b-2 cluster on chromosome 1 [181]. The miR-29s are known to be regulated in multiple tissues. Hyperinsulinemia dramatically reduces their expression, while hyperglycemia induces it. Experiments with MIN6 insulinoma beta cell line determined that miR-29 targets a member of the BCL2 family, an antiapoptotic protein, the MCL1 (myeloid cell leukemia 1) (MCL-1) gene. Interestingly, in humans, repression of MCL1 is related to diabetes mellitus-associated cardiomyocyte disorganization [182]. Since circulating miR-29 has been reported in newly diagnosed T2D patients and, furthermore, upregulation of miR-29 expression contributes to development of the first stage of type 1 diabetes mellitus in the T1D model of NOD mice [170], there is the possibility that miR-29 regulates MCL1 at different stages of the disease.

There are instances that indicate the miR-29 cluster family has a protective role against oxidative stress conditions. Its elevated expression has been associated with a compensatory mechanism for heart hypertrophy and fibrosis due to age increased oxidative stress, modulating targets such as DNA methylases and collagens [183]. A protective role in endothelial dysfunction in cardiometabolic disorders found in T2D has been reported. MiR-29 is upregulated in T2D arterioles to compensate for endothelial dysfunction. Specifically, miR-29 targets Lypla 1 (lysophospholipase I), a gene that negatively regulates production of NO, required for vasodilation. Lypla 1 depalmitoylates eNOS (nitric oxide synthase), reducing NO in endothelial cells [173].

The expression of miR-29a and miR-29c in skeletal muscle of patients with type 2 diabetes are upregulated which suppresses glucose and lipid metabolism possibly by targeting insulin receptor substrate 1 (IRS1) and phosphoinositide 3 kinase (PI3K). Both genes are involved in glucose insulin regulation, moreover they control lipid oxidation by targeting peroxisome activated receptor gamma coactivator1alpha (PGC1alpha). In vivo overexpression of miR-29 in mouse tibias anterior muscle resulted in a decrease of glucose uptake and glycogen content. MiR-29 acts as an important regulator of insulin stimulated glucose metabolism [171].

6.2. Retina Cells

Oxidative stress and hypoxia cause retinopathy by induction of miR-7 that negatively regulates the RAPGEF3/EPAC-1 (rap guanine nucleotide exchange factor 3). EPAC-1 is an accessory protein for cAMP activation and stimulation for survival and growth in response to extracellular signals [134]. MiR-7-mediated decrease of EPAC1 expression results in endothelial hyperpermeability and loss of (endothelial nitric oxide synthase) eNOS activity in murine experimental retinopathy. EPAC-1 is associated with cAMP-induced vascular relaxation in endothelial cells via eNOS and amelioration of endothelial hyperpermeability induced by inflammatory mediators [134]. Development of retinopathy in T2D is associated with miR-15 as well. This miRNA is mostly found in the pancreas, where it plays an important role in beta cell insulin secretion. Interestingly, miR-15 has been detected in the plasma of T2D patients, where its amount corelated with the severity of the disease. Experiments with the rat beta cell line INS1 showed that the concentration of miR-15 in the cells increases when cultured in high glucose media. Coculture of INS1 insulinoma cells with Muller cells (retinal glial cells) showed a clear transfer of miR-15 into Muller cells, and the transfer was achieved by exosomes. The deleterious effect of miR-15 in the retina is via targeting AKT3, an isoform of the AKT gene (serine/threonine kinase 1). Loss of AKT3 in the tissue increases intracellular content of ROS, leading to cellular apoptosis. These results also prove that under pathological conditions some miRNAs can travel from tissue to tissue through exosome transfer [154]. Incidentally, persistent exposure to high glucose causes intracellular accumulation of insulin in beta cells mediated by suppression of the UPC2 gene by miR-15a. High glucose treatment for a short time induces miR-15a, while longer exposure suppresses the expression. It has been found that inhibition of UPC2 by miR-15a increases O_2 consumption beta cell function and insulin synthesis [122].

Oxidative stress in retinal glial Muller cells induces upregulation of miR-365 causing damage by targeting TIMP3, the protein that inhibits matrix metalloproteinases and has antioxidative properties [130]. MiR-455-5p may have a positive role in diabetic retinopathy. Upregulation of miR-455-5p attenuates high glucose-triggered oxidative stress injury by targeting SOCS3 (suppressor of cytokine signaling 3) mRNA. SOCS3 downregulation decreases production of intracellular ROS, malondialdehyde (MDA) content, and NADPH oxidase 4 expression, while enhancing superoxide dismutase, catalase, and GPX activities [120].

6.3. Diabetic Wound

Moreover, the miR-200 family has an effect on the pathology of diabetic skin ulcers by targeting the angiogenic factor angiopoietin 1 (ANGPT1), resulting in disrupted angiogenesis. In diabetic wound healing, hyperglycemia-mediated oxidative stress produces an unmodulated, persistent unfolded protein response (UPR), generating deficiency in inositol-requiring enzyme 1 (IRE1α), a primary UPR transducer that modulates expression of mRNAs and miRNAs. This deficiency leads to the upregulation of the miR-200 family and miR-466, both targeting ANGPT1. Angiogenesis may be rescued by upregulation of IRE-1a, which attenuates maturation of both miRNAs [166].

6.4. Kidney Tissues and Functions

Another miRNA that interferes with ROS homeostasis in diabetes via targeting NRF2 is miR-27a. The adipokine omentin 1 restores renal function of type 2 diabetic db/db mice through suppression of miR-27a, which upregulates NRF2 and decreases oxidative stress [155]. NRF2/KEAP1 is a master antioxidant pathway regulating redox under nonstressed and stressed conditions. Under nonstressed conditions, NRF2 is anchored by a repressor KEAP1 in cytoplasm. A stressed situation releases KEAP1 and the stabilized NRF2 relocates to nucleus, where it binds to the antioxidant response element (ARE) activating transcription of antioxidant proteins [184]. In experiments with mice rendered diabetic with streptozotocin, hyperglycemia activates the polyol pathway in renal mesangial cells. The polyol pathway is involved in microvascular damage to retina in diabetes. On the one hand, activation of the polyol pathway increases the activity of aldose reductase which in turn decreases expression of miR-200a and miR-141. These miRNAs are regulators of KEAP-1. Their low expression enhances suppressive activity of KEAP-1 on NRF2. The suppressed transcription factor, NRF2, cannot activate transcription of antioxidant genes resulting in an increase of ROS and oxidative stress. On the other hand, aldose reductase deficiency in the renal cortex upregulates miR-200 and miR-141, which releases the KEAP-1 suppression of NRF2 and ameliorates the oxidative stress and downregulates TGF-beta, preventing kidney fibrosis [164]. The NRF2/KEAP1 pathway is also regulated in other organs under oxidative stress damage, such as in the pathological process of liver injury in T2DM. In this case, miR-233 targets KEAP1 allowing the released NFR2 to migrate to the nucleus and activate synthesis of antioxidative mRNAs and proteins such as SOD and HO-1 [125].

Endothelial dysfunction in cardiovascular disease is also affected by CKD (chronic kidney disease). CKD is caused by the accumulation of uremic toxin which upregulates miR-92a. The miRNA can be detected in the patient's serum, which could be useful for diagnostic purposes. Uremic toxins generated oxidative stress results in downregulation of endothelial protective factors such as SIRT1 and eNOS [144]. At this time, it is not known if this is through direct or indirect regulation. Additionally, miR-92a is upregulated in diabetic aortic endothelium of C57BL-db/db mice and in renal arteries from human diabetic subjects. MiR-92a downregulates expression of heme oxygenase 1 (HO-1), an endothelial protective enzyme synthesized through NRF2 binding to the ARE sequence in the nucleus. The resulting oxidative stress impairs endothelium dependent relaxation. The suppression of miR-92 restores the endothelial function and the expression of HO-1 [145]. The expression of miR-25 in diabetic mouse kidneys and in human peripheral blood of patients with diabetes is much lower than in non-diabetic subjects. MiR-25 has a protective role in ROS-mediated diabetic kidney disease, by direct regulation of the Ras-related gene CDC42. The CDC42 gene belongs to the family of Rho small GTPases which are central regulators of actin reorganization and have a role in nephrotic pathogenesis. An increase of miR-25 expression represses glomerular fibrosis [139]. Some of the intracellular effects of ROS are mediated by regulation of the PTEN/PI3K/AKT pathway [185]. Blood samples and kidney tissue from diabetic subjects show downregulation of miR-25. Gain and loss of function performed with the human kidney cell line HK2 confirmed the crucial role of miR-25 protection against dysfunction and apoptosis of renal tubular epithelial cells. MiR-25 inhibits the apoptotic effect of hyperglycemia-mediated ROS in renal tubular epithelial cells by targeting PTEN. Knockout of PTEN activates the PI3K/AKT. PTEN is a dual protein and lipid phosphatase whose main substrate is phosphatidyl-inositol,3,4,5 triphosphate (PIP3). PTEN catalysis dephosphorization of PIP3 to PIP2 which represses the antiapoptotic signaling pathway of PI3k/AKT. Knockout of PTEN by miR-25 activates the AKT pathway ameliorating ROS and apoptosis [140]. Some miRNAs exert their antioxidative role by regulating the expression of UCP2 (uncoupling protein 2) which attenuates ROS activity in mitochondria. In HK2 (kidney cortex and proximal tubule cell line), it has been shown that miR-214 suppresses oxidative stress in diabetic nephropathy via the ROS/Akt/mTOR signaling pathway and enhancing UCP2 expression [121]. On the other hand, an experiment in a diabetic mouse model showed that miR-30e targets directly UCP2 in kidney cells, thus mediating the TGF-β1-induced epithelial-mesenchymal transition and kidney fibrosis [123]. In diabetic nephropathy, miRNA-29c contributes to the progression of the disease by regulating proinflammatory cytokines via targeting tristetraprolin (TTP) mRNA [172]. Experiments were performed in kidney tissues from DN patients and controls. TTP has anti-inflammatory effects by enhancing the decay of mRNAs bearing the adenosine/uridine-rich element (ARE) present in the 3′UTR of cytokine transcripts such as Il-6 and TNF alpha. Additional experiments with cultured podocytes confirmed the findings. Finally, miR-21, a diabetes-related miRNA, described above, has a role in diabetic nephropathy by regulating TIMP3, an inhibitor of extracellular matrix degradation [176], involved in mesangial expansion characteristic of diabetic nephropathy.

6.5. Diabetic Neuropathy

In the case of diabetic peripheral neuropathy, PKC activity is linked to a protective role of miR-25. MiR-25 downregulates production of AGE and RAGE, reduces activation of PKC, and reduces NAPDH oxidase activity probably via regulation of NOX4, an isoform of the NOX family. NOX4 protects vasculature against inflammatory stress. Experiments to clarify the protective role of miR-25 in diabetic neuropathy were done with sciatic nerve from db/db diabetic mouse model and BALB/c healthy counterparts. The conclusions were confirmed with cultured Schwann cells [138]. Modulation of the PTEN/AKT pathway is also critical to attenuate the oxidative stress mediated by extracellular amyloid-β (Aβ) peptides in diabetic neurotoxicity. Activation of the AKT pathway through direct targeting of PTEN by miR-302 attenuates amyloid beta induced toxicity in neurons and activated AKT signaling, which subsequently stabilizes NRF2 and synthesis of cytoprotective protein HO-1 [135].

Finally, as stated above, we have not included in this review the miRNAs involved in the oxidative stress caused by the effect of proinflammatory cytokines in beta cells. However, beta cells are also the target of other oxidative sources such as oxidized LDL (low density of lipoproteins). Oxidative stress induced the generation of oxidized LDL in hyperlipidemia conditions. Oxidized LDL enhances the activity of LPS (lysophosphatidylcholine) increasing the expression of miR-155-5p in murine pancreatic beta cells. MiR-155 targets MAFB (v-maf musculoaponeurotic fibrosarcoma oncogene family, protein B), enhancing the transcription of IL-6 that stimulates the production of GLP-1 in alpha cells, which suppresses glucagon secretion from alpha cells and stimulates insulin secretion from beta cells in a glucose-dependent manner. Through this mechanism, miR-155-5p improves the adaptation of beta cells to insulin resistance and protection of islets from stress [169].

6.6. Gestational Diabetes

As discussed previously, the miR-29 family is regulated in multiple tissues. Although in most cases it has a deleterious and proinflammatory effect, in some organs the effect of miR-29 alleviates symptoms. In rats, miR-29b has a positive effect on gestational diabetes mellitus by targeting PI3K/Akt signal. Administration of miR-29 mimics reduced markers indicating oxidative stress, increased super oxide dismutase (SOD), catalase [165], and decreased malondialdehyde (MDA) in liver tissues of GDM rats [186]. Maternal diabetes and hyperglycemia dysregulate mitochondrial function through activation of protein kinase C (PKC) isoforms that have a role in the diabetic embryopathy. One of the isoforms of PKCα upregulates expression of miR-129-2, which targets the PGC-1α, the ligand of PPAR alpha (peroxisome proliferator activated receptor alpha). PGC1 alpha is a positive regulator of mitochondrial function and its downregulation by miR-129-2 mediates teratogenicity of hyperglycemia leading to NTDs (Neural tube defects in embryos) [131]. On the other hand, in the case of oxidative stress induced in embryo by maternal diabetes, inhibition of miR-27a increases NRF2 expression, which restores the homeostasis [129].

More recently, specific circular RNAs (circRNAs) interacting with miRNAs were identified in placentas from women with gestation diabetes mellitus that may regulate the AGE–RAGE interaction [174]. The circRNAs have their 5' end and 3' end covalently bond and are generated by a process known as back splicing, in which an upstream splice acceptor is joined to a downstream splice donor. They are expressed in various types of cells and tissues and, although little is known about their biological role, some act as gene regulators. In particular, several circRNAs have been described as acting as miRNA silencers or "sponges" by containing miRNA target sequences, in different type of cells including beta cells [187,188]. The differentially expressed circRNAs have been analyzed by Kyoto Encyclopedia of Genes and Genomes (KEGG) enrichment and circRNA–miRNA interaction, according to the sponge molecular interaction. The KEGG analysis predicted that circRNAs are likely to be involved in advanced glycation end products receptor for advanced glycation end products, AGE-RAGE, signaling pathways in diabetic complications. The expression of three circRNAs, circ-5824, circ-3636, and circ-0395, are downregulated in placentas of GDM. The circRNA–miRNA interaction analysis showed that miR-1273g-3p activated by acute glucose fluctuation is also involved in the progression of several complications caused by diabetes and it could be a potential gene of interest in GDM [174].

Figure 2 shows a scheme depicting the group of selected miRNAs described above and in Table 1 with their role in regulation of oxidative stress in diabetes

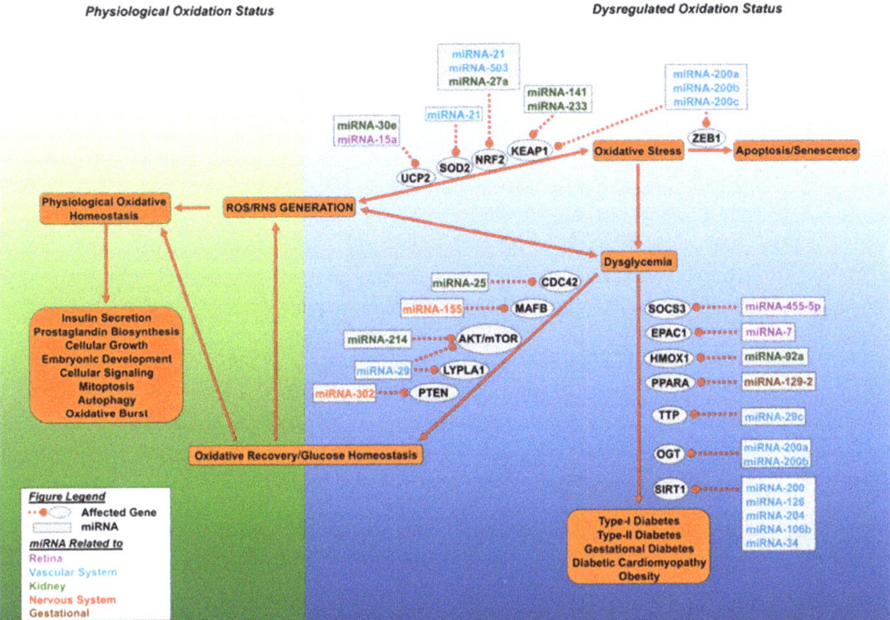

Figure 2. Dysregulated oxidative stress and microRNAs result in loss of glucose homeostasis. This figure outlines the effect of aberrant accumulation of cellular reactive oxygen species (ROS) and reactive nitrogen species (RNS). Cellular oxidative status is maintained by SOD2, NRF2, and UCP2, which allows for a spectrum of physiological functions carried out by the cell. Excessive ROS and RNS generation led to dysglycemia or cellular senescence. The miRNA molecules can target NRF2 (miRNA-21, miRNA-27a, miRNA-503, miRNA-233), SOD2 (miRNA-21), and UCP2 (miR-30e and miR-15a), leading to loss of oxidative regulation and the initiation of oxidative stress. Cellular oxidative stress can lead to either dysglycemia or cellular senescence. Cellular senescence is mediated by the inhibition of zinc finger E-box binding homeobox 1 (ZEB1) by miR-200 family miRNAs. Dysglycemia develops when O-linked β-N-acetylglucosamine transferase (OGT) and NAD-dependent deacetylase sirtuin-1 (SIRT1) are targeted by specific miRNAs. Oxidative stress driven dysglycemia rapidly initiates the expression of miRNA molecules which target suppressor of cytokine signaling 3 (SOCS3), exchange factor directly activated by cAMP 1 (EPAC1), and heme oxygenase (decycling) 1 (HMOX1), Peroxisome proliferator-activated receptor alpha (PPARA), mitochondrial uncoupling protein 2 (UCP2), and tristetraprolin (TTP), leading to decreased expression of these genes and the advance of diabetes. Alternatively, recovery can occur by miRNA directed targeting of genes involved in dysglycemia, they include: Cell division control protein 42 homolog (CDC42), V-maf musculoaponeurotic fibrosarcoma oncogene homolog B (MAFB), protein kinase B and mammalian target of rapamycin (AKT/mTOR), acyl-protein thioesterase 1 (LYPLA1) and phosphatase and tensin homolog (PTEN). Recovery of glucose homeostasis results in oxidative normalization and cellular homeostasis. Different colors of miRNA denote affected organ.

7. Conclusions

In diabetes, hyperglycemia induces intense oxidative stress that can no longer be modulated by the cellular antioxidative response, thus leading to accumulation of ROS. Overall, this process causes pancreatic beta cell dysfunction and unpaired glucose tolerance response, both of which have a deleterious effect on many types of cells and tissues. miRNAs have a critical role in the molecular mechanism involved in this process. Many of the studies reviewed here were performed in in vitro with animal cell lines or primary cells, in animal models (some in combination with tissues), some *in silico*, and a few cases in human tissues. It is expected that the development of new transgenic mice to study the role of miRNAs in oxidative stress will be useful to confirm or even discover novel potential targets and cellular pathways. However, the real challenge is the translation of all the in vitro, *in silico*, and animal model discovery to human diabetes. Although animal models, especially rodents, have been very useful for obtaining the basic information on the mechanism of several diseases, it is also true that the translation to human disease is not always straightforward. Specifically, many strategies were successful in treating autoimmune diabetes in rodent models, but none of them had been successful in treating human T1D. Furthermore, human basic and clinical research should aim at developing new strategies focusing on miRNAs and their target genes to cure diabetes and its complications. One of the emerging strategies is the use of a combination of human primary cells derived from human stem cell differentiation and organoid cultures plus genome editing alternatives to investigate the causes and role of miRNAs in oxidative stress in diabetes, as well as to screen for potential drugs to treat or alleviate its effects. However, it is important to remember that, currently, therapeutic approaches based on manipulation of miRNA expression are more effective in vitro than in vivo because of difficulties with specific delivery. As we have presented in this review, miRNAs are of variable nature, depending very much on the external and internal triggers. Therefore, it is of utmost importance to determine their specific targets and approach the treatment from that direction.

Author Contributions: M.M.F.Q., conceptualization, writing and design of figures; D.K., writing and critical reading; S.Á.-C., writing and critical reading; J.D.-B., writing and critical reading; R.L.P., conceptualization, writing, and revision.

Funding: The corresponding authors are funded by the Diabetes Research Institute Foundation (DRIF), the Inserra family, the Fred and Mabel R. Parks Foundation, the Tonkinson Foundation, ADA grant #1-19-ICTS-078, and NIH grants 1R43DK105655-01, 2R44 DK105655-02, and U01DK120393 (Human Islets Research Consortium (HIRN)). The MMFQ is funded by an International Fulbright predoctoral fellowship and grant administered by the Foreign Fulbright Scholarship Board and the International Institute of Education.

Conflicts of Interest: The authors do not report any conflict of interest.

References

1. World Health Organization. *Global Report on Diabetes*; World Health Organization: Geneva, Switzerland, 2018.

2. Saito, K.; Iwama, N.; Takahashi, T. Morphometrical Analysis on Topographical Difference in Size Distribution, Number and Volume of Islets in the Human Pancreas. *Tohoku J. Exp. Med.* **1978**, *124*, 177–186. [CrossRef] [PubMed]

3. Burrack, A.L.; Martinov, T.; Fife, B.T. T Cell-Mediated Beta Cell Destruction: Autoimmunity and Alloimmunity in the Context of Type 1 Diabetes. *Front. Endocrinol.* **2017**, *8*, 343. [CrossRef] [PubMed]

4. Cantley, J.; Ashcroft, F.M. Q&A: Insulin secretion and type 2 diabetes: Why do beta-cells fail? *BMC Biol.* **2015**, *13*, 33.

5. Meier, J.J.; Bonadonna, R.C. Role of reduced beta-cell mass versus impaired beta-cell function in the pathogenesis of type 2 diabetes. *Diabetes Care* **2013**, *36*, S113–S119. [CrossRef] [PubMed]

6. Kaufman, F.R. Type 1 diabetes mellitus. *Pediatr. Rev.* **2003**, *24*, 291–300. [CrossRef]

7. Soltész, G. Diabetes in the young: A paediatric and epidemiological perspective. *Diabetologia* **2003**, *46*, 447–454. [CrossRef]

8. Butler, P.C. The replication of beta cells in normal physiology, in disease and for therapy. *Nat. Clin. Pract. Endocrinol. Metab.* **2007**, *3*, 758–768. [CrossRef]

9. Gregg, B.E. Formation of a human beta-cell population within pancreatic islets is set early in life. *J. Clin. Endocrinol. Metab.* **2012**, *97*, 3197–3206. [CrossRef]

10. Chen, C.; Cohrs, C.M.; Stertmann, J.; Bozsak, R.; Speier, S. Human beta cell mass and function in diabetes: Recent advances in knowledge and technologies to understand disease pathogenesis. *Mol. Metab.* **2017**, *6*, 943–957. [CrossRef]

11. Ginsberg, H.N.; Maccallum, P.R. The obesity, metabolic syndrome, and type 2 diabetes mellitus pandemic: Part I. Increased cardiovascular disease risk and the importance of atherogenic dyslipidemia in persons with the metabolic syndrome and type 2 diabetes mellitus. *J. CardioMetabolic Syndr.* **2009**, *4*, 113–119. [CrossRef]

12. Shin, J.-A.; Lee, J.-H.; Lim, S.-Y.; Ha, H.-S.; Kwon, H.-S.; Park, Y.; Lee, W.-C.; Kang, M.-I.; Yim, H.-W.; Yoon, K.-H.; et al. Metabolic syndrome as a predictor of type 2 diabetes, and its clinical interpretations and usefulness. *J. Diabetes Investig.* **2013**, *4*, 334–343. [CrossRef] [PubMed]

13. Bartel, D.P. MicroRNAs: Genomics, biogenesis, mechanism, and function. *Cell* **2004**, *116*, 281–297. [CrossRef]

14. Reinhart, B.J.; Slack, F.J.; Basson, M.; Pasquinelli, A.E.; Bettinger, J.C.; Rougvie, A.E.; Horvitz, H.R.; Ruvkun, G. The 21-nucleotide let-7 RNA regulates developmental timing in Caenorhabditis elegans. *Nature* **2000**, *403*, 901–906. [CrossRef] [PubMed]

15. Wightman, B.; Ha, I.; Ruvkun, G. Posttranscriptional regulation of the heterochronic gene lin-14 by lin-4 mediates temporal pattern formation in C. elegans. *Cell* **1993**, *75*, 855–862. [CrossRef]

16. Lee, R.; Gowrishankar, T.; Basch, R.; Patel, P.; Golan, D. Cell shape-dependent rectification of surface receptor transport in a sinusoidal electric field. *Biophys. J.* **1993**, *64*, 44–57. [CrossRef]

17. Pasquinelli, A.E.; Reinhart, B.J.; Slack, F.; Martindale, M.Q.; Kuroda, M.I.; Maller, B.; Hayward, D.C.; Ball, E.E.; Degnan, B.; Müller, P.; et al. Conservation of the sequence and temporal expression of let-7 heterochronic regulatory RNA. *Nature* **2000**, *408*, 86–89. [CrossRef]

18. Davis, C.A. The Encyclopedia of DNA elements (ENCODE): Data portal update. *Nucleic Acids Res.* **2018**, *46*, D794–D801. [CrossRef]

19. Ambros, V. microRNAs: Tiny regulators with great potential. *Cell* **2001**, *107*, 823–826. [CrossRef]

20. Hinske, L.C.G.; Galante, P.A.; Kuo, W.P.; Ohno-Machado, L. A potential role for intragenic miRNAs on their hosts' interactome. *BMC Genom.* **2010**, *11*, 533. [CrossRef]

21. Cifuentes, D.; Xue, H.; Taylor, D.W.; Patnode, H.; Mishima, Y.; Cheloufi, S.; Ma, E.; Mane, S.; Hannon, G.J.; Lawson, N.D.; et al. A novel miRNA processing pathway independent of Dicer requires Argonaute2 catalytic activity. *Science* **2010**, *328*, 1694–1698. [CrossRef]

22. Liu, Y.P.; Karg, M.; Harwig, A.; Herrera-Carrillo, E.; Jongejan, A.; Van Kampen, A.; Berkhout, B. Mechanistic insights on the Dicer-independent AGO2-mediated processing of AgoshRNAs. *RNA Boil.* **2015**, *12*, 92–100. [CrossRef] [PubMed]

23. Barad, O.; Mann, M.; Chapnik, E.; Shenoy, A.; Blelloch, R.; Barkai, N.; Hornstein, E. Efficiency and specificity in microRNA biogenesis. *Nat. Struct. Mol. Boil.* **2012**, *19*, 650–652. [CrossRef] [PubMed]

24. Hammond, S.M. An overview of microRNAs. *Adv. Drug Deliv. Rev.* **2015**, *87*, 3–14. [CrossRef] [PubMed]

25. Yekta, S.; Shih, I.-H.; Bartel, D.P. MicroRNA-Directed Cleavage of HOXB8 mRNA. *Science* **2004**, *304*, 594–596. [CrossRef]

26. Sarshad, A.A.; Juan, A.H.; Muler, A.I.C.; Anastasakis, D.G.; Wang, X.; Genzor, P.; Feng, X.; Tsai, P.-F.; Sun, H.-W.; Haase, A.D.; et al. Argonaute-miRNA Complexes Silence Target mRNAs in the Nucleus of Mammalian Stem Cells. *Mol. Cell* **2018**, *71*, 1040–1050. [CrossRef]

27. Ohanian, M.; Humphreys, D.T.; Anderson, E.; Preiss, T.; Fatkin, D. A heterozygous variant in the human cardiac miR-133 gene, MIR133A2, alters miRNA duplex processing and strand abundance. *BMC Genet.* **2013**, *14*, 18. [CrossRef]

28. Meijer, H.A.; Smith, E.M.; Bushell, M. Regulation of miRNA strand selection: Follow the leader? *Biochem. Soc. Trans.* **2014**, *42*, 1135–1140. [CrossRef]

29. Friedman, R.C. Most mammalian mRNAs are conserved targets of microRNAs. *Genome Res.* **2009**, *19*, 92–105. [CrossRef]

30. Schirle, N.T.; Sheu-Gruttadauria, J.; Macrae, I.J. Structural basis for microRNA targeting. *Science* **2014**, *346*, 608–613. [CrossRef]

31. Kozomara, A.; Birgaoanu, M.; Griffiths-Jones, S. miRBase: From microRNA sequences to function. *Nucleic Acids Res.* **2019**, *47*, D155–D162. [CrossRef]

32. Ambros, V. The functions of animal microRNAs. *Nature* **2004**, *431*, 350–355. [CrossRef] [PubMed]

33. Bernstein, E.; Kim, S.Y.; Carmell, M.A.; Murchison, E.P.; Alcorn, H.; Li, M.Z.; Mills, A.A.; Elledge, S.J.; Anderson, K.V.; Hannon, G.J. Dicer is essential for mouse development. *Nat. Genet.* **2003**, *35*, 215–217. [CrossRef] [PubMed]

34. Sun, X.; Zhang, Z.; McManus, M.T.; Harfe, B.D.; Harris, K.S. Dicer function is essential for lung epithelium morphogenesis. *Dev. Boil.* **2006**, *295*, 460. [CrossRef]

35.	Kosik, K.S. The neuronal microRNA system. *Nat. Rev. Neurosci.* **2006**, *7*, 911–920. [CrossRef] [PubMed]

36.	Guay, C.; Regazzi, R. Circulating microRNAs as novel biomarkers for diabetes mellitus. *Nat. Rev. Endocrinol.* **2013**, *9*, 513–521. [CrossRef] [PubMed]

37.	Guay, C.; Regazzi, R. Exosomes as new players in metabolic organ cross-talk. *Diabetes Obes. Metab.* **2017**, *19*, 137–146. [CrossRef] [PubMed]

38.	Baroukh, N. MicroRNA-124a regulates Foxa2 expression and intracellular signaling in pancreatic beta-cell lines. *J. Biol. Chem.* **2007**, *282*, 19575–19588. [CrossRef]

39.	Joglekar, M.V.; Joglekar, V.M.; Hardikar, A.A. Expression of islet-specific microRNAs during human pancreatic development. *Gene Expr. Patterns* **2009**, *9*, 109–113. [CrossRef]

40.	Lynn, F.C.; Skewes-Cox, P.; Kosaka, Y.; McManus, M.; Harfe, B.D.; German, M.S. MicroRNA Expression Is Required for Pancreatic Islet Cell Genesis in the Mouse. *Diabetes* **2007**, *56*, 2938–2945. [CrossRef]

41.	Poy, M.N.; Eliasson, L.; Krutzfeldt, J.; Kuwajima, S.; Ma, X.; Macdonald, P.E.; Pfeffer, S.; Tuschl, T.; Rajewsky, N.; Rorsman, P.; et al. A pancreatic islet-specific microRNA regulates insulin secretion. *Nature* **2004**, *432*, 226–230. [CrossRef]

42.	Kaspi, H.; Pasvolsky, R.; Hornstein, E. Could microRNAs contribute to the maintenance of beta cell identity? *Trends Endocrinol. Metab.* **2014**, *25*, 285–292. [CrossRef] [PubMed]

43.	Kālis, M.; Bolmeson, C.; Esguerra, J.L.S.; Gupta, S.; Edlund, A.; Tormo-Badia, N.; Speidel, D.; Holmberg, D.; Mayans, S.; Khoo, N.K.S.; et al. Beta-Cell Specific Deletion of Dicer1 Leads to Defective Insulin Secretion and Diabetes Mellitus. *PLoS ONE* **2011**, *6*, e29166. [CrossRef] [PubMed]

44.	Mandelbaum, A.D.; Melkman-Zehavi, T.; Oren, R.; Kredo-Russo, S.; Nir, T.; Dor, Y.; Hornstein, E. Dysregulation of Dicer1 in Beta Cells Impairs Islet Architecture and Glucose Metabolism. *Exp. Diabetes Res.* **2012**, *2012*, 1–8. [CrossRef] [PubMed]

45.	Berry, C.; Lal, M.; Binukumar, B.K. Crosstalk Between the Unfolded Protein Response, MicroRNAs, and Insulin Signaling Pathways: In Search of Biomarkers for the Diagnosis and Treatment of Type 2 Diabetes. *Front. Endocrinol.* **2018**, *9*, 210. [CrossRef] [PubMed]

46.	Dotta, F. MicroRNAs: Markers of beta-cell stress and autoimmunity. *Curr. Opin. Endocrinol. Diabetes Obes.* **2018**, *25*, 237–245. [CrossRef]

47.	Feng, J.; Xing, W.; Xie, L. Regulatory Roles of MicroRNAs in Diabetes. *Int. J. Mol. Sci.* **2016**, *17*, 1729. [CrossRef]

48.	Lapierre, M.P.; Stoffel, M. MicroRNAs as stress regulators in pancreatic beta cells and diabetes. *Mol. Metab.* **2017**, *6*, 1010–1023. [CrossRef]

49.	Bravo-Egana, V. Quantitative differential expression analysis reveals miR-7 as major islet microRNA. *Biochem. Biophys. Res. Commun.* **2008**, *366*, 922–926. [CrossRef]

50.	Correa-Medina, M.; Bravo-Egana, V.; Rosero, S.; Ricordi, C.; Edlund, H.; Diez, J.; Pastori, R.L. MicroRNA miR-7 is preferentially expressed in endocrine cells of the developing and adult human pancreas. *Gene Expr. Patterns* **2009**, *9*, 193–199. [CrossRef]

51.	Rosero, S.; Bravo-Egana, V.; Jiang, Z.; Khuri, S.; Tsinoremas, N.; Klein, D.; Sabates, E.; Correa-Medina, M.; Ricordi, C.; Domínguez-Bendala, J.; et al. MicroRNA signature of the human developing pancreas. *BMC Genom.* **2010**, *11*, 509. [CrossRef]

52.	Klein, D.; Misawa, R.; Bravo-Egana, V.; Vargas, N.; Rosero, S.; Piroso, J.; Ichii, H.; Umland, O.; Zhijie, J.; Tsinoremas, N.; et al. MicroRNA Expression in Alpha and Beta Cells of Human Pancreatic Islets. *PLoS ONE* **2013**, *8*, e55064. [CrossRef] [PubMed]

53.	Kloosterman, W.P.; Lagendijk, A.K.; Ketting, R.F.; Moulton, J.D.; Plasterk, R.H.A. Targeted inhibition of miRNA maturation with morpholinos reveals a role for miR-375 in pancreatic islet development. *PLoS Boil.* **2007**, *5*, e203. [CrossRef] [PubMed]

54.	Kredo-Russo, S.; Ness, A.N.; Mandelbaum, A.D.; Walker, M.D.; Hornstein, E. Regulation of Pancreatic microRNA-7 Expression. *Exp. Diabetes Res.* **2012**, *2012*, 1–7. [CrossRef] [PubMed]

55.	Latreille, M. MicroRNA-7a regulates pancreatic beta cell function. *J. Clin. Investig.* **2014**, *124*, 2722–2735. [CrossRef] [PubMed]

56.	Nieto, M.; Hevia, P.; Garcia, E.; Klein, D.; Álvarez-Cubela, S.; Bravo-Egana, V.; Rosero, S.; Molano, R.D.; Vargas, N.; Ricordi, C.; et al. Antisense miR-7 Impairs Insulin Expression in Developing Pancreas and in Cultured Pancreatic Buds. *Cell Transplant.* **2012**, *21*, 1761–1774. [CrossRef] [PubMed]

57. Poy, M.N.; Hausser, J.; Trajkovski, M.; Braun, M.; Collins, S.; Rorsman, P.; Zavolan, M.; Stoffel, M. miR-375 maintains normal pancreatic alpha- and beta-cell mass. *Proc. Natl. Acad. Sci. USA* **2009**, *106*, 5813–5818. [CrossRef]

58. López-Beas, J.; Capilla-Gonzalez, V.; Aguilera, Y.; Mellado, N.; Lachaud, C.C.; Martín, F.; Smani, T.; Soria, B.; Hmadcha, A. miR-7 Modulates hESC Differentiation into Insulin-Producing Beta-like Cells and Contributes to Cell Maturation. *Mol. Ther. Nucleic Acids* **2018**, *12*, 463–477. [CrossRef]

59. Nathan, G. MiR-375 promotes redifferentiation of adult human beta cells expanded in vitro. *PLoS ONE* **2015**, *10*, e0122108. [CrossRef]

60. Shaer, A.; Azarpira, N.; Karimi, M.H. Differentiation of Human Induced Pluripotent Stem Cells into Insulin-Like Cell Clusters with miR-186 and miR-375 by using chemical transfection. *Appl. Biochem. Biotechnol.* **2014**, *174*, 242–258. [CrossRef]

61. Wei, R.; Yang, J.; Liu, G.-Q.; Gao, M.-J.; Hou, W.-F.; Zhang, L.; Gao, H.-W.; Liu, Y.; Chen, G.-A.; Hong, T.-P. Dynamic expression of microRNAs during the differentiation of human embryonic stem cells into insulin-producing cells. *Gene* **2013**, *518*, 246–255. [CrossRef]

62. Martinez-Sanchez, A.; Rutter, G.A.; Latreille, M. MiRNAs in beta-Cell Development, Identity, and Disease. *Front. Genet.* **2016**, *7*, 226. [PubMed]

63. Sies, H. Oxidative stress: Oxidants and antioxidants. *Exp. Physiol.* **1997**, *82*, 291–295. [CrossRef] [PubMed]

64. Sies, H.; Cadenas, E. Oxidative stress: Damage to intact cells and organs. *Philos Trans. R Soc. Lond. B Biol. Sci.* **1985**, *311*, 617–631. [CrossRef] [PubMed]

65. Cadenas, E.; Davies, K.J. Mitochondrial free radical generation, oxidative stress, and aging. *Free. Radic. Boil. Med.* **2000**, *29*, 222–230. [CrossRef]

66. Cheifetz, S.; Li, I.W.; McCulloch, C.A.; Sampath, K.; Sodek, J. Influence of osteogenic protein-1 (OP-1;BMP-7) and transforming growth factor-beta 1 on bone formation in vitro. *Connect. Tissue Res.* **1996**, *35*, 71–78. [CrossRef] [PubMed]

67. Atari-Hajipirloo, S.; Valizadeh, N.; Khadem-Ansari, M.-H.; Rasmi, Y.; Kheradmand, F. Altered Concentrations of Copper, Zinc, and Iron are Associated With Increased Levels of Glycated Hemoglobin in Patients With Type 2 Diabetes Mellitus and Their First-Degree Relatives. *Int. J. Endocrinol. Metab.* **2016**, *14*, 33273. [CrossRef]

68. Giulivi, C.; Boveris, A.; Cadenas, E. Hydroxyl Radical Generation during Mitochondrial Electron-Transfer and the Formation of 8-Hydroxydesoxyguanosine in Mitochondrial-DNA. *Arch. Biochem. Biophys.* **1995**, *316*, 909–916. [CrossRef]

69. Pi, J.; Bai, Y.; Zhang, Q.; Wong, V.; Floering, L.M.; Daniel, K.; Reece, J.M.; Deeney, J.T.; Andersen, M.E.; Corkey, B.E.; et al. Reactive Oxygen Species as a Signal in Glucose-Stimulated Insulin Secretion. *Diabetes* **2007**, *56*, 1783–1791. [CrossRef]

70. Roy, J.; Galano, J.-M.; Durand, T.; Le Guennec, J.-Y.; Lee, J.C.-Y. Physiological role of reactive oxygen species as promoters of natural defenses. *FASEB J.* **2017**, *31*, 3729–3745. [CrossRef]

71. Pizzino, G.; Irrera, N.; Cucinotta, M.; Pallio, G.; Mannino, F.; Arcoraci, V.; Squadrito, F.; Altavilla, D.; Bitto, A. Oxidative Stress: Harms and Benefits for Human Health. *Oxidative Med. Cell. Longev.* **2017**, *2017*, 1–13. [CrossRef]

72. Di Meo, S.; Reed, T.T.; Venditti, P.; Victor, V.M. Role of ROS and RNS Sources in Physiological and Pathological Conditions. *Oxidative Med. Cell. Longev.* **2016**, *2016*, 1–44. [CrossRef] [PubMed]

73. Barrera, G.; Pizzimenti, S.; Daga, M.; Dianzani, C.; Arcaro, A.; Cetrangolo, G.P.; Giordano, G.; Cucci, M.A.; Graf, M.; Gentile, F. Lipid Peroxidation-Derived Aldehydes, 4-Hydroxynonenal and Malondialdehyde in Aging-Related Disorders. *Antioxidants* **2018**, *7*, 102. [CrossRef] [PubMed]

74. Vaca, C.; Wilhelm, J.; Harms-Ringdahl, M. Interaction of lipid peroxidation products with DNA. A review. *Mutat. Res. Genet. Toxicol.* **1988**, *195*, 137–149. [CrossRef]

75. Ayala, A.; Muñoz, M.F.; Argüelles, S. Lipid Peroxidation: Production, Metabolism, and Signaling Mechanisms of Malondialdehyde and 4-Hydroxy-2-Nonenal. *Oxidative Med. Cell. Longev.* **2014**, *2014*, 1–31. [CrossRef] [PubMed]

76. Schieber, M.; Chandel, N.S. ROS function in redox signaling and oxidative stress. *Curr. Boil.* **2014**, *24*, R453–R462. [CrossRef]

77. Gaschler, M.M.; Stockwell, B.R. Lipid peroxidation in cell death. *Biochem. Biophys. Res. Commun.* **2017**, *482*, 419–425. [CrossRef]

78. Cadet, J.; Wagner, J.R. DNA Base Damage by Reactive Oxygen Species, Oxidizing Agents, and UV Radiation. *Cold Spring Harb. Perspect. Boil.* **2013**, *5*, a012559. [CrossRef]

79. Lenzen, S. Chemistry and biology of reactive species with special reference to the antioxidative defence status in pancreatic beta-cells. *Biochim. Biophys. Acta Gen. Subj.* **2017**, *1861*, 1929–1942. [CrossRef]

80. Shen, X.; Zheng, S.; Metreveli, N.S.; Epstein, P.N. Protection of cardiac mitochondria by overexpression of MnSOD reduces diabetic cardiomyopathy. *Diabetes* **2006**, *55*, 798–805. [CrossRef]

81. Ye, G.; Metreveli, N.S.; Donthi, R.V.; Xia, S.; Xu, M.; Carlson, E.C.; Epstein, P.N. Catalase protects cardiomyocyte function in models of type 1 and type 2 diabetes. *Diabetes* **2004**, *53*, 1336–1343. [CrossRef]

82. Brand, M.D.; Esteves, T.C. Physiological functions of the mitochondrial uncoupling proteins UCP2 and UCP3. *Cell Metab.* **2005**, *2*, 85–93. [CrossRef] [PubMed]

83. Pierelli, G.; Stanzione, R.; Forte, M.; Migliarino, S.; Perelli, M.; Volpe, M.; Rubattu, S. Uncoupling Protein 2: A Key Player and a Potential Therapeutic Target in Vascular Diseases. *Oxidative Med. Cell. Longev.* **2017**, *2017*, 1–11. [CrossRef] [PubMed]

84. Giralt, M.; Villarroya, F. Mitochondrial Uncoupling and the Regulation of Glucose Homeostasis. *Curr. Diabetes Rev.* **2017**, *13*, 386–394. [CrossRef] [PubMed]

85. Tonelli, C.; Chio, I.I.C.; Tuveson, D.A. Transcriptional Regulation by Nrf2. *Antioxid. Redox Signal* **2018**, *29*, 1727–1745. [CrossRef] [PubMed]

86. Assar, M.E.; Angulo, J.; Rodriguez-Manas, L. Diabetes and ageing-induced vascular inflammation. *J. Physiol.* **2016**, *594*, 2125–2146. [CrossRef]

87. Rabbani, P.S. Dysregulation of Nrf2/Keap1 Redox Pathway in Diabetes Affects Multipotency of Stromal Cells. *Diabetes* **2019**, *68*, 141–155. [CrossRef] [PubMed]

88. Simmons, K.M.; Michels, A.W. Type 1 diabetes: A predictable disease. *World J. Diabetes* **2015**, *6*, 380–390. [CrossRef]

89. Veld, P.I. Insulitis in human type 1 diabetes: A comparison between patients and animal models. *Semin. Immunopathol.* **2014**, *36*, 569–579. [CrossRef]

90. Nakayama, M. Insulin as a key autoantigen in the development of type 1 diabetes. *Diabetes Metab. Res. Rev.* **2011**, *27*, 773–777. [CrossRef]

91. Delmastro, M.M.; Piganelli, J.D. Oxidative Stress and Redox Modulation Potential in Type 1 Diabetes. *Clin. Dev. Immunol.* **2011**, *2011*, 1–15. [CrossRef]

92. Atkinson, M.A.; Eisenbarth, G.S. Type 1 diabetes: New perspectives on disease pathogenesis and treatment. *Lancet* **2001**, *358*, 221–229. [CrossRef]

93. Padgett, L.E.; Broniowska, K.A.; Hansen, P.A.; Corbett, J.A.; Tse, H.M. The role of reactive oxygen species and proinflammatory cytokines in type 1 diabetes pathogenesis. *Ann. N. Y. Acad. Sci.* **2013**, *1281*, 16–35. [CrossRef] [PubMed]

94. Feduska, J.M.; Tse, H.M. The proinflammatory effects of macrophage-derived NADPH oxidase function in autoimmune diabetes. *Free. Radic. Boil. Med.* **2018**, *125*, 81–89. [CrossRef] [PubMed]

95. Ali, M.K. Diabetes: An Update on the Pandemic and Potential Solutions. In *Cardiovascular, Respiratory, and Related Disorders*; World Bank and Oxford University Press: Washington, DC, USA, 2017.

96. Heitmeier, M.R. Pancreatic beta-cell damage mediated by beta-cell production of interleukin-1. A novel mechanism for virus-induced diabetes. *J. Biol. Chem.* **2001**, *276*, 11151–11158. [CrossRef]

97. Meares, G.P. Differential responses of pancreatic beta-cells to ROS and RNS. *Am. J. Physiol. Endocrinol. Metab.* **2013**, *304*, E614–E622. [CrossRef] [PubMed]

98. Oleson, B.J. Nitric Oxide Suppresses beta-Cell Apoptosis by Inhibiting the DNA Damage Response. *Mol. Cell Biol.* **2016**, *36*, 2067–2077. [CrossRef] [PubMed]

99. Pugliese, A. Insulitis in the pathogenesis of type 1 diabetes. *Pediatr. Diabetes* **2016**, *17*, 31–36. [CrossRef]

100. Matteucci, E.; Cinapri, V.; Quilici, S.; Forotti, G.; Giampietro, O. Oxidative stress in families of type 1 diabetic patients. *Diabetes Res. Clin. Pr.* **2000**, *50*, 308. [CrossRef]

101. Saklayen, M.G. The Global Epidemic of the Metabolic Syndrome. *Curr. Hypertens. Rep.* **2018**, *20*, 12. [CrossRef]

102. Newsholme, P. Oxidative stress pathways in pancreatic beta cells and insulin sensitive cells and tissues-importance to cell metabolism, function and dysfunction. *Am. J. Physiol. Cell Physiol.* **2019**. [CrossRef]

103. Evans, J.L.; Goldfine, I.D.; Maddux, B.A.; Grodsky, G.M. Are oxidative stress-activated signaling pathways mediators of insulin resistance and beta-cell dysfunction? *Diabetes* **2003**, *52*, 1–8. [CrossRef] [PubMed]

104. Gao, D.; Nong, S.; Huang, X.; Lu, Y.; Zhao, H.; Lin, Y.; Man, Y.; Wang, S.; Yang, J.; Li, J. The Effects of Palmitate on Hepatic Insulin Resistance Are Mediated by NADPH Oxidase 3-derived Reactive Oxygen Species through JNK and p38MAPK Pathways*. *J. Boil. Chem.* **2010**, *285*, 29965–29973. [CrossRef] [PubMed]

105. Guo, X.-X.; An, S.; Yang, Y.; Liu, Y.; Hao, Q.; Tang, T.; Xu, T.-R. Emerging role of the Jun N-terminal kinase interactome in human health. *Cell Boil. Int.* **2018**, *42*, 756–768. [CrossRef] [PubMed]

106. Patel, S.; Santani, D. Role of NF-kappa B in the pathogenesis of diabetes and its associated complications. *Pharmacol. Rep.* **2009**, *61*, 595–603. [CrossRef]

107. Chilelli, N.C.; Burlina, S.; Lapolla, A. AGEs, rather than hyperglycemia, are responsible for microvascular complications in diabetes: A "glycoxidation-centric" point of view. *Nutr. Metab. Cardiovasc. Dis.* **2013**, *23*, 913–919. [CrossRef] [PubMed]

108. Jud, P.; Sourij, H.; Philipp, J.; Harald, S. Therapeutic options to reduce advanced glycation end products in patients with diabetes mellitus: A review. *Diabetes Res. Clin. Pr.* **2019**, *148*, 54–63. [CrossRef] [PubMed]

109. Brandon, A.E.; Liao, B.M.; Diakanastasis, B.; Parker, B.L.; Raddatz, K.; McManus, S.A.; O'Reilly, L.; Kimber, E.; Van Der Kraan, A.G.; Hancock, D.; et al. Protein Kinase C Epsilon Deletion in Adipose Tissue, but Not in Liver, Improves Glucose Tolerance. *Cell Metab.* **2019**, *29*, 183–191. [CrossRef]

110. Fleming, A.K.; Storz, P. Protein kinase C isoforms in the normal pancreas and in pancreatic disease. *Cell. Signal.* **2017**, *40*, 1–9. [CrossRef]

111. Moser, B.; Schumacher, C.; Von Tscharner, V.; Clark-Lewis, I.; Baggiolini, M. Neutrophil-activating peptide 2 and gro/melanoma growth-stimulatory activity interact with neutrophil-activating peptide 1/interleukin 8 receptors on human neutrophils. *J. Boil. Chem.* **1991**, *266*, 10666–10671.

112. Giacco, F.; Brownlee, M. Oxidative stress and diabetic complications. *Circ. Res.* **2010**, *107*, 1058–1070. [CrossRef]

113. Yan, L.-J. Redox imbalance stress in diabetes mellitus: Role of the polyol pathway. *Anim. Model. Exp. Med.* **2018**, *1*, 7–13. [CrossRef] [PubMed]

114. Isaacs, S.R. MicroRNAs in Type 1 Diabetes: Complex Interregulation of the Immune System, beta Cell Function and Viral Infections. *Curr. Diabetes Rep.* **2016**, *16*, 133. [CrossRef] [PubMed]

115. Pileggi, A.; Klein, D.; Fotino, C.; Bravo-Egana, V.; Rosero, S.; Doni, M.; Podetta, M.; Ricordi, C.; Molano, R.D.; Pastori, R.L. MicroRNAs in islet immunobiology and transplantation. *Immunol. Res.* **2013**, *57*, 185–196. [CrossRef] [PubMed]

116. Zheng, Y.; Wang, Z.; Zhou, Z. miRNAs: Novel regulators of autoimmunity-mediated pancreatic beta-cell destruction in type 1 diabetes. *Cell Mol. Immunol.* **2017**, *14*, 488–496. [CrossRef]

117. Engedal, N.; Žerovnik, E.; Rudov, A.; Galli, F.; Olivieri, F.; Procopio, A.D.; Rippo, M.R.; Monsurrò, V.; Betti, M.; Albertini, M.C. From Oxidative Stress Damage to Pathways, Networks, and Autophagy via MicroRNAs. *Oxidative Med. Cell. Longev.* **2018**, *2018*, 4968321. [CrossRef]

118. Yang, X.; Li, X.; Lin, Q.; Xu, Q. Up-regulation of microRNA-203 inhibits myocardial fibrosis and oxidative stress in mice with diabetic cardiomyopathy through the inhibition of PI3K/Akt signaling pathway via PIK3CA. *Gene* **2019**, *715*, 143995. [CrossRef]

119. Dieter, C.; Assmann, T.S.; Costa, A.R.; Canani, L.H.; De Souza, B.M.; Bauer, A.C.; Crispim, D. MiR-30e-5p and MiR-15a-5p Expressions in Plasma and Urine of Type 1 Diabetic Patients With Diabetic Kidney Disease. *Front. Genet.* **2019**, *10*, 563. [CrossRef]

120. Chen, P.; Miao, Y.; Yan, P.; Wang, X.J.; Jiang, C.; Lei, Y. MiR-455-5p ameliorates HG-induced apoptosis, oxidative stress and inflammatory via targeting SOCS3 in retinal pigment epithelial cells. *J. Cell. Physiol.* **2019**, *234*, 21915–21924. [CrossRef]

121. Yang, S.; Fei, X.; Lu, Y.; Xu, B.; Ma, Y.; Wan, H. miRNA-214 suppresses oxidative stress in diabetic nephropathy via the ROS/Akt/mTOR signaling pathway and uncoupling protein 2. *Exp. Ther. Med.* **2019**, *17*, 3530–3538. [CrossRef]

122. Sun, L.-L.; Jiang, B.-G.; Li, W.-T.; Zou, J.-J.; Shi, Y.-Q.; Liu, Z.-M. MicroRNA-15a positively regulates insulin synthesis by inhibiting uncoupling protein-2 expression. *Diabetes Res. Clin. Pr.* **2011**, *91*, 94–100. [CrossRef]

123. Jiang, L. A microRNA-30e/mitochondrial uncoupling protein 2 axis mediates TGF-beta1-induced tubular epithelial cell extracellular matrix production and kidney fibrosis. *Kidney Int.* **2013**, *84*, 285–296. [CrossRef] [PubMed]

124. Yin, Z. MiR-30c/PGC-1beta protects against diabetic cardiomyopathy via PPARalpha. *Cardiovasc. Diabetol.* **2019**, *18*, 7. [CrossRef] [PubMed]

125. Ding, X.; Jian, T.; Wu, Y.; Zuo, Y.; Li, J.; Lv, H.; Ma, L.; Ren, B.; Zhao, L.; Li, W.; et al. Ellagic acid ameliorates oxidative stress and insulin resistance in high glucose-treated HepG2 cells via miR-223/keap1-Nrf2 pathway. *Biomed. Pharmacother.* **2019**, *110*, 85–94. [CrossRef]

126. Kubota, K.; Nakano, M.; Kobayashi, E.; Mizue, Y.; Chikenji, T.; Otani, M.; Nagaishi, K.; Fujimiya, M. An enriched environment prevents diabetes-induced cognitive impairment in rats by enhancing exosomal miR-146a secretion from endogenous bone marrow-derived mesenchymal stem cells. *PLoS ONE* **2018**, *13*, e0204252. [CrossRef] [PubMed]

127. Wan, R.J.; Li, Y.H. MicroRNA-146a/NAPDH oxidase4 decreases reactive oxygen species generation and inflammation in a diabetic nephropathy model. *Mol. Med. Rep.* **2018**, *17*, 4759–4766. [CrossRef] [PubMed]

128. Miao, Y.; Wan, Q.; Liu, X.; Wang, Y.; Luo, Y.; Liu, D.; Lin, N.; Zhou, H.; Zhong, J. miR-503 Is Involved in the Protective Effect of Phase II Enzyme Inducer (CPDT) in Diabetic Cardiomyopathy via Nrf2/ARE Signaling Pathway. *BioMed Res. Int.* **2017**, *2017*, 1–10. [CrossRef] [PubMed]

129. Zhao, Y. Oxidative stress-induced miR-27a targets the redox gene nuclear factor erythroid 2-related factor 2 in diabetic embryopathy. *Am. J. Obstet. Gynecol.* **2018**, *218*, 136 e1–136 e10. [CrossRef]

130. Wang, J.; Zhang, J.; Chen, X.; Yang, Y.; Wang, F.; Li, W.; Awuti, M.; Sun, Y.; Lian, C.; Li, Z.; et al. miR-365 promotes diabetic retinopathy through inhibiting Timp3 and increasing oxidative stress. *Exp. Eye Res.* **2018**, *168*, 89–99. [CrossRef]

131. Wang, F.; Xu, C.; Reece, E.A.; Li, X.; Wu, Y.; Harman, C.; Yu, J.; Dong, D.; Wang, C.; Yang, P.; et al. Protein kinase C-alpha suppresses autophagy and induces neural tube defects via miR-129-2 in diabetic pregnancy. *Nat. Commun.* **2017**, *8*, 15182. [CrossRef]

132. Chen, D.-L.; Yang, K.-Y.; Chen, D.; Yang, K. Berberine Alleviates Oxidative Stress in Islets of Diabetic Mice by Inhibiting miR-106b Expression and Up-Regulating SIRT1. *J. Cell. Biochem.* **2017**, *118*, 4349–4357. [CrossRef]

133. Wu, Y. MiR-106a Associated with Diabetic Peripheral Neuropathy through the Regulation of 12/15-LOX-meidiated Oxidative/Nitrative Stress. *Curr. Neurovasc. Res.* **2017**, *14*, 117–124. [CrossRef] [PubMed]

134. Garcia-Morales, V.; Friedrich, J.; Jorna, L.M.; Campos-Toimil, M.; Hammes, H.-P.; Schmidt, M.; Krenning, G. The microRNA-7-mediated reduction in EPAC-1 contributes to vascular endothelial permeability and eNOS uncoupling in murine experimental retinopathy. *Acta Diabetol.* **2017**, *54*, 581–591. [CrossRef] [PubMed]

135. Li, H.H. miR-302 Attenuates Amyloid-beta-Induced Neurotoxicity through Activation of Akt Signaling. *J. Alzheimers Dis.* **2016**, *50*, 1083–1098. [CrossRef] [PubMed]

136. Xiao, D.; Zhou, T.; Fu, Y.; Wang, R.; Zhang, H.; Li, M.; Lin, Y.; Li, Z.; Xu, C.; Yang, B.; et al. MicroRNA-17 impairs glucose metabolism in insulin-resistant skeletal muscle via repressing glucose transporter 4 expression. *Eur. J. Pharmacol.* **2018**, *838*, 170–176. [CrossRef] [PubMed]

137. Hui, Y.; Yin, Y. MicroRNA-145 attenuates high glucose-induced oxidative stress and inflammation in retinal endothelial cells through regulating TLR4/NF-kappaB signaling. *Life Sci.* **2018**, *207*, 212–218. [CrossRef] [PubMed]

138. Zhang, Y.; Song, C.; Liu, J.; Bi, Y.; Li, H. Inhibition of miR-25 aggravates diabetic peripheral neuropathy. *NeuroReport* **2018**, *29*, 945–953. [CrossRef] [PubMed]

139. Liu, Y.; Li, H.; Liu, J.; Han, P.; Li, X.; Bai, H.; Zhang, C.; Sun, X.; Teng, Y.; Zhang, Y.; et al. Variations in MicroRNA-25 Expression Influence the Severity of Diabetic Kidney Disease. *J. Am. Soc. Nephrol.* **2017**, *28*, 3627–3638. [CrossRef]

140. Li, H.; Zhu, X.; Zhang, J.; Shi, J. MicroRNA-25 inhibits high glucose-induced apoptosis in renal tubular epithelial cells via PTEN/AKT pathway. *Biomed. Pharmacother.* **2017**, *96*, 471–479. [CrossRef]

141. Filios, S.R.; Xu, G.; Chen, J.; Hong, K.; Jing, G.; Shalev, A. MicroRNA-200 Is Induced by Thioredoxin-interacting Protein and Regulates Zeb1 Protein Signaling and Beta Cell Apoptosis*. *J. Boil. Chem.* **2014**, *289*, 36275–36283. [CrossRef]

142. Zhang, H.; Liu, J.; Qu, D.; Wang, L.; Luo, J.-Y.; Lau, C.W.; Gao, Z.; Tipoe, G.L.; Lee, H.K.; Ng, C.F.; et al. Inhibition of miR-200c restores endothelial function in diabetic mice through suppression of COX-2. *Diabetes* **2016**, *65*, 1196–1207. [CrossRef]

143. Lo, W.-Y.; Yang, W.-K.; Peng, C.-T.; Pai, W.-Y.; Wang, H.-J. MicroRNA-200a/200b Modulate High Glucose-Induced Endothelial Inflammation by Targeting O-linked N-Acetylglucosamine Transferase Expression. *Front. Physiol.* **2018**, *9*, 355. [CrossRef] [PubMed]

144. Shang, F.; Wang, S.-C.; Hsu, C.-Y.; Miao, Y.; Martin, M.; Yin, Y.; Wu, C.-C.; Wang, Y.-T.; Wu, G.; Chien, S.; et al. MicroRNA-92a Mediates Endothelial Dysfunction in CKD. *J. Am. Soc. Nephrol.* **2017**, *28*, 3251–3261. [CrossRef] [PubMed]

145. Gou, L.; Zhao, L.; Song, W.; Wang, L.; Liu, J.; Zhang, H.; Huang, Y.; Lau, C.W.; Yao, X.; Tian, X.Y.; et al. Inhibition of miR-92a Suppresses Oxidative Stress and Improves Endothelial Function by Upregulating Heme Oxygenase-1 in db/db Mice. *Antioxid. Redox Signal.* **2018**, *28*, 358–370. [CrossRef] [PubMed]

146. Reddy, M.A.; Jin, W.; Villeneuve, L.; Wang, M.; Lanting, L.; Todorov, I.; Kato, M.; Natarajan, R. Pro-inflammatory role of microrna-200 in vascular smooth muscle cells from diabetic mice. *Arter. Thromb. Vasc. Boil.* **2012**, *32*, 721–729. [CrossRef]

147. Carlomosti, F.; D'Agostino, M.; Beji, S.; Torcinaro, A.; Rizzi, R.; Zaccagnini, G.; Maimone, B.; Di Stefano, V.; De Santa, F.; Cordisco, S.; et al. Oxidative Stress-Induced miR-200c Disrupts the Regulatory Loop Among SIRT1, FOXO1, and eNOS. *Antioxid. Redox Signal.* **2017**, *27*, 328–344. [CrossRef]

148. Togliatto, G. Unacylated ghrelin induces oxidative stress resistance in a glucose intolerance and peripheral artery disease mouse model by restoring endothelial cell miR-126 expression. *Diabetes* **2015**, *64*, 1370–1382. [CrossRef]

149. Mahavadi, S.; Sriwai, W.; Manion, O.; Grider, J.R.; Murthy, K.S. Diabetes-induced oxidative stress mediates upregulation of RhoA/Rho kinase pathway and hypercontractility of gastric smooth muscle. *PLoS ONE* **2017**, *12*, 0178574. [CrossRef]

150. La Sala, L.; Mrakic-Sposta, S.; Tagliabue, E.; Prattichizzo, F.; Micheloni, S.; Sangalli, E.; Specchia, C.; Uccellatore, A.C.; Lupini, S.; Spinetti, G.; et al. Circulating microRNA-21 is an early predictor of ROS-mediated damage in subjects with high risk of developing diabetes and in drug-naïve T2D. *Cardiovasc. Diabetol.* **2019**, *18*, 18. [CrossRef]

151. La Sala, L.; Mrakic-Sposta, S.; Micheloni, S.; Prattichizzo, F.; Ceriello, A. Glucose-sensing microRNA-21 disrupts ROS homeostasis and impairs antioxidant responses in cellular glucose variability. *Cardiovasc. Diabetol.* **2018**, *17*, 105. [CrossRef]

152. Gao, L.; Liu, Y.; Guo, S.; Xiao, L.; Wu, L.; Wang, Z.; Liang, C.; Yao, R.; Zhang, Y. LAZ3 protects cardiac remodeling in diabetic cardiomyopathy via regulating miR-21/PPARa signaling. *Biochim. Biophys. Acta BBA Mol. Basis Dis.* **2018**, *1864*, 3322–3338. [CrossRef]

153. Murray, A.R.; Chen, Q.; Takahashi, Y.; Zhou, K.K.; Park, K.; Ma, J.-X. MicroRNA-200b Downregulates Oxidation Resistance 1 (Oxr1) Expression in the Retina of Type 1 Diabetes Model. *Investig. Opthalmol. Vis. Sci.* **2013**, *54*, 1689–1697. [CrossRef] [PubMed]

154. Kamalden, T.A.; Macgregor-Das, A.M.; Kannan, S.M.; Dunkerly-Eyring, B.; Khaliddin, N.; Xu, Z.; Fusco, A.P.; Abu Yazib, S.; Chow, R.C.; Duh, E.J.; et al. Exosomal MicroRNA-15a Transfer from the Pancreas Augments Diabetic Complications by Inducing Oxidative Stress. *Antioxid. Redox Signal.* **2017**, *27*, 913–930. [CrossRef] [PubMed]

155. Song, J.; Zhang, H.; Sun, Y.; Guo, R.; Zhong, D.; Xu, R.; Song, M. Omentin-1 protects renal function of mice with type 2 diabetic nephropathy via regulating miR-27a-Nrf2/Keap1 axis. *Biomed. Pharmacother.* **2018**, *107*, 440–446. [CrossRef] [PubMed]

156. Li, Q.; Kim, Y.-R.; Vikram, A.; Kumar, S.; Kassan, M.; Gabani, M.; Lee, S.K.; Jacobs, J.S.; Irani, K. P66Shc-Induced MicroRNA-34a Causes Diabetic Endothelial Dysfunction by Downregulating Sirtuin1. *Arter. Thromb. Vasc. Boil.* **2016**, *36*, 2394–2403. [CrossRef]

157. Kassan, M.; Vikram, A.; Li, Q.; Kim, Y.-R.; Kumar, S.; Gabani, M.; Liu, J.; Jacobs, J.S.; Irani, K. MicroRNA-204 promotes vascular endoplasmic reticulum stress and endothelial dysfunction by targeting Sirtuin1. *Sci. Rep.* **2017**, *7*, 9308. [CrossRef]

158. Li, X.; Wang, H.; Yao, B.; Xu, W.; Chen, J.; Zhou, X. lncRNA H19/miR-675 axis regulates cardiomyocyte apoptosis by targeting VDAC1 in diabetic cardiomyopathy. *Sci. Rep.* **2016**, *6*, 36340. [CrossRef]

159. Zheng, D.; Ma, J.; Yu, Y.; Li, M.; Ni, R.; Wang, G.; Chen, R.; Li, J.; Fan, G.-C.; Lacefield, J.C.; et al. Silencing of miR-195 reduces diabetic cardiomyopathy in C57BL/6 mice. *Diabetologia* **2015**, *58*, 1949–1958. [CrossRef]

160. Zhang, R.; Garrett, Q.; Zhou, H.; Wu, X.; Mao, Y.; Cui, X.; Xie, B.; Liu, Z.; Cui, D.; Jiang, L.; et al. Upregulation of miR-195 accelerates oxidative stress-induced retinal endothelial cell injury by targeting mitofusin 2 in diabetic rats. *Mol. Cell. Endocrinol.* **2017**, *452*, 33–43. [CrossRef]

161. Gu, H. The miR-322-TRAF3 circuit mediates the pro-apoptotic effect of high glucose on neural stem cells. *Toxicol. Sci.* **2015**, *144*, 186–196. [CrossRef]

162. Wu, K.; Yang, Y.; Zhong, Y.; Ammar, H.M.; Zhang, P.; Guo, R.; Liu, H.; Cheng, C.; Koroscil, T.M.; Chen, Y.; et al. The effects of microvesicles on endothelial progenitor cells are compromised in type 2 diabetic patients via downregulation of the miR-126/VEGFR2 pathway. *Am. J. Physiol. Metab.* **2016**, *310*, E828–E837. [CrossRef]

163. Wang, J.M. MicroRNA miR-27b rescues bone marrow-derived angiogenic cell function and accelerates wound healing in type 2 diabetes mellitus. *Arter. Thromb. Vasc. Biol.* **2014**, *34*, 99–109. [CrossRef] [PubMed]

164. Wei, J. Aldose reductase regulates miR-200a-3p/141-3p to coordinate Keap1-Nrf2, Tgfbeta1/2, and Zeb1/2 signaling in renal mesangial cells and the renal cortex of diabetic mice. *Free Radic. Biol. Med.* **2014**, *67*, 91–102. [CrossRef] [PubMed]

165. Yildirim, S.S.; Akman, D.; Catalucci, D.; Turan, B. Relationship Between Downregulation of miRNAs and Increase of Oxidative Stress in the Development of Diabetic Cardiac Dysfunction: Junctin as a Target Protein of miR-1. *Cell Biophys.* **2013**, *67*, 1397–1408. [CrossRef] [PubMed]

166. Wang, J.M. Inositol-Requiring Enzyme 1 Facilitates Diabetic Wound Healing Through Modulating MicroRNAs. *Diabetes* **2017**, *66*, 177–192. [CrossRef] [PubMed]

167. Jeyabal, P.; Thandavarayan, R.A.; Joladarashi, D.; Babu, S.S.; Krishnamurthy, S.; Bhimaraj, A.; Youker, K.A.; Kishore, R.; Krishnamurthy, P. MicroRNA-9 inhibits hyperglycemia-induced pyroptosis in human ventricular cardiomyocytes by targeting ELAVL1. *Biochem. Biophys. Res. Commun.* **2016**, *471*, 423–429. [CrossRef]

168. Zhang, Z.-W.; Guo, R.-W.; Lv, J.-L.; Wang, X.-M.; Ye, J.-S.; Lu, N.-H.; Liang, X.; Yang, L.-X. MicroRNA-99a inhibits insulin-induced proliferation, migration, dedifferentiation, and rapamycin resistance of vascular smooth muscle cells by inhibiting insulin-like growth factor-1 receptor and mammalian target of rapamycin. *Biochem. Biophys. Res. Commun.* **2017**, *486*, 414–422. [CrossRef]

169. Zhu, M. Hyperlipidemia-Induced MicroRNA-155-5p Improves beta-Cell Function by Targeting Mafb. *Diabetes* **2017**, *66*, 3072–3084. [CrossRef]

170. Roggli, E. Changes in microRNA expression contribute to pancreatic beta-cell dysfunction in prediabetic NOD mice. *Diabetes* **2012**, *61*, 1742–1751. [CrossRef]

171. Massart, J.; Sjögren, R.J.; Lundell, L.S.; Mudry, J.M.; Franck, N.; O'Gorman, D.J.; Egan, B.; Zierath, J.R.; Krook, A. Altered miR-29 Expression in Type 2 Diabetes Influences Glucose and Lipid Metabolism in Skeletal Muscle. *Diabetes* **2017**, *66*, 1807–1818. [CrossRef]

172. Guo, J.; Li, J.; Zhao, J.; Yang, S.; Wang, L.; Cheng, G.; Liu, D.; Xiao, J.; Liu, Z.; Zhao, Z. MiRNA-29c regulates the expression of inflammatory cytokines in diabetic nephropathy by targeting tristetraprolin. *Sci. Rep.* **2017**, *7*, 2314. [CrossRef]

173. Widlansky, M.E.; Jensen, D.M.; Wang, J.; Liu, Y.; Geurts, A.M.; Kriegel, A.J.; Liu, P.; Ying, R.; Zhang, G.; Casati, M.; et al. miR-29 contributes to normal endothelial function and can restore it in cardiometabolic disorders. *EMBO Mol. Med.* **2018**, *10*, e8046. [CrossRef] [PubMed]

174. Wang, H.; She, G.; Zhou, W.; Liu, K.; Miao, J.; Yu, B. Expression profile of circular RNAs in placentas of women with gestational diabetes mellitus. *Endocr. J.* **2019**, *66*, 431–441. [CrossRef] [PubMed]

175. Lakhter, A.J.; Pratt, R.E.; Moore, R.E.; Doucette, K.K.; Maier, B.F.; DiMeglio, L.A.; Sims, E.K. Beta cell extracellular vesicle miR-21-5p cargo is increased in response to inflammatory cytokines and serves as a biomarker of type 1 diabetes. *Diabetologia* **2018**, *61*, 1124–1134. [CrossRef] [PubMed]

176. Piperi, C.; Goumenos, A.; Adamopoulos, C.; Papavassiliou, A.G. AGE/RAGE signalling regulation by miRNAs: Associations with diabetic complications and therapeutic potential. *Int. J. Biochem. Cell Boil.* **2015**, *60*, 197–201. [CrossRef] [PubMed]

177. Magenta, A.; Ciarapica, R.; Capogrossi, M.C. The Emerging Role of miR-200 Family in Cardiovascular Diseases. *Circ. Res.* **2017**, *120*, 1399–1402. [CrossRef]

178. Magenta, A.; Cencioni, C.; Fasanaro, P.; Zaccagnini, G.; Greco, S.; Sarra-Ferraris, G.; Antonini, A.; Martelli, F.; Capogrossi, M.C. miR-200c is upregulated by oxidative stress and induces endothelial cell apoptosis and senescence via ZEB1 inhibition. *Cell Death Differ.* **2011**, *18*, 1628–1639. [CrossRef]

179. Belgardt, B.-F.; Ahmed, K.; Spranger, M.; Latreille, M.; Denzler, R.; Kondratiuk, N.; Von Meyenn, F.; Villena, F.N.; Herrmanns, K.; Bosco, D.; et al. The microRNA-200 family regulates pancreatic beta cell survival in type 2 diabetes. *Nat. Med.* **2015**, *21*, 619–627. [CrossRef]

180. Kitada, M.; Ogura, Y.; Monno, I.; Koya, D. Sirtuins and Type 2 Diabetes: Role in Inflammation, Oxidative Stress, and Mitochondrial Function. *Front. Endocrinol.* **2019**, *10*, 187. [CrossRef]

181. Kwon, J.J.; Factora, T.D.; Dey, S.; Kota, J. A Systematic Review of miR-29 in Cancer. *Mol. Ther. Oncol.* **2019**, *12*, 173–194. [CrossRef]

182. Slusarz, A.; Pulakat, L. The two faces of miR-29. *J. Cardiovasc. Med.* **2015**, *16*, 480–490. [CrossRef]

183. Heid, J.; Cencioni, C.; Ripa, R.; Baumgart, M.; Atlante, S.; Milano, G.; Scopece, A.; Kuenne, C.; Guenther, S.; Azzimato, V.; et al. Age-dependent increase of oxidative stress regulates microRNA-29 family preserving cardiac health. *Sci. Rep.* **2017**, *7*, 16839. [CrossRef] [PubMed]

184. David, J.A.; Rifkin, W.J.; Rabbani, P.S.; Ceradini, D.J. The Nrf2/Keap1/ARE Pathway and Oxidative Stress as a Therapeutic Target in Type II Diabetes Mellitus. *J. Diabetes Res.* **2017**, *2017*, 1–15. [CrossRef] [PubMed]

185. Matsuda, S.; Nakagawa, Y.; Kitagishi, Y.; Nakanishi, A.; Murai, T. Reactive Oxygen Species, Superoxide Dimutases, and PTEN-p53-AKT-MDM2 Signaling Loop Network in Mesenchymal Stem/Stromal Cells Regulation. *Cells* **2018**, *7*, 36. [CrossRef] [PubMed]

186. Zong, H.-Y.; Wang, E.-L.; Han, Y.-M.; Wang, Q.-J.; Wang, J.-L.; Wang, Z. Effect of miR-29b on rats with gestational diabetes mellitus by targeting PI3K/Akt signal. *Eur. Rev. Med Pharmacol. Sci.* **2019**, *23*, 2325–2331.

187. Stoll, L. Circular RNAs as novel regulators of beta-cell functions in normal and disease conditions. *Mol. Metab.* **2018**, *9*, 69–83. [CrossRef]

188. Hansen, T.B.; Jensen, T.I.; Clausen, B.H.; Bramsen, J.B.; Finsen, B.; Damgaard, C.K.; Kjems, J. Natural RNA circles function as efficient microRNA sponges. *Nature* **2013**, *495*, 384–388. [CrossRef]

International Journal of
Molecular Sciences

Review

The Yin-Yang Regulation of Reactive Oxygen Species and MicroRNAs in Cancer

Kamesh R. Babu [1] and Yvonne Tay [1,2,*]

[1] Cancer Science Institute of Singapore, National University of Singapore, Singapore 117599, Singapore; csirbk@nus.edu.sg

[2] Department of Biochemistry, Yong Loo Lin School of Medicine, National University of Singapore, Singapore 117597, Singapore

[*] Correspondence: yvonnetay@nus.edu.sg; Tel.: +65-6516-7756

Received: 25 September 2019; Accepted: 24 October 2019; Published: 26 October 2019

Abstract: Reactive oxygen species (ROS) are highly reactive oxygen-containing chemical species formed as a by-product of normal aerobic respiration and also from a number of other cellular enzymatic reactions. ROS function as key mediators of cellular signaling pathways involved in proliferation, survival, apoptosis, and immune response. However, elevated and sustained ROS production promotes tumor initiation by inducing DNA damage or mutation and activates oncogenic signaling pathways to promote cancer progression. Recent studies have shown that ROS can facilitate carcinogenesis by controlling microRNA (miRNA) expression through regulating miRNA biogenesis, transcription, and epigenetic modifications. Likewise, miRNAs have been shown to control cellular ROS homeostasis by regulating the expression of proteins involved in ROS production and elimination. In this review, we summarized the significance of ROS in cancer initiation, progression, and the regulatory crosstalk between ROS and miRNAs in cancer.

Keywords: ROS; oxidative stress; antioxidants; miRNA; cancer

---

1. Introduction

Reactive oxygen species (ROS) are free radicals, ions, or molecules with a single unpaired electron. ROS, including hydrogen peroxide (H_2O_2), hydroxyl radicals (OH^-), nitric oxide (NO), and superoxide radicals (O_2^-) are highly reactive and generated as a byproduct during metabolic processes in various subcellular compartments of a cell [1]. Mitochondria are the main cellular source of ROS. However, ROS are also generated in other cellular organelles including endoplasmic reticulum, lysosomes, and peroxisomes [2]. At lower concentrations, ROS play significant roles in various physiological functions including gene activation, cell growth, proliferation, survival, apoptosis, chemical reaction modulation, blood pressure control, prostaglandin biosynthesis, embryonic development, cognitive function, and immune response [3,4]. However, at higher concentrations, ROS can cause oxidative damage via oxidation of macromolecules such as DNA, RNA, proteins, and lipids that can contribute to the pathogenesis of various diseases including cancer [5–9]. Elevated ROS production is associated with tumorigenesis and suggested to be a hallmark of cancer. Nevertheless, the molecular mechanisms responsible for sustained high ROS levels in cancer is not well understood.

MicroRNAs (miRNAs) are a class of small non-coding RNAs that are approximately 22 nucleotides long and regulate gene expression at the post-transcriptional level [10]. They regulate gene expression by binding to the target messenger RNA (mRNA) transcript which activates either degradation or translation suppression based on the extent of basepairing. However, several studies reported that miRNAs can also target and regulate the stability of non-coding RNAs. Studies have demonstrated that the deregulation of miRNA expression is associated with cancer development, and miRNAs may function as potential oncogenes or tumor suppressors [11]. Surprisingly, studies show the existence of a regulatory connection between ROS and miRNA. For example, H_2O_2 treatment has been shown to dysregulate the expression of certain miRNAs in vascular smooth muscle cells and macrophage cells [12,13]. Another study has shown that miR-30e regulates oxidative stress and ROS levels by targeting SNAI1 mRNA in human umbilical endothelial vein cells [14]. These findings suggest that ROS and miRNAs may co-regulate each other in cancer to maintain cellular ROS levels that support cancer development. In this review, we discuss the significance of ROS in cancer development, as well as the crosstalk between ROS and miRNAs in the regulation of redox homeostasis and cancer progression.

2. Significance of ROS in Cancer Development

ROS are required by cells to carry out physiological cellular functions and this is also true in the case of cancer cells. However, cancer cells show elevated levels of ROS when compared to normal cells, which is mainly due to persistent and high metabolic rate in mitochondria, endoplasmic reticulum (ER). and cell membranes. In this section, we discuss how ROS play a significant role in the whole process of cancer development, including initiation, promotion, and progression.

2.1. ROS in Cancer Initiation

ROS are potent mutagens that can stimulate cancer initiation. High levels of ROS oxidize DNA bases resulting in DNA lesions including base damage, strand breaks, and mutations, which are usually repaired by the endogenous DNA repair enzymes of the base excision repair, nuclear excision repair, or mismatch repair pathways [15]. Cells unable to repair DNA lesions undergo apoptosis to prevent the passage of DNA mutations to progeny cells. However, under certain conditions, cells harboring DNA lesions evade apoptosis, which eventually leads to cancer. In a similar fashion to DNA, RNA also undergoes oxidation under oxidative stress that results in strand breaks and oxidative base modifications. Oxidized mRNA can cause several defects during protein translation, which include synthesis of truncated, mutated, or non-functional proteins, ribosome stalling, and ribosome dysfunction [6]. Oxidized RNA can promote the pathogenesis of chronic degenerative diseases including cancer [7]. For example, oxidation of tumor suppressor mRNAs results in the synthesis of mutated or truncated proteins that lack proper function, and this may lead to carcinogenesis. It is important to note that RNA oxidation is not limited only to mRNA as all RNA species including non-coding RNAs are subjected to oxidative damage. Since several studies have shown the significant participation of non-coding RNAs, including miRNAs and long non-coding RNAs (lncRNAs) in cancer development [16], oxidative modification of non-coding RNAs may also promote cancer initiation. ROS-induced mutation or modification is not only restricted to nucleotides, but even protein molecules are also susceptible to such modifications. Oxidation of proteins by ROS results in amino acid modification, protein carbonylation, nitration of tyrosine and phenylalanine residues, protein degradation, or formation of cross-linked proteins or glycated proteins [17,18]. Oxidized amino acid residues can affect their protein activity. For example, oxidation of DNA polymerase affects its fidelity during replication/synthesis, transcription, or DNA repair activity, which is closely associated with cancer initiation [19]. Finally, ROS can also damage polyunsaturated or polydesaturated fatty acids by the process of lipid peroxidation which generates various toxic molecules including malondialdehyde, 2-alkenals, 4-hydroxynonenal (HNE), and lipoperoxyl radical (LOO^-) [9,20]. The LOO^- reacts with the lipids to generate lipid peroxides, which are unstable and can produce new peroxyl and alkoxy radicals. These radicals may further increase the oxidation of macromolecules. Furthermore, HNE is a chemically reactive molecule that can react with macromolecules and form covalent modifications, which has been proposed as the mechanism to induce carcinogenesis [20]. These studies indicate that higher levels of ROS are detrimental to cells and can increase the risk of developing cancer.

2.2. ROS in Cancer Cell Proliferation

ROS function as secondary messengers in cellular signaling and activate ROS-sensitive signaling pathways by regulating protein activity through the reversible oxidation of target proteins. Redox-sensitive signaling pathways, including the mitogen-activated protein kinase (MAPK)/extracellular signal-regulated kinase (ERK), phosphoinositide-3 kinase (PI3K)/protein kinase B (AKT), and nuclear factor κ-B (NF-κB) signaling pathways, are constantly upregulated in various cancer subtypes, where they play a pivotal role in cell proliferation, growth, protein synthesis, glucose metabolism, cell survival, and inflammation [21]. Activation of MAPK/ERK signaling has been shown to increase anchorage-independent growth, cell survival, and motility of many cancer subtypes including breast cancer, leukemia, melanoma, and ovarian cancer. Studies have shown that high ROS levels in cancer cells can elevate MAPK/ERK signaling and can increase cancer cell proliferation [19]. Analogously, high levels of ROS, either produced endogenously or added exogenously, have shown to increase the activation PI3K/AKT signaling pathway in breast and ovarian cancer. Furthermore, studies show that elevated ROS levels can activate the transcription factor NF-κB. Oxidative stress-induced through the exogenous treatment of sodium arsenite, rotenone, H_2O_2, or through inhibition of endogenous antioxidants elevated the NF-κB activation and increased cancer cell proliferation [19,22]. Moreover, ROS play a significant role in the cell cycle by regulating mRNA levels of cyclins that promote G1 to S phase transition, which include cyclin B2, cyclin D3, cyclin E1, and cyclin E3 [23]. In breast cancer cells, ROS generated by sodium arsenite treatment promote S phase transition and aberrant cell proliferation [24], whereas reduction of ROS levels through antioxidant N-acetyl cysteine (NAC) treatment reduces cyclin D1 levels and slowed the G1 to S phase transition in the non-cancerous human breast epithelial cells [25]. All these studies suggest that besides being a highly reactive mutagen, ROS can also function as a secondary messenger that mediate physiological signaling pathways involved in cell proliferation, thus higher ROS production in cancer cells favor cancer progression through elevated and sustained activation of these pathways.

2.3. ROS in Cancer Metastasis

Metastasis is a multistep process that involves the spread of cancer cells from its original site to distal parts of the body, the process comprises migration, invasion, intravasation into the blood, anchorage-independent survival in the blood, and extravasation into distal organs [26]. Several studies show that ROS levels are increased in cells that undergo metastasis, and they play a significant role in the cancer cell metastasis. A study has shown that endogenous ROS levels are increased in circulating melanoma cells and metastasis nodules of xenografted mice compared to primary subcutaneous tumors [27]. Importantly, cancer cells treated with H_2O_2 have shown high metastasis upon injected intravenously into mice. Likewise, a sub-population of the breast cancer cells that has elevated intracellular ROS levels compared to the parental cells exhibits high motility and metastasized to distant organs including lung, liver, and spleen [19]. It is noteworthy that the levels and activity of endogenous antioxidants are decreased in metastatic cancer cells. For example, the levels and catalytic activity of manganese-dependent superoxide dismutase (MnSOD) are lower in highly invasive pancreatic cancer cells and metastatic breast cancer cells [28,29]. Cancer cells go through epithelial to mesenchymal transition (EMT) before migrating to distant sites of the body. During the EMT process, expression of matrix metalloproteinases (MMPs) is increased to mediate degradation and reorganization of extracellular matrix and their elevated activation is associated with tumor growth, angiogenesis, invasion, and metastasis [30]. ROS play a significant role in the EMT process in which they regulate the expression of MMPs and their inhibitors tissue inhibitor of metalloproteinases (TIMP) [31]. A study has shown that treatment of MMP-3, a stromal protease whose expression is upregulated in mammary tumors, has increased cellular ROS and induced EMT in murine mammary epithelial cells. In contrast, scavenging cellular ROS through NAC treatment abrogated MMP-3-induced EMT, suggesting that high levels of cellular ROS can lead to malignant transformation [32]. Moreover, ROS also facilitate metastasis by increasing vascular permeability through various mechanisms. Oxidative stress in endothelial cells mediate Rac-1-induced loss of cell–cell adhesion and loosens the endothelium integrity, which favors the cancer cell intravasation [33]. ROS regulate the expression of IL-8 and intracellular adhesion protein 1 (ICAM-1) via NF-κB activation. Both IL-8 and ICAM-1 regulate transendothelial migration of tumor cells [34,35]. Furthermore, ROS induce actin reorganization in vascular endothelial cells through p38-mediated phosphorylation of the heat shock protein Hsp27, which may contribute to promote invasive processes [36]. Taken altogether, these studies suggest that ROS has a versatile role in the pathogenesis of cancer, therefore it would be interesting to identify further novel roles of ROS in other physiological processes that could possibly support the process of cancer development.

2.4. ROS in Cancer Stem Cells

Cancer stem cells (CSCs) are a subset of tumorigenic cells that possess similar characteristics as normal stem cells, in particular the capabilities of self-renewal or differentiation. Interestingly, CSCs have been shown to have a high capacity to grow into tumors. Similar to cancer cells, ROS also play an important role in CSCs. However, in contrast to cancer cells in which ROS levels are elevated, CSCs exhibit lower levels of ROS. This is similar to the levels found in normal stem cells [37]. The lower cellular ROS levels in CSCs are associated with increased expression of ROS scavenging systems and are essential for the maintenance of self-renewal and stemness. A study has shown that pharmacological depletion of ROS scavengers in breast CSCs reduces their clonogenicity and results in radiosensitization [38]. Conversely, ovarian CSCs exhibit higher mitochondrial ROS production, and inhibition of the mitochondrial respiratory chain in CSCs results in apoptosis [39]. Furthermore, a study has shown that the population of hematopoietic stem cells (HSCs) with higher ROS levels possess higher myeloid differentiation potential compared to the HSC fraction with lower ROS levels [40]. These findings suggest that ROS levels in CSCs are crucial for their survival and differentiation. Nevertheless, the effects of ROS and the regulation of ROS levels in CSCs have not been studied extensively. Future investigations may unravel the molecular mechanisms behind the regulation of redox homeostasis in CSCs.

3. ROS Regulate MiRNA Expression

Accumulating studies show functional regulatory links between ROS and miRNAs in carcinogenesis. ROS also contribute to cancer development by regulating the expression of miRNAs that target genes responsible for enhancing or suppressing carcinogenesis. In this section, we discuss how the ROS affect the miRNA expression in cancer via different mechanisms including alteration of epigenetic signatures, transcription, and biogenesis.

3.1. Regulation of MiRNA Expression via Epigenetic Modifications

Dysregulated miRNA expression in cancer is associated with altered DNA methylation and histone modifications such as acetylation, methylation, and phosphorylation. ROS can regulate miRNA expression by altering the epigenetic signatures including DNA methylation or histone modifications (Figure 1a). For example, ROS inhibit the expression of miR-199a and miR-125b in ovarian cancer cells via increasing promoter methylation of the miR-199a and miR-125b genes, which is mediated by the DNA methyltransferase 1 (DNMT1) [41]. Interestingly, overexpression of miR-199a and miR-125b in ovarian cancer cells decreased the expression of the hypoxia-inducible factor 1-alpha (HIF-1α) and vascular endothelial growth factor, which suppressed tumor-induced angiogenesis [42]. Histone modifications play an important role in chromatin remodeling in order to regulate gene transcription. Histone acetylation is a type of histone modification in which the lysine residues of histone are acetylated to relax the chromatin structure for gene transcription. In contrast, deacetylation of lysine residues catalyzed by the histone deacetylases (HDACs) causes chromatin condensation and transcriptional gene silencing [43]. ROS can regulate the activity of HDACs. For example, the Cys667 and Cys669 amino acid residues of HDAC4 are oxidized to form an intramolecular dis-sulfide bond, which promotes its nuclear export [44]. Cancer cells promote nuclear translocation of HDAC4 by increasing endogenous antioxidants, which decreases miR-206 expression through deacetylation of its promoter and promotes cancer progression [45,46]. Furthermore, oxidative stress-induced by glucose depletion increases the expression of miR-466h-5p by inhibiting HDAC2 activity, which results in increased apoptosis due to the fact that miR-466h-5p directly targets and downregulates many anti-apoptotic proteins including BCL212, DAD1, BIRC6, STAT5A, and SMO [47,48]. These findings suggest that ROS can affect the epigenetic status of miRNA genes thereby regulating its expression in cancer. It is important to note that ROS-mediated regulation of DNMT1 and HDACs in cancer may change its global epigenetic signature, therefore the expression of other genes including oncogenes and tumor suppressors can also be activated or silenced.

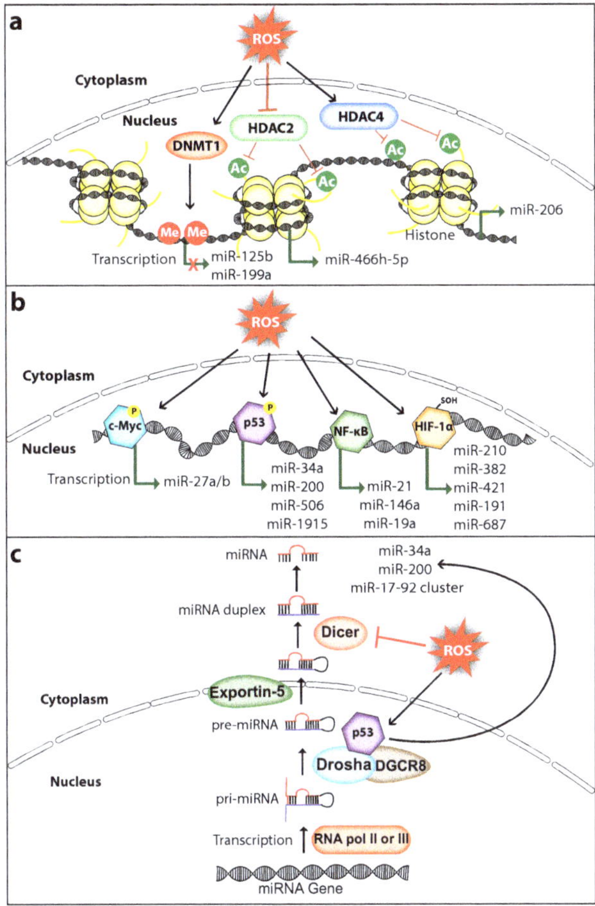

Figure 1. ROS regulate miRNA expression and processing in cancer. (**a**) ROS induce epigenetic modifications to regulate miRNA expression. ROS affect DNMT1 (DNA methyltransferase 1), HDAC2 (Histone deacetylase 2), or HDAC4 (Histone deacetylase 4) to either inhibit or activate miRNA expression. (**b**) ROS can induce miRNA transcription through activating transcription factors c-Myc, p53, NF-κB (nuclear factor κ-B), or HIF-1α (hypoxia-inducible factor 1-alpha). (**c**) ROS affect miRNA biogenesis and maturation through regulating the activity and expression of miRNA processing enzymes Drosha and Dicer, respectively. Me, methyl group; Ac, acetyl group; P, phosphoryl group (The black arrow indicates upmodulation, the red T arrow indicates inhibition, the green arrow indicates transcription activation, the green arrow with red cross indicates transcription inhibition).

3.2. Regulation of MiRNA Expression via Transcription Factors

ROS can also control miRNA expression by regulating the ROS-responsive transcription factors that activate miRNA transcription (Figure 1b). ROS regulate the activation of transcription factors through direct or indirect mechanisms. The activated transcription factor binds to the target miRNA promoter and upregulates miRNA transcription.

3.2.1. C-Myc

C-Myc is a well-studied transcription factor and characterized as an oncoprotein whose expression is elevated in a wide range of tumors. It promotes tumorigenesis by activating the transcription of several oncogenes including the miR-17-92 cluster, or by inhibiting the transcription of tumor-suppressors including let-7a which functions as a negative regulator of CSC features by regulating PTEN and Lin28b expression in pancreatic and prostate cancer [49]. c-Myc is a redox-sensitive transcription factor. Under oxidative stress, ROS cause ERK-dependent phosphorylation at the Ser62 amino acid residue of c-Myc which enhances the c-Myc recruitment to the promoter of gamma-glutamyl-cysteine synthetase, the rate-limiting enzyme catalyzing glutathione (GSH) synthesis. The c-Myc phosphorylation-dependent activation of GSH promotes the survival of cancer cells under oxidative-stress conditions [50]. Lithocholic acid (LCA)-induced ROS increased c-Myc expression in the human hepatocellular carcinoma (HCC) cells and in mouse liver. Importantly, LCA mediated c-Myc overexpression activates the expression of miR-27a/b that promotes HCC proliferation [51]. miR-27a/b directly targets and suppresses the expression of nuclear factor-erythroid 2-related factor 2 (NRF2) and prohibitin 1 (PHB1), a mitochondrial chaperone function as a tumor suppressor in liver cancer [52], whereas knockdown of c-Myc or miR-27a/b in Huh-7 cells rescued the LCA-mediated suppression of NRF2 and PHB1. This suggests that the interplay of ROS, c-Myc, and miR-27 has a significant role in HCC progression.

3.2.2. P53

The tumor suppressor protein p53 maintains genome integrity by inducing antiproliferative programs such as cell cycle arrest, senescence, and apoptosis through differential activation of key effector genes including the tumor suppressor miR-34a [53,54]. p53 is an oxidative stress-responsive transcription factor whose expression can be induced by ROS to protect genome stability via selectively activating its target genes [55]. Furthermore, the transcriptional activity of p53 is affected by oxidative stress, as the endogenous antioxidants thioredoxin (TRX) and GSH modify the cysteine amino acids of p53, which affects p53 activity including DNA binding capacity, activation of target gene transcription, and apoptosis induction [56–58]. A study has shown that H_2O_2 treatment phosphorylates the Ser33 amino acid residue of p53 in hepatic cells, which promotes miR-200 transcription and cell death [59]. Interestingly, p53 knockdown reversed the H_2O_2 mediated miR-200 expression [60], confirming that miR-200 expression under oxidative stress is p53-dependent. Importantly, miR-200 has shown to function as a tumor suppressor by inhibiting the CSC self-renewal potential and EMT process in various cancer subtypes including bladder cancer, gastric cancer, ovarian cancer, pancreatic cancer, and prostate cancer [49,61,62]. Furthermore, ROS mediated p53 activation also upregulates the expression of miR-506, which inhibited the growth of lung tumor in-vitro and in-vivo [63]. In addition, expression of miR-34a-5p and miR-1915 is regulated by p53 in HCC cells during oxidative stress [64]. Moreover, miR-34 inhibits pancreatic CSC proliferation, self-renewal, and induces apoptosis and cell cycle arrest [49]. Altogether, these studies strongly suggest that p53 mediates anticancer roles through promoting the expression of tumor suppressor miRNAs in a redox-dependent fashion.

3.2.3. NFκB

NF-κB is an inducible transcription factor that plays a pivotal role in DNA transcription, cytokine production, cell proliferation, survival, differentiation, cell cycle regulation, and especially in inflammation [65]. The activity of NF-κB is inhibited by its inhibitor IκB which sequesters NF-κB in the cytosol to prevent its translocation to the nucleus. The canonical NF-κB activation is mediated through the degradation of IκB, induced via site-specific phosphorylation by NF-κB-inducing kinase (NIK) and IκB kinase (IKK) protein complex, consisting of IKKα, IKKβ, and NF-κB essential modulator. ROS activate the NF-κB pathway by activating NIK through oxidative inhibition of regulatory phosphatases, and through tyrosine phosphorylation of IκBα [22]. NFκB mediates transcription of several miRNAs including let-7, miR-21, and miR-146 [66]. miR-21 is a well-studied oncomiR which mediates pro-survival and anti-proliferative effects through directly targeting and suppressing the expression of tumor suppressors such as PTEN, PDCD4, IGFBP3, and MKK3 [67–70]. Overexpression of miR-21 is associated with the progression of many cancer types and considered as a biomarker and target for cancer treatment [71]. Interestingly, miR-21 is elevated in breast CSC subpopulations and regulates the EMT phenotype [49]. ROS-induced miR-21 expression has been shown to contribute to the invasion and metastasis of prostate cancer [72]. NFκB activates miR-21 transcription by directly binding to the promoter of the miR-21 gene [73]. Likewise, ROS-activated NFκB can also upregulate miR-146a transcription, which suppresses the progression of acute myeloid leukemia (AML) [74]. In contrast, berberine-treatment-induced oxidative stress, suppressed miR-21 expression by inhibiting the nuclear translocation of NFκB in human multiple myeloma cells, which induces apoptosis [75]. In addition, oxidative stress deactivated NFκB activity that downregulated miR-19a transcription and activated apoptosis of the pheochromocytoma cells [76]. These findings suggest that the transcription factor NFκB can be either activated or inhibited under oxidative stress.

3.2.4. HIF-1α

HIF-1α is a subunit of heterodimeric transcription factor hypoxia-inducible factor 1, which regulates the expression of genes involved in the process of angiogenesis and erythropoiesis, which is important for blood vessel formation and the survival of cells under hypoxic condition [77,78]. Under hypoxia, HIF-1α activates the transcription of certain miRNAs called hypoxamiRs, which function as key regulators of the cell against decreased oxygen tension [79]. miR-210 is one such miRNA whose transcription is activated through direct binding of HIF-1α to the hypoxia-responsive element located within its promoter. Interestingly, miR-210 can negatively regulate HIF-1α expression by directly targeting its mRNA forming a negative-feedback loop, and disruption of this loop has been implicated in autoimmune diseases and tumor initiation [79,80]. Studies have shown that miR-210 promotes CSC proliferation, migration, metastasis, and self-renewal [49]. Furthermore, HIF-1α activates the transcription of many other miRNAs including miR-382, miR-421, miR-191, and miR-687 that promote migration, angiogenesis, metastasis, tumor growth, or drug resistance in cancer [81–84]. ROS regulate HIF-1α directly by oxidizing the Cys533 amino acid residue of HIF-1α, which increases the HIF-1α protein stability under oxidative stress [85]. In addition, ROS can activate HIF-1α indirectly through downregulating SIRT1 deacetylase, which results in acetylation at the Lys647 amino acid residue of HIF-1α [86]. This strongly suggests that ROS may regulate the expression of a broad range of miRNA genes in cancer by regulating the redox-sensitive HIF-1α transcription factor.

3.3. Regulation of MiRNA Processing

ROS can also affect miRNA expression by regulating proteins involved in miRNA processing. Generally, miRNAs are transcribed as primary miRNA (pri-miRNA) transcripts by RNA polymerase II or RNA polymerase III. Pri-miRNAs are then processed into premature miRNA (pre-miRNA) transcripts that are approximately 60–70 nucleotide long by the RNA-specific RNAse III type ribonuclease Drosha and DGCR8 protein complex. The pre-miRNA hairpins are then exported to the cytoplasm by the Exportin-5 and are processed into mature miRNA duplex by the ribonuclease Dicer [87] (Figure 1c). Interestingly, p53 regulates the processing of pri-miRNA to pre-miRNA by interacting with the Drosha processing complex via the association with DEAD-box RNA helicase p68 (DDX5), thus indirectly inducing the transcription of miR-34a, miR-200c, and miR-17-92 cluster [88]. A study has demonstrated that H_2O_2 treatment in endothelial cells decreased the expression of Dicer which in turn downregulated the majority of miRNAs that are normally expressed in cerebromicrovascular endothelial cells [89]. Strikingly, ROS production is also regulated by cellular Dicer levels. A study has shown that dicer knockdown downregulated miRNA expression and decreased the production of ROS in human microvascular endothelial cells [90]. Although this has been investigated in non-cancerous cell models, it would be interesting to analyze whether this phenomenon also exists in cancer cells. Furthermore, NFκB can also regulate miRNA expression indirectly by expressing proteins involved in miRNA processing. A study has shown that NFκB activates the transcription of the miRNA processing inhibitor Lin28, which decreased the let-7 levels rapidly leading to Src-induced cellular transformation [91]. Moreover, ROS not only affect miRNA expression but also modify miRNAs directly through oxidation. A study has shown that upon oxidative modification, miR-184 can target the 3′UTR of antiapoptotic proteins BCL-XL and BCL-W, which are non-native targets of miR-184. Oxidized miR-184 induces apoptosis through downregulating the expression of BCL-XL and BCL-W in the rat heart cell line H9c2 [92]. Altogether, these studies indicate that ROS promote cancer progression through controlling miRNA expression, and the mechanisms involved in the ROS-mediated miRNA expression are not limited. Therefore, more novel mechanisms involved in ROS-dependent miRNA regulation continue to be unraveled in future studies.

4. MiRNAs Regulate ROS Homeostasis

MiRNAs can affect cellular redox homeostasis by regulating the expression of endogenous ROS producers and antioxidants. They usually manipulate ROS levels by directly targeting the genes involved in ROS production or elimination processes (Figure 2). In this section, we discuss how miRNAs control cellular ROS levels in cancer by targeting genes involved in redox homeostasis.

4.1. Regulation of ROS Producer

Studies have shown that miRNAs can affect the expression and function of endogenous ROS producers through functional interactions, thereby controlling cellular ROS production in cancer cells. The membrane-bound enzyme NADPH oxidases (NOXs) produce O_2^- through catalyzing the reduction of O_2 by transferring an electron from NADPH [93]. The tumor suppressor miR-34a regulates NOX2, the catalytic subunit of NADPH oxidase and overexpression of miR-34a in glioma cells induced apoptosis through NOX2 mediated ROS production [94]. Proline oxidase (POX) is a p53-activated ROS producer whose expression is decreased in human cancer tissues including renal cancer. POX is a direct target of miR-23b, and knockdown of miR-23b promotes ROS production and apoptosis thereby inhibiting kidney tumor growth [95]. Knockdown of dicer in mouse endothelial cells increased the activity of miR-21a-3p targeting NOX4 3′UTR, which resulted in decreased cellular ROS production and endothelial cell tumor formation [96]. These findings indicate that ROS can act as a double-edged sword, thus both overproduction or inhibition of ROS can have a significant effect on cancer progression.

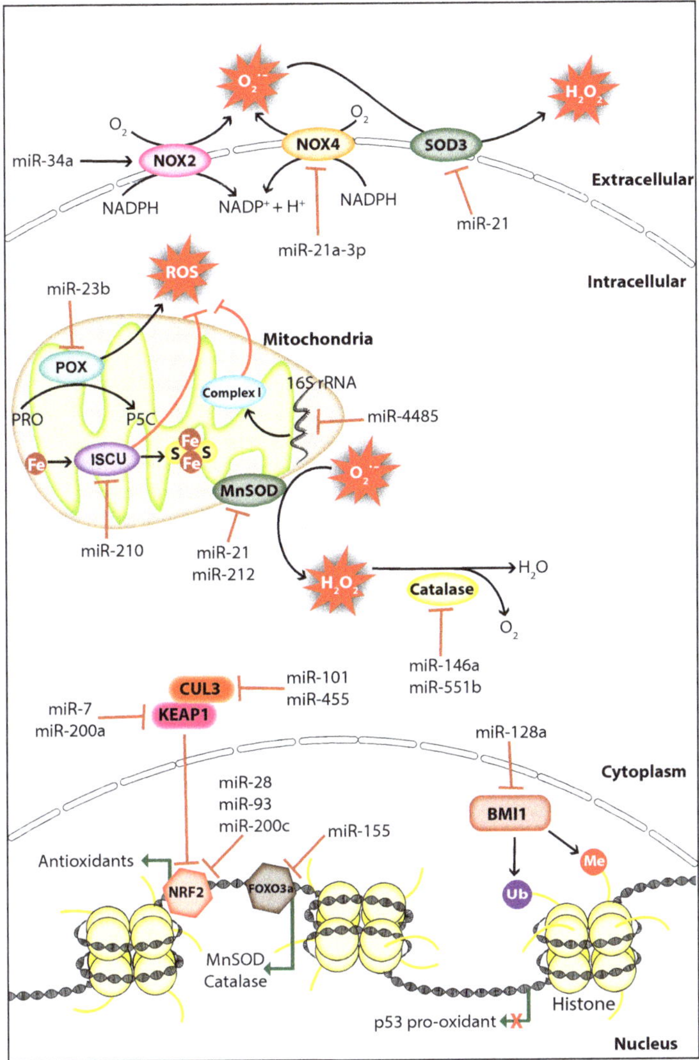

Figure 2. MiRNAs regulate ROS levels in cancer. MiRNAs regulate ROS levels in cancer by inhibiting the expression of ROS producers or antioxidants. MiRNAs decrease ROS levels through inhibiting ROS producers NOX2 (NADPH oxidase 2), NOX4 (NADPH oxidase 4), POX (Proline oxidase), or indirectly by inhibiting the polycomb complex protein BMI1 which repress p53 pro-oxidant expression. ROS levels are elevated by miRNAs through direct or indirect inhibition of antioxidants including catalase, SOD3 (Superoxide dismutase 3), MnSOD (Manganese-dependent superoxide dismutase), and proteins involved in mitochondrial function including mitochondrial complex I (NADH Coenzyme Q reductase) and ISCU (Iron-sulfur cluster assembly enzyme). BMI1, B lymphoma Mo-MLV insertion region 1 homolog; CUL3, cullin-3, Fe, iron; FOXO3a, forkhead box O3; KEAP1, Kelch-like ECH-associated protein 1, NRF2, nuclear factor-erythroid 2-related factor 2; H$_2$O$_2$, hydrogen peroxide; Me, methyl group; NADPH, nicotinamide adenine dinucleotide phosphate; O$_2$⁻, superoxide; PRO, proline; P5C, 1-pyrroline-5-carboxylate; S, sulfur; Ub, ubiquitin (The black arrow indicates upmodulation, the red T arrow indicates inhibition, the green arrow indicates transcription activation, the green arrow with red cross indicates transcription inhibition).

4.2. Regulation of Mitochondrial Functions

Mitochondria are the major site for ROS production, and redox homeostasis in mitochondria is crucial for normal cellular processes. MiRNAs have been shown to affect the ROS production of mitochondria in cancer cells by regulating genes associated with mitochondrial function. The hypoxia-induced miR-210 promotes ROS production by repressing the iron-sulfur cluster assembly enzyme (ISCU) which is essential for the assembly of iron-sulfur (Fe-S) cluster and mitochondria respiratory activity [97]. However, a study suggests that miR-210 mediated ROS accumulation may be due to the repression of other gene targets since the ISCU knockdown in colon cancer cells does not increase ROS levels significantly [98]. A study has shown that miR-128a promotes intracellular ROS levels and cellular senescence in medulloblastoma cells by directly targeting the polycomb complex protein BMI-1 which is involved in the maintenance of mitochondrial activities and redox homeostasis [99]. Surprisingly, a study demonstrated that miRNAs regulate ROS production by targeting non-coding RNAs, during cellular stress miR-4485 translocates to mitochondria and directly targets mitochondrial 16S ribosomal RNA (rRNA), thus modulates mitochondrial function and subsequent ROS accumulation (Figure 2). Importantly, miR-4485 levels are decreased in human breast cancer tissues and overexpression of miR-4485 suppressed breast cancer tumorigenesis in-vitro and in-vivo [100]. These findings strongly suggest that hindrance in mitochondrial metabolism can promote carcinogenesis through ROS accumulation.

4.3. Regulation of Antioxidants

Antioxidant enzymes and non-enzymatic antioxidants mediate the detoxification of ROS to protect cells from oxidative damage. Superoxide dismutase (SOD) is an antioxidant metalloenzyme expressed in both eukaryotes and prokaryotes, which utilizes the metal ions including copper, iron, manganese, and zinc as cofactors to catalyze the dismutation of O_2^- into molecular oxygen (O_2) and H_2O_2. Similarly, catalase is an antioxidant enzyme located mostly in the cytosol and peroxisomes scavenge ROS through catalyzing the conversion of H_2O_2 into water (H_2O) and O_2 [101]. Several studies have shown that miRNAs can upregulate cellular ROS levels in cancer cells by inhibiting antioxidants including SOD and catalase. The oncomiR miR-21 promotes tumorigenesis through increasing cellular ROS levels by directly targeting the SOD3 or by targeting TNFα that results in MnSOD downregulation (Figure 2) [102]. Furthermore, the miR-212 which is downregulated in human colorectal cancer (CRC) can regulate MnSOD by directly targeting its mRNA, and overexpression of miR-212 inhibited metastasis of CRC cells by suppressing MnSOD expression [103]. In cancer cells, catalase expression is regulated by miR-551b and miR-146a, and inhibition of catalase by these miRNAs promotes ROS accumulation [104,105]. Interestingly, miRNAs can also control the expression of antioxidants indirectly through targeting transcription factors that promote the transcription of antioxidants. For example, K-Ras-induced miR-155 increases ROS levels by directly targeting FOXO3a, a transcription factor that activates the transcription of antioxidants MnSOD and catalase (Figure 2) [106]. These findings suggest that the endogenous expression of endogenous antioxidants is crucial for the prevention of cellular ROS accumulation, which is manipulated by miRNAs in cancer cells to support cancer progression.

4.4. Regulation of NRF2/KEAP1 System

Cellular redox homeostasis is controlled by the nuclear factor-erythroid 2-related factor 2 (NRF2)/ Kelch-like ECH-associated protein 1 (KEAP1) system. NRF2 is a transcriptional factor, which activates the transcription of genes that encode antioxidant enzymes and non-enzymatic antioxidants in response to oxidative stress. Under normal conditions, NRF2 is inactivated by the KEAP1-cullin3 (CUL3) complex, which sequesters NRF2 in the cytoplasm and promotes NRF2 degradation through ubiquitination. During oxidative stress, NRF2 is dissociated from the KEAP1-CUL3 complex caused by the rapid oxidation on cys151 residue of KEAP1 [107]. In cancer, miRNAs can affect the cellular redox homeostasis by targeting genes involved in the NRF2/KEAP1 regulatory system. Overexpression of miR-200c in lung cancer cells increases ROS levels through suppressing the expression of proteins involved in oxidative stress defense including peroxiredoxin 2, NRF2, and Sestrin 1 [108]. A study has shown that miR-28 decreases NRF2 expression by directly targeting its 3'UTR, which increased the colony formation capacity in breast cancer cells [109]. Similarly, miR-93 regulates NRF2 and is associated with breast cancer development [110]. Moreover, a bioinformatic prediction showed that about 85 miRNAs may negatively regulate NRF2 expression by directly targeting its mRNA [111]. miRNAs also regulate NRF2 activity indirectly through targeting its inhibitors KEAP1 and CUL3 (Figure 2). miR-7 and miR-200a target KEAP1 mRNA and decrease its protein expression thereby mediating NRF2 nuclear localization and target gene transcription in neuroblastoma and breast cancer cells, respectively [112,113]. Likewise, miR-101 and miR-455 target CUL3 mRNA, which promotes NRF2 nuclear localization that leads to angiogenesis and oxidative stress protection, respectively [114,115]. Altogether, these studies strongly suggest that cancer cells manipulate ROS levels by controlling miRNA expression to support their survival and promotion.

5. The Interplay of ROS and MiRNAs in Cancer

Oxidative stress induces DNA damage or mutation that may affect the expression and function of genes associated with the damaged genomic loci, and can eventually cause cancer initiation. ROS may also affect miRNA expression and function directly by causing oxidative damage-induced mutation on miRNA genes and mature miRNA sequences, or indirectly by altering its epigenetic signature or biogenesis pathway. Deregulated miRNA expression caused through genomic deletion, epigenetic silencing, or overexpression can contribute to cancer initiation and progression by controlling oncogenes and tumor suppressor genes. Therefore, ROS can regulate miRNA-mediated carcinogenesis. Elevated ROS production is observed in various cancer types and high cellular ROS can activate oncogenic signaling pathways that support cancer progression. MiRNAs are able to control the cellular ROS levels by targeting genes involved in ROS production and elimination, thus miRNA can control ROS-mediated carcinogenesis. These facts suggest that ROS and miRNAs can function synergistically in the process of cancer development (Figure 3). ROS upregulate the expression of the oncomiR miR-21 and miR-146a through activating NFκB, and these miRNAs can increase cellular ROS levels by downregulating endogenous antioxidants [73,102,105]. Similarly, the miR-210 expression is upregulated through ROS-mediated activation of HIF-1α, and the miR-210 has been shown to increase ROS production by negatively regulating ISCU [97,116]. Interestingly, ROS can also upregulate the expression of the tumor suppressor miR-34 through p53 activation, whereas the miR-34 has been shown to increase ROS production by upregulating the expression of NOX2 [64,94]. These studies strongly suggest that ROS and miRNAs crosstalk in cancer cells to orchestrate the ROS production to activate and promote cancer development. Furthermore, it is of importance to investigate whether the mRNA of genes involved in miRNA expression, ROS production, and detoxification would function as potential competing endogenous RNAs (ceRNAs) which can co-regulate each other's expression by competing for binding to shared miRNAs [117]. For example, miR-210 can directly target the mRNA of HIF-1α and ISCU, suggesting that HIF-1α and ISCU could function as potential ceRNAs [80,97]. Likewise, miR-21 has been shown to downregulate the expression of antioxidants SOD3 and MnSOD [102]. However, miR-21 targets only the mRNA of SOD3 but not the MnSOD. Therefore, it would be interesting to investigate whether the MnSOD mRNA encompasses a binding site for miR-21 or any other miRNA that can target SOD3. Nevertheless, more studies should be done in this perspective to unravel the complete regulatory network between miRNA and ROS in cancer.

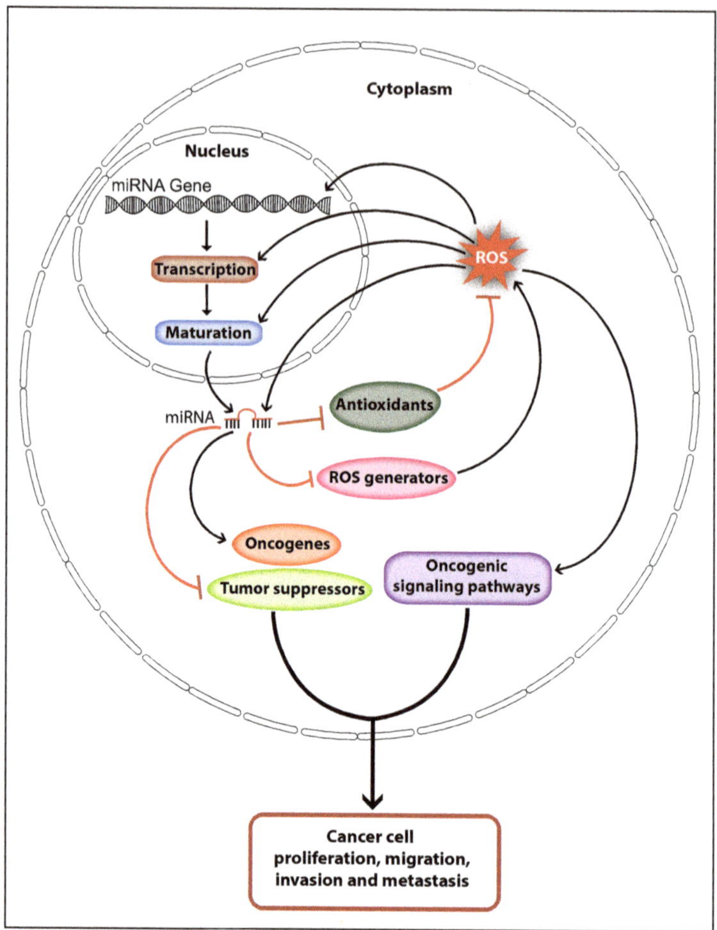

Figure 3. Schematic model illustrates the interplay of ROS and miRNAs in cancer progression. ROS regulate miRNA expression and function by altering miRNA transcription, maturation, or sequences. Dysregulated miRNA expression promotes cancer progression by regulating oncogenes and tumor suppressors. MiRNAs control cellular ROS levels by regulating endogenous antioxidants and ROS generators, which favor cancer development through activating oncogenic signaling pathways (The black arrow indicates upmodulation, the red T arrow indicates inhibition).

6. Challenges in Using Antioxidants for Anti-Cancer Therapy

Since ROS mediate cellular damage and oncogenic mutations, usage of dietary supplement with antioxidants was proposed to prevent or treat cancer. Dietary supplement containing antioxidants such as selenium, vitamin E, and β-carotene was tested to reduce the occurrence of cancer in individuals with a history of cancer. This resulted in a significant decrease in total cancer occurrence and overall mortality. Conversely, studies also show that nutritional supplements of antioxidants may promote cancer incidence and mortality [118–120]. Moreover, the usage of antioxidants as additional therapy in cancer treatment failed to show beneficiary effect, supplementing breast and colorectal cancer patients with ascorbate/vitamin C does not improve overall or progression-free survival [121]. Even though antioxidants are often ineffective for cancer prevention/treatment in humans with a high risk of cancer, it was shown that antioxidant treatment might suppress cancer risk in mice with certain genetic modifications. NAC treatment reduces ROS generation, DNA damage, and cancer occurrence in mice deficient of ATM and p53 [122,123]. However, this was not consistent since another study demonstrated that treating mouse models of lung cancer with antioxidants NAC and vitamin E promotes tumor progression and decreases mouse survival [124]. The underlying cause of antioxidants promoting cancer progression may be due to the fact that cancer cells are more susceptible to oxidative stress when compared to normal cells. Therefore, cancer cells depend on endogenous antioxidants including GSH, TRX, NRF2, thioredoxin-like 2, SOD, MnSOD, and glutamate-cysteine ligase to protect them from oxidative stress during cancer development [125–130]. In cancer cells, several oncogenes increase NRF2 transcription to promote ROS detoxification and tumorigenesis, whereas the deletion of NRF2 promotes DNA damage and suppresses tumorigenesis in pancreatic cancer cells [131]. In some cancers, ROS levels are suppressed by continuous activation of NRF2 achieved via mutations in NRF2 or its inhibitors KEAP1 that prevents NRF2 translocation from nucleus to cytoplasm [132]. Oxidative stress also limits metastasis by melanoma cells, whereas antioxidant treatment in a mouse model of malignant melanoma promotes the distant metastasis without affecting the growth of primary subcutaneous tumors [133]. Furthermore, cancer cells manage ROS levels by increasing NAPDH generation through accelerating multiple metabolic pathways including the pentose phosphate, folate, and malic enzyme pathway [134–136]. These studies suggest that antioxidant treatments are beneficial for cancer progression instead of being detrimental to cancer cells. Importantly, the inconsistent outcome from the clinical trials and experimental mouse models suggest that the application of antioxidants for anti-cancer therapy may not be a promising approach.

On the other hand, miRNAs are suggested as promising therapeutic agents for cancer treatment. In recent years, several studies proposed many novel miRNA-based cancer therapies that have significantly improved the survival of cancer patients [137]. The application of miRNAs as therapeutic agents has many potential advantages. Basically, miRNAs are highly conserved endogenous small RNA molecules with known sequences which may simplify the process of designing therapeutic agents with less off-target effects. A single miRNA can potentially regulate multiple target genes associated with single or multiple pathways which could be a very efficient way to treat multi-pathway diseases including cancer. For anti-cancer therapy, two miRNA-based strategies are applied. MiRNA replacement therapy is applied to either induce apoptosis or suppress the proliferation of cancer cells by using exogenous tumor suppressor miRNA mimics. MiRNA reduction therapy is applied to inhibit the function of oncogenic miRNAs by using antagomiRs or locked-nucleic acids antisense oligonucleotides (LNAs) [138,139]. To date, there is no miRNA-based drug available for cancer treatment. However, some miRNA drug candidates have entered into the early phase of human clinical trials. These include MesomiR-1, the miRNA mimic of tumor-suppressing miR-16 for treating lung cancer; MRX34, the miRNA mimic of tumor-suppressing miR-34 for treating liver cancer, lymphoma and melanoma; and MRG106, the LNA-modified anti-miR of miR-155 for treating T-cell lymphoma [140]. Although miRNA-based therapy has made progress, still there are some challenges ahead to become an efficient therapeutic approach. The adverse effect is one of the major challenges encountered by this therapy. For example, MRX34 has been withdrawn from entering phase 2 trials due to the serious immune response observed in some patients during phase 1 trials [140]. There are limitations in the efficiency of in-vivo delivery of miRNA mimics and antagomiRs as the oligonucleotides are degraded by the endonucleases in the blood. Understanding the regulatory network of miRNA and ROS production in cancer would further help to develop an alternative effective therapeutic approach to treat cancer. One such approach would be aggravating oxidative stress in cancer cells through miRNA-based therapy that either enhance ROS production or inhibit endogenous antioxidant system.

7. Concluding Remarks

ROS function as a mediator of cellular signaling pathways involved in proliferation, growth, survival, and apoptosis, and the redox homeostasis is actively maintained by endogenous antioxidant systems. Cancer cells manipulate the cellular ROS levels to favor their proliferation, survival, and metastasis. ROS levels are regulated via fine-tuning the expression of ROS producers and scavengers by miRNAs. On the other hand, ROS regulate miRNA expression by altering the activity of proteins involved in miRNA transcription and maturation. The regulatory network of ROS and miRNAs is orchestrated in cancer to promote cancer progression and to cope with oxidative stress. Identification of regulatory crosstalk between miRNA and redox signaling opens up new horizons for using miRNAs as potential therapeutic targets in cancer treatment. However, further understanding of the miRNA-ROS regulatory network is needed for the application of miRNAs to augment ROS-mediated cancer cell death.

Author Contributions: K.R.B. wrote the manuscript. K.R.B. and Y.T. edited and revised the manuscript.

Funding: This work was supported by the RNA Biology Center at the Cancer Science Institute of Singapore, National University of Singapore (NUS), as part of funding under the Singapore Ministry of Education's Tier 3 grants, grant number MOE2014-T3-1-006.

Acknowledgments: We apologize to all colleagues whose work could not be cited due to space constraints. We thank Yvonne Tay lab members (Cancer Science Institute of Singapore) for critical reading of the manuscript. Yvonne Tay was supported by a Singapore National Research Foundation Fellowship and a National University of Singapore President's Assistant Professorship.

Conflicts of Interest: The authors declare no conflict of interest.

References

1. Snezhkina, A.V.; Kudryavtseva, A.V.; Kardymon, O.L.; Savvateeva, M.V.; Melnikova, N.V.; Krasnov, G.S.; Dmitriev, A.A. ROS Generation and antioxidant defense systems in normal and malignant cells. *Oxidative Med. Cell. Longev.* **2019**, *2019*, 6175804. [CrossRef] [PubMed]

2. Di Meo, S.; Reed, T.T.; Venditti, P.; Victor, V.M. Role of ROS and RNS sources in physiological and pathological conditions. *Oxidative Med. Cell. Longev.* **2016**, *2016*, 1245049. [CrossRef] [PubMed]

3. Brieger, K.; Schiavone, S.; Miller, F.J., Jr.; Krause, K.H. Reactive oxygen species: From health to disease. *Swiss Med Wkly.* **2012**, *142*, 13659. [CrossRef] [PubMed]

4. Schieber, M.; Chandel, N.S. ROS function in redox signaling and oxidative stress. *Curr. Biol.* **2014**, *24*, 453–462. [CrossRef] [PubMed]

5. Jena, N.R. DNA damage by reactive species: Mechanisms, mutation and repair. *J. Biosci.* **2012**, *37*, 503–517. [CrossRef] [PubMed]

6. Poulsen, H.E.; Specht, E.; Broedbaek, K.; Henriksen, T.; Ellervik, C.; Mandrup-Poulsen, T.; Tonnesen, M.; Nielsen, P.E.; Andersen, H.U.; Weimann, A. RNA modifications by oxidation: A novel disease mechanism? *Free. Radic. Biol. Med.* **2012**, *52*, 1353–1361. [CrossRef] [PubMed]

7. Fimognari, C. Role of oxidative RNA damage in chronic-degenerative diseases. *Oxidative Med. Cell. Longev.* **2015**, *2015*, 358713. [CrossRef]

8. Ma, Y.; Zhang, L.; Rong, S.; Qu, H.; Zhang, Y.; Chang, D.; Pan, H.; Wang, W. Relation between gastric cancer and protein oxidation, DNA damage, and lipid peroxidation. *Oxidative Med. Cell. Longev.* **2013**, *2013*, 543760. [CrossRef]

9. Barrera, G. Oxidative stress and lipid peroxidation products in cancer progression and therapy. *ISRN Oncol.* **2012**, *2012*, 137289. [CrossRef]

10. Bartel, D.P. Metazoan MicroRNAs. *Cell* **2018**, *173*, 20–51. [CrossRef]

11. Peng, Y.; Croce, C.M. The role of MicroRNAs in human cancer. *Signal Transduct. Target. Ther.* **2016**, *1*, 15004. [CrossRef] [PubMed]

12. Thulasingam, S.; Massilamany, C.; Gangaplara, A.; Dai, H.; Yarbaeva, S.; Subramaniam, S.; Riethoven, J.J.; Eudy, J.; Lou, M.; Reddy, J. MiR-27b, an oxidative stress-responsive microRNA modulates nuclear factor-kB pathway in RAW 264.7 cells. *Mol. Cell. Biochem.* **2011**, *352*, 181–188. [CrossRef] [PubMed]

13. Peng, J.; He, X.; Zhang, L.; Liu, P. MicroRNA26a protects vascular smooth muscle cells against H_2O_2 induced injury through activation of the PTEN/AKT/mTOR pathway. *Int. J. Mol. Med.* **2018**, *42*, 1367–1378. [CrossRef] [PubMed]

14. Cheng, Y.; Zhou, M.; Zhou, W. MicroRNA-30e regulates TGF-beta-mediated NADPH oxidase 4-dependent oxidative stress by Snail in atherosclerosis. *Int. J. Mol. Med.* **2019**, *43*, 1806–1816. [CrossRef]

15. Van Houten, B.; Santa-Gonzalez, G.A.; Camargo, M. DNA repair after oxidative stress: Current challenges. *Curr. Opin. Toxicol.* **2018**, *7*, 9–16. [CrossRef]

16. Anastasiadou, E.; Jacob, L.S.; Slack, F.J. Non-coding RNA networks in cancer. *Nat. Rev. Cancer* **2018**, *18*, 5–18. [CrossRef]

17. Davies, M.J. Protein oxidation and peroxidation. *Biochem. J.* **2016**, *473*, 805–825. [CrossRef]

18. Gangwar, A.; Paul, S.; Ahmad, Y.; Bhargava, K. Competing trends of ROS and RNS-mediated protein modifications during hypoxia as an alternate mechanism of NO benefits. *Biochimie* **2018**, *148*, 127–138. [CrossRef]

19. Liou, G.Y.; Storz, P. Reactive oxygen species in cancer. *Free Radic. Res.* **2010**, *44*, 479–496. [CrossRef]

20. Zhong, H.; Yin, H. Role of lipid peroxidation derived 4-hydroxynonenal (4-HNE) in cancer: Focusing on mitochondria. *Redox Biol.* **2015**, *4*, 193–199. [CrossRef]

21. Liao, Z.; Chua, D.; Tan, N.S. Reactive oxygen species: A volatile driver of field cancerization and metastasis. *Mol. Cancer* **2019**, *18*, 65. [CrossRef] [PubMed]

22. Lingappan, K. NF-kappaB in oxidative stress. *Curr. Opin. Toxicol.* **2018**, *7*, 81–86. [CrossRef] [PubMed]

23. Verbon, E.H.; Post, J.A.; Boonstra, J. The influence of reactive oxygen species on cell cycle progression in mammalian cells. *Gene* **2012**, *511*, 1–6. [CrossRef] [PubMed]

24. Ruiz-Ramos, R.; Lopez-Carrillo, L.; Rios-Perez, A.D.; De Vizcaya-Ruiz, A.; Cebrian, M.E. Sodium arsenite induces ROS generation, DNA oxidative damage, HO-1 and c-Myc proteins, NF-kappaB activation and cell proliferation in human breast cancer MCF-7 cells. *Mutat. Res.* **2009**, *674*, 109–115. [CrossRef] [PubMed]

25. Menon, S.G.; Coleman, M.C.; Walsh, S.A.; Spitz, D.R.; Goswami, P.C. Differential susceptibility of nonmalignant human breast epithelial cells and breast cancer cells to thiol antioxidant-induced G(1)-delay. *Antioxid. Redox Signal.* **2005**, *7*, 711–718. [CrossRef] [PubMed]

26. Vanharanta, S.; Massague, J. Origins of metastatic traits. *Cancer cell* **2013**, *24*, 410–421. [CrossRef]

27. Piskounova, E.; Agathocleous, M.; Murphy, M.M.; Hu, Z.; Huddlestun, S.E.; Zhao, Z.; Leitch, A.M.; Johnson, T.M.; DeBerardinis, R.J.; Morrison, S.J. Oxidative stress inhibits distant metastasis by human melanoma cells. *Nature* **2015**, *527*, 186–191. [CrossRef]

28. Hitchler, M.J.; Oberley, L.W.; Domann, F.E. Epigenetic silencing of SOD2 by histone modifications in human breast cancer cells. *Free Radic. Biol. Med.* **2008**, *45*, 1573–1580. [CrossRef]

29. Lewis, A.; Du, J.; Liu, J.; Ritchie, J.M.; Oberley, L.W.; Cullen, J.J. Metastatic progression of pancreatic cancer: Changes in antioxidant enzymes and cell growth. *Clin. Exp. Metastasis* **2005**, *22*, 523–532. [CrossRef]

30. Roy, R.; Morad, G.; Jedinak, A.; Moses, M.A. Metalloproteinases and their roles in human cancer. *Anat. Rec. (Hoboken)* **2019**. [CrossRef]

31. Kang, K.A.; Ryu, Y.S.; Piao, M.J.; Shilnikova, K.; Kang, H.K.; Yi, J.M.; Boulanger, M.; Paolillo, R.; Bossis, G.; Yoon, S.Y.; et al. DUOX2-mediated production of reactive oxygen species induces epithelial mesenchymal transition in 5-fluorouracil resistant human colon cancer cells. *Redox. Biol.* **2018**, *17*, 224–235. [CrossRef] [PubMed]

32. Radisky, D.C.; Levy, D.D.; Littlepage, L.E.; Liu, H.; Nelson, C.M.; Fata, J.E.; Leake, D.; Godden, E.L.; Albertson, D.G.; Nieto, M.A.; et al. Rac1b and reactive oxygen species mediate MMP-3-induced EMT and genomic instability. *Nature* **2005**, *436*, 123–127. [CrossRef] [PubMed]

33. van Wetering, S.; van Buul, J.D.; Quik, S.; Mul, F.P.; Anthony, E.C.; ten Klooster, J.P.; Collard, J.G.; Hordijk, P.L. Reactive oxygen species mediate Rac-induced loss of cell-cell adhesion in primary human endothelial cells. *J. Cell Sci.* **2002**, *115*, 1837–1846. [PubMed]

34. Kuai, W.X.; Wang, Q.; Yang, X.Z.; Zhao, Y.; Yu, R.; Tang, X.J. Interleukin-8 associates with adhesion, migration, invasion and chemosensitivity of human gastric cancer cells. *World J. Gastroenterol.* **2012**, *18*, 979–985. [CrossRef]

35. Ghislin, S.; Obino, D.; Middendorp, S.; Boggetto, N.; Alcaide-Loridan, C.; Deshayes, F. LFA-1 and ICAM-1 expression induced during melanoma-endothelial cell co-culture favors the transendothelial migration of melanoma cell lines in vitro. *BMC Cancer* **2012**, *12*, 455. [CrossRef] [PubMed]

36. Sawada, J.; Li, F.; Komatsu, M. R-Ras Inhibits VEGF-Induced p38MAPK Activation and HSP27 Phosphorylation in Endothelial Cells. *J. Vasc. Res.* **2015**, *52*, 347–359. [CrossRef]

37. Shi, X.; Zhang, Y.; Zheng, J.; Pan, J. Reactive oxygen species in cancer stem cells. *Antioxid. Redox Signal.* **2012**, *16*, 1215–1228. [CrossRef]

38. Diehn, M.; Cho, R.W.; Lobo, N.A.; Kalisky, T.; Dorie, M.J.; Kulp, A.N.; Qian, D.; Lam, J.S.; Ailles, L.E.; Wong, M.; et al. Association of reactive oxygen species levels and radioresistance in cancer stem cells. *Nature* **2009**, *458*, 780–783. [CrossRef]

39. Pasto, A.; Bellio, C.; Pilotto, G.; Ciminale, V.; Silic-Benussi, M.; Guzzo, G.; Rasola, A.; Frasson, C.; Nardo, G.; Zulato, E.; et al. Cancer stem cells from epithelial ovarian cancer patients privilege oxidative phosphorylation, and resist glucose deprivation. *Oncotarget* **2014**, *5*, 4305–4319. [CrossRef]

40. Jang, Y.Y.; Sharkis, S.J. A low level of reactive oxygen species selects for primitive hematopoietic stem cells that may reside in the low-oxygenic niche. *Blood* **2007**, *110*, 3056–3063. [CrossRef]

41. He, J.; Xu, Q.; Jing, Y.; Agani, F.; Qian, X.; Carpenter, R.; Li, Q.; Wang, X.R.; Peiper, S.S.; Lu, Z.; et al. Reactive oxygen species regulate ERBB2 and ERBB3 expression via miR-199a/125b and DNA methylation. *EMBO Rep.* **2012**, *13*, 1116–1122. [CrossRef] [PubMed]

42. He, J.; Jing, Y.; Li, W.; Qian, X.; Xu, Q.; Li, F.S.; Liu, L.Z.; Jiang, B.H.; Jiang, Y. Roles and mechanism of miR-199a and miR-125b in tumor angiogenesis. *PLoS ONE* **2013**, *8*, e56647. [CrossRef] [PubMed]

43. Li, Y.; Seto, E. HDACs and HDAC inhibitors in cancer development and therapy. *Cold Spring Harb. Perspect. Med.* **2016**, *6*. [CrossRef] [PubMed]

44. Ago, T.; Liu, T.; Zhai, P.; Chen, W.; Li, H.; Molkentin, J.D.; Vatner, S.F.; Sadoshima, J. A redox-dependent pathway for regulating class II HDACs and cardiac hypertrophy. *Cell* **2008**, *133*, 978–993. [CrossRef] [PubMed]

45. Ciesla, M.; Marona, P.; Kozakowska, M.; Jez, M.; Seczynska, M.; Loboda, A.; Bukowska-Strakova, K.; Szade, A.; Walawender, M.; Kusior, M.; et al. Heme oxygenase-1 controls an HDAC4-miR-206 pathway of oxidative stress in rhabdomyosarcoma. *Cancer Res.* **2016**, *76*, 5707–5718. [CrossRef]

46. Singh, A.; Happel, C.; Manna, S.K.; Acquaah-Mensah, G.; Carrerero, J.; Kumar, S.; Nasipuri, P.; Krausz, K.W.; Wakabayashi, N.; Dewi, R.; et al. Transcription factor NRF2 regulates miR-1 and miR-206 to drive tumorigenesis. *J. Clin. Investig.* **2013**, *123*, 2921–2934. [CrossRef]

47. Druz, A.; Betenbaugh, M.; Shiloach, J. Glucose depletion activates mmu-miR-466h-5p expression through oxidative stress and inhibition of histone deacetylation. *Nucleic Acids Res.* **2012**, *40*, 7291–7302. [CrossRef]

48. Druz, A.; Chu, C.; Majors, B.; Santuary, R.; Betenbaugh, M.; Shiloach, J. A novel microRNA mmu-miR-466h affects apoptosis regulation in mammalian cells. *Biotechnol. Bioeng.* **2011**, *108*, 1651–1661. [CrossRef]

49. Dando, I.; Cordani, M.; Dalla Pozza, E.; Biondani, G.; Donadelli, M.; Palmieri, M. Antioxidant Mechanisms and ROS-Related MicroRNAs in Cancer Stem Cells. *Oxidative Med. Cell. Longev.* **2015**, *2015*, 425708. [CrossRef]

50. Benassi, B.; Fanciulli, M.; Fiorentino, F.; Porrello, A.; Chiorino, G.; Loda, M.; Zupi, G.; Biroccio, A. c-Myc phosphorylation is required for cellular response to oxidative stress. *Mol. cell* **2006**, *21*, 509–519. [CrossRef]

51. Huang, S.; He, X.; Ding, J.; Liang, L.; Zhao, Y.; Zhang, Z.; Yao, X.; Pan, Z.; Zhang, P.; Li, J.; et al. Upregulation of miR-23a approximately 27a approximately 24 decreases transforming growth factor-beta-induced tumor-suppressive activities in human hepatocellular carcinoma cells. *Int. J. Cancer* **2008**, *123*, 972–978. [CrossRef] [PubMed]

52. Yang, H.; Li, T.W.; Zhou, Y.; Peng, H.; Liu, T.; Zandi, E.; Martinez-Chantar, M.L.; Mato, J.M.; Lu, S.C. Activation of a novel c-Myc-miR27-prohibitin 1 circuitry in cholestatic liver injury inhibits glutathione synthesis in mice. *Antioxid. Redox Signal.* **2015**, *22*, 259–274. [CrossRef] [PubMed]

53. Yeo, C.Q.X.; Alexander, I.; Lin, Z.; Lim, S.; Aning, O.A.; Kumar, R.; Sangthongpitag, K.; Pendharkar, V.; Ho, V.H.B.; Cheok, C.F. p53 Maintains genomic stability by preventing interference between transcription and replication. *Cell Rep.* **2016**, *15*, 132–146. [CrossRef] [PubMed]

54. Navarro, F.; Lieberman, J. miR-34 and p53: New insights into a complex functional relationship. *PLoS ONE* **2015**, *10*, e0132767. [CrossRef]

55. Chen, Y.; Liu, K.; Shi, Y.; Shao, C. The tango of ROS and p53 in tissue stem cells. *Cell Death Differ.* **2018**, *25*, 639–641. [CrossRef]

56. Peuget, S.; Bonacci, T.; Soubeyran, P.; Iovanna, J.; Dusetti, N.J. Oxidative stress-induced p53 activity is enhanced by a redox-sensitive TP53INP1 SUMOylation. *Cell Death Differ.* **2014**, *21*, 1107–1118. [CrossRef]

57. Maillet, A.; Pervaiz, S. Redox regulation of p53, redox effectors regulated by p53: A subtle balance. *Antioxid. Redox Signal.* **2012**, *16*, 1285–1294. [CrossRef]

58. Haffo, L.; Lu, J.; Bykov, V.J.N.; Martin, S.S.; Ren, X.; Coppo, L.; Wiman, K.G.; Holmgren, A. Inhibition of the glutaredoxin and thioredoxin systems and ribonucleotide reductase by mutant p53-targeting compound APR-246. *Sci. Rep.* **2018**, *8*, 12671. [CrossRef]

59. Xiao, Y.; Yan, W.; Lu, L.; Wang, Y.; Lu, W.; Cao, Y.; Cai, W. p38/p53/miR-200a-3p feedback loop promotes oxidative stress-mediated liver cell death. *Cell Cycle* **2015**, *14*, 1548–1558. [CrossRef]

60. Magenta, A.; Cencioni, C.; Fasanaro, P.; Zaccagnini, G.; Greco, S.; Sarra-Ferraris, G.; Antonini, A.; Martelli, F.; Capogrossi, M.C. miR-200c is upregulated by oxidative stress and induces endothelial cell apoptosis and senescence via ZEB1 inhibition. *Cell Death Differ.* **2011**, *18*, 1628–1639. [CrossRef]

61. Gibbons, D.L.; Lin, W.; Creighton, C.J.; Rizvi, Z.H.; Gregory, P.A.; Goodall, G.J.; Thilaganathan, N.; Du, L.; Zhang, Y.; Pertsemlidis, A.; et al. Contextual extracellular cues promote tumor cell EMT and metastasis by regulating miR-200 family expression. *Genes Dev.* **2009**, *23*, 2140–2151. [CrossRef] [PubMed]

62. Feng, X.; Wang, Z.; Fillmore, R.; Xi, Y. MiR-200, a new star miRNA in human cancer. *Cancer Lett.* **2014**, *344*, 166–173. [CrossRef] [PubMed]

63. Yin, M.; Ren, X.; Zhang, X.; Luo, Y.; Wang, G.; Huang, K.; Feng, S.; Bao, X.; Huang, K.; He, X.; et al. Selective killing of lung cancer cells by miRNA-506 molecule through inhibiting NF-kappaB p65 to evoke reactive oxygen species generation and p53 activation. *Oncogene* **2015**, *34*, 691–703. [CrossRef]

64. Wan, Y.; Cui, R.; Gu, J.; Zhang, X.; Xiang, X.; Liu, C.; Qu, K.; Lin, T. Identification of Four Oxidative Stress-Responsive MicroRNAs, miR-34a-5p, miR-1915-3p, miR-638, and miR-150-3p, in Hepatocellular Carcinoma. *Oxidative Med. Cell. Longev.* **2017**, *2017*, 5189138. [CrossRef] [PubMed]

65. Liu, T.; Zhang, L.; Joo, D.; Sun, S.C. NF-kappaB signaling in inflammation. *Signal Transduct. Target. Ther.* **2017**, *2*. [CrossRef] [PubMed]

66. Markopoulos, G.S.; Roupakia, E.; Tokamani, M.; Alabasi, G.; Sandaltzopoulos, R.; Marcu, K.B.; Kolettas, E. Roles of NF-kappaB Signaling in the regulation of miRNAs impacting on inflammation in cancer. *Biomedicines* **2018**, *6*, 40. [CrossRef] [PubMed]

67. Li, X.; Xin, S.; He, Z.; Che, X.; Wang, J.; Xiao, X.; Chen, J.; Song, X. MicroRNA-21 (miR-21) post-transcriptionally downregulates tumor suppressor PDCD4 and promotes cell transformation, proliferation, and metastasis in renal cell carcinoma. *Cell. Physiol. Biochem.* **2014**, *33*, 1631–1642. [CrossRef]

68. Li, Z.; Deng, X.; Kang, Z.; Wang, Y.; Xia, T.; Ding, N.; Yin, Y. Elevation of miR-21, through targeting MKK3, may be involved in ischemia pretreatment protection from ischemia-reperfusion induced kidney injury. *J. Nephrol.* **2016**, *29*, 27–36. [CrossRef]

69. Yang, C.H.; Yue, J.; Pfeffer, S.R.; Fan, M.; Paulus, E.; Hosni-Ahmed, A.; Sims, M.; Qayyum, S.; Davidoff, A.M.; Handorf, C.R.; et al. MicroRNA-21 promotes glioblastoma tumorigenesis by down-regulating insulin-like growth factor-binding protein-3 (IGFBP3). *J. Biol. Chem.* **2014**, *289*, 25079–25087. [CrossRef]

70. Peralta-Zaragoza, O.; Deas, J.; Meneses-Acosta, A.; De la, O.G.F.; Fernandez-Tilapa, G.; Gomez-Ceron, C.; Benitez-Boijseauneau, O.; Burguete-Garcia, A.; Torres-Poveda, K.; Bermudez-Morales, V.H.; et al. Relevance of miR-21 in regulation of tumor suppressor gene PTEN in human cervical cancer cells. *BMC Cancer* **2016**, *16*, 215. [CrossRef]

71. Feng, Y.H.; Tsao, C.J. Emerging role of microRNA-21 in cancer. *Biomed. Rep.* **2016**, *5*, 395–402. [CrossRef] [PubMed]

72. Jajoo, S.; Mukherjea, D.; Kaur, T.; Sheehan, K.E.; Sheth, S.; Borse, V.; Rybak, L.P.; Ramkumar, V. Essential role of NADPH oxidase-dependent reactive oxygen species generation in regulating microRNA-21 expression and function in prostate cancer. *Antioxid. Redox Signal.* **2013**, *19*, 1863–1876. [CrossRef] [PubMed]

73. Ling, M.; Li, Y.; Xu, Y.; Pang, Y.; Shen, L.; Jiang, R.; Zhao, Y.; Yang, X.; Zhang, J.; Zhou, J.; et al. Regulation of miRNA-21 by reactive oxygen species-activated ERK/NF-kappaB in arsenite-induced cell transformation. *Free Radic. Biol. Med.* **2012**, *52*, 1508–1518. [CrossRef] [PubMed]

74. Mohr, S.; Doebele, C.; Comoglio, F.; Berg, T.; Beck, J.; Bohnenberger, H.; Alexe, G.; Corso, J.; Strobel, P.; Wachter, A.; et al. Hoxa9 and Meis1 cooperatively induce addiction to syk signaling by suppressing miR-146a in acute myeloid leukemia. *Cancer Cell* **2017**, *31*, 549–562. [CrossRef]

75. Hu, H.Y.; Li, K.P.; Wang, X.J.; Liu, Y.; Lu, Z.G.; Dong, R.H.; Guo, H.B.; Zhang, M.X. Set9, NF-kappaB, and microRNA-21 mediate berberine-induced apoptosis of human multiple myeloma cells. *Acta Pharmacol. Sin.* **2013**, *34*, 157–166. [CrossRef]

76. Hong, J.; Wang, Y.; Hu, B.C.; Xu, L.; Liu, J.Q.; Chen, M.H.; Wang, J.Z.; Han, F.; Zheng, Y.; Chen, X.; et al. Transcriptional downregulation of microRNA-19a by ROS production and NF-kappaB deactivation governs resistance to oxidative stress-initiated apoptosis. *Oncotarget* **2017**, *8*, 70967–70981. [CrossRef]

77. Gerri, C.; Marin-Juez, R.; Marass, M.; Marks, A.; Maischein, H.M.; Stainier, D.Y.R. Hif-1alpha regulates macrophage-endothelial interactions during blood vessel development in zebrafish. *Nat. Commun.* **2017**, *8*, 15492. [CrossRef]

78. Koh, M.Y.; Lemos, R. Jr.; Liu, X.; Powis, G. The hypoxia-associated factor switches cells from HIF-1alpha- to HIF-2alpha-dependent signaling promoting stem cell characteristics, aggressive tumor growth and invasion. *Cancer Res.* **2011**, *71*, 4015–4027. [CrossRef]

79. Serocki, M.; Bartoszewska, S.; Janaszak-Jasiecka, A.; Ochocka, R.J.; Collawn, J.F.; Bartoszewski, R. miRNAs regulate the HIF switch during hypoxia: A novel therapeutic target. *Angiogenesis* **2018**, *21*, 183–202. [CrossRef]

80. Wang, H.; Flach, H.; Onizawa, M.; Wei, L.; McManus, M.T.; Weiss, A. Negative regulation of Hif1a expression and TH17 differentiation by the hypoxia-regulated microRNA miR-210. *Nat. Immunol.* **2014**, *15*, 393–401. [CrossRef]

81. Seok, J.K.; Lee, S.H.; Kim, M.J.; Lee, Y.M. MicroRNA-382 induced by HIF-1alpha is an angiogenic miR targeting the tumor suppressor phosphatase and tensin homolog. *Nucleic Acids Res.* **2014**, *42*, 8062–8072. [CrossRef] [PubMed]

82. Ge, X.; Liu, X.; Lin, F.; Li, P.; Liu, K.; Geng, R.; Dai, C.; Lin, Y.; Tang, W.; Wu, Z.; et al. MicroRNA-421 regulated by HIF-1alpha promotes metastasis, inhibits apoptosis, and induces cisplatin resistance by targeting E-cadherin and caspase-3 in gastric cancer. *Oncotarget* **2016**, *7*, 24466–24482. [CrossRef] [PubMed]

83. Nagpal, N.; Ahmad, H.M.; Chameettachal, S.; Sundar, D.; Ghosh, S.; Kulshreshtha, R. HIF-inducible miR-191 promotes migration in breast cancer through complex regulation of TGFbeta-signaling in hypoxic microenvironment. *Sci. Rep.* **2015**, *5*, 9650. [CrossRef] [PubMed]

84. Bhatt, K.; Wei, Q.; Pabla, N.; Dong, G.; Mi, Q.S.; Liang, M.; Mei, C.; Dong, Z. MicroRNA-687 Induced by hypoxia-inducible factor-1 targets phosphatase and tensin homolog in renal ischemia-reperfusion injury. *JASN* **2015**, *26*, 1588–1596. [CrossRef] [PubMed]

85. Li, F.; Sonveaux, P.; Rabbani, Z.N.; Liu, S.; Yan, B.; Huang, Q.; Vujaskovic, Z.; Dewhirst, M.W.; Li, C.Y. Regulation of HIF-1alpha stability through S-nitrosylation. *Mol. Cell* **2007**, *26*, 63–74. [CrossRef]

86. Lim, J.H.; Lee, Y.M.; Chun, Y.S.; Chen, J.; Kim, J.E.; Park, J.W. Sirtuin 1 modulates cellular responses to hypoxia by deacetylating hypoxia-inducible factor 1alpha. *Mol. Cell* **2010**, *38*, 864–878. [CrossRef]

87. Macfarlane, L.A.; Murphy, P.R. MicroRNA: Biogenesis, function and role in cancer. *Curr. Genom.* **2010**, *11*, 537–561. [CrossRef]

88. Abdi, J.; Rastgoo, N.; Li, L.; Chen, W.; Chang, H. Role of tumor suppressor p53 and micro-RNA interplay in multiple myeloma pathogenesis. *J. Hematol. Oncol.* **2017**, *10*, 169. [CrossRef]

89. Ungvari, Z.; Tucsek, Z.; Sosnowska, D.; Toth, P.; Gautam, T.; Podlutsky, A.; Csiszar, A.; Losonczy, G.; Valcarcel-Ares, M.N.; Sonntag, W.E.; et al. Aging-induced dysregulation of dicer1-dependent microRNA expression impairs angiogenic capacity of rat cerebromicrovascular endothelial cells. *J. Gerontol. Ser. A Biol. Sci. Med Sci.* **2013**, *68*, 877–891. [CrossRef]

90. Shilo, S.; Roy, S.; Khanna, S.; Sen, C.K. Evidence for the involvement of miRNA in redox regulated angiogenic response of human microvascular endothelial cells. *Arterioscler. Thromb. Vasc. Biol.* **2008**, *28*, 471–477. [CrossRef]

91. Iliopoulos, D.; Hirsch, H.A.; Struhl, K. An epigenetic switch involving NF-kappaB, Lin28, Let-7 MicroRNA, and IL6 links inflammation to cell transformation. *Cell* **2009**, *139*, 693–706. [CrossRef] [PubMed]

92. Wang, J.X.; Gao, J.; Ding, S.L.; Wang, K.; Jiao, J.Q.; Wang, Y.; Sun, T.; Zhou, L.Y.; Long, B.; Zhang, X.J.; et al. Oxidative modification of miR-184 enables it to target Bcl-xL and Bcl-w. *Mol. Cell* **2015**, *59*, 50–61. [CrossRef] [PubMed]

93. Panday, A.; Sahoo, M.K.; Osorio, D.; Batra, S. NADPH oxidases: An overview from structure to innate immunity-associated pathologies. *Cell. Mol. Immunol.* **2015**, *12*, 5–23. [CrossRef] [PubMed]

94. Li, S.Z.; Hu, Y.Y.; Zhao, J.; Zhao, Y.B.; Sun, J.D.; Yang, Y.F.; Ji, C.C.; Liu, Z.B.; Cao, W.D.; Qu, Y.; et al. MicroRNA-34a induces apoptosis in the human glioma cell line, A172, through enhanced ROS production and NOX2 expression. *Biochem. Biophys. Res. Commun.* **2014**, *444*, 6–12. [CrossRef]

95. Liu, W.; Zabirnyk, O.; Wang, H.; Shiao, Y.H.; Nickerson, M.L.; Khalil, S.; Anderson, L.M.; Perantoni, A.O.; Phang, J.M. miR-23b targets proline oxidase, a novel tumor suppressor protein in renal cancer. *Oncogene* **2010**, *29*, 4914–4924. [CrossRef]

96. Gordillo, G.M.; Biswas, A.; Khanna, S.; Pan, X.; Sinha, M.; Roy, S.; Sen, C.K. Dicer knockdown inhibits endothelial cell tumor growth via microRNA 21a-3p targeting of Nox-4. *J. Biol. Chem.* **2014**, *289*, 9027–9038. [CrossRef]

97. Kim, J.H.; Park, S.G.; Song, S.Y.; Kim, J.K.; Sung, J.H. Reactive oxygen species-responsive miR-210 regulates proliferation and migration of adipose-derived stem cells via PTPN2. *Cell Death Dis.* **2013**, *4*, e588. [CrossRef]

98. Chen, Z.; Li, Y.; Zhang, H.; Huang, P.; Luthra, R. Hypoxia-regulated microRNA-210 modulates mitochondrial function and decreases ISCU and COX10 expression. *Oncogene* **2010**, *29*, 4362–4368. [CrossRef]

99. Venkataraman, S.; Alimova, I.; Fan, R.; Harris, P.; Foreman, N.; Vibhakar, R. MicroRNA 128a increases intracellular ROS level by targeting Bmi-1 and inhibits medulloblastoma cancer cell growth by promoting senescence. *PLoS ONE* **2010**, *5*, e10748. [CrossRef]

100. Sripada, L.; Singh, K.; Lipatova, A.V.; Singh, A.; Prajapati, P.; Tomar, D.; Bhatelia, K.; Roy, M.; Singh, R.; Godbole, M.M.; et al. hsa-miR-4485 regulates mitochondrial functions and inhibits the tumorigenicity of breast cancer cells. *J. Mol. Med.* **2017**, *95*, 641–651. [CrossRef]

101. Glorieux, C.; Calderon, P.B. Catalase, a remarkable enzyme: Targeting the oldest antioxidant enzyme to find a new cancer treatment approach. *Biol. Chem.* **2017**, *398*, 1095–1108. [CrossRef] [PubMed]

102. Zhang, X.; Ng, W.L.; Wang, P.; Tian, L.; Werner, E.; Wang, H.; Doetsch, P.; Wang, Y. MicroRNA-21 modulates the levels of reactive oxygen species by targeting SOD3 and TNFalpha. *Cancer Res.* **2012**, *72*, 4707–4713. [CrossRef] [PubMed]

103. Meng, X.; Wu, J.; Pan, C.; Wang, H.; Ying, X.; Zhou, Y.; Yu, H.; Zuo, Y.; Pan, Z.; Liu, R.Y.; et al. Genetic and epigenetic down-regulation of microRNA-212 promotes colorectal tumor metastasis via dysregulation of MnSOD. *Gastroenterology* **2013**, *145*, 426–436. [CrossRef] [PubMed]

104. Xu, X.; Wells, A.; Padilla, M.T.; Kato, K.; Kim, K.C.; Lin, Y. A signaling pathway consisting of miR-551b, catalase and MUC1 contributes to acquired apoptosis resistance and chemoresistance. *Carcinogenesis* **2014**, *35*, 2457–2466. [CrossRef] [PubMed]

105. Wang, Q.; Chen, W.; Bai, L.; Chen, W.; Padilla, M.T.; Lin, A.S.; Shi, S.; Wang, X.; Lin, Y. Receptor-interacting protein 1 increases chemoresistance by maintaining inhibitor of apoptosis protein levels and reducing reactive oxygen species through a microRNA-146a-mediated catalase pathway. *J. Biol. Chem.* **2014**, *289*, 5654–5663. [CrossRef] [PubMed]

106. Wang, P.; Zhu, C.F.; Ma, M.Z.; Chen, G.; Song, M.; Zeng, Z.L.; Lu, W.H.; Yang, J.; Wen, S.; Chiao, P.J.; et al. Micro-RNA-155 is induced by K-Ras oncogenic signal and promotes ROS stress in pancreatic cancer. *Oncotarget* **2015**, *6*, 21148–21158. [CrossRef] [PubMed]

107. Ma, Q. Role of nrf2 in oxidative stress and toxicity. *Annu. Rev. Pharmacol. Toxicol.* **2013**, *53*, 401–426. [CrossRef]

108. Cortez, M.A.; Valdecanas, D.; Zhang, X.; Zhan, Y.; Bhardwaj, V.; Calin, G.A.; Komaki, R.; Giri, D.K.; Quini, C.C.; Wolfe, T.; et al. Therapeutic delivery of miR-200c enhances radiosensitivity in lung cancer. *J. Am. Soc. Gene Ther.* **2014**, *22*, 1494–1503. [CrossRef]

109. Yang, M.; Yao, Y.; Eades, G.; Zhang, Y.; Zhou, Q. MiR-28 regulates Nrf2 expression through a Keap1-independent mechanism. *Breast Cancer Res. Treat.* **2011**, *129*, 983–991. [CrossRef]

110. Singh, B.; Ronghe, A.M.; Chatterjee, A.; Bhat, N.K.; Bhat, H.K. MicroRNA-93 regulates NRF2 expression and is associated with breast carcinogenesis. *Carcinogenesis* **2013**, *34*, 1165–1172. [CrossRef]

111. Papp, D.; Lenti, K.; Modos, D.; Fazekas, D.; Dul, Z.; Turei, D.; Foldvari-Nagy, L.; Nussinov, R.; Csermely, P.; Korcsmaros, T. The NRF2-related interactome and regulome contain multifunctional proteins and fine-tuned autoregulatory loops. *FEBS Lett.* **2012**, *586*, 1795–1802. [CrossRef] [PubMed]

112. Kabaria, S.; Choi, D.C.; Chaudhuri, A.D.; Jain, M.R.; Li, H.; Junn, E. MicroRNA-7 activates Nrf2 pathway by targeting Keap1 expression. *Free Radic. Biol. Med.* **2015**, *89*, 548–556. [CrossRef] [PubMed]

113. Eades, G.; Yang, M.; Yao, Y.; Zhang, Y.; Zhou, Q. miR-200a regulates Nrf2 activation by targeting Keap1 mRNA in breast cancer cells. *J. Biol. Chem.* **2011**, *286*, 40725–40733. [CrossRef] [PubMed]

114. Kim, J.H.; Lee, K.S.; Lee, D.K.; Kim, J.; Kwak, S.N.; Ha, K.S.; Choe, J.; Won, M.H.; Cho, B.R.; Jeoung, D.; et al. Hypoxia-responsive microRNA-101 promotes angiogenesis via heme oxygenase-1/vascular endothelial growth factor axis by targeting cullin 3. *Antioxid. Redox Signal.* **2014**, *21*, 2469–2482. [CrossRef] [PubMed]

115. Xu, D.; Zhu, H.; Wang, C.; Zhu, X.; Liu, G.; Chen, C.; Cui, Z. microRNA-455 targets cullin 3 to activate Nrf2 signaling and protect human osteoblasts from hydrogen peroxide. *Oncotarget* **2017**, *8*, 59225–59234. [CrossRef] [PubMed]

116. Li, L.; Huang, K.; You, Y.; Fu, X.; Hu, L.; Song, L.; Meng, Y. Hypoxia-induced miR-210 in epithelial ovarian cancer enhances cancer cell viability via promoting proliferation and inhibiting apoptosis. *Int. J. Oncol.* **2014**, *44*, 2111–2120. [CrossRef] [PubMed]

117. Tay, Y.; Rinn, J.; Pandolfi, P.P. The multilayered complexity of ceRNA crosstalk and competition. *Nature* **2014**, *505*, 344–352. [CrossRef]

118. Vinceti, M.; Filippini, T.; Del Giovane, C.; Dennert, G.; Zwahlen, M.; Brinkman, M.; Zeegers, M.P.; Horneber, M.; D'Amico, R.; Crespi, C.M. Selenium for preventing cancer. *Cochrane Database Syst. Rev.* **2018**, *1*, CD005195. [CrossRef]

119. Harvie, M. Nutritional supplements and cancer: Potential benefits and proven harms. In *American Society of Clinical Oncology Educational Book*; American Society of Clinical Oncology: Alexandria, VA, USA, 2014. [CrossRef]

120. Druesne-Pecollo, N.; Latino-Martel, P.; Norat, T.; Barrandon, E.; Bertrais, S.; Galan, P.; Hercberg, S. Beta-carotene supplementation and cancer risk: A systematic review and metaanalysis of randomized controlled trials. *Int. J. Cancer* **2010**, *127*, 172–184. [CrossRef]

121. Jacobs, C.; Hutton, B.; Ng, T.; Shorr, R.; Clemons, M. Is there a role for oral or intravenous ascorbate (vitamin C) in treating patients with cancer? A systematic review. *Oncologist* **2015**, *20*, 210–223. [CrossRef] [PubMed]

122. Sablina, A.A.; Budanov, A.V.; Ilyinskaya, G.V.; Agapova, L.S.; Kravchenko, J.E.; Chumakov, P.M. The antioxidant function of the p53 tumor suppressor. *Nature Med.* **2005**, *11*, 1306–1313. [CrossRef] [PubMed]

123. Schubert, R.; Erker, L.; Barlow, C.; Yakushiji, H.; Larson, D.; Russo, A.; Mitchell, J.B.; Wynshaw-Boris, A. Cancer chemoprevention by the antioxidant tempol in Atm-deficient mice. *Hum. Mol. Genet.* **2004**, *13*, 1793–1802. [CrossRef] [PubMed]

124. Sayin, V.I.; Ibrahim, M.X.; Larsson, E.; Nilsson, J.A.; Lindahl, P.; Bergo, M.O. Antioxidants accelerate lung cancer progression in mice. *Sci. Transl. Med.* **2014**, *6*, 221ra215. [CrossRef] [PubMed]

125. Harris, I.S.; Treloar, A.E.; Inoue, S.; Sasaki, M.; Gorrini, C.; Lee, K.C.; Yung, K.Y.; Brenner, D.; Knobbe-Thomsen, C.B.; Cox, M.A.; et al. Glutathione and thioredoxin antioxidant pathways synergize to drive cancer initiation and progression. *Cancer Cell* **2015**, *27*, 211–222. [CrossRef]

126. Wang, H.; Liu, X.; Long, M.; Huang, Y.; Zhang, L.; Zhang, R.; Zheng, Y.; Liao, X.; Wang, Y.; Liao, Q.; et al. NRF2 activation by antioxidant antidiabetic agents accelerates tumor metastasis. *Sci. Transl. Med.* **2016**, *8*, 334ra351. [CrossRef]

127. Qu, Y.; Wang, J.; Ray, P.S.; Guo, H.; Huang, J.; Shin-Sim, M.; Bukoye, B.A.; Liu, B.; Lee, A.V.; Lin, X.; et al. Thioredoxin-like 2 regulates human cancer cell growth and metastasis via redox homeostasis and NF-kappaB signaling. *J. Clin. Investig.* **2011**, *121*, 212–225. [CrossRef]

128. Kamarajugadda, S.; Cai, Q.; Chen, H.; Nayak, S.; Zhu, J.; He, M.; Jin, Y.; Zhang, Y.; Ai, L.; Martin, S.S.; et al. Manganese superoxide dismutase promotes anoikis resistance and tumor metastasis. *Cell Death Dis.* **2013**, *4*, e504. [CrossRef]

129. Glasauer, A.; Sena, L.A.; Diebold, L.P.; Mazar, A.P.; Chandel, N.S. Targeting SOD1 reduces experimental non-small-cell lung cancer. *J. Clin. Investig.* **2014**, *124*, 117–128. [CrossRef]

130. Nguyen, A.; Loo, J.M.; Mital, R.; Weinberg, E.M.; Man, F.Y.; Zeng, Z.; Paty, P.B.; Saltz, L.; Janjigian, Y.Y.; de Stanchina, E.; et al. PKLR promotes colorectal cancer liver colonization through induction of glutathione synthesis. *J. Clin. Investig.* **2016**, *126*, 681–694. [CrossRef]

131. DeNicola, G.M.; Karreth, F.A.; Humpton, T.J.; Gopinathan, A.; Wei, C.; Frese, K.; Mangal, D.; Yu, K.H.; Yeo, C.J.; Calhoun, E.S.; et al. Oncogene-induced Nrf2 transcription promotes ROS detoxification and tumorigenesis. *Nature* **2011**, *475*, 106–109. [CrossRef] [PubMed]

132. Kitamura, H.; Motohashi, H. NRF2 addiction in cancer cells. *Cancer Sci.* **2018**, *109*, 900–911. [CrossRef] [PubMed]

133. Le Gal, K.; Ibrahim, M.X.; Wiel, C.; Sayin, V.I.; Akula, M.K.; Karlsson, C.; Dalin, M.G.; Akyurek, L.M.; Lindahl, P.; Nilsson, J.; et al. Antioxidants can increase melanoma metastasis in mice. *Sci. Transl. Med.* **2015**, *7*, 308re308. [CrossRef] [PubMed]

134. Patra, K.C.; Hay, N. The pentose phosphate pathway and cancer. *Trends Biochem. Sci.* **2014**, *39*, 347–354. [CrossRef]

135. Fan, J.; Ye, J.; Kamphorst, J.J.; Shlomi, T.; Thompson, C.B.; Rabinowitz, J.D. Quantitative flux analysis reveals folate-dependent NADPH production. *Nature* **2014**, *510*, 298–302. [CrossRef]
136. Ren, J.G.; Seth, P.; Clish, C.B.; Lorkiewicz, P.K.; Higashi, R.M.; Lane, A.N.; Fan, T.W.; Sukhatme, V.P. Knockdown of malic enzyme 2 suppresses lung tumor growth, induces differentiation and impacts PI3K/AKT signaling. *Sci. Rep.* **2014**, *4*, 5414. [CrossRef]
137. Mollaei, H.; Safaralizadeh, R.; Rostami, Z. MicroRNA replacement therapy in cancer. *J. Cell. Physiol.* **2019**, *234*, 12369–12384. [CrossRef]
138. Bader, A.G.; Brown, D.; Winkler, M. The promise of microRNA replacement therapy. *Cancer Res.* **2010**, *70*, 7027–7030. [CrossRef]
139. Ling, H.; Fabbri, M.; Calin, G.A. MicroRNAs and other non-coding RNAs as targets for anticancer drug development. *Nature reviews. Drug Discov.* **2013**, *12*, 847–865. [CrossRef]
140. Hanna, J.; Hossain, G.S.; Kocerha, J. The Potential for microRNA Therapeutics and Clinical Research. *Front. Genet.* **2019**, *10*, 478. [CrossRef]

International Journal of
Molecular Sciences

Review

Interplay Between MicroRNAs and Oxidative Stress in Ovarian Conditions with a Focus on Ovarian Cancer and Endometriosis

Josep Marí-Alexandre [1,*,†], Antonio Pellín Carcelén [2,†], Cristina Agababyan [1,3], Andrea Moreno-Manuel [4,5], Javier García-Oms [1,3], Silvia Calabuig-Fariñas [4,5,6,7] and Juan Gilabert-Estellés [1,3,8]

[1] Research Laboratory in Biomarkers in Reproduction, Gynaecology and Obstetrics, Fundación Hospital General Universitario de Valencia, 46014 València, Spain; dra.kristina.agababyan@gmail.com (C.A.); javiergoms@yahoo.es (J.G.-O.); juangilaeste@yahoo.es (J.G.-E.)
[2] Department of Physiology, Universitat de València, 46010 València, Spain; apellincarcelen@gmail.com
[3] Comprehensive Multidisciplinary Endometriosis Unit, Consorcio Hospital General Universitario de València, 46014 València, Spain
[4] Molecular Oncology Laboratory, Fundación para la Investigación del Hospital General Universitario de València, 46014, València, Spain; andrea.morenomanuel@gmail.com (A.M.-M.); calabuix_sil@gva.es (S.C.-F.)
[5] TRIAL Mixed Unit, Centro de Investigación Príncipe Felipe-Fundación para la Investigación del Hospital General Universitario de València, 46014 València, Spain
[6] Department of Pathology, Universitat de València, 46010 València, Spain
[7] Centro de Investigación Biomédica en Red en Cáncer (CIBERONC), 46014 València, Spain
[8] Department of Paediatrics, Obstetrics and Gynaecology, University of València, 46010 València, Spain
* Correspondence: mari_josale@gva.es; Tel.: +34-96-313-1893 (ext. 437211)
† These authors contributed equally to this work.

Received: 23 September 2019; Accepted: 23 October 2019; Published: 25 October 2019

Abstract: Ovarian cancer and endometriosis are two distinct gynaecological conditions that share many biological aspects incuding proliferation, invasion of surrounding tissue, inflammation, inhibition of apoptosis, deregulation of angiogenesis and the ability to spread at a distance. miRNAs are small non-coding RNAs (19–22 nt) that act as post-transcriptional modulators of gene expression and are involved in several of the aforementioned processes. In addition, a growing body of evidence supports the contribution of oxidative stress (OS) to these gynaecological diseases: increased peritoneal OS due to the decomposition of retrograde menstruation blood facilitates both endometriotic lesion development and fallopian tube malignant transformation leading to high-grade serous ovarian cancer (HGSOC). Furthermore, as HGSOC develops, increased OS levels are associated with chemoresistance. Finally, continued bleeding within ovarian endometrioma raises OS levels and contributes to the development of endometriosis-associated ovarian cancer (EAOC). Therefore, this review aims to address the need for a better understanding of the dialogue between miRNAs and oxidative stress in the pathophysiology of ovarian conditions: endometriosis, EAOC and HGSOC.

Keywords: oxidative stress; miRNAs; endometriosis; high-grade serous ovarian cancer; endometriosis-associated ovarian cancer; epithelial-to-mesenchymal transition; chemoresistance

---

1. Introduction

The Great Oxidative Event occurred between 2.4 and 2.1 billion years ago when high O_2 concentrations appeared in the Earth atmosphere as a metabolic product of cyanobacteria oxygenic photosynthesis [1]. This phenomenon is considered a breakthrough for life on Earth, as living organisms had to develop an arsenal of antioxidant strategies to adapt to this powerful compound. However, in some circumstances the balance between oxidants and antioxidants might be shifted in favour of the former, giving raise to oxidative stress (OS). Remarkably, high partial oxygen pressures might also lead to OS through the formation of highly reactive oxygen species (ROS), including superoxide anion radical ($O_2^{-\bullet}$), hydroxyl radical ($OH^\bullet$), hydrogen peroxide (H_2O_2), and peroxyl radical ($ROO^\bullet$) (Figure 1).

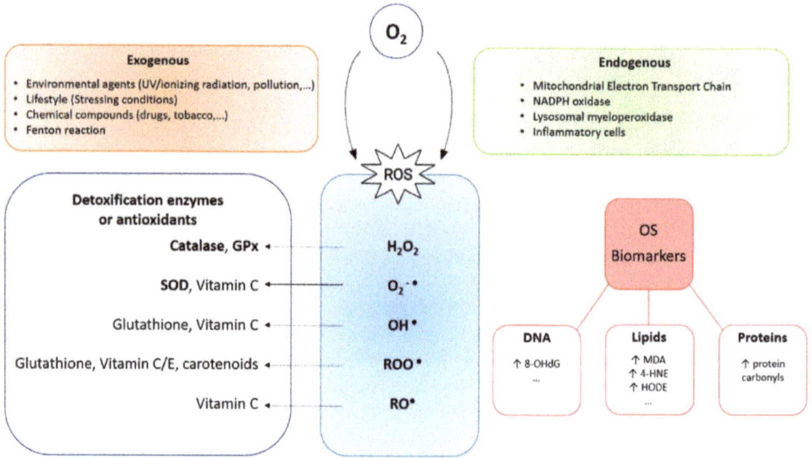

Figure 1. Diagram representing reactive oxygen species (ROS) production, their detoxification mechanisms and the oxidative stress alterations produced by their action, which can serve as oxidative stress (OS) biomarkers. Abbreviations: NADPH, nicotinamide adenine dinucleotide phosphate; GPx, glutathione peroxidase; SOD, superoxide dismutase; H_2O_2, hydrogen peroxide; $O_2^{-\bullet}$, superoxide anion radical; $OH^\bullet$, hydroxyl radical; $ROO^\bullet$, p eroxyl radical; $RO^\bullet$, alcoxyl radical; 8-OHdG, 8-hydroxy-2′-deoxyguanosine; MDA, malondialdehyde; 4-HNE, 4-hydroxynonenal; HODE, hydroxyoctadecadienoic acid.

Certain levels of ROS can be produced endogenously as a sub-product of countless chemical reactions essential for cell life (including those mediated by the mitochondrial electron transport chain reactions and nicotinamide adenine dinucleotide phosphate (NADPH) oxidase), playing an important role in the regulation of cellular signalling processes [2]. However, an excess of ROS provokes the disruption of redox signalling and control and/or molecular damage [3] (Figure 1). On the other hand, ROS can also be generated exogenously, as a result of the exposure of biological systems to environmental agents (i.e., ultraviolet or ionizing radiation) or by the action of free iron (via the Fenton reaction) [4,5] (Figure 1).

The antioxidant mechanisms in charge of avoiding oxidative damage to cells include: a) enzymes (such as superoxide dismutase, glutathione peroxidase, catalase and metal binding proteins), b) non-enzymatic protectors (such as glutathione, vitamin E, vitamin C, uric acid, bilirubin and albumin) and c) repairers of damaged molecules (such as DNA repair enzymes, methionine, sulfoxide reductase) [6] (Figure 1). Contrary to the popular belief, growing evidence suggests a predominant role of antioxidant enzymes over dietary antioxidants in protection against OS [7]. When these antioxidant defences are overwhelmed (due to an excess of prooxidant substances, a deficiency of antioxidant agents or both [8]) OS produces damage to biomolecules essential for life, as lipids (malondialdehyde) [6,9], proteins (protein carbonyls) [3,9] and DNA (8-hydroxy-2′-deoxyguanosine (8-OHdG) [6,7,10] (Figure 1). This contributes to the pathophysiology of many pathological conditions such as Alzheimer's disease [11], frailty [9], ovarian cancer [4], and endometriosis [1].

On the other hand, epigenetics refers to the heritable changes in gene function that cannot be explained by changes in the DNA sequence [12]. These changes are produced through four epigenetic mechanisms that are dynamic and reversible, and include: DNA methylation, histone modifications, chromatin remodelling, and the expression of non-coding RNAs, including miRNAs [13]. A growing body of evidence suggests that epigenetics could be involved in the pathophysiology of endometriosis [14] and that carcinogenesis cannot be explained only by DNA mutations, but also that epigenetic alterations need to be included in the equation [15]. In this respect, distinct epigenetic mechanisms could contribute to carcinogenesis either by repressing the expression of tumour suppressor genes (TSG) (i.e., DNA hypermethylation at gene promoters, over-expression of miRNA targeting TSG, histone modifications, and heterochromatin conformation at TSG coding regions) or allowing the activation of oncogenes (OG) (i.e., global DNA hypomethylation, down-regulation of miRNA targeting OG, histone modifications, and euchromatin conformation at OG coding regions).

Importantly, great research endeavours have been conducted to decipher the role of miRNAs in these pathologies. miRNAs are small (19–22 nt) non-coding RNAs that can act as post-transcriptional regulators of gene expression, reducing the expression of their target mRNAs either by inhibiting its translation or by promoting its degradation. Thus, the levels of their target mRNAs are opposed to those of their targeting miRNAs (Figure 2B). It is worth mentioning that several miRNAs can target a given mRNA and a single miRNA can target several mRNAs, increasing the complexity of the regulatory mechanism mediated by these molecules [16–19]. miRNAs are involved in pivotal biological processes including development, differentiation, apoptosis, and proliferation. Remarkably, miRNAs themselves can also act as OG or TSG, depending on their targets [20]. Extensive literature supports the role of miRNAs in the development of endometriosis (reviewed in [21–24]) as well as in endometriosis-associated ovarian cancer (EAOC) [25,26] and in high-grade serous ovarian cancer (HGSOC) [27–29], as hereafter described.

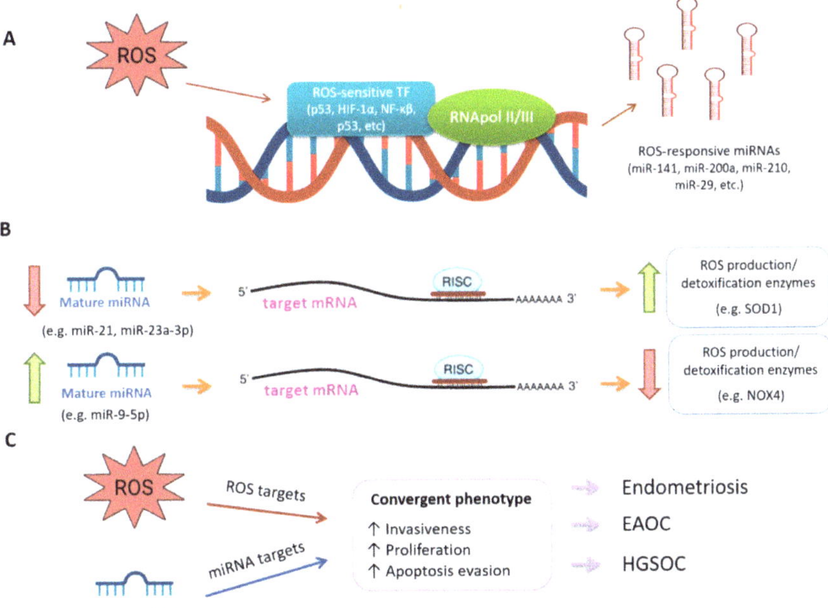

Figure 2. Schematic representation of the possible mechanisms of interplay between miRNAs and oxidative stress. (**A**) ROS can activate ROS-sensitive transcription factors to induce the transcription of specific primary miRNA (pri-miRNA) that will conduce to mature miRNAs; (**B**) The levels of a given miRNA inversely correlate with those of their target mRNAs, that could belong to ROS production/detoxification enzymes; (**C**) ROS and miRNAs can produce separate effects that converge

in a common phenotype, leading to endometriosis, EAOC and HGOSC. Abbreviations: ROS, reactive oxygen species; TF, transcription factors; RNApol II/III, RNA polymerase II or III; HIF-1α, hypoxia-inducible factor 1α; NF-κβ, nuclear factor κβ; RISC, RNA-induced silencing complex; SOD1, superoxide dismutase 1; NOX4, NADPH oxygenase 4; EAOC, endometriosis-associated ovarian cancer; HGSOC, high-grade serous ovarian cancer.

As for the goal of this review, the complex interplay between OS and miRNAs represents an active research area in different pathologies that develops on the basis of several premises: (1) OS triggers the expression of responsive miRNAs through ROS-sensitive transcription factors (reviewed in [30,31]) (Figure 2A); (2) miRNAs can regulate the expression of enzymes involved in ROS production or detoxification (reviewed in [31,32]) (Figure 2B); (3) miRNAs and OS independently act to reach convergent phenotypes (Figure 2C).

Therefore, the aim of the present review is to compile existing evidence on how miRNAs and OS interact through the aforementioned mechanisms into the pathophysiology of three important gynaecological diseases: endometriosis, EAOC, and HGSOC.

2. Endometriosis

Endometriosis is an oestrogen-dependent inflammatory disorder defined by the presence of endometrial-like tissue in ectopic locations, which limits the quality of life of affected women [33–35]. This pathology affects 10% of reproductive-aged women from all ethnic and social groups, although the prevalence in those patients experiencing pain, infertility, or both is as high as 35%–50% [36], being the estimated prevalence of this condition around 176 million worldwide. The most frequent involvement is in the peritoneum (superficial and deep endometriotic implants) and ovaries (ovarian endometrioma (OMA) or endometriotic cysts), although cases of pulmonary [37] and cerebral endometriosis [38] have also been documented.

While a unifying theory regarding the exact aetiopathogenic mechanism of endometriosis is still lacking [13], nowadays the most widely accepted theory is Sampson's proposal of retrograde menstruation and implantation [39]. This theory postulates that desquamated endometrial cells reach the peritoneal cavity by retrograde flow through the fallopian tubes, where they are able to implant and survive (Figure 3). Since retrograde menstruation occurs in 90% of healthy women of reproductive age with patent fallopian tubes [40], the fact that only a small percentage develops the disease suggests that there must be additional mechanisms that allow the migrated tissue to implant and survive [41].

miRNAs and Oxidative Stress in Endometriosis

Sampson's proposal not only postulates an origin for the endometrial-like ectopic tissue, but also provides a mechanism for the action of OS in the pathogenesis of endometriosis. Remarkably, compelling evidence demonstrates an increase in OS markers in several fluids (serum, peritoneal and follicular fluid) and tissues in women with endometriosis (reviewed in [42,43]).

From the point of view of the desquamated cells, several studies including ours have observed that eutopic endometria expresses higher levels of vascular endothelial growth factor A (VEGF-A) (the main pro-angiogenic factor) [44,45], urokinase-type plasminogen activator (uPA), and matrix metallopeptidase 3 (MMP-3) (proteolytic factors) [46] in comparison to control endometria, a process that could be mediated by miRNAs [47]. Upon menstruation, endometrial cells lose their blood supply, entering into a hypoxic state by the time they reach the peritoneal cavity. This upregulates the expression of hypoxia-inducible factor 1-alpha (HIF-1α), which further stimulates the transcription of several hypoxia-inducible miRNAs or hypoxamiRs [30,48], as the prototypical hypoxamiR, miR-210, which is overexpressed in OMA tissues, promoting cell survival [49] (Figure 2A).

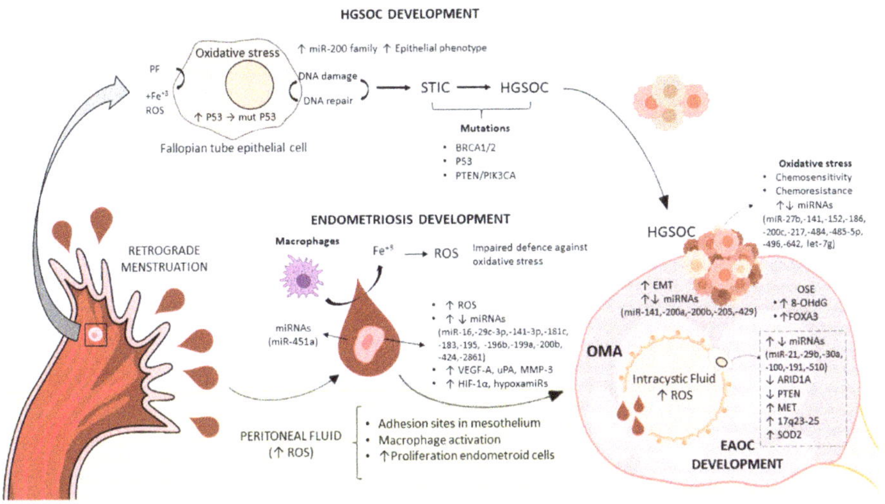

Figure 3. Global vision of aetiopathogenic mechanisms leading to endometriosis, EAOC and HGSOC development. Regarding HGSOC development, oxidative stress contributes to fallopian tube epithelial cells alterations, through the action of ROS. Repeated cycles of DNA damage and repair produce mutations in driver genes *BRCA1/2*, *P53*, *PTEN* and *PIK3CA*. Additionally, miRNA deregulation contributes to tumour progression. Once the malignant lesion is established in the ovary, oxidative stress is initially involved in first-line chemotherapy mechanism of action, although excessive oxidative stress is linked to tumour chemoresistance. Regarding endometriosis development, refluxed endometrial cells from patients show some features predisposing them to the development of this condition (i.e., increased angiogenesis and proteolysis, disbalanced miRNAs profile, etc). Upon menstruation, endometrial cells lose their blood supply and activate hypoxia-responsive miRNAs (hypoxamiRs) that together with erythrocyte-derived miRNAs contribute to the development of the condition. Blood decomposition by pelvic macrophages contribute to ROS production, which alters the peritoneal microenvironment to enhance endometrial cells attachment and proliferation. Finally, the intra-cystic fluid of OMAs presents with higher levels of ROS, triggering subsequent events such as miRNAs disbalance, decreased expression of ARID1A and PTEN, amplification of *MET* and 17q24–25, and increased generation of SOD2, all of which enhance the development of EAOC. Abbreviations: EAOC, endometriosis-associated ovarian cancer; HGSOC, High-Grade Serous Ovarian Cancer; STIC, Serous Tubal Intraepithelial Carcinoma; OSE, Ovarian Surface Epithelium; OMA, Ovarian endometrioma; ROS, Reactive Oxygen Species; PF, Peritoneal Fluid; BRCA, Breast Cancer gene; PTEN, Phosphatase and tensin homolog; PIK3CA, Phosphatidylinostil-4,5 Biphosphonate 3-Kinase Catalytic Subunit Alpha; VEGF-A, Vascular Endothelial Growth Factor A; uPA, Urokinase-type plasminogen Activator; MMP-3; Matrix Metallopeptidase 3; HIF-1α; Hypoxia-inducible factor 1-alpha; ARID1A, AT-Rich Interaction Domain A; MET, mesenchymal-to-epithelial transcription factor; EMT, Epithelial-Mesenchymal Transition; SOD2, Superoxide Dismutase 2; FOXA3, Hepatocyte Nuclear Factor 3-gamma; 8-OHdG, 8-Oxo-2′-deoxyguanosine.

Once in the peritoneal cavity, endometrial cells floating in a mixture of blood and peritoneal fluid (PF) need to undergo a continuum of important events if they are to implant and survive. These events include the attachment to ectopic sites, extracellular matrix degradation, invasion, and angiogenesis (a complex and sequential process devoted to the formation of new blood vessels from pre-existing ones, to assure oxygen and nutrient supply for proliferation and survival [50]). Notably, in vitro studies revealed that several miRNAs are involved in distinct processes leading to the establishment and survival of the endometriotic lesions, including invasiveness (miR-200b [51], miR-183 [52], miR-199a [53]), proliferation (miR-210 [54], miR-200b [51], miR-2861 [55], miR-195 [56], miR-196b [57]), apoptosis evasion (miR-181c [58], miR-141-3p [59], miR-2861 [55], miR-195 [56], miR-196b [57], miR-210 [54], and increased angiogenesis (miR-16, miR-29c-3p, miR-424 [60]) (Table 1). A myriad of studies reveals the putative role of miRNAs in endometriosis, recently reviewed by Panir and collaborators [61]. Among these miRNAs are miR-21, miR-23a-3p, and miR-9-5p, which have been linked to the regulation or redox enzymes [30] (Figure 2B).

Table 1. Deregulated miRNAs in selected in vitro studies in endometriosis.

References	Main Biological Function Promoted	Experimental Design	Main Deregulated miRNAs in Patients
[49]	Cell survival	OMA cell line under hypoxia	↑ miR-210
[51]	Invasiveness	Immortalized endometriotic cell line 12Z, the stromal cell line ST-T1b and primary endometriotic stromal cells	↓ miR-200b
[52]		Primary eutopic and control stromal cells	↓ miR-183
[53]		Primary ectopic, eutopic and control stromal cells	↓ miR-199a
[54]		Primary ectopic and control stromal cells	↑ miR-210
[51]	Proliferation	Immortalized endometriotic cell line 12Z, the stromal cell line ST-T1b and primary endometriotic stromal cells	↓ miR-200b
[55]		Ectopic endometrial cells	↓ miR-2861
[56]		Primary ectopic, eutopic and control stromal cells	↓ miR-195
[57]		Primary ectopic and control stromal cells	↓ mi-196b
[58]	Apoptosis evasion	Endometrial cell lines	↑ miR-181c
[59]		Ectopic endometrial stromal cells	↓ miR-143-3p
[55]		Ectopic endometrial cells	↓ miR-2861
[56]		Primary ectopic, eutopic and control stromal cells	↓ miR-195
[57]		Primary ectopic and control stromal cells	↓ mi-196b
[49] [54]		OMA cell line under hypoxia Primary ectopic and control stromal cells	↑ miR-210
[60]	Angiogenesis	Primary ectopic, eutopic and control stromal cells	↓ miR-16, ↓ miR-29c-3p, ↓ miR-424

↑, up-regulated levels; ↓ down-regulated levels; OMA: ovarian endometrioma.

From the point of view of the milieu into which these cells arrive, PF from patients might favour endometriosis development through several mechanisms: (a) Firstly, increased OS in PF might create adhesion sites for the migrated cells by damaging the mesothelial wall [1]; (b) acting on endometrial cells, PF from patients is involved in the over-expression of the proteolytic factors uPA and MMP-3 [62], and not yet resolved components in this biofluid increase angiogenesis by down-regulation of angiogenesis-related miRNAs (miR-16-5p, miR-29c-3p, and miR-424-5p), mainly in eutopic cells from patients, favouring their survival [60]. The putative role of OS in mediating these effects is reinforced by recent results from Wright and co-workers [63], who observed that global down-regulation in miRNAs could be recapitulated by stimulating endometrial cells with oxidized low-density lipoprotein (LDL) (an OS marker present in patient's PF and associated with pain). These in vitro findings have also been observed in tissues, since several authors, including us, have reported an increase in VEGF-A in endometriotic tissues, which might also be regulated by miRNAs such as miR-16, miR-29c-3p, and miR-424 [45,60,64]; and (c) accompanying erythrocytes might be a source not only of OS but also of miRNAs [65], as observed by the presence of miR-451, the most abundant miRNA in erythrocytes, in PF [47]. Surprisingly, in vitro and in vivo evidence suggests that miR-451 is uptaken by endometriotic tissues, correlating its expression with survival status of the lesions [66] (Table 2, Figure 3).

Table 2. Deregulated miRNAs in selected studies considering distinct biofluids from patients with endometriosis, EAOC or HGSOC compared to control women.

miRNAs in Biofluids			
Reference	Gynaecological Condition	Biofluid Specimen	Main Deregulated miRNAs in Patients
[47]	Endometriosis	Peritoneal fluid	↑ miR-106b-3p, miR-451a and miR-486-5p
[67]		Serum	↓ let-7b and miR-135
[68]		Serum	↓ miR-9 *, miR-141 *, miR-145 * and miR-542-3p ↑ miR-122 and miR-199a
[69]		Serum	↑ miR-122 and miR-199a
[70]		Serum	↓ miR-30c-5p, miR-127-3p, miR-99b-5p, miRNA-15b-5p and miRNA-20a-5p ↑ miR-424-3p and miR-185-5p
[71]		Plasma	↓ miR-17-5p, miR-20a and miR-22
[72]		Plasma	↓ miR-200a-3p, miR-200b-3p and miR-414-3p
[73]		Plasma	↑ miR-154-5p
[74]	EAOC	Plasma	Three distinct miRNA signatures, including ↑ miR-15b, miR-16, miR-21, and miR-195
[75]	HGSOC	Serum	↑ miR-1290
[76]		Serum	↓ miR-375 + CA-125 levels
[77]		Serum	↑ miR-1246
[78]		Serum	↑ miR-200b, miR-200c

↑, up-regulated levels; ↓ down-regulated levels; CA-125, cancer antigen 125.

Provided that several authors do not distinguish the type of ectopic lesions into their analyses, the precise effect of OS in different endometriotic lesions is difficult to evaluate. In spite of this, some conclusions can be drawn for the OMA landscape which will be commented on from the extra- to the intra-cystic space. Firstly, several authors observed an increase in 8-OHdG [79,80], forkhead box A3 (FOXA3) and advanced glycation end products [80] in the normal ovarian cortex surrounding OMAs in comparison to the normal ovarian cortex surrounding benign ovarian cysts, which might postulate OS as a specific mechanism in endometriosis. Secondly, Ngô and co-workers [81] observed that the pro-oxidant/antioxidant balance is shifted towards enhanced OS in both epithelial and stromal cells within the cyst wall, which increased cell proliferation through ERK1/2 pathway activation. Additionally, Chen and co-workers [82] observed an increase in generation of both ROS and of superoxide dismutase 2 (SOD2) by mitochondria in stromal cells, which may support the development of the disease by allowing a high metabolic rate within these lesions. Finally, the increased ROS production in endometriotic cells might also be a consequence of the pro-oxidative inner cyst fluid stimulation [83]. Interestingly, stimulation of immortalized ovarian surface epithelium (OSE) and endometrial glandular cells with endometriotic cyst content produced more ROS than treatment with non-endometriotic cyst content [84]. Altogether, the increased OS might have consequences in gene expression, rgarding the down-regulation of the tumour suppressor gene AT-rich interaction domain A (*ARID1A*) via promoter hypermethylation [85], which is considered an early event in endometriosis malignant transformation (Figure 3).

Opposite to this evidence, Santulli and collaborators [86] did not find any significant differences in protein OS markers in the PF of women with ovarian or peritoneal endometriosis when compared with control PF, in contrast to women with deep infiltrating endometriosis. This counterintuitive finding might find a rationale either when considering OMA as an encapsulated lesion within the ovary unlikely to influence PF composition, or a possible bias in these observations due to the inclusion of patients with benign pathologies in the control population.

3. Endometriosis-Associated Ovarian Cancer

Endometriosis malignant transformation occurs in OMA at a higher rate than in other endometriotic lesions [87], producing the so-called EAOC. This clinical entity might be the result of a sequential process of malignant transformation from endometriotic lesions through atypical endometriosis [88] and, finally, to ovarian cancer, especially to the endometrioid (EOC) and clear cell (OCCC) histological subtypes [89]. Importantly, the estimated risk of malignancy of ovarian endometriosis may be close to 1% [90,91]. This enhanced rate of transformation of OMA, together with the specific histotypes to which it leads, points to unique carcinogenic processes within these endometriotic lesions, with a plausible involvement of OS and miRNAs.

miRNAs and Oxidative Stress in Endometriosis-Associated Ovarian Cancer

The intra-cystic fluid within OMA represents a unique milieu that may underlay the initiation of the malignant transformation process [92], probably triggered by the release of free iron as a result of monthly bleeding and the subsequent raise in OS [84] (Figure 3). At this respect, endometriotic cysts fluid present with higher levels of free iron, lactate dehydrogenase, lipid peroxide, 8-OHdG and potential antioxidant in comparison to non-endometriotic cysts and OCCC. Also, higher levels of 8-OHdG are reportedly associated with OCCC, being almost negative in EOC [93]. In agreement with these results, Fujimoto and co-workers [94] found increased levels of 8-OHdG and heme-oxygenase 1 in cyst fluid from OMA when compared to EAOC (mainly OCCC). Additionally, the expression of 8-OHdG is decreased in EAOC tissues (OCCC and EOC) when compared to paired adjacent endometriotic tissue and OMA [95]. Therefore, one might conclude that the levels of OS are higher in OMA and decrease in OCCC, and further in EOC, which suggests their involvement mainly in the initiation process of EAOC.

Regarding antioxidants, decreased MnSOD expression and increased malondialdehyde (MDA) expression has been observed in OMA and EOAC in comparison to non-EAOC and control endometria, corroborating the increased OS levels commonly found in these tissues [96]. In vitro studies denoted that the over-expression of the antioxidant lipocalin2 increases intracellular iron concentrations in OCCC cell lines but reduces the levels of ROS and DNA damage, probably through increasing glutathione, xCT (a cystine transporter protein) and CD44v (a stem cell marker), resulting in reduced apoptosis and prolonged cell survival of OCCC [97]. Altogether, it seems plausible to acknowledge that OS is involved in early steps of the malignant transformation of endometriosis.

In this context, some of the specific alterations produced by this pro-oxidant milieu have been unravelled. Opposed to HGSOC, EAOC is characterized by mutations in several genes, including the TSG *ARID1A*, *PTEN*, and the OG phosphatidylinostil-4,5 biphosphonate 3-kinase catalytic subunit alpha *(PIK3CA)* [98]. Winarto and collaborators [96] investigated tissue samples from patients with endometriosis, EAOC, or non-EAOC to observe that *ARID1A* expression decreases with increased OS a finding also corroborated in vitro. Interestingly, an epigenetic modification (*ARID1A* promoter hypermethylation in OMA) might underlay this observation [85]. In addition, chromosomal aberrations are more frequently found in EOC over OMA and less frequently in extragonadal endometriosis, which might reflect a clonal expansion of aberrant OMA cells produced as a result of the harmful intra-cystic milieu [87]. In addition, copy number variation in OCCC has been identified by several authors: for instance, mesenchymal-to-epithelial transcription factor (*MET*) gene amplification is a frequent event in OCCC and this genomic amplification can be recapitulated in an ROS-induced rat carcinoma model [99]. Furthermore, amplification at loci 17q23-25 occurs in approximately 40% of OCCC, which might over-express the encoded miR-21, decreasing the expression of their target PTEN. Therefore, both OS and miRNAs converge in some cases with genetic alterations to provoke the loss of function of important tumour suppressor genes in EAOC carcinogenesis (Figure 3).

Apart from its role in tumour initiation, a growing body of evidence has shed light into the role of deregulated miRNAs in other processes linked to EAOC carcinogenesis, as cell proliferation [100–102], migration [100,101], invasion [100,102], and epithelial-to-mesenchymal transition [100,103–106].

However, differences in study design and methodological approaches render low overlapping between reported deregulated miRNAs (Table 3), that is limited to miR-21 [106,107], miR-510 [104,105], miR-29b [108,109], miR-191 [102,110] and miR-30a [111,112].

Table 3. Deregulated miRNAs in selected studies in EAOC.

Reference	Effect	Experimental Design	Main Deregulated miRNAs in Patients
[100]	Promoted proliferation, migration, invasion	OCCC and adjacent non-tumor tissues	↓ miR-424
[101]	Increased cell motility, growth and colony formation	OCCC and EOC cell lines and OMA primary stromal cells	↓ miR-381
[102]	Increased cell proliferation and invasion	EOC, OCCC, OMA and control endometria tissues	↑ miR-191
[103]	Increased MET phenotype and good prognosis	HGSOC, EOC, OCCC and mucinous ovarian cancer tissues	↓ miR-506
[104]	Increased EMT phenotype	HGSOC and OSE tissues	↑ miR-205-5p
[104]	EMT (miR-200s), poor PFS and OS (miR-200c -3p)	HGSOC, OCCC and OSE tissues	↑ miR-200s, miR-182-5p ↓miR-383
[104]	Hystology differentiatiors	OCCC and HGSOC tissues	↑ miR-509-3-5p, miR-509-3p, miR-509-5p, miR-510
[105]	Poor overall survival	OCCC and HGSOC tissues	↓ miR-510, miR-129-3p
[106]	Increased EMT phenotype	OCCC and HGSOC tissues	↑ miR-9
[107]	Down-regulation of the TSG PTEN	OCCC tissues	↑ miR-21
[108]	Increased paclitaxel chemosensitivity	OCCC cell lines	↑ miR-29b
[109]	Poor prognosis	OCCC, HGSOC, mucinous ovarian cancer and control tissues	↓ miR-29b
[110]	Increased apoptosis evasion	EOC, OMA and control tissues	↑ miR-191
[111]	Hystology differentiatiors	OCCC, EOC, HGSOC and mucinous ovarian cancer	↑ miR-30a and miR-30a *
[112]	Poor overall survival in ovarian papillary serous carcinoma tissues	OCCC and ovarian papillary serous carcinoma tissues	↓ miR-30a, miR-30e and miR-505
[113]	Enhanced sensitivity to everolimus	OCCC and OSE cell lines	↓ miR-100

↑, up-regulated levels; ↓, down-regulated levels; EOC, endometrioid ovarian cancer; OCCC, ovarian clear cell carcinoma; OSE, ovarian surface epithelium; HGSOC, high-grade serous ovarian cancer; EMT, epithelial-to-mesenchymal transition; MET, mesenchymal-to-epithelial transition. PTEN, phosphatase and tensin homologue; TSG, tumour suppressor gene; PFS, progression-free survival.

Regarding treatment, first-line chemotherapy agents in ovarian cancer (EAOC and HGSOC) are represented by taxanes and carboplatin, which exert their antitumour effect partly mediated by an increase in OS (Figure 3). A very recent study from Amano and co-workers [114] observed an association between higher mitochondrial SOD2 levels in EAOC and poor prognosis, without any difference regarding histology (OCCC vs. EOC), which could reflect an improved scavenging of platinum-mediated ROS production during EAOC treatment. On the other hand, the acquisition of chemoresistance by both EAOC and HGSOC is directly linked to the poor overall results in these patients. Specifically, the clear cell subtype is known to be more chemorresistant and associated to worse prognosis than the endometrioid subtype.

In this context, Sugio and co-workers [108] observed that increased miR-29b levels correlate with progression-free survival in OCCC patients. Of note, the down-regulation of the protein Bcl2-associated athanogene 3 (BAG3) seemed to induce the miR-29b expression, which finally sensitized cells to paclitaxel. In an attempt to overwhelm chemoresistance in OCCC, new therapeutic approaches have been developed, including the use of the rapamycin analogue everolimus, for which resistances have also been documented [115]. However, Nagaraja and collaborators [113] found that induced over-expression of miR-100 increased the sensitivity to everolimus in OCCC cell lines, an effect mediated by the inhibition of mammalian target of ramapycin (mTOR) signalling (Table 3). All in all, these results might pave the way for a miRNA-based therapy to overcome chemoresistance in EAOC.

Therefore, there exists compelling evidence that miRNAs and OS cooperate in the distinct steps of EAOC carcinogenesis to accomplish the malignant transformation of endometriosis and that both of them influence therapeutic outcomes in these conditions.

4. High-Grade Serous Ovarian Cancer

Ovarian cancer is the fifth cause of cancer death in women and the most lethal of gynaecological malignancies [116]. Although this term designates a group of multiple malignant diseases sharing the same anatomical location and with different histological types [117], the poor prognosis data is mainly related to the most frequent (70%) histological subtype, HGSOC. Actually, HGSOC causes 70%–80% of gynaecological cancer-related deaths [118]. Regarding prognosis, more than 80% of HGSOC patients are diagnosed in advanced stages and the 5-year survival rates are below 50% [119,120]. This poor prognosis in HGSOC may be attributable to the asymptomatic nature, the lack of diagnostic methods in initial stages, and the presence of metastases at the time of diagnosis, although major determinants of HGSOC-related deaths are the frequent recurrences and the acquisition of chemoresistance by the tumour [121]. To this respect, slight changes have been produced in the medical treatment algorithm for HGSOC in the last decades, which relies on taxanes and cisplatin derivatives.

Despite its clinical importance, the exact aetiopathogenic mechanism of HGSOC remains elusive, since an unsolved debate on whether HGSOC arises from the ovary itself or from the fimbriae of the fallopian tubes still exists. Initially, the OSE was proposed as the focus for epithelial ovarian cancers, postulating the repeated ovulation and inclusion of cysts during regeneration as an initiation mechanism. The later theory of the fallopian tube epithelium (FTE) as a source of this tumour has gained growing attention [122], considering the premalignant lesion serous tubal intraepithelial carcinoma (STIC) as an early event. Accordingly, the tumour formed in the fallopian tube spreads to the ovary helped by the retrograde menstruation, where it is more capable of metastasizing [123] (Figure 3).

miRNAs and Oxidative Stress in High-Grade Serous Ovarian Cancer

Multiple studies based on both OS and miRNAs have reinforced an FTE origin of HGSOC. Vercellini and co-workers [124] proposed the "incessant menstruation" hypothesis, which states that pelvic macrophages decompose menstrual blood in the peritoneal cavity, and the released free-iron can damage epithelial fimbriae via ROS production. Specifically, the action of ROS on the FTE provokes a continued process of DNA damage and repair, which permits the sequential acquisition of mutations, and genomic instability. This genomic instability is represented by very frequent structural and numerical aberrations in chromosomes 3, 8, 11, 17, and 21 [125] and the mutations in HGSOC-driver genes, breast cancer gene 1 and 2 (*BRCA1/2*), *TP53*, or *PTEN/PIK3CA* [126] (Figure 3). Additionally, the loss of function of these genes has been involved with an impaired defence against OS in these tumours [127], what determines a positive feed-back loop. Nevertheless, we cannot withstand that 17% of HGSOC patients carry germline mutations in BRCA1/2 [128].

From an epigenetic standpoint, DNA methylation [129] and miRNA [130] studies endorse the fallopian tube origin of HGSOC. For the former, Klinkebiel and co-workers [129] examined a small cohort of paired HGSOC, FTE, and OSE found that DNA methylomes are more highly conserved between HGSOC and FTE than between HGSOC and OSE. For the latter, Yang and co-workers [130] examined the expression of the miR-200 family (i.e., miR-200a, -200b, -141, and -429) and miR-205, their target genes and downstream effectors in a panel of HGSOC, STIC, FTE, and OSE tissues. As a result, the authors observed an over-expression of miR-200 family in HGSOC, STIC, and FTE and an increase of the epithelial phenotype through down-regulation of the target genes *ZEB1*, *ZEB2*, *TGFβ1*, and *TGFβ2*. These effects were not observed in OSE. Interestingly, pre-miR-200 transfection in FTE cells increased the levels of CA-125, recapitulating the high expression of this mucin in HGSOC.

Compelling literature shows evidence that epithelial-mesenchymal transition (EMT), a process by which epithelial cells lose their characteristic organization and acquire the motility of mesenchymal cells, plays a central role in HGSOC tumour progression and chemoresistance acquisition [131,132]. Notably, several miRNAs are among their master regulators. Boac and collaborators [133] serially treated four ovarian cancer cell lines (A2780CP, A2780S, IGROV1, and OVCAR5) with six cycles of cisplatin, and assessed the miRNA patterns in each of the treatment-recovery cycles. They identified five known miRNAs positively (namely miR-496, miR-485-5p, let-7g, and miR-152) or negatively (miR-27b) correlated with cisplatin chemoresistance, being the modulated pathways mainly involved in EMT regulation. Zhu and collaborators [134] demonstrated that decreased expression of miR-186 is associated with increased cisplatin resistance in HGSOC patients. Remarkably, decreased miR-186 levels up-regulate those of their target Twist1, an EMT driver, promoting the mesenchymal phenotype. Hellman and collaborators [135] analysed nine studies involving gene data sets to discover pathways associated with platinum resistance in ovarian cancer. Interestingly, despite the low degree of gene overlapping due to study design heterogeneity and technology employed, pathways related to OS ("oxidative stress", "oxidative stress response mediated by nuclear factor (NF)-E2-related factor 2") and to EMT ("TGFbeta signalling", "cell migration", "cellular movement", and "cell-to-cell signalling") were among the most over-represented in the studied datasets. In addition, components of the miR-17-92 cluster, which down-regulates two key TFGβ signalling molecules, and let-7 family members were also associated with platinum resistance in these analyses. Finally, Brozovic and collaborators [136] found that decreased miR-200s (miR-200a, miR200b, miR-200c, miR-429, and miR-141) expression is associated with a partial EMT phenotype in the ovarian cancer paclitaxel resistant cell lines OVCAR-3/TP and MES-OV/TP. Consistently, miR-200c and miR-141 inhibition increased the mesenchymal phenotype and the resistance to paclitaxel in non-resistant OVCAR-3 cell lines. As expected, miR-200c and miR-141 over-expression sensitized MES-OV/TP cells to paclitaxel through a mesenchymal-to-epithelial transition, and increased the levels of a set of redox enzymes, mainly reductases. It is important to mention that oxidative stress induces the expression of both miR-141 and miR-200c [137] (Table 4, Figure 2A). As can be observed, the vast majority of studies regarding the miRNA regulation of EMT have been developed in cell cultures. Although they represent a valuable source of information, established cell lines do not completely mirror the biological complexity of a tumour tissue sample. At the light of the importance of the EMT phenomenon in HGSOC patients' prognosis, it becomes clear that there is a need for major number of studies in HGSOC tissue specimens that would increase the knowledge about the miRNA regulation of the EMT.

Table 4. Deregulated miRNAs in selected studies in HGSOC.

Reference	Effect	Experimental Design	Main Deregulated miRNAs in Patients
[130]	Susceptibility to oncogenic mutations and histologic differentiation. FTE cells increase CA-125 upon pre-miR-200 transfection.	HGSOC, STIC and FTE vs. OSE	↑ miR-200a, miR-200b, miR-141 and miR-429 and miR-205
[133]	Five miRNAs associated with cisplatin resistance EMT phenotype associated with higher chemoresistance Two pathways associated with overall patient survival (TGF/WNT and Regulation of EMT)	Four ovarian cancer cell lines, public ovarian cancer dataset	Positively correlated namely miR-496, miR-485-5p, let-7g and miR-152 Negatively correlated miR-27b
[134]	EMT phenotype, cisplatin resistance and worse prognosis	HGSOC tissue and ovarian cancer cell lines (chemosensitive and chemoresistant)	↓ miR-186, ↑ miR-200 family (significantly miR-141 and miR-200a)
[135]	Platinum resistance, related to EMT and stemness	Exploratory study based on nine published gene sets associated with platinum resistance in ovarian cancer.	↓ miR-17-92 cluster, let-7 family members

Table 4. *Cont.*

Reference	Effect	Experimental Design	Main Deregulated miRNAs in Patients
[136]	Stronger EMT phenotype and paclitaxel resistance	Two ovarian cancer cell lines (sensitive and resistant to paclitaxel and carboplatin)	↓ miR-200s (miR-200a, miR200b, miR-200c, miR-429 and miR-141)
[138]	Increased chemoresistance by regulation of the VEGFB and VEGFR2 pathway	198 serous epithelial ovarian carcinomas, six epithelial ovarian carcinoma cell lines	↓ miR-484 (tumour angiogenesis), miR-642, miR-217
[139]	Poor prognosis, increased placlitaxel resistance	HGSOC tissues relative to normal control tissues. Placlitaxel resistant ovarian cell lines.	↓ miR-136
[140]	Decreased cisplatin resistance by PARP1 regulation	Cisplatin-resistant and cisplatin-sensitive ovarian cancer cell lines	↓ miR-216b
[141]	Longer progression-free survival (PFS), increased platinum sensitivity to cisplatin and PARP inhibitors by directly targeting BRCA1	Serous ovarian cancer patients and tumour xenografts	↑ miR-9

↑, up-regulated levels; ↓, down-regulated levels; BRCA1, Breast Cancer type 1 susceptibility protein; EMT, epithelial-to-mesenchymal transition; VEGFB, vascular endothelial growth factor B; VEGFR2, vascular endothelial growth factor receptor 2; FTE, Fallopian Tube Epithelial; HGSOC, high-grade serous ovarian cancer; OSE, ovarian surface epithelium; PARP1, Poly [ADP-ribose] polymerase 1; STIC, Serous Tubal Intraepithelial Carcinoma, TGF, Transforming Growth Factor; Wnt, Wingless-related integration site.

Apart from EMT, Vecchione and collaborators [138] analysed the miRNA expression profiles in 198 HGSOC patients and validated a signature of three miRNAs (miR-484, miR-642 and miR-217) involved in chemoresistance, of which, miR-484 was associated with angiogenesis regulation. In addition, Jeong and co-workers [139] observed that miR-136 behaves as a TSG and that miR-136 down-regulation is associated with poor overall results in HGSOC patients. Specifically, miR-136 targets Notch3, and miR-136 over-expression re-sensitized paclitaxel-resistant ovarian cancer cells and significantly reduced cell viability, proliferation, cancer stem cell spheroid formation, and angiogenesis, as well as increased apoptosis when compared with the effects of isolated paclitaxel treatment.

As aforementioned, first-line chemotherapy schemes (cis-platin derivatives and taxanes) in HGSOC exert their anticancer effect partially mediated by increased oxidative stress. Accordingly, Ayyagari and co-workers [142] showed a synergistical effect on reduced cell viability and increased apoptosis when ovarian cancer cell lines were simultaneously treated with the anti-parasitic drug bithionol and paclitaxel, and this effect was attributable to an increase in intracellular ROS production. These findings are in agreement with previous studies from the same research group considering the combination of biothionol and cis-platin [143].

On the other hand, several PARP inhibitors (namely olaparib, niraparib, and rucaparib) have been approved for the treatment of HGSOC. These drugs act by preventing the poly [ADP-ribose] polymerase (PARP)-mediated repair of DNA damage and are especially effective in *BRCA1/2* mutation carriers. In this respect, Hou and collaborators [144] observed that the anti-tumour effect of PARP is mediated by increased ROS production and that antioxidant treatment with *N*-acetylcysteine rescued the effect. Notably, several miRNAs have been associated with PARP inhibitors effectiveness. In vitro studies have linked miR-622 and miR-493-5p over-expression with platinum and PARP inhibitor resistance in *BRCA1* and *BRCA2* mutated ovarian cancer cell lines, respectively [145,146].

A growing body of evidence suggests that OS is involved in the acquisition of chemoresistance in HGSOC as the tumour develops (Figure 3). Belotte and co-workers [147] reported that chemorresistant MDAH-2774 and SKOV-3 ovarian cancer cell lines display a pro-oxidant state, with reduced expression of the antioxidant enzyme glutathione reductase and increased expression of reactive nitrogen species nitrate/nitrite and their synthetizing enzyme iNOs. One step further, Fletcher and co-workers [148] observed that first-line chemotherapy agents induce point mutations in key redox enzymes, allowing a pro-oxidant state in ovarian cancer cells and favouring chemoresistance. Specifically, the authors observed decreased levels of SOD2, cytochrome b-245 alpha chain (CYBA, a NADPH oxidase subunit) and glutathione reductase and an increase in iNOS activity, nitrate/nitrite levels and glutathione peroxidase in chemoresistant cells. As expected, the combination of SOD with chemotherapy (cisplatin or taxanes) significantly increased the sensitivity to chemotherapy. In addition, miR-216b increases cisplatin sensitivity by directly targeting PARP1 [140] whilst miR-9 increases sensitivity to cisplatin and PARP inhibitor by directly targeting BRCA1 [141].

Regarding the antioxidant response, Pei and co-workers [149] observed a significant inhibition of cell adhesion, migration, invasion, metastasis, and oxidative stress levels in SKOV-3 cells when treated with the antioxidant bisdemethoxycurcumin. In an interesting approach, Pons and collaborators [150] wondered whether the initial activation state of the antioxidant response could influence patients' outcomes. The authors observed a significant reduction of the antioxidant enzymes glutathione reductase and catalase, as well as of the uncoupling proteins (UCP) UCP2 and UCP5 in HGSOC patients resistant to carboplatin/paclitaxel. Nevertheless, the small cohort assessed and the few markers of OS preclude further conclusions on this study.

Altogether, it seems clear that whichever the origin of HGSOC, both OS and miRNAs play a crucial role in its initiation, promotion, and progression, including chemotherapy outcomes.

5. Potential Role of miRNAs and Oxidative Stress in Diagnosis and Treatment of Endometriosis, EAOC and HGSOC

As expected, the opportunity that alterations in both miRNAs and OS markers confer as potential biomarkers of disease has not been overlooked by researchers. However, the small number of patients included in the majority of studies and the high variability in the assessed analytes preclude any molecule being proposed as a reliable biomarker at this point. A common finding in studies evaluating OS as biomarkers for endometriosis is an increase in either plasma or serum oxidative stress biomarkers [151,152] and reduced levels of thiols [153,154] in patients in comparison to control women. In addition, other authors evaluated urine as a source of biomarkers, observing higher concentration of metabolites related to inflammation and oxidative stress (namely N(1)-methyl-4-pyridone-5-carboxamide, guanidinosuccinate, creatinine, taurine, valine, and 2-hydroxyisovalerate) in patients with endometriosis [155]. In EAOC, two studies demonstrated the utility of examining OS-related molecules as prognosis biomarkers. Amano and collaborators [114] observed that increased levels of SOD2 in EAOC specimens associated with worse prognosis (overall survival and progression-free survival). Additionally, protease-activated receptor-2 (PAR-2) expression, which is up-regulated by OS, correlated with shorter survival in OCCC specimens [93]. Further immunohistochemical analyses also revealed that higher 8-OHdG levels is associated with poor differentiation, higher stage, and non-optimal surgical outcomes in epithelial ovarian cancer, including HGSOC, EOC, and OCCC specimens [156]. Importantly, these observations might have a counterpart in peripheral markers, since higher serum 8-OHdG levels were associated with poor prognosis and platinum resistance in epithelial ovarian cancers, especially in EOC patients [157]. Similar results were also obtained in stage I-II epithelial ovarian cancer studies [158].

Interestingly, miRNAs can be found as circulating miRNAs in a number of biofluids, including blood, urine, and peritoneal fluid [159]. Although far from the scope of this review, several mechanisms explain the higher stability of circulating miRNAs in these biofluids [160], which make them attractive biomarkers in a myriad of pathologies, including gynaecological conditions. A limited number of studies have explored the potential of circulating miRNAs as non-invasive biomarkers for endometriosis [67–73]. As a matter of fact, slight reproducibility has been found among studies and neither a single miRNA nor a combination of them has demonstrated a higher performance when compared to current diagnostic techniques. These differences in results might find a rationale in the type of blood sample analyzed (either serum or plasma), the distinct stages of endometriosis considered and the heterogeneous cohort considered as control. Cho and co-workers [67] found decreased levels of let-7b and miR-135 in serum samples from patients with endometriosis. Wang and co-workers [68] observed decreased levels of miR-9 *, miR-141 *, miR-145 * and miR-542-3p * and increased levels of miR-122 and miR-199a in sera from patients in comparison to control women. Interestingly, these results considering miR-122 and miR-199a have been recently corroborated by other authors [69]. Wang and co-workers [70] performed a deep sequencing approach and qRT-PCR validation to determine that down-regulated miR-30c-5p, miR127-3p, miR-99b-5p, miRNA-15b-5p, and miRNA-20a-5p and up-regulated miR-424-3p and miR-185-5p could be putative biomarkers of the disease. Regarding plasma samples, decreased levels of miR-17-5p, miR-20a, and miR-22 [71], decreased levels of miR-200a-3p, miR-200b-3p, miR-414-3p [72] and increased levels of miR-154-5p [73] have been proposed as biomarkers of endometriosis. Perhaps due to the scarcity of cases, only one study reports circulating miRNA profiles in EAOC patients [74]. Suryawanshi and collaborators identified three distinct miRNA signatures in plasma capable of discriminating among patients with EAOC, endometriosis and healthy individuals. Interestingly, four miRNAs distinguishing EAOC patients from healthy women (namely miR-15b, -16, -21, and -195) also discriminated cancer and control mice in a pre-clinical murine model. Finally, several recent studies have evidenced the utility of miRNAs as biomarkers in HGSOC. Kobayashi and co-workers [75] observed that miR-1290 is elevated in sera from HGSOC patients in comparison to control women, but not in ovarian cancers of other histological types, and that its expression correlates with tumour burden. Shah and co-workers [76] observed that the combination of sera miR-375 and CA-125 is a diagnostic biomarker of HGSOC. Todeschini and collaborators [77] employed two independent patient cohorts to identify miR-1246 as the best diagnostic marker in HGSOC. Finally, Kan and collaborators [78] built a predictive model including sera levels of miR-200b and miR-200c, with an area under the curve of 0.784, distinguishing HGSOC patients from control individuals (Table 2).

With respect to treatment, a number of studies have considered the antioxidant treatment in endometriosis, based on supplementation with Vitamin C and E [161], or preparations with different antioxidants [162,163]. As a result, peripheral oxidative stress markers diminished [161], and the pain symptoms were improved [162,163] albeit without any benefit on pregnancy rates [161]. In the last 10 years, a few clinical trials have evaluated the utility of oxidative-stress based treatments for HGSOC, either aiming to increase oxidative stress or employing antioxidants. Monk and co-workers report a very recent two-stage phase II clinical trial, which failed to provide any benefit in objective tumour responses in a panel of platinum-resistant recurrent ovarian, tubal and peritoneal cancer patients treated with i.v. elesclomol, a ROS inducer, plus weekly paclitaxel in comparison to paclitaxel alone [164]. This observation might be in agreement with the involvement of ROS in the acquisition of chemoresistance by the tumour. In agreement with this rationale, a recent phase II trial evaluated the combination of the vitamin E analogue, delta tocotrienol, with bevacizumab in 23 patients with refractory ovarian cancer. The combination produced an improvement in progression-free survival (median 6.9 months) and overall survival (median 10.9 months) with regards to the data in current literature [165]. Trudel and co-workers [166] performed a two-stage, single-arm, phase II study (NCT00721890) of the tea drink enriched with the polyphenol epigallocatechin gallate (EGCG) for maintenance treatment in advanced ovarian cancer. Sixteen participants (13 HGSOC and three EOC) were included in the study since they were in complete remission after completion of their first line treatment and were followed for 18 months thereafter. Unfortunately, daily nutritional intervention with 500mL of the drink failed to prove any benefit regarding recurrence improvement at 18 months.

Regarding miRNAs, the vast literature involving the outstanding role of miRNAs in the three considered gynaecological conditions pave the way for a miRNA-based therapy in order to restore deregulated miRNA levels. Indeed, pathologically down- or over-expressed levels of a given miRNA could be in vitro restored by the employment of miRNA mimics or antimiRs, respectively. Although a limited number of miRNAs has been tested in clinical trials (i.e., anti-miR-122 for Hepatitis C, antimir-103/107 for type 2 diabetes and non-alcoholic fatty liver diseases, antimir-155 for cutaneous T cell lymphoma and mycosis fungoides, miR-29 mimic for scleroderma, miR-16 mimic for mesothelioma and non-small cell lung cancer, and miR-34 mimics for multiple solid tumors (reviewed in [167]), none of them have reached phase III clinical trials and there is no single ongoing trial considering gynecological diseases, to the best of our knowledge. Several limitations might provide a rationale for the lack of translation of miRNA-based therapeutics, as the high probability of off-targets cellular and systemic effects (provided the multiple mRNAs targeted by a single miRNA and the difficulty of delivering miRNA therapies to a specific organ or even to a specific cellular type, respectively). However, a major limitation might lay in the lack of knowledge of the dynamic expression of miRNA patterns and the overall effect produced due to the interaction among them, since researchers usually only have access to a still picture of disease specimens.

6. Conclusions

In conclusion, extensive literature shows that there exists an interplay between miRNAs and oxidative stress in several gynaecological conditions, highlighting endometriosis, EAOC and HGSOC. In endometriosis, oxidative stress and miRNAs contribute to the establishment and development of endometriotic lesions. Regarding OMA, repeated ROS stimulation and miRNA deregulation on the cells of the cystic wall seem to play an important role in the malignant transformation of endometriosis. Besides, a mechanistic model for oxidative stress and miRNAs in OMA malignant transformation may be established: Within endometrial cysts, oxidative stress might be involved in the carcinogenesis of EAOC. Initially, repeated intra-cystic bleeding generates ROS and OS that act on the cystic wall cells. To cope with this adverse event, OMA epithelial cells increase their antioxidant response, but eventually are overwhelmed by repeated ROS, producing genetic and epigenetic alterations in crucial tumour suppressor genes. Additionally, miRNAs can contribute not only to the loss of function of these tumour suppressor genes but also to important carcinogenic events. Yet, once EAOC is established, this increased antioxidant response diminishes the platinum-mediated ROS injury and the efficiency of the chemotherapy treatment, which is also modulated by specific miRNAs. Finally, oxidative stress and miRNA deregulation is involved in the carcinogenesis of HGSOC and crucially influences the response to first-line chemotherapeutics, both regarding initial treatment outcomes and acquisition of chemoresistance.

With respect to diagnosis, increased circulating levels of OS markers have been involved with endometriosis diagnosis, and increased circulating levels of SOD and OS markers have been associated with poor prognosis in EAOC and HGSOC. Additionally, several studies have proven the putative role of miRNAs as biomarkers of these three gynaecological conditions. Regarding therapeutics, antioxidant treatment in endometriosis seems to improve the associated pain and the combination of antioxidants with bevacizumab is a promising approach in refractory ovarian cancer patients. On the other hand, miRNA treatment is still in its infancy, with few ongoing clinical trials and none of them in gynaecological diseases. The difficulty in developing miRNA-based therapies could be related to the likely off-target effects and the lack of knowledge of the precise dynamics and interactions of miRNAs throughout the disease development.

Altogether, future research endeavours are guaranteed to enlarge the knowledge on the action of miRNAs and oxidative stress in the pathophysiology of these important gynaecological pathologies and to propose targeted therapeutic strategies to deal with their pernicious effect.

Author Contributions: J.M.-A. conceived and designed the manuscript, revised the literature and wrote the manuscript. A.P.C. revised the literature and drafted the manuscript. C.A. and J.G.-O. contributed to literature revision and revised the manuscript. A.M.-M. and S.C.-F. prepared the figures and tables and revised the manuscript. J.G.-E. has written the manuscript.

Funding: This work was supported by grants from ISCIII and FEDER (PI17/01945). J. M-A. is supported by a post-doctoral grant (APOSTD/2019/087) and A.M-M. is supported by PhD scholarship from Asociación Española Contra el Cáncer (AECC).

Conflicts of Interest: The authors declare no conflict of interest.

References

1. Donnez, J.; Binda, M.M.; Donnez, O.; Dolmans, M.M. Oxidative stress in the pelvic cavity and its role in the pathogenesis of endometriosis. *Fertil. Steril.* **2016**, *106*, 1011–1017. [CrossRef] [PubMed]

2.   Dröge, W. Free radicals in the physiological control of cell function. *Physiol. Rev.* **2002**, *82*, 47–95. [CrossRef] [PubMed]

3.   Sies, H. Oxidative stress: A concept in redox biology and medicine. *Redox Biol.* **2015**, *4*, 180–183. [CrossRef] [PubMed]

4.   Rockfield, S.; Raffel, J.; Mehta, R.; Rehman, N.; Nanjundan, M. Iron overload and altered iron metabolism in ovarian cancer. *Biol. Chem.* **2017**, *398*, 995–1007. [CrossRef]

5.   Dixon, S.J.; Stockwell, B.R. The role of iron and reactive oxygen species in cell death. *Nat. Chem. Biol.* **2014**, *10*, 9–17. [CrossRef] [PubMed]

6.   Sies, H.; Berndt, C.; Jones, D.P. Oxidative Stress. *Annu. Rev. Biochem.* **2017**, *86*, 715–748. [CrossRef] [PubMed]

7.   Bokov, A.; Chaudhuri, A.; Richardson, A. The role of oxidative damage and stress in aging. *Mech. Ageing Dev.* **2004**, *125*, 811–826. [CrossRef] [PubMed]

8.   Sies, H. *Oxidative Stress: Introductory Remarks*; Academic Press: London, UK, 1985; pp. 1–8.

9.   Inglés, M.; Gambini, J.; Carnicero, J.A.; García-García, F.J.; Rodríguez-Mañas, L.; Olaso-González, G.; Dromant, M.; Borrás, C.; Viña, J. Oxidative stress is related to frailty, not to age or sex, in a geriatric population: Lipid and protein oxidation as biomarkers of frailty. *J. Am. Geriatr. Soc.* **2014**, *62*, 1324–1328. [CrossRef]

10.   Sies, H.; Jones, D. Oxidative stress. In *Encyclopedia of Stress*, 2nd ed.; Fink, G., Ed.; Elsevier: Amsterdam, The Netherland, 2007; pp. 45–48.

11.   Smith, M.A.; Perry, G.; Richey, P.L.; Sayre, L.M.; Anderson, V.E.; Beal, M.F.; Kowall, N. Oxidative damage in Alzheimer's. *Nature* **1996**, *382*, 120–121. [CrossRef]

12.   Fincham, J. *Epigenetic Mechanisms of Gene Regulation*; Russo, V.E.A., Martienssen, R.A., Riggs, A.D., Eds.; Cold Spring Harbor Laboratory: Cold Spring Harbor, NY, USA, 1997; pp. 159–162.

13.   Portela, A.; Esteller, M. Epigenetic modifications and human disease. *Nat. Biotechnol.* **2010**, *28*, 1057–1068. [CrossRef]

14.   Guo, S.W. Epigenetics of endometriosis. *Mol. Hum. Reprod.* **2009**, *15*, 587–607. [CrossRef] [PubMed]

15.   Esteller, M. Epigenetics in cancer. *N. Engl. J. Med.* **2008**, *358*, 1148–1159. [CrossRef] [PubMed]

16.   Lee, R.C.; Feinbaum, R.L.; Ambros, V. The C. elegans heterochronic gene lin-4 encodes small RNAs with antisense complementarity tolin-14. *Cell* **1993**, *75*, 843–854. [CrossRef]

17.   Lee, Y.; Ahn, C.; Han, J.; Choi, H.; Kim, J.; Yim, J.; Lee, J.; Provost, P.; Rådmark, O.; Kim, S.; et al. The nuclear RNase III Drosha initiates microRNA processing. *Nature* **2003**, *425*, 415–419. [CrossRef]

18.   Bartel, D.P. MicroRNAs: Genomics, biogenesis, mechanism, and function. *Cell* **2004**, *116*, 281–297. [CrossRef]

19.   Caporali, A.; Emanueli, C. MicroRNA regulation in angiogenesis. *Vascul. Pharmacol.* **2011**, *55*, 79–86. [CrossRef]

20.   Gailhouste, L.; Ochiya, T. Cancer-related microRNAs and their role as tumor suppressors and oncogenes in hepatocellular carcinoma. *Histol. Histopathol.* **2013**, *28*, 437–451.

21.   Santamaria, X.; Taylor, H. MicroRNA and gynecological reproductive diseases. *Fertil. Steril.* **2014**, *101*, 1545–1551. [CrossRef]

22.   Marí-Alexandre, J.; Sánchez-Izquierdo, D.; Gilabert-Estellés, J.; Barceló-Molina, M.; Braza-Boïls, A.; Sandoval, J. miRNAs Regulation and Its Role as Biomarkers in Endometriosis. *Int. J. Mol. Sci.* **2016**, *17*, E93. [CrossRef]

23.   Borghese, B.; Zondervan, K.T.; Abrao, M.S.; Chapron, C.; Vaiman, D. Recent insights on the genetics and epigenetics of endometriosis. *Clin. Genet.* **2017**, *91*, 254–264. [CrossRef]

24.   Saare, M.; Rekker, K.; Laisk-Podar, T.; Rahmioglu, N.; Zondervan, K.; Salumets, A.; Götte, M.; Peters, M. Challenges in endometriosis miRNA studies-From tissue heterogeneity to disease specific miRNAs. *Biochim. Biophys. Acta* **2017**, *1863*, 2282–2292. [CrossRef] [PubMed]

25.   He, J.; Chang, W.; Feng, C.; Cui, M.; Xu, T. Endometriosis Malignant Transformation: Epigenetics as a Probable Mechanism in Ovarian Tumorigenesis. *Int. J. Genom.* **2018**, *2018*, 1465348. [CrossRef] [PubMed]

26.   Siufi Neto, J.; Kho, R.M.; Siufi, D.F.; Baracat, E.C.; Anderson, K.S.; Abrão, M.S. Cellular, histologic, and molecular changes associated with endometriosis and ovarian cancer. *J. Minim. Invasive Gynecol.* **2014**, *21*, 55–63. [CrossRef] [PubMed]

27.   Sun, Y.; Guo, F.; Bagnoli, M.; Xue, F.X.; Sun, B.C.; Shmulevich, I.; Mezzanzanica, D.; Chen, K.X.; Sood, A.K.; Yang, D.; et al. Key nodes of a microRNA network associated with the integrated mesenchymal subtype of high-grade serous ovarian cancer. *Chin. J. Cancer.* **2015**, *34*, 28–40. [CrossRef]

28. Mittempergher, L. Genomic Characterization of High-Grade Serous Ovarian Cancer: Dissecting Its Molecular Heterogeneity as a Road Towards Effective Therapeutic Strategies. *Curr. Oncol. Rep.* **2016**, *18*, 44. [CrossRef]

29. Samuel, P.; Pink, R.C.; Brooks, S.A.; Carter, D.R. miRNAs and ovarian cancer: A miRiad of mechanisms to induce cisplatin drug resistance. *Expert Rev. Anticancer Ther.* **2016**, *16*, 57–70. [CrossRef]

30. Lan, J.; Huang, Z.; Han, J.; Shao, J.; Huang, C. Redox regulation of microRNAs in cancer. *Cancer Lett.* **2018**, *418*, 250–259. [CrossRef]

31. He, J.; Jiang, B.H. Interplay between Reactive oxygen Species and MicroRNAs in Cancer. *Curr. Pharmacol. Rep.* **2016**, *2*, 82–90. [CrossRef]

32. Cheng, X.; Ku, C.H.; Siow, R.C. Regulation of the Nrf2 antioxidant pathway by microRNAs: New players in micromanaging redox homeostasis. *Free Radic. Biol. Med.* **2013**, *64*, 4–11. [CrossRef]

33. Giudice, L.C.; Kao, L.C. Endometriosis. *Lancet* **2004**, *364*, 1789–1799. [CrossRef]

34. Giudice, L.C. Endometriosis. *N. Engl. J. Med.* **2010**, *362*, 2389–2398. [CrossRef] [PubMed]

35. Tamaresis, J.S.; Irwin, J.C.; Goldfien, G.A.; Rabban, J.T.; Burney, R.O.; Nezhat, C.; DePaolo, L.V.; Giudice, L.C. Molecular classification of endometriosis and disease stage using high-dimensional genomic data. *Endocrinology* **2014**, *155*, 4986–4999. [CrossRef] [PubMed]

36. Burney, R.O.; Giudice, L.C. Pathogenesis and pathophysiology of endometriosis. *Fertil. Steril.* **2012**, *98*, 511–519. [CrossRef] [PubMed]

37. Huang, H.; Li, C.; Zarogoulidis, P.; Darwiche, K.; Machairiotis, N.; Yang, L.; Simoff, M.; Celis, E.; Zhao, T.; Zarogoulidis, K.; et al. Endometriosis of the lung: Report of a case and literature review. *Eur. J. Med. Res.* **2013**, *18*, 13. [CrossRef] [PubMed]

38. Thibodeau, L.L.; Prioleau, G.R.; Manuelidis, E.E.; Merino, M.J.; Heafner, M.D. Cerebral endometriosis. Case report. *J. Neurosurg.* **1987**, *66*, 609–610. [CrossRef] [PubMed]

39. Sampson, J.A. Metastatic or Embolic Endometriosis, due to the Menstrual Dissemination of Endometrial Tissue into the Venous Circulation. *Am. J. Pathol.* **1927**, *3*, 93–110.

40. Halme, J.; Hammond, M.G.; Hulka, J.F.; Raj, S.G.; Talbert, L.M. Retrograde menstruation in healthy women and in patients with endometriosis. *Obstet. Gynecol.* **1984**, *64*, 151–154.

41. Koninckx, P.R.; Kennedy, S.H.; Barlow, D.H. Pathogenesis of endometriosis: The role of peritoneal fluid. *Gynecol. Obstet. Investig.* **1999**, *47*, 23–33. [CrossRef]

42. Scutiero, G.; Iannone, P.; Bernardi, G.; Bonaccorsi, G.; Spadaro, S.; Volta, C.A.; Greco, P.; Nappi, L. Oxidative Stress and Endometriosis: A Systematic Review of the Literature. *Oxid. Med. Cell. Longev.* **2017**, *2017*, 7265238. [CrossRef]

43. Carvalho, L.F.; Samadder, A.N.; Agarwal, A.; Fernandes, L.F.; Abrão, M.S. Oxidative stress biomarkers in patients with endometriosis: Systematic review. *Arch. Gynecol. Obstet.* **2012**, *286*, 1033–1040. [CrossRef]

44. Ramón, L.A.; Braza-Boïls, A.; Gilabert-Estellés, J.; Gilabert, J.; España, F.; Chirivella, M.; Estellés, A. microRNAs expression in endometriosis and their relation to angiogenic factors. *Hum. Reprod.* **2011**, *26*, 1082–1090. [CrossRef] [PubMed]

45. Braza-Boïls, A.; Marí-Alexandre, J.; Gilabert, J.; Sánchez-Izquierdo, D.; España, F.; Estellés, A.; Gilabert-Estellés, J. MicroRNA expression profile in endometriosis: Its relation to angiogenesis and fibrinolytic factors. *Hum. Reprod.* **2014**, *29*, 978–988. [CrossRef] [PubMed]

46. Gilabert-Estellés, J.; Ramón, L.A.; España, F.; Gilabert, J.; Vila, V.; Réganon, E.; Castelló, R.; Chirivella, M.; Estellés, A. Expression of angiogenic factors in endometriosis: Its relation to fibrinolytic and metalloproteinase (MMP) systems. *Hum. Reprod.* **2007**, *22*, 2120–2127. [CrossRef] [PubMed]

47. Marí-Alexandre, J.; Barceló-Molina, M.; Belmonte-López, E.; García-Oms, J.; Estellés, A.; Braza-Boïls, A.; Gilabert-Estellés, J. Micro-RNA profile and proteins in peritoneal fluid from women with endometriosis: Their relationship with sterility. *Fertil. Steril.* **2018**, *109*, 675–684. [CrossRef]

48. Gee, H.E.; Ivan, C.; Calin, G.A.; Ivan, M. HypoxamiRs and cancer: From biology to targeted therapy. *Antioxid. Redox Signal.* **2014**, *21*, 1220–1238. [CrossRef]

49. Xu, T.X.; Zhao, S.Z.; Dong, M.; Yu, X.R. Hypoxia responsive miR-210 promotes cell survival and autophagy of endometriotic cells in hypoxia. *Eur. Rev. Med. Pharmacol. Sci.* **2016**, *20*, 399–406.

50. Augustin, H.G. Vascular morphogenesis in the ovary. *Baillieres Best Pract. Res. Clin. Obstet. Gynaecol.* **2000**, *14*, 867–882. [CrossRef]

51. Eggers, J.C.; Martino, V.; Reinbold, R.; Schäfer, S.D.; Kiesel, L.; Starzinski-Powitz, A.; Schüring, A.N.; Kemper, B.; Greve, B.; Götte, M. microRNA miR-200b affects proliferation, invasiveness and stemness of endometriotic cells by targeting ZEB1, ZEB2 and KLF4. *Reprod. Biomed. Online* **2016**, *32*, 434–445. [CrossRef]

52. Chen, J.; Gu, L.; Ni, J.; Hu, P.; Hu, K.; Shi, Y.L. MiR-183 Regulates ITGB1P Expression and Promotes Invasion of Endometrial Stromal Cells. *Biomed. Res. Int.* **2015**, *2015*, 340218. [CrossRef]

53. Dai, L.; Gu, L.; Di, W. MiR-199a attenuates endometrial stromal cell invasiveness through suppression of the IKKβ/NF-κB pathway and reduced interleukin-8 expression. *Mol. Hum. Reprod.* **2012**, *18*, 136–145. [CrossRef]

54. Okamoto, M.; Nasu, K.; Abe, W.; Aoyagi, Y.; Kawano, Y.; Kai, K.; Moriyama, M.; Narahara, H. Enhanced miR-210 expression promotes the pathogenesis of endometriosis through activation of signal transducer and activator of transcription 3. *Hum. Reprod.* **2015**, *30*, 632–641. [CrossRef]

55. Yu, H.; Zhong, Q.; Xia, Y.; Li, E.; Wang, S.; Ren, R. MicroRNA-2861 targets STAT3 and MMP2 to regulate the proliferation and apoptosis of ectopic endometrial cells in endometriosis. *Pharmazie* **2019**, *74*, 243–249. [PubMed]

56. Wang, Y.; Chen, H.; Fu, Y.; Ai, A.; Xue, S.; Lyu, Q.; Kuang, Y. MiR-195 inhibits proliferation and growth and induces apoptosis of endometrial stromal cells by targeting FKN. *Int. J. Clin. Exp. Pathol.* **2013**, *15*, 2824–2834.

57. Abe, W.; Nasu, K.; Nakada, C.; Kawano, Y.; Moriyama, M.; Narahara, H. miR-196b targets c-myc and Bcl-2 expression, inhibits proliferation and induces apoptosis in endometriotic stromal cells. *Hum. Reprod.* **2013**, *28*, 750–761. [CrossRef] [PubMed]

58. Zhao, Q.; Ye, M.; Yang, W.; Wang, M.; Li, M.; Gu, C.; Zhao, L.; Zhang, Z.; Han, W.; Fan, W.; et al. Effect of Mst1 on Endometriosis Apoptosis and Migration: Role of Drp1-Related Mitochondrial Fission and Parkin-Required Mitophagy. *Cell Physiol. Biochem.* **2018**, *45*, 1172–1190. [CrossRef] [PubMed]

59. Zhang, Y.; Yan, J.; Pan, X. miR-141-3p affects apoptosis and migration of endometrial stromal cells by targeting KLF-12. *Pflugers Arch.* **2019**, *471*, 1055–1063. [CrossRef]

60. Braza-Boïls, A.; Salloum-Asfar, S.; Marí-Alexandre, J.; Arroyo, A.B.; González-Conejero, R.; Barceló-Molina, M.; García-Oms, J.; Vicente, V.; Estellés, A.; Gilabert-Estellés, J.; et al. Peritoneal fluid modifies the microRNA expression profile in endometrial and endometriotic cells from women with endometriosis. *Hum. Reprod.* **2015**, *30*, 2292–2302. [CrossRef]

61. Panir, K.; Schjenken, J.E.; Robertson, S.A.; Hull, M.L. Non-coding RNAs in endometriosis: A narrative review. *Hum. Reprod. Update* **2018**, *24*, 497–515. [CrossRef]

62. Cosín, R.; Gilabert-Estellés, J.; Ramón, L.A.; Gómez-Lechón, M.J.; Gilabert, J.; Chirivella, M.; Braza-Boïls, A.; España, F.; Estellés, A. Influence of peritoneal fluid on the expression of angiogenic and proteolytic factors in cultures of endometrial cells from women with endometriosis. *Hum. Reprod.* **2010**, *25*, 398–405. [CrossRef]

63. Wright, K.R.; Mitchell, B.; Santanam, N. Redox regulation of microRNAs in endometriosis-associated pain. *Redox Biol.* **2017**, *12*, 956–966. [CrossRef]

64. Yang, R.Q.; Teng, H.; Xu, X.H.; Liu, S.Y.; Wang, Y.H.; Guo, F.J.; Liu, X.J. Microarray analysis of microRNA deregulation and angiogenesis-related proteins in endometriosis. *Genet. Mol. Res.* **2016**, *15*, 1–8. [CrossRef] [PubMed]

65. Teruel-Montoya, R.; Kong, X.; Abraham, S.; Ma, L.; Kunapuli, S.P.; Holinstat, M.; Shaw, C.A.; McKenzie, S.E.; Edelstein, L.C.; Bray, P.F. MicroRNA expression differences in human hematopoietic cell lineages enable regulated transgene expression. *PLoS ONE* **2014**, *9*, e102259. [CrossRef] [PubMed]

66. Nothnick, W.B.; Swan, K.; Flyckt, R.; Falcone, T.; Graham, A. Human endometriotic lesion expression of the miR-144-3p/miR-451a cluster, its correlation with markers of cell survival and origin of lesion content. *Sci. Rep.* **2019**, *9*, 8823. [CrossRef] [PubMed]

67. Cho, S.; Mutlu, L.; Grechukhina, O.; Taylor, H.S. Circulating microRNAs as potential biomarkers for endometriosis. *Fertil Steril.* **2015**, *103*, 1252–1260. [CrossRef] [PubMed]

68. Wang, W.T.; Zhao, Y.N.; Han, B.W.; Hong, S.J.; Chen, Y.Q. Circulating microRNAs identified in a genome-wide serum microRNA expression analysis as non-invasive biomarkers for endometriosis. *J. Clin. Endocrinol. Metab.* **2013**, *98*, 281–289. [CrossRef]

69. Maged, A.M.; Deeb, W.S.; El Amir, A.; Zaki, S.S.; El Sawah, H.; Al Mohamady, M.; Metwally, A.A.; Katta, M.A. Diagnostic accuracy of serum miR-122 and miR-199a in women with endometriosis. *Int. J. Gynaecol. Obstet.* **2018**, *141*, 14–19. [CrossRef]

70. Wang, L.; Huang, W.; Ren, C.; Zhao, M.; Jiang, X.; Fang, X.; Xia, X. Analysis of Serum microRNA Profile by Solexa Sequencing in Women with Endometriosis. *Reprod. Sci.* **2016**, *23*, 1359–1370. [CrossRef]

71. Jia, S.Z.; Yang, Y.; Lang, J.; Sun, P.; Leng, J. Plasma miR-17-5p, miR20a and miR-22 are down-regulated in women with endometriosis. *Hum. Reprod.* **2013**, *28*, 322–330. [CrossRef]

72. Rekker, K.; Saare, M.; Roost, A.M.; Kaart, T.; Sõritsa, D.; Karro, H.; Sõritsa, A.; Simón, C.; Salumets, A.; Peters, M. Circulating miR-200-family microRNAs have altered plasma levels in patients with endometriosis and vary with blood collection time. *Fertil Steril.* **2015**, *104*, 938–946. [CrossRef]

73. Pateisky, P.; Pils, D.; Szabo, L.; Kuessel, L.; Husslein, H.; Schmitz, A.; Wenzl, R.; Yotova, I. hsa-miRNA-154-5p expression in plasma of endometriosis patients is a potential diagnostic marker for the disease. *Reprod. Biomed. Online* **2018**, *37*, 449–466. [CrossRef]

74. Suryawanshi, S.; Vlad, A.M.; Lin, H.M.; Mantia-Smaldone, G.; Laskey, R.; Lee, M.; Lin, Y.; Donnellan, N.; Klein-Patel, M. Plasma microRNAs as novel biomarkers for endometriosis and endometriosis-associated ovarian cancer. *Clin. Cancer Res.* **2013**, *19*, 1213–1224. [CrossRef] [PubMed]

75. Kobayashi, M.; Sawada, K.; Nakamura, K.; Yoshimura, A.; Miyamoto, M.; Shimizu, A.; Ishida, K.; Nakatsuka, E.; Kodama, M.; Hashimoto, K.; et al. Exosomal miR-1290 is a potential biomarker of high-grade serous ovarian carcinoma and can discriminate patients from those with malignancies of other histological types. *J. Ovarian Res.* **2018**, *11*, 81. [CrossRef] [PubMed]

76. Shah, J.S.; Gard, G.B.; Yang, J.; Maidens, J.; Valmadre, S.; Soon, P.S.; Marsh, D.J. Combining serum microRNA and CA-125 as prognostic indicators of preoperative surgical outcome in women with high-grade serous ovarian cancer. *Gynecol. Oncol.* **2018**, *148*, 181–188. [CrossRef] [PubMed]

77. Todeschini, P.; Salviato, E.; Paracchini, L.; Ferracin, M.; Petrillo, M.; Zanotti, L.; Tognon, G.; Gambino, A.; Calura, E.; Caratti, G.; et al. Circulating miRNA landscape identifies miR-1246 as promising diagnostic biomarker in high-grade serous ovarian carcinoma: A validation across two independent cohorts. *Cancer Lett.* **2017**, *388*, 320–327. [CrossRef]

78. Kan, C.W.; Hahn, M.A.; Gard, G.B.; Maidens, J.; Huh, J.Y.; Marsh, D.J.; Howell, V.M. Elevated levels of circulating microRNA-200 family members correlate with serous epithelial ovarian cancer. *BMC Cancer* **2012**, *12*, 627. [CrossRef]

79. Matsuzaki, S.; Schubert, B. Oxidative stress status in normal ovarian cortex surrounding ovarian endometriosis. *Fertil. Steril.* **2010**, *93*, 2431–2432. [CrossRef]

80. Di Emidio, G.; D'Alfonso, A.; Leocata, P.; Parisse, V.; Di Fonso, A.; Artini, P.G.; Patacchiola, F.; Tatone, C.; Carta, G. Increased levels of oxidative and carbonyl stress markers in normal ovarian cortex surrounding endometriotic cysts. *Gynecol. Endocrinol.* **2014**, *30*, 808–812. [CrossRef]

81. Ngô, C.; Chéreau, C.; Nicco, C.; Weill, B.; Chapron, C.; Batteux, F. Reactive oxygen species controls endometriosis progression. *Am. J. Pathol.* **2009**, *175*, 225–234. [CrossRef]

82. Chen, C.; Zhou, Y.; Hu, C.; Wang, Y.; Yan, Z.; Li, Z.; Wu, R. Mitochondria and oxidative stress in ovarian endometriosis. *Free Radic. Biol. Med.* **2019**, *136*, 22–34. [CrossRef]

83. Kobayashi, Y.; Osanai, K.; Tanaka, K.; Nishigaya, Y.; Matsumoto, H.; Momomura, M.; Hashiba, M.; Mita, S.; Kyo, S.; Iwashita, M. Endometriotic cyst fluid induces reactive oxygen species (ROS) in human immortalized epithelial cells derived from ovarian endometrioma. *Redox Rep.* **2017**, *22*, 361–366. [CrossRef]

84. Yamaguchi, K.; Mandai, M.; Toyokuni, S.; Hamanishi, J.; Higuchi, T.; Takakura, K.; Fujii, S. Contents of endometriotic cysts, especially the high concentration of free iron, are a possible cause of carcinogenesis in the cysts through the iron-induced persistent oxidative stress. *Clin. Cancer Res.* **2008**, *14*, 32–40. [CrossRef] [PubMed]

85. Xie, H.; Chen, P.; Huang, H.W.; Liu, L.P.; Zhao, F. Reactive oxygen species downregulate ARID1A expression via its promoter methylation during the pathogenesis of endometriosis. *Eur. Rev. Med. Pharmacol. Sci.* **2017**, *21*, 4509–4515. [PubMed]

86. Santulli, P.; Chouzenoux, S.; Fiorese, M.; Marcellin, L.; Lemarechal, H.; Millischer, A.E.; Batteux, F.; Borderie, D.; Chapron, C. Protein oxidative stress markers in peritoneal fluids of women with deep infiltrating endometriosis are increased. *Hum Reprod.* **2015**, *30*, 49–60. [CrossRef] [PubMed]

87. Körner, M.; Burckhardt, E.; Mazzucchelli, L. Higher frequency of chromosomal aberrations in ovarian endometriosis compared to extragonadal endometriosis: A possible link to endometrioid adenocarcinoma. *Mod. Pathol.* **2006**, *19*, 1615–1623. [CrossRef] [PubMed]

88. Ñiguez-Sevilla, I.; Machado-Linde, F.; Marín-Sánchez, M.D.P.; Arense, J.J.; Torroba, A.; Nieto-Díaz, A.; Sánchez Ferrer, M.L. Prognostic importance of atypical endometriosis with architectural hyperplasia versus cytologic atypia in endometriosis-associated ovarian cancer. *J. Gynecol. Oncol.* **2019**, *30*, e63. [CrossRef]

89. Pearce, C.L.; Templeman, C.; Rossing, M.A.; Lee, A.; Near, A.M.; Webb, P.M.; Nagle, C.M.; Doherty, J.A.; Cushing-Haugen, K.L.; Wicklund, K.G.; et al. Association between endometriosis and risk of histological subtypes of ovarian cancer: A pooled analysis of case-control studies. *Lancet Oncol.* **2012**, *13*, 385–394. [CrossRef]

90. Guo, S.W. Endometriosis and ovarian cancer: Potential benefits and harms of screening and risk-reducing surgery. *Fertil Steril.* **2015**, *104*, 813–830. [CrossRef]

91. Poole, E.M.; Lin, W.T.; Kvaskoff, M.; De Vivo, I.; Terry, K.L.; Missmer, S.A. Endometriosis and risk of ovarian and endometrial cancers in a large prospective cohort of U.S. nurses. *Cancer Causes Control.* **2017**, *28*, 437–445. [CrossRef]

92. Mandai, M.; Matsumura, N.; Baba, T.; Yamaguchi, K.; Hamanishi, J.; Konishi, I. Ovarian clear cell carcinoma as a stress-responsive cancer: Influence of the microenvironment on the carcinogenesis and cancer phenotype. *Cancer Lett.* **2011**, *310*, 129–133. [CrossRef]

93. Aman, M.; Ohishi, Y.; Imamura, H.; Shinozaki, T.; Yasutake, N.; Kato, K.; Oda, Y. Expression of protease-activated receptor-2 (PAR-2) is related to advanced clinical stage and adverse prognosis in ovarian clear cell carcinoma. *Hum Pathol.* **2017**, *64*, 156–163. [CrossRef]

94. Fujimoto, Y.; Imanaka, S.; Yamada, Y.; Ogawa, K.; Ito, F.; Kawahara, N.; Yoshimoto, C.; Kobayashi, H. Comparison of redox parameters in ovarian endometrioma and its malignant transformation. *Oncol. Lett.* **2018**, *16*, 5257–5264. [CrossRef] [PubMed]

95. Sova, H.; Kangas, J.; Puistola, U.; Santala, M.; Liakka, A.; Karihtala, P. Down-regulation of 8-hydroxydeoxyguanosine and peroxiredoxin II in the pathogenesis of endometriosis-associated ovarian cancer. *Anticancer Res.* **2012**, *32*, 3037–3044. [PubMed]

96. Winarto, H.; Tan, M.I.; Sadikin, M.; Wanandi, S.I. ARID1A Expression is Down-Regulated by Oxidative Stress in Endometriosis and Endometriosis-Associated Ovarian Cancer. *Transl. Oncogenom.* **2017**, *24*, 9.

97. Yamada, Y.; Miyamoto, T.; Kashima, H.; Kobara, H.; Asaka, R.; Ando, H.; Higuchi, S.; Ida, K.; Shiozawa, T. Lipocalin 2 attenuates iron-related oxidative stress and prolongs the survival of ovarian clear cell carcinoma cells by up-regulating the CD44 variant. *Free Radic Res.* **2016**, *50*, 414–425. [CrossRef] [PubMed]

98. Mabuchi, S.; Sugiyama, T.; Kimura, T. Clear cell carcinoma of the ovary: Molecular insights and future therapeutic perspectives. *J. Gynecol. Oncol.* **2016**, *27*, e31. [CrossRef] [PubMed]

99. Yamashita, Y.; Akatsuka, S.; Shinjo, K.; Yatabe, Y.; Kobayashi, H.; Seko, H.; Kajiyama, H.; Kikkawa, F.; Takahashi, T.; Toyokuni, S. Met is the most frequently amplified gene in endometriosis-associated ovarian clear cell adenocarcinoma and correlates with worsened prognosis. *PLoS ONE* **2013**, *8*, e57724. [CrossRef]

100. Wu, X.; Ruan, Y.; Jiang, H.; Xu, C. MicroRNA-424 inhibits cell migration, invasion, and epithelial mesenchymal transition by downregulating doublecortin-like kinase 1 in ovarian clear cell carcinoma. *Int. J. Biochem. Cell Biol.* **2017**, *85*, 66–74. [CrossRef]

101. Hsu, C.Y.; Hsieh, T.H.; Er, T.K.; Chen, H.S.; Tsai, C.C.; Tsai, E.M. MiR-381 regulates cell motility, growth and colony formation through PIK3CA in endometriosis-associated clear cell and endometrioid ovarian cancer. *Oncol. Rep.* **2018**, *40*, 3734–3742. [CrossRef]

102. Dong, M.; Yang, P.; Hua, F. MiR-191 modulates malignant transformation of endometriosis through regulating TIMP3. *Med. Sci. Monit.* **2015**, *21*, 915–920.

103. Sun, Y.; Hu, L.; Zheng, H.; Bagnoli, M.; Guo, Y.; Rupaimoole, R.; Rodriguez-Aguayo, C.; Lopez-Berestein, G.; Ji, P.; Chen, K.; et al. MiR-506 inhibits multiple targets in the epithelial-to-mesenchymal transition network and is associated with good prognosis in epithelial ovarian cancer. *J. Pathol.* **2015**, *235*, 25–36. [CrossRef]

104. Vilming-Elgaaen, B.; Olstad, O.K.; Haug, K.B.; Brusletto, B.; Sandvik, L.; Staff, A.C.; Gautvik, K.M.; Davidson, B. Global miRNA expression analysis of serous and clear cell ovarian carcinomas identifies differentially expressed miRNAs including miR-200c-3p as a prognostic marker. *BMC Cancer.* **2014**, *14*, 80. [CrossRef] [PubMed]

105. Zhang, X.; Guo, G.; Wang, G.; Zhao, J.; Wang, B.; Yu, X.; Ding, Y. Profile of differentially expressed miRNAs in high-grade serous carcinoma and clear cell ovarian carcinoma, and the expression of miR-510 in ovarian carcinoma. *Mol. Med. Rep.* **2015**, *12*, 8021–8031. [CrossRef] [PubMed]

106. Yanaihara, N.; Noguchi, Y.; Saito, M.; Takenaka, M.; Takakura, S.; Yamada, K.; Okamoto, A. MicroRNA Gene Expression Signature Driven by miR-9 Overexpression in Ovarian Clear Cell Carcinoma. *PLoS ONE* **2016**, *11*, e0162584. [CrossRef] [PubMed]

107. Hirata, Y.; Murai, N.; Yanaihara, N.; Saito, M.; Saito, M.; Urashima, M.; Murakami, Y.; Matsufuji, S.; Okamoto, A. MicroRNA-21 is a candidate driver gene for 17q23-25 amplification in ovarian clear cell carcinoma. *BMC Cancer* **2014**, *14*, 799. [CrossRef]

108. Sugio, A.; Iwasaki, M.; Habata, S.; Mariya, T.; Suzuki, M.; Osogami, H.; Tamate, M.; Tanaka, R.; Saito, T. BAG3 upregulates Mcl-1 through downregulation of miR-29b to induce anticancer drug resistance in ovarian cancer. *Gynecol. Oncol.* **2014**, *134*, 615–623. [CrossRef]

109. Dai, F.; Zhang, Y.; Chen, Y. Involvement of miR-29b signaling in the sensitivity to chemotherapy in patients with ovarian carcinoma. *Hum. Pathol.* **2014**, *45*, 1285–1293. [CrossRef]

110. Tian, X.; Xu, L.; Wang, P. MiR-191 inhibits TNF-α induced apoptosis of ovarian endometriosis and endometrioid carcinoma cells by targeting DAPK1. *Int. J. Clin. Exp. Pathol.* **2015**, *8*, 4933–4942.

111. Calura, E.; Fruscio, R.; Paracchini, L.; Bignotti, E.; Ravaggi, A.; Martini, P.; Sales, G.; Beltrame, L.; Clivio, L.; Ceppi, L.; et al. MiRNA landscape in stage I epithelial ovarian cancer defines the histotype specificities. *Clin. Cancer Res.* **2013**, *19*, 4114–4123. [CrossRef]

112. Zhao, H.; Ding, Y.; Tie, B.; Sun, Z.F.; Jiang, J.Y.; Zhao, J.; Lin, X.; Cui, S. miRNA expression pattern associated with prognosis in elderly patients with advanced OPSC and OCC. *Int. J. Oncol.* **2013**, *43*, 839–849. [CrossRef]

113. Nagaraja, A.K.; Creighton, C.J.; Yu, Z.; Zhu, H.; Gunaratne, P.H.; Reid, J.G.; Olokpa, E.; Itamochi, H.; Ueno, N.T.; Hawkins, S.M.; et al. A link between mir-100 and FRAP1/mTOR in clear cell ovarian cancer. *Mol. Endocrinol.* **2010**, *24*, 447–463. [CrossRef]

114. Amano, T.; Chano, T.; Isono, T.; Kimura, F.; Kushima, R.; Murakami, T. Abundance of mitochondrial superoxide dismutase is a negative predictive biomarker for endometriosis-associated ovarian cancers. *World J. Surg. Oncol.* **2019**, *17*, 24. [CrossRef] [PubMed]

115. Ho, C.M.; Lee, F.K.; Huang, S.H.; Cheng, W.F. Everolimus following 5-aza-2-deoxycytidine is a promising therapy in paclitaxel-resistant clear cell carcinoma of the ovary. *Am. J. Cancer Res.* **2018**, *8*, 56–69. [PubMed]

116. Siegel, R.L.; Miller, K.D.; Jemal, A. Cancer Statistics, 2019. *CA Cancer J. Clin.* **2019**, *69*, 7–34. [CrossRef] [PubMed]

117. Vaughan, S.; Coward, J.; Bast, R.C., Jr.; Berchuck, A.; Berek, J.S.; Brenton, J.D.; Coukos, G.; Crum, C.C.; Drapkin, R.; Etemadmoghadam, D. Rethinking ovarian cancer: Recommendations for improving outcomes. *Nat. Rev. Cancer* **2011**, *11*, 719–725. [CrossRef] [PubMed]

118. Lisio, M.A.; Fu, L.; Goyeneche, A.; Gao, Z.H.; Telleria, C. High-Grade Serous Ovarian Cancer: Basic Sciences, Clinical and Therapeutic Standpoints. *Int. J. Mol. Sci.* **2019**, *20*, E952. [CrossRef] [PubMed]

119. Gilabert-Estelles, J.; Braza-Boils, A.; Ramon, L.A.; Zorio, E.; Medina, P.; España, F.; Estelles, A. Role of microRNAs in gynecological pathology. *Curr. Med. Chem.* **2012**, *19*, 2406–2413. [CrossRef]

120. Llueca, A.; Escrig, J. Prognostic value of peritoneal cancer index in primary advanced ovarian cancer. *Eur. J. Surg. Oncol.* **2018**, *44*, 163–169. [CrossRef]

121. Arend, R.C.; Londoño-Joshi, A.I.; Straughn, J.M.; Buchsbaum, D.J. The Wnt/β-catenin pathway in ovarian cancer: A review. *Gynecol. Oncol.* **2013**, *131*, 772–779. [CrossRef]

122. Kim, J.; Park, E.Y.; Kim, O.; Schilder, J.M.; Coffey, D.M.; Cho, C.H.; Bast, R.C., Jr. Cell Origins of High-Grade Serous Ovarian Cancer. *Cancers* **2018**, *10*, E433. [CrossRef]

123. Labidi-Galy, S.I.; Papp, E.; Hallberg, D.; Niknafs, N.; Adleff, V.; Noe, M.; Bhattacharya, R.; Novak, M.; Jones, S.; Phallen, J.; et al. High grade serous ovarian carcinomas originate in the fallopian tube. *Nat. Commun.* **2017**, *8*, 1093. [CrossRef]

124. Vercellini, P.; Crosignani, P.; Somigliana, E.; Viganò, P.; Buggio, L.; Bolis, G.; Fedele, L. The 'incessant menstruation' hypothesis: A mechanistic ovarian cancer model with implications for prevention. *Hum. Reprod.* **2011**, *26*, 2262–2273. [CrossRef] [PubMed]

125. Bayani, J.; Brenton, J.D.; Macgregor, P.F.; Beheshti, B.; Albert, M.; Nallainathan, D.; Karaskova, J.; Rosen, B.; Murphy, J.; Laframboise, S.; et al. Parallel Analysis of Sporadic Primary Ovarian Carcinomas by Spectral Karyotyping, Comparative Genomic Hybridization, and Expression Microarrays. *Cancer Res.* **2002**, *62*, 3466–3476.

126. Kobayashi, H.; Ogawa, K.; Kawahara, N.; Iwai, K.; Niiro, E.; Morioka, S.; Yamada, Y. Sequential molecular changes and dynamic oxidative stress in high-grade serous ovarian carcinogenesis. *Free Radic Res.* **2017**, *51*, 755–764. [CrossRef]

127. Martinez-Outschoorn, U.E.; Balliet, R.M.; Lin, Z.; Whitaker-Menezes, D.; Howell, A.M.; Sotgia, F.; Lisanti, M.P. Hereditary ovarian cancer and two-compartment tumor metabolism: Epithelial loss of BRCA1 induces hydrogen peroxide production, driving oxidative stress and NFκB activation in the tumor stroma. *Cell Cycle* **2012**, *11*, 4152–4166. [CrossRef]

128. Alsop, K.; Fereday, S.; Meldrum, C.; DeFazio, A.; Emmanuel, C.; George, J.; Dobrovic, A.; Birrer, M.J.; Webb, P.M.; Stewart, C.; et al. BRCA mutation frequency and patterns of treatment response in BRCA mutation-positive women with ovarian cancer: A report from the Australian Ovarian Cancer Study Group. *J. Clin. Oncol.* **2012**, *30*, 2654–2663. [CrossRef] [PubMed]

129. Klinkebiel, D.; Zhang, W.; Akers, S.N.; Odunsi, K.; Karpf, A.R. DNA Methylome Analyses Implicate Fallopian Tube Epithelia as the Origin for High-Grade Serous Ovarian Cancer. *Mol. Cancer Res.* **2016**, *14*, 787–794. [CrossRef] [PubMed]

130. Yang, J.; Zhou, Y.; Ng, S.K.; Huang, K.C.; Ni, X.; Choi, P.W.; Hasselblatt, K.; Muto, M.G.; Welch, W.R.; Berkowitz, R.S.; et al. Characterization of MicroRNA-200 pathway in ovarian cancer and serous intraepithelial carcinoma of fallopian tube. *BMC Cancer* **2017**, *17*, 422. [CrossRef]

131. Katsuno, Y.; Lamouille, S.; Derynck, R. TGF-β signaling and epithelial-mesenchymal transition in cancer progression. *Curr. Opin. Oncol.* **2013**, *25*, 76–84. [CrossRef]

132. Prieto-García, E.; Díaz-García, C.V.; García-Ruiz, I.; Agulló-Ortuño, M.T. Epithelial-to-mesenchymal transition in tumor progression. *Med. Oncol.* **2017**, *34*, 122. [CrossRef]

133. Boac, B.M.; Xiong, Y.; Marchion, D.C.; Abbasi, F.; Bush, S.H.; Ramirez, I.J.; Khulpateea, B.R.; Clair-McClung, E.; Berry, A.L.; Bou-Zgheib, N.; et al. Micro-RNAs associated with the evolution of ovarian cancer cisplatin resistance. *Gynecol. Oncol.* **2016**, *140*, 259–263. [CrossRef]

134. Zhu, X.; Shen, H.; Yin, X.; Long, L.; Xie, C.; Liu, Y.; Hui, L.; Lin, X.; Fang, Y.; Cao, Y.; et al. miR-186 regulation of Twist1 and ovarian cancer sensitivity to cisplatin. *Oncogene* **2016**, *35*, 323–332. [CrossRef] [PubMed]

135. Helleman, J.; Smid, M.; Jansen, M.P.; van der Burg, M.E.; Berns, E.M. Pathway analysis of gene lists associated with platinum-based chemotherapy resistance in ovarian cancer: The big picture. *Gynecol Oncol.* **2010**, *17*, 170–176. [CrossRef] [PubMed]

136. Brozovic, A.; Duran, G.E.; Wang, Y.C.; Francisco, E.B.; Sikic, B.I. The miR-200 family differentially regulates sensitivity to paclitaxel and carboplatin in human ovarian carcinoma OVCAR-3 and MES-OV cells. *Mol Oncol.* **2015**, *9*, 1678–1693. [CrossRef] [PubMed]

137. Mateescu, B.; Batista, L.; Cardon, M.; Gruosso, T.; de Feraudy, Y.; Mariani, O.; Nicolas, A.; Meyniel, J.P.; Cottu, P.; Sastre-Garau, X.; et al. miR-141 and miR-200a act on ovarian tumorigenesis by controlling oxidative stress response. *Nat Med.* **2011**, *17*, 1627–1635. [CrossRef] [PubMed]

138. Vecchione, A.; Belletti, B.; Lovat, F.; Volinia, S.; Chiappetta, G.; Giglio, S.; Sonego, M.; Cirombella, R.; Onesti, E.C.; Pellegrini, P.; et al. A microRNA signature defines chemoresistance in ovarian cancer through modulation of angiogenesis. *Proc. Natl. Acad. Sci. USA* **2013**, *110*, 9845–9850. [CrossRef]

139. Jeong, J.Y.; Kang, H.; Kim, T.H.; Kim, G.; Heo, J.H.; Kwon, A.Y.; Kim, S.; Jung, S.G.; An, H.J. MicroRNA-136 inhibits cancer stem cell activity and enhances the anti-tumor effect of paclitaxel against chemoresistant ovarian cancer cells by targeting Notch3. *Cancer Lett.* **2017**, *386*, 168–178. [CrossRef]

140. Liu, Y.; Niu, Z.; Lin, X.; Tian, Y. MiR-216b increases cisplatin sensitivity in ovarian cancer cells by targeting PARP1. *Cancer Gene Ther.* **2017**, *24*, 208–214. [CrossRef]

141. Sun, C.; Li, N.; Yang, Z.; Zhou, B.; He, Y.; Weng, D.; Fang, Y.; Wu, P.; Chen, P.; Yang, X.; et al. miR-9 regulation of BRCA1 and ovarian cancer sensitivity to cisplatin and PARP inhibition. *J. Natl. Cancer Inst.* **2013**, *105*, 1750–1758. [CrossRef]

142. Ayyagari, V.N.; Diaz-Sylvester, P.L.; Hsieh, T.J.; Brard, L. Evaluation of the cytotoxicity of the Bithionol-paclitaxel combination in a panel of human ovarian cancer cell lines. *PLoS ONE* **2017**, *12*, e0185111. [CrossRef]

143. Ayyagari, V.N.; Brard, L. Bithionol inhibits ovarian cancer cell growth in vitro—Studies on mechanism(s) of action. *BMC Cancer* **2014**, *14*, 61. [CrossRef]

144. Hou, D.; Liu, Z.; Xu, X.; Liu, Q.; Zhang, X.; Kong, B.; Wei, J.J.; Gong, Y.; Shao, C. Increased oxidative stress mediates the antitumor effect of PARP inhibition in ovarian cancer. *Redox Biol.* **2018**, *17*, 99–111. [CrossRef] [PubMed]

145. Choi, Y.E.; Meghani, K.; Brault, M.E.; Leclerc, L.; He, Y.J.; Day, T.A.; Elias, K.M.; Drapkin, R.; Weinstock, D.M.; Dao, F.; et al. Platinum and PARP Inhibitor Resistance Due to Overexpression of MicroRNA-622 in BRCA1-Mutant Ovarian Cancer. *Cell Rep.* **2016**, *14*, 429–439. [CrossRef] [PubMed]

146. Meghani, K.; Fuchs, W.; Detappe, A.; Drané, P.; Gogola, E.; Rottenberg, S.; Jonkers, J.; Matulonis, U.; Swisher, E.M.; Konstantinopoulos, P.A.; et al. Multifaceted Impact of MicroRNA 493-5p on Genome-Stabilizing Pathways Induces Platinum and PARP Inhibitor Resistance in BRCA2-Mutated Carcinomas. *Cell Rep.* **2018**, *23*, 100–111. [CrossRef] [PubMed]

147. Belotte, J.; Fletcher, N.M.; Awonuga, A.O.; Alexis, M.; Abu-Soud, H.M.; Saed, M.G.; Diamond, M.P.; Saed, G.M. The role of oxidative stress in the development of cisplatin resistance in epithelial ovarian cancer. *Reprod. Sci.* **2014**, *21*, 503–508. [CrossRef]

148. Fletcher, N.M.; Belotte, J.; Saed, M.G.; Memaj, I.; Diamond, M.P.; Morris, R.T.; Saed, G.M. Specific point mutations in key redox enzymes are associated with chemoresistance in epithelial ovarian cancer. *Free Radic Biol. Med.* **2017**, *102*, 122–132. [CrossRef]

149. Pei, H.; Yang, Y.; Cui, L.; Yang, J.; Li, X.; Yang, Y.; Duan, H. Bisdemethoxycurcumin inhibits ovarian cancer via reducing oxidative stress mediated MMPs expressions. *Sci. Rep.* **2016**, *6*, 28773. [CrossRef]

150. Pons, D.G.; Sastre-Serra, J.; Nadal-Serrano, M.; Oliver, A.; García-Bonafé, M.; Bover, I.; Roca, P.; Oliver, J. Initial activation status of the antioxidant response determines sensitivity to carboplatin/paclitaxel treatment of ovarian cancer. *Anticancer Res.* **2012**, *32*, 4723–4728.

151. Montoya-Estrada, A.; Coria-García, C.F.; Cruz-Orozco, O.P.; Aguayo-González, P.; Torres-Ramos, Y.D.; Flores-Herrera, H.; Hicks, J.J.; Medina-Navarro, R.; Guzmán-Grenfell, A.M. Increased systemic and peritoneal oxidative stress biomarkers in endometriosis are not related to retrograde menstruation. *Redox Rep.* **2019**, *24*, 51–55. [CrossRef]

152. Lambrinoudaki, I.V.; Augoulea, A.; Christodoulakos, G.E.; Economou, E.V.; Kaparos, G.; Kontoravdis, A.; Papadias, C.; Creatsas, G. Measurable serum markers of oxidative stress response in women with endometriosis. *Fertil Steril.* **2009**, *91*, 46–50. [CrossRef]

153. Turkyilmaz, E.; Yildirim, M.; Cendek, B.D.; Baran, P.; Alisik, M.; Dalgaci, F.; Yavuz, A.F. Evaluation of oxidative stress markers and intra-extracellular antioxidant activities in patients with endometriosis. *Eur. J. Obstet. Gynecol. Reprod. Biol.* **2016**, *199*, 164–168. [CrossRef]

154. Rosa e Silva, J.C.; do Amara, V.F.; Mendonça, J.L.; Rosa e Silva, A.C.; Nakao, L.S.; Poli Neto, O.B.; Ferriani, R.A. Serum markers of oxidative stress and endometriosis. *Clin. Exp. Obstet. Gynecol.* **2014**, *41*, 371–374. [PubMed]

155. Vicente-Muñoz, S.; Morcillo, I.; Puchades-Carrasco, L.; Payá, V.; Pellicer, A.; Pineda-Lucena, A. Nuclear magnetic resonance metabolomic profiling of urine provides a noninvasive alternative to the identification of biomarkers associated with endometriosis. *Fertil Steril.* **2015**, *104*, 1202–1209. [CrossRef] [PubMed]

156. Karihtala, P.; Soini, Y.; Vaskivuo, L.; Bloigu, R.; Puistola, U. DNA adduct 8-hydroxydeoxyguanosine, a novel putative marker of prognostic significance in ovarian carcinoma. *Int. J. Gynecol. Cancer.* **2009**, *19*, 1047–1051. [CrossRef] [PubMed]

157. Pylväs-Eerola, M.; Karihtala, P.; Puistola, U. Preoperative serum 8-hydroxydeoxyguanosine is associated with chemoresistance and is a powerful prognostic factor in endometrioid-type epithelial ovarian cancer. *BMC Cancer* **2015**, *15*, 493. [CrossRef]

158. Pylväs, M.; Puistola, U.; Laatio, L.; Kauppila, S.; Karihtala, P. Elevated serum 8-OHdG is associated with poor prognosis in epithelial ovarian cancer. *Anticancer Res.* **2011**, *31*, 1411–1415.

159. Weber, J.A.; Baxter, D.H.; Zhang, S.; Huang, D.Y.; Huang, K.H.; Lee, M.J.; Galas, D.J.; Wang, K. The microRNA spectrum in 12 body fluids. *Clin. Chem.* **2010**, *56*, 1733–1741. [CrossRef]

160. Shah, M.Y.; Calin, G.A. The mix of two worlds: Non-coding RNAs and hormones. *Nucl. Acid Ther.* **2013**, *23*, 2–8. [CrossRef]

161. Mier-Cabrera, J.; Genera-García, M.; De la Jara-Díaz, J.; Perichart-Perera, O.; Vadillo-Ortega, F.; Hernández-Guerrero, C. Effect of vitamins C and E supplementation on peripheral oxidative stress markers and pregnancy rate in women with endometriosis. *Int. J. Gynaecol. Obstet.* **2008**, *100*, 252–256. [CrossRef]

162. Lete, I.; Mendoza, N.; de la Viuda, E.; Carmona, F. Effectiveness of an antioxidant preparation with N-acetyl cysteine, alpha lipoic acid and bromelain in the treatment of endometriosis-associated pelvic pain: LEAP study. *Eur. J. Obstet. Gynecol. Reprod. Biol.* **2018**, *228*, 221–224. [CrossRef]
163. Ray, K.; Fahrmann, J.; Mitchell, B.; Paul, D.; King, H.; Crain, C.; Cook, C.; Golovko, M.; Brose, S.; Golovko, S.; et al. Oxidation-sensitive nociception involved in endometriosis-associated pain. *Pain* **2015**, *156*, 528–539. [CrossRef]
164. Monk, B.J.; Kauderer, J.T.; Moxley, K.M.; Bonebrake, A.J.; Dewdney, S.B.; Secord, A.A.; Ueland, F.R.; Johnston, C.M.; Aghajanian, C. A phase II evaluation of elesclomol sodium and weekly paclitaxel in the treatment of recurrent or persistent platinum-resistant ovarian, fallopian tube or primary peritoneal cancer: An NRG oncology/gynecologic oncology group study. *Gynecol. Oncol.* **2018**, *151*, 422–427. [CrossRef] [PubMed]
165. Thomsen, C.B.; Andersen, R.F.; Steffensen, K.D.; Adimi, P.; Jakobsen, A. Delta tocotrienol in recurrent ovarian cancer. A phase II trial. *Pharmacol. Res.* **2019**, *141*, 392–396. [CrossRef] [PubMed]
166. Trudel, D.; Labbé, D.P.; Araya-Farias, M.; Doyen, A.; Bazinet, L.; Duchesne, T.; Plante, M.; Grégoire, J.; Renaud, M.C.; Bachvarov, D.; et al. A two-stage, single-arm, phase II study of EGCG-enriched green tea drink as a maintenance therapy in women with advanced stage ovarian cancer. *Gynecol. Oncol.* **2013**, *131*, 357–361. [CrossRef] [PubMed]
167. Rupaimoole, R.; Slack, F.J. MicroRNA therapeutics: Towards a new era for the management of cancer and other diseases. *Nat. Rev. Drug Discov.* **2017**, *16*, 203–222. [CrossRef] [PubMed]

International Journal of
Molecular Sciences

Review

The Crosstalk of miRNA and Oxidative Stress in the Liver: From Physiology to Pathology and Clinical Implications

Eckhard Klieser [1,2,*], Christian Mayr [3,4], Tobias Kiesslich [3,4], Till Wissniowski [5], Pietro Di Fazio [6], Daniel Neureiter [1,2] and Matthias Ocker [7,8]

[1]   Institute of Pathology, Paracelsus Medical University/Salzburger Landeskliniken (SALK),
     5020 Salzburg, Austria; d.neureiter@salk.at
[2]   Cancer Cluster Salzburg, 5020 Salzburg, Austria
[3]   Department of Internal Medicine I, Paracelsus Medical University/Salzburger Landeskliniken (SALK),
     5020 Salzburg, Austria; christian.mayr@pmu.ac.at (C.M.); t.kiesslich@salk.at (T.K.)
[4]   Institute of Physiology and Pathophysiology, Paracelsus Medical University/Salzburger
     Landeskliniken (SALK), 5020 Salzburg, Austria
[5]   Department of Gastroenterology and Endocrinology, Philipps University Marburg,
     35043 Marburg, Germany; wissniowski@me.com
[6]   Department of Visceral, Thoracic and Vascular Surgery, Philipps University Marburg,
     35043 Marburg, Germany; difazio@med.uni-marburg.de
[7]   Translational Medicine Oncology, Bayer AG, 13353 Berlin, Germany;
     Matthias.Ocker@bayer.com or Matthias.Ocker@charite.de
[8]   Department of Gastroenterology CBF, Charité University Medicine Berlin, 12200 Berlin, Germany
[*]   Correspondence: e.klieser@salk.at; Tel.: +43-662-57255-29030

Received: 30 September 2019; Accepted: 21 October 2019; Published: 23 October 2019

Abstract: The liver is the central metabolic organ of mammals. In humans, most diseases of the liver are primarily caused by an unhealthy lifestyle–high fat diet, drug and alcohol consumption- or due to infections and exposure to toxic substances like aflatoxin or other environmental factors. All these noxae cause changes in the metabolism of functional cells in the liver. In this literature review we focus on the changes at the miRNA level, the formation and impact of reactive oxygen species and the crosstalk between those factors. Both, miRNAs and oxidative stress are involved in the multifactorial development and progression of acute and chronic liver diseases, as well as in viral hepatitis and carcinogenesis, by influencing numerous signaling and metabolic pathways. Furthermore, expression patterns of miRNAs and antioxidants can be used for biomonitoring the course of disease and show potential to serve as possible therapeutic targets.

Keywords: microRNA; oxidative stress; metabolism; physiology; ASH; NAFLD; NASH; HCC; HCV; HBV

---

1. Introduction

The liver is the central metabolic organ in the human body. It serves as a storage organ for e.g., glycogen, lipoproteins, vitamins, iron and blood, synthesizes important proteins such as albumins, transferrin and coagulation factors as well as fats and lipoproteins. Enzyme systems that are necessary for the metabolism of fat–for example–are involved in the formation of reactive oxygen species (ROS) and reactive nitrogen species (RNS), which in turn play a role in the development of non-tumours and tumorous liver diseases like non-alcoholic fatty liver disease (NAFLD) and liver cancer (hepatocellular carcinoma (HCC)).

In the human body there is normally a balance between antioxidants and ROS. If, however, the metabolic situation changes in favor of ROS, then oxidative stress (OS) is present in the cell [1].

ROS are almost always considered to be purely toxic, although ROS also have important regulatory tasks in various signaling pathways [2–4]. After growth factor stimulation an increase of ROS is necessary for the increase of tyrosine phosphorylation, which is needed for downstream signaling [5,6]. In the bone marrow, ROS play an important role in the redox regulation of stem cells and progenitor cells of hematopoiesis [7]. Besides these beneficial tasks, ROS are involved in aging and carcinogenesis, although they also play an ambivalent role herein [8–11]. This also applies to antioxidants which degrade ROS and thus ensure the stability of the genome but that also prevent apoptosis of damaged cells [12]. Aging and carcinogenesis are also part of the spectrum of activity of micro-RNAs (miRNAs) [13–16]. miRNAs do not code for proteins but play a role in the regulation of the expression of genes that are involved in regulation of diverse biological pathways. miRNAscause either degradation (if they match perfectly to messenger RNA) or prevention of translation (imperfect match) of the respective mRNAs [17]. The main task of miRNAs is gene regulation. They are a pillar of self-regulation, but also interact with other mechanisms of epigenetics such as histone modification [18]. Inferred from this, they have a decisive role not only in malignant diseases, but also in physiologic conditions and in metabolic diseases like NAFLD [19]. This review is intended to provide a comprehensive overview of the (inter)actions of oxidative stress and miRNA in pathological processes of the liver.

2. Physiology and Metabolism

In the context of oxidative stress, ROS are a group of chemically reactive, intracellular compounds containing oxygen and include the superoxide anion (O_2^-), hydrogen peroxide (H_2O_2) and the hydroxyl radical (HO•)–each with different chemical properties such as reactivity, half-life, diffusion distance, and permeability through cellular membranes [20–22]. While these species have physiologic functions in cell signaling and regulation ("oxidative eustress"), supraphysiologic oxidative levels may cause damage to biomolecules and cells, i.e., "oxidative distress". The general concept of oxidative stress is defined as a dysbalance favoring oxidants (ROS as well as RNS) over antioxidants thus disrupting redox signaling and control and/or inducing molecular damage [23,24] (see Figure 1).

In non-phagocytic cells, mitochondria represent the main sources of ROS produced within the steps of oxidative phosphorylation. Catalyzed by NADP(H) or xanthine oxidase, about 1% of the mitochondrial electron flow contributes to generation of superoxide anion. Importantly, at physiologic levels, free radicals play a role in the cell's signal transduction [25], regulation of gene expression and defense against pathogens [1]. As reviewed by Dickinson and Chang, other cellular sources of ROS include the endoplasmic reticulum during oxidative protein folding mechanisms (post-translational protein disulfide bond formation) and NADPH oxidases (NOX) located at various cellular membranes [26]. In the context of an immune response, NADPH-dependent enzymes such as NOX2 seem indispensable [27]. In the gut, bacteria stimulate ROS production via NOX1 and DUOX2 and ROS promote intestinal stem cell proliferation [12,27–29].

Besides side reactions in the electron flow within the oxidative phosphorylation pathway, (ethanol-inducible) cytochrome P450 enzymes (CYP2E1) represent a non-mitochondrial source of ROS in the liver [1,30]. As reviewed by Li et al. [31], other sources of ROS include hepatic metabolization of drugs, environmental pollutants and other factors such as radiation, temperature, high fat or high salt diet. As a tissue characterized by high metabolic activity, the liver parenchyma is equipped with several ROS scavenging mechanisms: besides non-enzymatic factors (α-tocopherol, glutathione (GSH), β-carotene, bilirubin, flavonoids, and plasma proteins [22]), the nuclear factor erythroid 2 like 2 (Nrf2) is a cellular redox sensor which—induced by elevated levels of ROS—is released from sequestration via the cytoplasmic cytoskeletal-anchoring protein Kelch-like ECH-associated protein 1 (Keap1) and, in turn, promotes transcription of ROS-protective genes. Genes regulated by Nrf2 via antioxidant response elements (ARE) include ROS-relevant factors involved in GSH turnover (regeneration), reduction of oxidized protein thiol groups and NADPH-producing enzymes (required for drug-metabolizing enzymes and antioxidant systems)–for review see Hayes and Dinkova-Kostova [32]. In the context of hepatic pathology, Nrf2-mediated cytoprotective responses are involved in (counteracting) the development of various liver diseases including alcoholic and non-alcoholic liver diseases, viral hepatitis, fibrosis and HCC. Therefore, ROS are central factors in the pathogenesis of various hepatic diseases [1]—as summarized in Table 1.

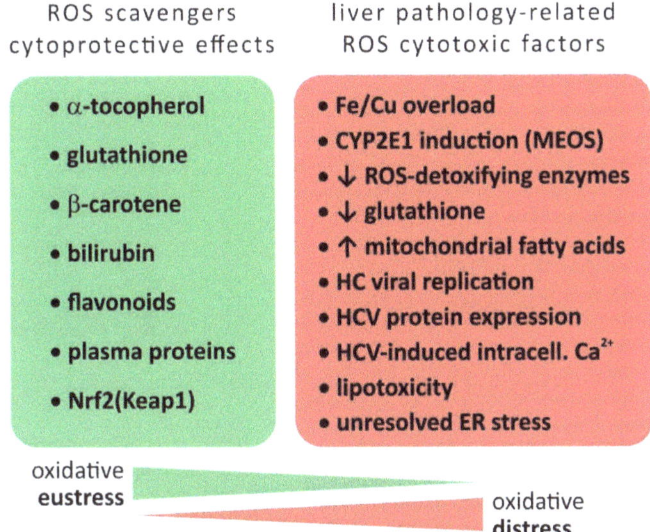

Figure 1. Disbalance between oxidative eustress and distress. Based on [1,22,30,33]. Abbreviations: Ca = Calcium, Cu = Copper, CYP = cytochrome P450, ER = endoplasmatic reticulum, Fe = Ferrum (Iron), GSH = Glutathione, HC = hepatitis C, HCV = hepatitis C virus, Keap1 = Kelch-like ECH-associated protein 1, MEOS = microsomal ethanol oxidizing system, Nrf2 = nuclear factor erythroid 2 like 2, ROS = reactive oxygen species, ↓ = downregulation/reduction, ↑ = upregulation/increase.

Most chronic liver diseases are characterized by deposition and accumulation of extracellular matrix components (collagens, fibronectin, elastin, laminin, hyaluronan, and proteoglycans), mostly secreted by hepatic stellate cells (HSC), resulting in up to six times more extracellular matrix (ECM) than normal in advanced stages of fibrosis [33]. Depending on genetic and environmental factors, fibro-proliferative disorders (i.e., NAFLD or NASH) may proceed to liver cirrhosis, which in its uncompensated form, is associated with acute and chronic liver failure, portal hypertension and often require liver transplantation [34]. The involvement of ROS during the development of liver fibrosis is evident based on several mechanisms [23] (see also [22,35,36] for reviews): i) ROS-based stimulation of collagen (Col1α1) synthesis by HSC, ii) ROS-based intracellular signaling of transforming growth factor β (TGFβ) as a major fibrogenic factor as well as its up-regulation by ROS, and, iii) ROS produced by NOX enzymes contributing to HSC activation.

Taken together, production of ROS and correlated tissue damage represent central aspects of various hepatic diseases. Therefore, understanding (epigenetic) regulation of factors involved in either generation of ROS or in their detoxification is necessary to fully comprehend pathophysiologic mechanisms of liver diseases as well as to develop new epigenetics-based therapeutic approaches.

Table 1. Involvement of ROS in various liver diseases. Based on [1,22,30].

Liver Disease	ROS-Production by	(Patho)Mechanism
Hemochromatosis, Wilson's disease	Iron/copper overload	Presence of metal catalyst for ROS production
Alcoholic liver disease (ALD)	CYP2E1 induction (MEOS)	High NADPH oxidase activity of CYP2E1 associated with production of O_2^- and H_2O_2
	Reduced expression of ROS-detoxifying enzymes	Alcohol-induced reduction of PPARγ coactivator 1α
Nonalcoholic steatohepatitis (NASH)	Increased concentration and metabolisms of fatty acids in mitochondria	Saturation of mitochondrial β-oxidation and H_2O_2 production through peroxisomal β-oxidation
	CYP2E1 (CYP4A) induction	See above
HCV infection	Reduction of ROS detoxification	Reduced levels of glutathione and its regeneration as well as ROS-detoxifying enzymes
	Increased mitochondrial ROS production due to viral replication or virus protein expression	Inhibition of mitochondrial electron transport chain
	Increased NADPH oxidase triggered by calcium	Virus-induced redistribution of cellular calcium

Abbreviations: ALD = alcoholic liver disease, CYP = cytochrome P450, HCV = hepatitis C virus, MEOS = microsomal ethanol oxidizing system, NASH = nonalcoholic steatohepatitis, PPARγ = peroxisome proliferator activated receptor gamma, ROS = reactive oxygen species.

3. Alcoholic Liver Disease (ALD) and Alcoholic Steatohepatitis (ASH)

ASH is a liver disease caused by high alcohol consumption. The accumulation of ethanol and its metabolic products lead to production of ROS that alter the hepatocyte function, finally leading to fibrosis, cirrhosis, and in 5% to 6% of patients, to the development of HCC. Beyond the genetic and metabolic alterations occurring during ASH, epigenetic modifications have been shown to exert a key role. Changes in DNA methylation at the promoter regions of several genes were discovered in ASH, as were changes in histone acetylation. Nonetheless, it has been found that also miRNAs are differentially expressed in patients affected by ASH [37].

Alcohol intake favors the hepatic accumulation of lipopolysaccharide (LPS), a bacterial antigen, thus mediating the activation of Toll Like Receptor 4 (TLR4). This promotes the transcriptional activity of Nuclear Factor kappa B (NFκB), leading to the expression of miR-155. The over-expression of miR-155 causes the release of tumor necrosis factor α (TNFα), ROS and oxidative stress in Kupffer cells (liver resident macrophages) and HSC [38,39]. miR-155 exerts a significant role in hepatocytes by suppressing peroxisome proliferator activated receptor α (PPARα). The down-regulation of this anti-oxidative enzyme causes the over-expression of genes involved in lipid metabolism and uptake, e.g., Fatty Acid Binding Protein 4 (FABP4), Acetyl-CoA-carboxylase 1 (ACC1) and Low-density Lipoprotein Receptor (LDLR) [40].

miR-181b-3p has also been found to be responsible for glucose and lipid homeostasis alterations as well as for liver injury and LPS-induced TLR4/NFκB activation in murine Kupffer cells [41]. Additionally, miR-291b expression is responsible for the suppression of Toll interacting protein (Tollip) in Kupffer cells, which is a down-regulator of the TLR4/NFκB pathway [42].

miR-34a, a member of the miR-34 family with known tumor suppressor activity because of its ability to promote p53-mediated apoptosis [43], has been found to correlate with alcoholic liver disease by targeting Sirtuin 1 (SIRT1) mRNA and inhibiting its protein coding [44]. Moreover, SIRT1 is a target of miR-217 during alcohol-associated inflammation [45]. Mice with hepatic deletion of miR-122 develop steatosis at birth, leading to fibrosis and HCC. Its expression is strongly down-regulated in alcohol fed mice as well as in patients affected by alcohol related cirrhosis [46]. Additionally, miR-122 down-regulation enables Hypoxia Inducible Factor 1 alpha (HIF1α) expression in ALD, which contributes to the development of hepatobiliary cancer [47]. It has also been reported that alcohol intake increases the level of miR-155 in Kupffer cells, triggering their sensitization to LPS produced by gut microbiota [48].

Alcohol consumption enhances the level of miR-21, which is overexpressed in several solid tumors including HCC, in hepatocytes and stellate cells [49]. However, its over-expression reduced ethanol-induced cell death, highlighting its role to protect the liver cells during injury [50]. miR-223 is responsible for the peripheral neutrophils activation and liver infiltration induced by ethanol. An increase in its level has been found in serum and neutrophils of patients with elevated alcohol consumption. Its over-expression could trigger NADPH oxidase, thus causing ROS production and liver cell death [51]. Alcohol is furthermore responsible for the suppression of miR-199 in human endothelial cells that leads to steatohepatitis in patients affected by cirrhosis by inducing HIF1α and endothelin-1 (ET-1) [52].

Alcohol-mediated miR-214 expression suppresses cytochrome P450 oxidoreductase (POR), CYP2E1 and glutathione reductase (GSR), which results in oxidative stress in the liver [53] and impairs alcohol metabolism [54]. Table 2 gives a short summary of the ASH associated miRNA and their relation to OS.

Table 2. Deregulated miRNAs and relation to OS in ALD/ASH.

miRNA	Evidence				Target Gene/Pathway	(Patho)Mechanism	References
	In Vitro	In Vivo	In Situ	In Silico			
155 ↑[1]	✓	✓	✓		TNFα ↑ PPARα ↓	LPS mediates the activation of NFκB. Increase of miR-155; release of TNFα, ROS and oxidative stress in Kupffer cells and hepatic stellate cells via suppression of PPARα causing overexpression of FABP4, ACC1 and LDLR	[38–40]
181b-3p ↑[2]	✓	✓			TLR4 ↑ NFκB ↑	Alterations in glucose and lipid homeostasis; activation of Kupffer cells	[41]
291b ↑[2]	✓	✓			Tollip ↓	Loss of downregulation of TLR4/NFκβ in Kupffer cells	[42]
34a ↑[2]	✓	✓			SIRT1 ↓	Inhibition of SIRT1 protein coding	[44]
217 ↑[2]	✓	✓			SIRT1 ↓	Alcohol-associated inflammation	[45]
122 ↓[2]		✓	✓		HIF1α ↑	miR-122 loss (deletion) or down-regulation (due to alcohol diet via GRHL2) leads to steathosis at birth, following fibrosis; miR-122 down-regulation enables HIF1α expression in ALD	[46,47]
21 ↑[2]	✓	✓			FASLG ↓ DR5 ↓	Reduced ethanol induced cell death in hepatocytes; stellate cells dysregulation via miR-21 in ethanol-induced altered extrinsic apoptotic signaling and its progression to ALD	[50]
223 ↑[1]					IL-6 ↑ p47^{phox} ↑	Peripheral neutrophils activation and liver infiltration induced by ethanol; triggering NADPH oxidase → ROS	[51]
199 ↓[1]	✓	✓			HIF1α ↑ ET-1 ↑	Leading to steatohepatitis in cirrhosis patients	[52]
214 ↑[1]	✓	✓		✓	CypP450 ↓ GSR ↓	Affecting alcohol metabolism and causing oxidative stress	[53,54]

Relation to oxidative stress: [1]: yes, [2]: no, [3]: not mentioned. Abbreviations: ACC1 = Acetyl-CoA carboxylase, ALD = Alcoholic liver disease, DR5 = Death receptor 5, ET-1 = endothelin 1, FABP4 = Fatty acid binding protein 4, FASLG = Fas ligand, GRHL2 = grainyhead like transcription factor 2, GSR = glutathione reductase, HIF1α = Hypoxia Inducible Factor 1 alpha, LDLR = Low-density Lipoprotein Receptor, NFκB = nuclear factor "kappa-light-chain-enhancer", PPARα = Peroxisome proliferator-activated receptor alpha, ROS = Reactive oxygen species, SIRT1 = Sirtuin 1, TLR4 = Toll-like receptor 4, TNFα = Tumor necrosis factor alpha, Tollip = Toll interacting protein, ↓ = downregulation/reduction, ↑ = upregulation/increase.

4. Nonalcoholic Fatty Liver Disease (NAFLD) and Nonalcoholic Steatohepatitis (NASH)

NAFLD is defined by fatty degeneration of hepatocytes comprising more than 5 to 10% of the liver and insulin resistance (IR) but without any history of alcohol abuse and/or other diseases that might lead to fatty liver disease [55]. One third of NAFLD patients progresses to nonalcoholic steatohepatitis (NASH) and fibrosis within 4 to 5 years, depending on the spectrum of lipotoxicity, cellular stress and inflammation [56,57]. NAFLD is caused by an imbalance of free fatty acid (FFA) uptake and *de novo* lipogenesis as well as fatty acid (FA) oxidation and formation of lipoproteins [58,59]. Oxidative stress is seen to be an important player leading to defective hepatocyte regeneration, development of NAFLD and progression to NASH [60]. Excessive nutrients intake, especially high fat diet, leads to excessive FA oxidation [61] and consequently to excessive generation of ROS that are either directly toxic or indirectly by depleting antioxidant reserves [60]. ROS can damage mitochondria, which leads to reduced FA oxidation and accumulation of FA, finally leading to lipotoxicity and release of proapoptotic factors [62]. In turn, lipotoxicity induces endoplasmic reticulum (ER) stress, impairs autophagy and promotes a sterile inflammatory response that aggravates liver cell injury and leads to death of liver cells [63]. Subsequently the unfolded protein response (UPR) is activated by toxic free cholesterol, FFA and diacylglyceride and induces upregulation of proapoptotic C/EBP homologous protein (CHOP) [64–67]. Usually, UPR induces antioxidant mechanisms by activation of Nrf2 via upregulation of ATF4 transcription factors to counteract the oxidative stress [32]. However, in contrast to normal physiology, NAFLD-related Nrf2 activity is impaired, which also leads to mitochondrial dysfunction and increased intracellular FFA [68–70].

Aberrant miRNA expression profiles have been shown to contribute to the development of metabolic syndrome and NAFLD [71]. As also many other genes and pathways that contribute to NAFLD and the progression to NASH are influenced by miRNA, we only provide an overview of the most relevant miRNAs.

miR-21 positively correlates with NAFLD and NASH severity [72]. In hepatocytes, unsaturated FFA increase miR-21 in a mTOR/NFκB dependent manner and inhibit phosphatase and tensin homolog (PTEN) that usually controls FA oxidation in the liver and stimulates glucose uptake in muscle cells [73]. Dattaroy et al. described in 2015 that in HSC NOX upregulates the levels of miR-21, which targets the TGFβ pathway and in turn causes activation of HSC and promotion of fibrogenesis via alpha-1 type I collagen (Col1α1) and alpha smooth muscle actin (α-SMA) upregulation [72]. In 2017 Rodrigues et al. were able to show that ablation of miR-21 results in a progressive decrease in steatosis, inflammation and lipoapoptosis with impaired fibrosis [74]. Fast food diet leads to increased miR-21 levels in liver and muscle of NASH mouse models with concomitantly decreased expression of PPARα, thereby promoting steatohepatitis [74,75].

The best characterized miRNA is miR-122 [76–78]. In cases of hepatocellular damage, miR-122 is secreted by damaged cells [79] and appears elevated in the serum during NAFLD. This correlates with disease severity [80,81], although it is contemporaneously reduced in liver tissue [78]. In the context of fibrogenesis, the protective actions of miR-122 are inhibited, which is mediated by long non-coding RNA Nuclear Enriched Abundant Transcript 1 (NEAT1) or via circRNA_002581 and subsequently triggers an increased expression of Kruppel-like factor 6 (KLF6) in HSC [82]. In addition, a miR-122 knockout leads to a higher accumulation of triglycerides (TG), micro steatosis, NASH and fibrosis [83].

Another miRNA that is upregulated in liver tissue and serum and which is integrated into the lipid metabolism is miR-34a [84,85]. Its targets are the transcription factors hepatocyte nuclear factor 4 alpha (HNF4α), PPARα, SIRT1 and p53, all in all leading to an accumulation of TG [86–89]. miR-34a inhibits SIRT1, which causes the inactivation of AMP-Kinase. This mechanism leads to an increase of hepatic cholesterol synthesis and activation of pro-apoptotic genes (p53 and P66SHC), which contributes to oxidative stress and apoptosis due to reduced β-oxidation resulting in restoration of nicotinamide phosphoribosyltransferase/nicotinamide-adenine-dinucleotide (NAMPT/NAD+) levels and therefore ameliorates hepatic steatosis and inflammation [86,88,90,91].

It was shown that miR-29 family (a, b, c) expression is altered in mice with liver fibrosis and in liver tissue of NASH patients [92,93]. miR-29a and c are downregulated in dietary induced NASH that is accompanied by an upregulation of HMG-CoA reductase (HMCGR), which in turn triggers severe hepatic steatosis and inflammation, probably via enhanced expression of lipoprotein lipase [94,95].

In contrast to that, Kurtz et al. demonstrated that blocking of miR-29 leads to significantly decreased plasma cholesterol and TG levels as a result of the inhibition of *de novo* hepatic lipid synthesis [96]. The reason for these contrary findings could be clarified by Mattis et al. who induced a conditional knockout mouse model and investigated the function of miR-29a [94,95] while Kurtz et al. used the LNA-29 inhibitor to deplete the entire miR-29 family [96]. Furthermore, miR-29b is downregulated in activated mouse HSC, leading to a loss of interaction with Col1α 3′-UTR, which stimulates the collagen production [93,97].

miR-155 is upregulated in a NASH mouse model induced via high fat diet [98]. iR-155 elevates the Forkhead-Box-Protein O3 (FOXO3a) expression thereby regulating the activation of that pathway, whose proteins are involved in the maintenance of the intercellular redox balance [99]. Additionally, miR-155 regulates lipid metabolism by modulating the protein expression of SREBP-1c and fatty acid synthase (FAS) resulting in increased intracellular lipid accumulation in hepatocytes [100]. Interestingly, decreased levels of miR-155 were shown by Csak et al. to be associated with fibrosis via dysregulation of HIF1α and vimentin [101]. This working group showed that a miR-155 knockout reduced steatosis and fibrosis in a mouse model fed with methionine-choline-deficient diet. This leads to the conclusion that miR-155 expression might be stage relevant. In high fat fed mice, miR-155 might exert a protective feedback regulation of the SERBP-1 pathway in order to suppress *de novo* lipid synthesis and reduce lipid load in the hepatocytes [102]. Furthermore, it has been shown that seven miRNAs belonging to the miRNA cluster located at chromosome locus 14q32.2 maternally imprinted region are over-expressed in a NASH mouse model, which was characterized by genetic modification (leptin knock-out) and high fat diet. Therefore, they could represent valid biomarkers for NAFLD/NASH [103].

Many other miRNAs can be linked more directly to OS and ER stress. During OS, NADPH is responsible for an upregulation of miR-21 and miR-155, therefore influencing FOXO3a pathways and fibrosis [72,99]. Protein expression of CHOP can be induced and cells sensitized to apoptosis by miR-211, -689, -70, -711, -712, -762, -1897-3p, -2132, -2137 and inhibited by miR-322, -351, -503 [104,105]. OS related activation of transcription factor 6α (ATF6α) is pro-apoptotic, but can be inhibited by miR-702 [106,107]. Inhibition of miR-199a-5p results in increased ER stress-induced apoptosis [108]. In summary, both oxidative and ER stress as well as miRNAs make a decisive contribution to the development of NAFLD and the progression to NASH (summarized in Table 3). In particular, the combination of these two mechanisms provides information on pathophysiology and promises starting points for monitoring disease progression and therapy.

Table 3. Deregulated miRNAs and relation to OS in NAFLD/NASH.

miRNA	Evidence				Target Gene/Pathway	(Patho)Mechanism	References
	In Vitro	In Vivo	In Situ	In Silico			
21 ↑[3]	✓				PPARα ↓	Liver injury, inflammation and fibrosis	[75]
21 ↑[3]	✓	✓	✓		PTEN ↓	Development of steatosis	[73]
21 ↑[1]		✓	✓		TGFβ ↑	Induced collagen production and extracellular matrix formation fibrogenesis via increase of Col1α1 and α-SMA expression	[72]
122 ↑[2]	✓				KLF6 ↑	Activation of hepatic stellate cells and progression of liver fibrosis	[82]
34a ↑[2]	✓		✓		HNF4α ↓	Inhibition of very low-density lipoprotein secretion and promotion of liver steatosis and hypolipidemia	[89]
34a ↑[1]	✓	✓			PPARα ↓	Loss of regulation genes encoding fatty acid metabolizing enzymes and mitochondrial fatty acid oxidation activity	[87]
34a ↑[1]		✓			SIRT1 ↓	Increase of hepatic cholesterol synthesis and activation of pro-apoptotic genes (p53, p66shc)	[88]
29a and c ↓[2]	✓	✓			SIRT1 ↓	Increased levels of free cholesterol	[94,95]
29 ↑[2]	✓	✓			Col1α1 ↓	Downregulation in activated hepatic stellate cells and therefore loss of interaction with Col1α1 → decreased collagen production	[97]

Table 3. *Cont.*

miRNA	Evidence				Target Gene/Pathway	(Patho)Mechanism	References
	In Vitro	In Vivo	In Situ	In Silico			
155 ↑[1]	✓	✓			AKT/ FOXO3a ↑	Regulates proliferation of hepatic stellate cells promotes liver fibrosis; FOXO3a proteins maintain intracellular redox balance and survival	[99]
155 ↑[2]		✓	✓		LXRα ↓	Decreased SREBP1 and FAS resulting in an increased intracellular lipid content	[100]
155 ↑[2]		✓	✓		HIF1α and vimentin ↑	NASH-induced liver fibrosis	[101]

Relation to oxidative stress: [1]: yes, [2]: no, [3]: not mentioned. Abbreviations: AKT = Protein kinase B, Col1α1 = Collagen type I alpha 1, FAS = Fatty acid synthase, FOXO3 = Forkhead-Box-Protein O3, HIF1α = Hypoxia-inducible factor 1-alpha, HNF4α = Hepatocyte nuclear factor 4 alpha, KLF6 = Krueppel-like factor 6, LXRα = Liver X receptor alpha, PPARa = Peroxisome proliferator-activated receptor alpha, PTEN = Phosphatase and Tensin homolog, SIRT1 = Sirtuin 1, SREBP1 = sterol regulatory element-binding protein, ↓ = downregulation/reduction, ↑ = upregulation/increase.

5. Viral Hepatitis

According to the WHO fact sheet, 257 million people were living with a chronic hepatits B virus (HBV) infection in 2015 with nearly 887,000 estimated deaths. Around 71 million people had a chronic hepatitis C virus (HCV) infection, resulting in an estimated 399,000 related deaths [109,110]. It has been shown that the immune system initiates the production of ROS and RNS in chronic hepatitis [111,112] and it seems that oxidative stress is important in the pathogenesis of viral hepatitis and some of these pathomechanisms are influenced by miRNAs.

Patients suffering from HCV infection produce more ROS compared to other types of virus associated hepatitis [113]. Hou et al. stated that miR-196 directly acts on Bach1 mRNA by repressing Bach1 expression and upregulating heme oxygenase 1 (HO1) leading to viral-induced oxidative stress [114]. Furthermore, miR-196 inhibits the HCV expression in HCV replicon cell lines, highlighting miR-196 as a potential therapeutic target.

As also demonstrated in other liver diseases, miR-122 also plays an important role in HCV infection. Here, miR-122 directly binds to the viral genome and enhances viral RNA replication, thus resulting in reduced miR-122 expression within the cell [115,116]. The NFκB-inducing kinase (NIK) is usually a target of miR-122, but due to the decreased levels of miR-122, NIK is increased in HCV infection [117]. In addition, HNF4α, a transcriptional regulator of miR-122 expression and known for its OS-association [118], is downregulated in HCV infection, too [117]. Both effects result in disturbance of the NIK mediated lipid metabolism and HCV-induced lipogenesis and lipid droplet formation [117,119].

Moreover, miR-122 also contributes to the pathomechanisms of HBV infection where it inhibits the effects of p53 on HBV replication by initiating a cyclin G1-p53 complex [120]. Wójcik K and co-workers described a link to oxidative stress in HVB infection as well. In a gene expression study, a positive correlation between miR-122 and NAD(P)H quinone dehydrogenase 1(NQO1) was demonstrated and it is supposed that miR-122 directly limits OS by suppression of the HBV replication and as a consequence affects the balance between pro-oxidants and antioxidants [121].

In summary, miRNAs and especially miR-122 are involved in the pathogenesis of HBV and HCV infections (see Table 4) and represent a potential target for novel treatment options [122].

Table 4. Deregulated miRNAs and relation to OS in viral hepatitis.

miRNA	Evidence				Target Gene/Pathway	(Patho)Mechanism	References
	In Vitro	In Vivo	In Situ	In Silico			
196↓[1, C]	✓				Bach1/ HMOX1 ↓	Down-regulation of Bach1 gene expression, up-regulation of HMOX1 gene expression, a key cytoprotective enzyme	[114]
196↓[2, C]	✓				HCV NS5A gene ↓	miR-196 perfectly matches coding region of the HCV NS5A gene down-regulatory effect of miR-196 on HCV expression in the HCV J6/JFH1 cell culture system	[114]
122↓[2, C]	✓				HCV viral genome ↑	Enhances viral RNA replication	[115,116]
122↓[1, C]	✓	✓	✓		NIK ↑ and HNF4α ↑	Disturbance of the NIK mediated lipid metabolism → lipogenesis and lipid droplet formation → promotion of oxidative stress	[117]
122↓[2, B]			✓		cyclin G1-p53 complex ↑	Inhibits the effects of p53 on HBV replication	[120]
122↓[1, B]			✓		NQO1 ↑ and HO1 ↓	miR-122 affects balance between the pro-oxidants and antioxidants	[121]

Relation to oxidative stress: [1]: yes, [2]: no, [3]: not mentioned, [B]: Hepatitis B virus infection, [C]: Hepatitis C virus infection. Abbreviations: HBV = Hepatitis B virus, HCV = Hepatitis C virus, HO1 = Heme oxygenase 1, HNF4α = Hepatocyte nuclear factor 4 alpha, NIK = NFκB-inducing kinase, NQO1 = NAD(P)H quinone dehydrogenase 1, NS5A = Non-structural protein 5A, ↓ = downregulation/reduction, ↑ = upregulation/increase.

6. Hepatocellular Carcinoma (HCC)

Hepatocellular carcinoma (HCC) is the most common primary malignancy of the liver, representing about 85% of all cases. HCC is the 6th most common malignancy worldwide and is the 4th most common cause of cancer related deaths [123]. HCC usually develops on the basis of other (chronic) liver diseases, esp. chronic viral hepatitis B or C, aflatoxin intoxication or ALD. Recently, also NAFLD and NASH became more prevalent and are now considered as major causes for HCC development in developed countries [124]. All of these conditions lead to chronic inflammation, fibrosis and cirrhosis development being essentially associated with oxidative stress conditions [125]. Interestingly, genes involved in antioxidation like Nrf2 or Keap1 were found to be mutated in up 8% of HCCs, linking the chronic stress conditions to OS pathways but also to metabolic conditions and autophagy [126], which are themselves regulated by different mechanisms, including long non-coding RNA and miRNA [127]. Under metabolic stress conditions, ROS is produced as a by-product from elevated mitochondrial fatty acid oxidation or inadequate respiratory chain function, e.g., due to fructose overload or insulin resistance. This leads to lipid accumulation which can further promote ROS production via β-oxidation of FA [125,128,129]. Additional ROS and RNS are produced by inflammatory cells that are attracted under those conditions but are also activated in case of viral hepatitis [130–134]. ROS can increase activity and expression of cytokines (e.g., IL-1α, IL-1β, IL-6, IL-8, TNFα) and growth factors, lead to DNA damage and trigger persistent necro-inflammation and hepatocyte regeneration that is considered a key event for HCC pathogenesis [135,136]. This can initiate a vicious circle, as the same mediators are also pathophysiologic drivers of the potentially underlying chronic liver disease, e.g., steatohepatitis, fibrosis or chronic inflammation.

8-hydroxy-2′-deoxy-guanosine (8-OHdG) was shown to be a prognostic biomarker in HCC [137]. 8-OHdG also links OS to epigenetic regulation of gene expression via DNA methylation as it is an important co-factor for the ten-eleven translocation methylcytosine dioxygenase (TET) family of DNA demethylases [138].

miRNAs have been shown to regulate expression of oncogenes and tumorsupressor genes also in HCC and provide a mechanistic link between epigenetics, inflammation, viral infection and OS [139]. Various miRNAs have been shown to be affected by OS in HCC–summarized in Table 5, e.g., downregulation of miR-26 or upregulation of miR-155 [83,140,141]. Interestingly, miR-26 expression was shown to be under the control of TET and targets the histone lysine methyltransferase Enhancer of Zeste Homolog 2 (EZH2), which is involved in the epigenetic regulation of various cell cycle control genes [142,143]. TET1 expression, in return, was shown to be under the control of miR-29b, and found to be downregulated in a study with 25 HCC patients from China [144]. In other studies, several other miRNAs, e.g., miR-494 [145] or miR-520b [146], were also shown to regulate TET1 expression in HCC, confirming the "multiple targets, multiple hits" problem and context sensitivity when analyzing miRNA signaling.

Expression of miRNA and levels of 8-OHdG were analyzed in a study comparing 29 HCC tissue samples to 58 non-cancerous liver specimens (including viral and alcoholic hepatitis). Here, significantly elevated levels of 8-OHdG were found in HCC and non-cancerous cirrhotic tissue compared to chronic hepatitis without cirrhosis or normal liver tissue. This was paralleled by increased telomerase activity and inversely correlated to telomere length. Several miRNAs were differentially regulated and the miR-17-92 cluster was down-regulated in about 50% of the analyzed samples [147]. Interestingly, the epigenetic down-regulation of miRNAs belonging to the miRNA cluster 17–92 promoted cell death in HCC cells [148]. Additional experimental findings showed that ROS reduces the expression of this miRNA cluster [149]. In HCC patients, miR-222 was found to be overexpressed and the endogenous cell cycle regulator p27^{kip1} was identified as a predicted target gene of this miRNA and expression of p27 protein is significantly decreased in HCC tissue [150]. Additionally, the tumorsuppressor is responsible for the suppression of HMGA2 leading to cell cycle block and liver cancer cell death [151].

Table 5. Deregulated miRNAs and relation to OS in liver cancer.

miRNA	Evidence				Target Gene/Pathway	(Patho)Mechanism	References
	In Vitro	In Vivo	In Situ	In Silico			
26↓[1]	✓		✓		EZH2 ↑	Sequestration of miR-26 from its target EZH2, which released the suppression on EZH2, and thereby led to EZH2 overexpression in gastric cancer	[142]
29b↓[2]	✓		✓		TET1 ↓	Feedback of miRNA-29-TET1 downregulation in HCC development suggesting a potential target in identification of the prognosis and application of cancer therapy for HCC patients	[144]
494↑[2]	✓		✓		TET1 ↓	miR-494 inhibition or enforced TET1 expression is able to restore invasion-suppressor miRNAs and inhibit miR-494-mediated HCC cell invasion	[145]
520b↓[2]	✓				TET1 ↓	Depresses proliferation of liver cancer cells through targeting 3′UTR of TET1 mRNA	[146]
17-92 cluster↓[1]			✓		E2F family ↑	ROS-mediated oxidative DNA damage correlates with over-expression of miR-92–playing a role in both the apoptotic process and in cellular proliferation pathways	[147]

Relation to oxidative stress: [1]: yes, [2]: no, [3]: not mentioned. Abbreviations: E2F = E2F transcription factor family, EZH2 = Enhancer of zeste homolog 2, TET1 = Ten-eleven translocation methylcytosine dioxygenase 1, ↓ = downregulation/reduction, ↑ = upregulation/increase.

7. Clinical Implications/Studies

Translating molecular scientific findings into clinical practice is the final destination of life sciences. While numberless miRNAs have been identified to play central roles in regulating nearly all pathways in cell homeostasis, it seems that science got lost in translation. OS has a key role in chronic liver diseases as it is strongly linked to acute and chronic inflammation and is therefore a main driver of progressive organ fibrosis and cancer development [152]. In chronic HCV infection antioxidant supplementation attenuates OS and although no clear clinical studies are available they are also recommended for patients with NASH [153].

Therapeutic approaches to miRNA are rare in liver diseases. Most miRNA based drugs are assessing antagonism by inhibitory antisense miRNA or by application of miRNA [154,155]. More than 6000 patents in the US market and more than 3000 in the EU market were granted in 2016 for miRNA and siRNA therapeutics [154]. Anti-miRNA oligonucleotides, so called anti-miRs or antago-miRs, have been used in experimental settings to inhibit signaling of corresponding miRNAs. Improvement of chemical structures of these oligonucleotides, e.g., adding 2′-O-methyl or 2′-O-methoxyethyl groups, generated locked nucleid acid (LNA-) antimiRs with improved pharmacokinetic and pharmacodynamic properties [156,157]. Liver specific targeting of antimiRs was achieved by conjugating these oligonucleotides to N-acetylgalactosamine (GalNAc), which is recognized by the asialoglycoprotein receptor on hepatocytes [158]. However, the therapeutic application in clinical practice seems to be far away. Actually, there are no ongoing clinical trials addressing both OS and liver disease registered to clinicaltrials.gov in a therapeutic manner.

Several trials are evaluating miRNA as biomarkers for prognosis of liver diseases–e.g., fibrosis, survival, progression of HCC. Only 10 clinical trials are registered for recruiting patients addressing microRNA and OS conditions–none of them has a therapeutic approach by addressing miRNAs.

The miR-210 group seems to be promising as a biomarker and therapeutic target in hypoxia [159]. It is up-regulated in hypoxia-related activation of HIF1α, is a key factor in induction of (tumor) cell proliferation by targeting fibroblast growth factor receptor-like 1 (FGFRL1) [160] and modulates mitochondrial alterations due to hypoxia [161]. By regulating miR-210, it could be possible to attenuate hypoxic cell damage and tissue alteration due to reperfusion after revascularization procedures. A clinical trial NCT04089943 (clinicaltrials. gov) is evaluating patients with peripheral artery disease (PAD) for the expression of miR-210 in skeletal tissue. The miR-210 group could also serve as OS marker, which could be even measured in peripheral blood [159]. By the dependency to HIF1α it could serve as prognostic factor for determining the aggressiveness and/or early stage of HCC [160,162,163].

However, OS-related miRNAs are evaluated as therapeutic target and/or biomarker for the outcome of ischemic injury such as myocardial ischemia, ischemic central insults and development of metabolic disorders. Using miRNAs as biomarkers for disease development, risk scoring, prognostic factors and drug monitoring seems actually the best approach. Countless studies are evaluating whole panels of miRNA as biomarkers in nearly all conditions of diseases (Figure 2).

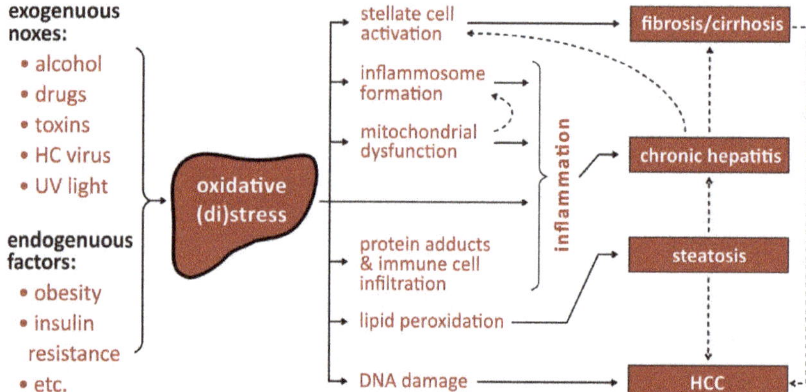

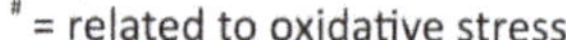

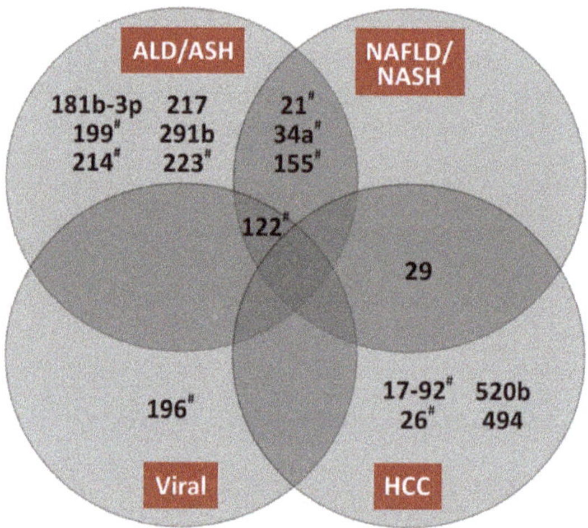

Figure 2. Oxidative stress- and microRNA-dependent liver pathogenesis. (**A**). Possible effects of oxidative stress and subsequent liver diseases; modified from [31]. (**B**). Venn diagram on known involvement of microRNAs in the four liver pathologies; for details, see text and Tables 2–5. Abbreviations: ALD = alcoholic liver disease, ASH = alcoholic steatohepatitis, HCC = hepatocellular carcinoma, NAFLD = nonalcoholic fatty liver disease, NASH = nonalcoholic steatohepatitis, # = related to oxidative stress.

7.1. ALD and ASH

Because of their altered expression, miRNAs could represent a valid diagnostic marker for patients affected by ALD. miR-192 and miR-30a serum levels have been correlated with ALD diagnosis [164]. Other examples are miR-103 and miR-107, which have been found to be strongly increased in the serum of patients affected by ALD and NAFLD. Their levels were low in healthy patients and in subjects affected by viral hepatitis [165]. Binge alcohol drinking caused an increase of miR-155 and miR-122 in healthy patients. Unfortunately, these miRNAs have been found over-expressed in several liver diseases and therefore could not be applied as a valid biomarker for ALD [166].

Targeting miRNAs could represent an effective therapeutic strategy for the treatment of ALD/ASH. Recently, the treatment with hyaluronic acid determined the stabilization of miR-181b-3p and importin α5 in mice fed with ethanol, thus protecting from the alcohol-derived liver damage [41]. Additionally, hyaluronic acid could normalize the level of miR-291b thus allowing the increase of Tollip and the consequent inhibition of the inflammatory pathway TLR4/NFκB [42]. Despite the contradictory role of miR-122 as determined by interrupting the cross-talk between hepatocytes and stromal cells [167], its suppression, mediated by the inhibitor Miravirsen, has shown a strong beneficial effect in chronic hepatitis suggesting a potential benefit for patients affected by ALD [167,168]. Beneficial effects for the treatment of ASH could be represented by the over-expression of liver protecting miRNAs. Unfortunately, no trial has been established to identify the clinical benefit of patients affected by ASH [168].

7.2. NAFLD/NASH

With rising incidence of NAFLD, obesity and diabetes in the Western and Asiatic world, NASH will be the most common cause for the development of liver cirrhosis and HCC [169]. Today, the definitive diagnosis of NASH requires a liver biopsy showing evidence with regard to steatosis, lobular or portal inflammation and ballooning of hepatocytes [170]. In NAFLD and NASH, different expression patterns of up to 44 miRNAs could be shown [78,171]. Latorre et al. and Su et al. described high serum levels of miR-451, -122, -34a and 21 in patients suffering from hepatic steatosis. miRNA-122 is elevated in the serum due to liver damage and levels are higher in severe steatosis than in mild and higher in severe fibrosis [78,79]. Furthermore, expression of miR-122 correlated positively with very low density lipoproteins (VLDL), free cholesterol and TGs [84]. With these properties miR-122 is suitable to act as a biomarker. Liu et al. was able to show that isochlorogenic acid B (ICAB) has a protective effect and is possibly associated with the ability to attenuate OS by up-regulating Nrf2 and suppressing fibrogenic factors through miR-122/HIF1α pathway [172]. Carnosic acid, an antioxidant, provides protection against NAFLD by decreasing miR-34a expression and stimulating the SIRT1/p66shc pathway [88]. In a mouse model Derdak et al. abrogated the overexpression of miR-34a with pifithrin-α *p*-nitro (PFT) and activated the SIRT1 pathway which ended up in diminished hepatic TG deposition and ameliorated the liver steatosis [91]. Kumar et al. treated mice with nanoparticles carrying a mimic of miR-29b1 which was able to significantly decrease collagen deposition in liver and serum in a liver fibrosis model. miR-29 has been associated with fibrosis in many different organs [79,173]. In a phase I trial patients suffering from fibrosis benefited from a miR-29 mimic [174]. In a transgenic mouse model overexpressing platelet derived growth factor C (PDGF C), LNA-antimiR-124 suppressed miR-124 signaling and expression of cognate target genes, leading to reduced hepatic fibrosis and even inhibited tumor formation [175]. Inhibition of miR-30b by lentiviral antimiR expression was able to reduce ER stress and improve insulin sensitivity in a high-fat dietary rat model of NAFLD [176]. These are encouraging further steps towards miRNA-directed therapies in the treatment of NASH and liver fibrosis.

7.3. HCC

In HCC patients, high levels of thioredoxin and manganese superoxide dismutase levels were detected and could be used as prognostic biomarkers [177–179]. In line with this, elevated levels of 8-OHdG, an established biomarker for oxidative stress conditions [180], were detected in various chronic liver diseases including HCV and HCC [181–184]. The miR-122 group could be another really promising candidate. It is involved in HCV related HCC progression and liver fibrosis. It targets most importantly mRNA is Aldolase A mRNA and MYC downstream regulated gene 3 [185]. Since Aldolase A is indirectly linked to hypoxia as downstream target of HIF1α [186] and its expression could be suppressed by miR-122, it could be possible to influence response to hypoxia-related survival of HCC by antagonizing miR-122. Interestingly, a nanoparticle-carrier based antimiR was able to suppress miR-122 expression for up 28 days in a murine HCV model [187]. However, in the next few years a

Int. J. Mol. Sci. **2019**, 20, 5266

wide range of patterns of miRNA will be available for clinical prognosis. miRNA-based drugs still need to be put into translation for clinical studies.

Besides its role in promoting tumorigenesis, OS has also been shown to exert anti-tumor effects in HCC. Downregulation or inhibition of thioredoxin reductase 1 (TXNRD1), a negative prognostic factor for HCC [188], suppressed growth of HCC models and induced sensitization to the current standard of care, sorafenib [189]. Sorafenib acts as a multi-tyrosine kinase inhibitor and impacts tumor growth by blocking the RAF/MEK/ERK pathway and by inhibiting angiogenesis [190,191]. In addition, sorafenib induced HCC cell death in vitro and in vivo also via induction of ROS production. This was linked to an increased median overall and progression free survival of patients showing higher levels of advanced oxidation protein products, which was used as a surrogate serum biomarker for OS in 26 patients [192]. Mechanistically, sorafenib blocks the mitochondrial respiratory chain and leads to disruption of the mitochondrial membranes which increases ROS production [193,194]. Resistance to sorafenib treatment is limiting its clinical efficacy. In a computational modelling approach, the miR-17-92 cluster was shown to be a key regulator of resistance to sorafenib via interaction with several components of the EGFR and IL-6 signaling pathways, including e.g., Januskinase/sterol regulatory element-binding proteins (JAK/STAT) signaling and induced myeloid leukemia cell differentiation protein (Mcl-1) function [195]. Altogether, these data confirm a complex interplay between chronic liver diseases, oxidative stress, miRNA expression, epigenetics and HCC pathogenesis.

8. Summary

The liver is the central metabolic organ and thus subjected to various potential external and internal factors. The increasing prevalence of NAFLD, which is projected to become the major causer of end-stage liver disease and liver transplantation, highlights the importance of understanding the pathophysiology of liver damaging conditions. While reactive (oxygen) species play a central role in normal tissue homeostasis and cellular signaling, these mediators can also contribute to acute and chronic injury of the liver, leading to fibrosis, cirrhosis and ultimately HCC formation. Recent studies demonstrated that ROS impacts lipid metabolism, detoxication, as well as central cellular survival and homeostasis processes like ER stress, calcium signaling and unfolded protein response. These pathways involve several genes that have been demonstrated to be regulated by miRNAs. While several miRNAs have now been identified to be involved in different liver diseases and some of these have been further associated to OS, we still do not fully understand the complex network of those signaling and regulatory pathways under distinct pathophysiologic conditions. Both axes, OS and miRNAs, represent potential biomarkers for surveillance, diagnosis and treatment response and may be used as novel therapeutic targets in the near future. Looking back from bed-side to bench, clinicians have to wait for stable formulations targeting miRNAs e.g., with antagomiRs enveloped into microparticles which are already available for siRNAs and being tested in clinical trials [154].

Author Contributions: Writing–Original Draft Preparation, E.K., C.M., T.K., P.D.F., T.W., D.N. and M.O.; Writing–Review and Editing E.K., T.K. and D.N.; Visualization T.K. and C.M.; Supervision D.N.; Project Administration E.K.; All authors have read and approved the final version of the manuscript.

Funding: This research received no external funding.

Conflicts of Interest: The authors declare no conflict of interest.

References

1. Cichoz-Lach, H.; Michalak, A. Oxidative stress as a crucial factor in liver diseases. *World J. Gastroenterol.* **2014**, 20, 8082–8091. [CrossRef] [PubMed]
2. Chandel, N.S.; Maltepe, E.; Goldwasser, E.; Mathieu, C.E.; Simon, M.C.; Schumacker, P.T. Mitochondrial reactive oxygen species trigger hypoxia-induced transcription. *Proc. Natl. Acad. Sci. USA* **1998**, 95, 11715–11720. [CrossRef] [PubMed]

3. Sena, L.A.; Chandel, N.S. Physiological roles of mitochondrial reactive oxygen species. *Mol. Cell* **2012**, *48*, 158–167. [CrossRef] [PubMed]

4. Nemoto, S.; Takeda, K.; Yu, Z.X.; Ferrans, V.J.; Finkel, T. Role for mitochondrial oxidants as regulators of cellular metabolism. *Mol. Cell. Biol.* **2000**, *20*, 7311–7318. [CrossRef] [PubMed]

5. Lo, Y.Y.; Cruz, T.F. Involvement of reactive oxygen species in cytokine and growth factor induction of c-fos expression in chondrocytes. *J. Biol. Chem.* **1995**, *270*, 11727–11730. [CrossRef] [PubMed]

6. Bae, Y.S.; Kang, S.W.; Seo, M.S.; Baines, I.C.; Tekle, E.; Chock, P.B.; Rhee, S.G. Epidermal growth factor (EGF)-induced generation of hydrogen peroxide. Role in EGF receptor-mediated tyrosine phosphorylation. *J. Biol. Chem.* **1997**, *272*, 217–221. [CrossRef]

7. Urao, N.; Ushio-Fukai, M. Redox regulation of stem/progenitor cells and bone marrow niche. *Free Radic. Biol. Med.* **2013**, *54*, 26–39. [CrossRef]

8. The Alpha-Tocopherol Beta Carotene Cancer Prevention Study Group. The effect of vitamin E and beta carotene on the incidence of lung cancer and other cancers in male smokers. *N. Engl. J. Med.* **1994**, *330*, 1029–1035. [CrossRef]

9. Watson, J. Oxidants, antioxidants and the current incurability of metastatic cancers. *Open Biol.* **2013**, *3*, 120144. [CrossRef]

10. Anastasiou, D.; Poulogiannis, G.; Asara, J.M.; Boxer, M.B.; Jiang, J.K.; Shen, M.; Bellinger, G.; Sasaki, A.T.; Locasale, J.W.; Auld, D.S.; et al. Inhibition of pyruvate kinase M2 by reactive oxygen species contributes to cellular antioxidant responses. *Science* **2011**, *334*, 1278–1283. [CrossRef]

11. Mathew, R.; White, E. Autophagy, stress, and cancer metabolism: What doesn't kill you makes you stronger. *Cold Spring Harb. Symp. Quant. Biol.* **2011**, *76*, 389–396. [CrossRef] [PubMed]

12. Holmstrom, K.M.; Finkel, T. Cellular mechanisms and physiological consequences of redox-dependent signalling. *Nat. Rev. Mol. Cell. Biol.* **2014**, *15*, 411–421. [CrossRef]

13. Calin, G.A.; Croce, C.M. MicroRNA signatures in human cancers. *Nat. Rev. Cancer* **2006**, *6*, 857–866. [CrossRef] [PubMed]

14. Pichler, M.; Calin, G.A. MicroRNAs in cancer: From developmental genes in worms to their clinical application in patients. *Br. J. Cancer* **2015**, *113*, 569–573. [CrossRef] [PubMed]

15. Kinser, H.E.; Pincus, Z. MicroRNAs as modulators of longevity and the aging process. In *Human Genetics*; Springer: Berlin, Germany, 2019.

16. Greco, S.; Gaetano, C.; Martelli, F. Long Noncoding Competing Endogenous RNA Networks in Age-Associated Cardiovascular Diseases. *Int. J. Mol. Sci* **2019**, *20*, 3079. [CrossRef] [PubMed]

17. Piontek, K.; Selaru, F.M. MicroRNAs in the biology and diagnosis of cholangiocarcinoma. *Semin. Liver Dis.* **2015**, *35*, 55–62. [CrossRef] [PubMed]

18. Swierczynski, S.; Klieser, E.; Illig, R.; Alinger-Scharinger, B.; Kiesslich, T.; Neureiter, D. Histone deacetylation meets miRNA: Epigenetics and post-transcriptional regulation in cancer and chronic diseases. *Expert Opin. Biol. Ther.* **2015**, *15*, 651–664. [CrossRef]

19. Gjorgjieva, M.; Sobolewski, C.; Dolicka, D.; Correia de Sousa, M.; Foti, M. miRNAs and NAFLD: From pathophysiology to therapy. *Gut* **2019**, *68*, 2065–2079. [CrossRef]

20. Collin, F. Chemical Basis of Reactive Oxygen Species Reactivity and Involvement in Neurodegenerative Diseases. *Int. J. Mol. Sci.* **2019**, *20*, 2407. [CrossRef]

21. Glasauer, A.; Chandel, N.S. Ros. *Curr. Biol.* **2013**, *23*, R100–R102. [CrossRef]

22. Parola, M.; Robino, G. Oxidative stress-related molecules and liver fibrosis. *J. Hepatol.* **2001**, *35*, 297–306. [CrossRef]

23. Fierro-Fernandez, M.; Miguel, V.; Lamas, S. Role of redoximiRs in fibrogenesis. *Redox Biol.* **2016**, *7*, 58–67. [CrossRef] [PubMed]

24. Sies, H. On the history of oxidative stress: Concept and some aspects of current development. *Curr. Opin. Toxicol.* **2018**, *7*, 122–126. [CrossRef]

25. Finkel, T. Signal transduction by reactive oxygen species. *J. Cell Biol.* **2011**, *194*, 7–15. [CrossRef]

26. Dickinson, B.C.; Chang, C.J. Chemistry and biology of reactive oxygen species in signaling or stress responses. *Nat. Chem. Biol.* **2011**, *7*, 504–511. [CrossRef]

27. Kumar, A.; Wu, H.; Collier-Hyams, L.S.; Hansen, J.M.; Li, T.; Yamoah, K.; Pan, Z.Q.; Jones, D.P.; Neish, A.S. Commensal bacteria modulate cullin-dependent signaling via generation of reactive oxygen species. *EMBO J.* **2007**, *26*, 4457–4466. [CrossRef]

28. Ha, E.M.; Oh, C.T.; Bae, Y.S.; Lee, W.J. A direct role for dual oxidase in Drosophila gut immunity. *Science* **2005**, *310*, 847–850. [CrossRef]

29. Jones, R.M.; Luo, L.; Ardita, C.S.; Richardson, A.N.; Kwon, Y.M.; Mercante, J.W.; Alam, A.; Gates, C.L.; Wu, H.; Swanson, P.A.; et al. Symbiotic lactobacilli stimulate gut epithelial proliferation via Nox-mediated generation of reactive oxygen species. *EMBO J.* **2013**, *32*, 3017–3028. [CrossRef]

30. Ceni, E.; Mello, T.; Galli, A. Pathogenesis of alcoholic liver disease: Role of oxidative metabolism. *World J. Gastroenterol.* **2014**, *20*, 17756–17772. [CrossRef]

31. Li, S.; Tan, H.Y.; Wang, N.; Zhang, Z.J.; Lao, L.; Wong, C.W.; Feng, Y. The Role of Oxidative Stress and Antioxidants in Liver Diseases. *Int. J. Mol. Sci.* **2015**, *16*, 26087–26124. [CrossRef]

32. Hayes, J.D.; Dinkova-Kostova, A.T. The Nrf2 regulatory network provides an interface between redox and intermediary metabolism. *Trends Biochem. Sci.* **2014**, *39*, 199–218. [CrossRef] [PubMed]

33. Bataller, R.; Brenner, D.A. Liver fibrosis. *J. Clin. Investig.* **2005**, *115*, 209–218. [CrossRef] [PubMed]

34. Francoz, C.; Belghiti, J.; Durand, F. Indications of liver transplantation in patients with complications of cirrhosis. *Best Pract. Res. Clin. Gastroenterol.* **2007**, *21*, 175–190. [CrossRef] [PubMed]

35. Poli, G. Pathogenesis of liver fibrosis: Role of oxidative stress. *Mol. Asp. Med.* **2000**, *21*, 49–98. [CrossRef]

36. Richter, K.; Kietzmann, T. Reactive oxygen species and fibrosis: Further evidence of a significant liaison. *Cell Tissue Res.* **2016**, *365*, 591–605. [CrossRef]

37. Meroni, M.; Longo, M.; Rametta, R.; Dongiovanni, P. Genetic and Epigenetic Modifiers of Alcoholic Liver Disease. *Int. J. Mol. Sci.* **2018**, *19*, 3857. [CrossRef]

38. Bala, S.; Marcos, M.; Kodys, K.; Csak, T.; Catalano, D.; Mandrekar, P.; Szabo, G. Up-regulation of microRNA-155 in macrophages contributes to increased tumor necrosis factor α (TNFα) production via increased mRNA half-life in alcoholic liver disease. *J. Biol. Chem.* **2011**, *286*, 1436–1444. [CrossRef]

39. Hritz, I.; Mandrekar, P.; Velayudham, A.; Catalano, D.; Dolganiuc, A.; Kodys, K.; Kurt-Jones, E.; Szabo, G. The critical role of toll-like receptor (TLR) 4 in alcoholic liver disease is independent of the common TLR adapter MyD88. *Hepatology* **2008**, *48*, 1224–1231. [CrossRef]

40. Bala, S.; Csak, T.; Saha, B.; Zatsiorsky, J.; Kodys, K.; Catalano, D.; Satishchandran, A.; Szabo, G. The pro-inflammatory effects of miR-155 promote liver fibrosis and alcohol-induced steatohepatitis. *J. Hepatol.* **2016**, *64*, 1378–1387. [CrossRef]

41. Saikia, P.; Bellos, D.; McMullen, M.R.; Pollard, K.A.; de la Motte, C.; Nagy, L.E. MicroRNA 181b-3p and its target importin alpha5 regulate toll-like receptor 4 signaling in Kupffer cells and liver injury in mice in response to ethanol. *Hepatology* **2017**, *66*, 602–615. [CrossRef]

42. Saikia, P.; Roychowdhury, S.; Bellos, D.; Pollard, K.A.; McMullen, M.R.; McCullough, R.L.; McCullough, A.J.; Gholam, P.; de la Motte, C.; Nagy, L.E. Hyaluronic acid 35 normalizes TLR4 signaling in Kupffer cells from ethanol-fed rats via regulation of microRNA291b and its target Tollip. *Sci. Rep.* **2017**, *7*, 15671. [CrossRef] [PubMed]

43. Raver-Shapira, N.; Marciano, E.; Meiri, E.; Spector, Y.; Rosenfeld, N.; Moskovits, N.; Bentwich, Z.; Oren, M. Transcriptional activation of miR-34a contributes to p53-mediated apoptosis. *Mol. Cell* **2007**, *26*, 731–743. [CrossRef] [PubMed]

44. Lee, J.; Padhye, A.; Sharma, A.; Song, G.; Miao, J.; Mo, Y.Y.; Wang, L.; Kemper, J.K. A pathway involving farnesoid X receptor and small heterodimer partner positively regulates hepatic sirtuin 1 levels via microRNA-34a inhibition. *J. Biol. Chem.* **2010**, *285*, 12604–12611. [CrossRef] [PubMed]

45. Yin, H.; Liang, X.; Jogasuria, A.; Davidson, N.O.; You, M. miR-217 regulates ethanol-induced hepatic inflammation by disrupting sirtuin 1-lipin-1 signaling. *Am. J. Pathol.* **2015**, *185*, 1286–1296. [CrossRef]

46. Satishchandran, A.; Ambade, A.; Rao, S.; Hsueh, Y.C.; Iracheta-Vellve, A.; Tornai, D.; Lowe, P.; Gyongyosi, B.; Li, J.; Catalano, D.; et al. MicroRNA 122, Regulated by GRLH2, Protects Livers of Mice and Patients from Ethanol-Induced Liver Disease. *Gastroenterology* **2018**, *154*, 238–252. [CrossRef]

47. Li, F.; Duan, K.; Wang, C.; McClain, C.; Feng, W. Probiotics and Alcoholic Liver Disease: Treatment and Potential Mechanisms. *Gastroenterol. Res. Pract.* **2016**, *2016*, 5491465. [CrossRef]

48. Szabo, G.; Bala, S.; Petrasek, J.; Gattu, A. Gut-liver axis and sensing microbes. *Dig. Dis.* **2010**, *28*, 737–744. [CrossRef]

49. Ng, R.; Song, G.; Roll, G.R.; Frandsen, N.M.; Willenbring, H. A microRNA-21 surge facilitates rapid cyclin D1 translation and cell cycle progression in mouse liver regeneration. *J. Clin. Investig.* **2012**, *122*, 1097–1108. [CrossRef]

50. Francis, H.; McDaniel, K.; Han, Y.; Liu, X.; Kennedy, L.; Yang, F.; McCarra, J.; Zhou, T.; Glaser, S.; Venter, J.; et al. Regulation of the extrinsic apoptotic pathway by microRNA-21 in alcoholic liver injury. *J. Biol. Chem.* **2014**, *289*, 27526–27539. [CrossRef]

51. Xu, T.; Li, L.; Hu, H.Q.; Meng, X.M.; Huang, C.; Zhang, L.; Qin, J.; Li, J. MicroRNAs in alcoholic liver disease: Recent advances and future applications. *J. Cell. Physiol.* **2018**, *234*, 382–394. [CrossRef]

52. Yeligar, S.; Tsukamoto, H.; Kalra, V.K. Ethanol-induced expression of ET-1 and ET-BR in liver sinusoidal endothelial cells and human endothelial cells involves hypoxia-inducible factor-1alpha and microrNA-199. *J. Immunol.* **2009**, *183*, 5232–5243. [CrossRef] [PubMed]

53. Dong, X.; Liu, H.; Chen, F.; Li, D.; Zhao, Y. MiR-214 promotes the alcohol-induced oxidative stress via down-regulation of glutathione reductase and cytochrome P450 oxidoreductase in liver cells. *Alcohol. Clin. Exp. Res.* **2014**, *38*, 68–77. [CrossRef] [PubMed]

54. Wang, Y.; Yu, D.; Tolleson, W.H.; Yu, L.R.; Green, B.; Zeng, L.; Chen, Y.; Chen, S.; Ren, Z.; Guo, L.; et al. A systematic evaluation of microRNAs in regulating human hepatic CYP2E1. *Biochem. Pharmacol.* **2017**, *138*, 174–184. [CrossRef] [PubMed]

55. Araujo, A.R.; Rosso, N.; Bedogni, G.; Tiribelli, C.; Bellentani, S. Global epidemiology of non-alcoholic fatty liver disease/non-alcoholic steatohepatitis: What we need in the future. *Liver Int.* **2018**, *38*, 47–51. [CrossRef]

56. Maher, J.J.; Leon, P.; Ryan, J.C. Beyond insulin resistance: Innate immunity in nonalcoholic steatohepatitis. *Hepatology* **2008**, *48*, 670–678. [CrossRef]

57. Wruck, W.; Graffmann, N.; Kawala, M.A.; Adjaye, J. Concise Review: Current Status and Future Directions on Research Related to Nonalcoholic Fatty Liver Disease. *Stem Cells* **2017**, *35*, 89–96. [CrossRef]

58. Manne, V.; Handa, P.; Kowdley, K.V. Pathophysiology of Nonalcoholic Fatty Liver Disease/Nonalcoholic Steatohepatitis. *Clin. Liver Dis.* **2018**, *22*, 23–37. [CrossRef]

59. Magee, N.; Zou, A.; Zhang, Y. Pathogenesis of Nonalcoholic Steatohepatitis: Interactions between Liver Parenchymal and Nonparenchymal Cells. *Biomed. Res. Int.* **2016**, *2016*, 5170402. [CrossRef]

60. Begriche, K.; Massart, J.; Robin, M.A.; Bonnet, F.; Fromenty, B. Mitochondrial adaptations and dysfunctions in nonalcoholic fatty liver disease. *Hepatology* **2013**, *58*, 1497–1507. [CrossRef]

61. Hu, F.; Liu, F. Mitochondrial stress: A bridge between mitochondrial dysfunction and metabolic diseases? *Cell. Signal.* **2011**, *23*, 1528–1533. [CrossRef]

62. Koliaki, C.; Szendroedi, J.; Kaul, K.; Jelenik, T.; Nowotny, P.; Jankowiak, F.; Herder, C.; Carstensen, M.; Krausch, M.; Knoefel, W.T.; et al. Adaptation of hepatic mitochondrial function in humans with non-alcoholic fatty liver is lost in steatohepatitis. *Cell Metab.* **2015**, *21*, 739–746. [CrossRef] [PubMed]

63. Ashraf, N.U.; Sheikh, T.A. Endoplasmic reticulum stress and Oxidative stress in the pathogenesis of Non-alcoholic fatty liver disease. *Free Radic. Res.* **2015**, *49*, 1405–1418. [CrossRef] [PubMed]

64. Kaufman, R.J. Stress signaling from the lumen of the endoplasmic reticulum: Coordination of gene transcriptional and translational controls. *Genes Dev.* **1999**, *13*, 1211–1233. [CrossRef] [PubMed]

65. Miyamoto, Y.; Mauer, A.S.; Kumar, S.; Mott, J.L.; Malhi, H. Mmu-miR-615-3p regulates lipoapoptosis by inhibiting C/EBP homologous protein. *PLoS ONE* **2014**, *9*, e109637. [CrossRef]

66. Park, S.W.; Zhou, Y.; Lee, J.; Lee, J.; Ozcan, U. Sarco(endo)plasmic reticulum Ca2+-ATPase 2b is a major regulator of endoplasmic reticulum stress and glucose homeostasis in obesity. *Proc. Natl. Acad. Sci. USA* **2010**, *107*, 19320–19325. [CrossRef]

67. Zhang, J.; Li, Y.; Jiang, S.; Yu, H.; An, W. Enhanced endoplasmic reticulum SERCA activity by overexpression of hepatic stimulator substance gene prevents hepatic cells from ER stress-induced apoptosis. *Am. J. Physiol. Cell Physiol.* **2014**, *306*, C279–C290. [CrossRef]

68. Chowdhry, S.; Nazmy, M.H.; Meakin, P.J.; Dinkova-Kostova, A.T.; Walsh, S.V.; Tsujita, T.; Dillon, J.F.; Ashford, M.L.; Hayes, J.D. Loss of Nrf2 markedly exacerbates nonalcoholic steatohepatitis. *Free Radic. Biol. Med.* **2010**, *48*, 357–371. [CrossRef]

69. Sugimoto, H.; Okada, K.; Shoda, J.; Warabi, E.; Ishige, K.; Ueda, T.; Taguchi, K.; Yanagawa, T.; Nakahara, A.; Hyodo, I.; et al. Deletion of nuclear factor-E2-related factor-2 leads to rapid onset and progression of nutritional steatohepatitis in mice. *Am. J. Physiol. Gastrointest Liver Physiol.* **2010**, *298*, G283–G294. [CrossRef] [PubMed]

70. Zhan, S.S.; Jiang, J.X.; Wu, J.; Halsted, C.; Friedman, S.L.; Zern, M.A.; Torok, N.J. Phagocytosis of apoptotic bodies by hepatic stellate cells induces NADPH oxidase and is associated with liver fibrosis in vivo. *Hepatology* **2006**, *43*, 435–443. [CrossRef]

71. Gallego-Duran, R.; Romero-Gomez, M. Epigenetic mechanisms in non-alcoholic fatty liver disease: An emerging field. *World J. Hepatol.* **2015**, *7*, 2497–2502. [CrossRef]

72. Dattaroy, D.; Pourhoseini, S.; Das, S.; Alhasson, F.; Seth, R.K.; Nagarkatti, M.; Michelotti, G.A.; Diehl, A.M.; Chatterjee, S. Micro-RNA 21 inhibition of SMAD7 enhances fibrogenesis via leptin-mediated NADPH oxidase in experimental and human nonalcoholic steatohepatitis. *Am. J. Physiol. Gastrointest Liver Physiol.* **2015**, *308*, G298–G312. [CrossRef] [PubMed]

73. Vinciguerra, M.; Sgroi, A.; Veyrat-Durebex, C.; Rubbia-Brandt, L.; Buhler, L.H.; Foti, M. Unsaturated fatty acids inhibit the expression of tumor suppressor phosphatase and tensin homolog (PTEN) via microRNA-21 up-regulation in hepatocytes. *Hepatology* **2009**, *49*, 1176–1184. [CrossRef] [PubMed]

74. Rodrigues, P.M.; Afonso, M.B.; Simao, A.L.; Carvalho, C.C.; Trindade, A.; Duarte, A.; Borralho, P.M.; Machado, M.V.; Cortez-Pinto, H.; Rodrigues, C.M.; et al. miR-21 ablation and obeticholic acid ameliorate nonalcoholic steatohepatitis in mice. *Cell Death Dis.* **2017**, *8*, e2748. [CrossRef] [PubMed]

75. Loyer, X.; Paradis, V.; Henique, C.; Vion, A.C.; Colnot, N.; Guerin, C.L.; Devue, C.; On, S.; Scetbun, J.; Romain, M.; et al. Liver microRNA-21 is overexpressed in non-alcoholic steatohepatitis and contributes to the disease in experimental models by inhibiting PPARalpha expression. *Gut* **2016**, *65*, 1882–1894. [CrossRef]

76. Sulaiman, S.A.; Muhsin, N.I.A.; Jamal, R. Regulatory Non-coding RNAs Network in Non-alcoholic Fatty Liver Disease. *Front. Physiol.* **2019**, *10*, 279. [CrossRef]

77. Brandt, S.; Roos, J.; Inzaghi, E.; Kotnik, P.; Kovac, J.; Battelino, T.; Cianfarani, S.; Nobili, V.; Colajacomo, M.; Kratzer, W.; et al. Circulating levels of miR-122 and nonalcoholic fatty liver disease in pre-pubertal obese children. *Pediatr. Obes.* **2018**, *13*, 175–182. [CrossRef]

78. Latorre, J.; Moreno-Navarrete, J.M.; Mercader, J.M.; Sabater, M.; Rovira, O.; Girones, J.; Ricart, W.; Fernandez-Real, J.M.; Ortega, F.J. Decreased lipid metabolism but increased FA biosynthesis are coupled with changes in liver microRNAs in obese subjects with NAFLD. *Int. J. Obes. (Lond.)* **2017**, *41*, 620–630. [CrossRef]

79. Su, Q.; Kumar, V.; Sud, N.; Mahato, R.I. MicroRNAs in the pathogenesis and treatment of progressive liver injury in NAFLD and liver fibrosis. *Adv. Drug Deliv. Rev.* **2018**, *129*, 54–63. [CrossRef]

80. Cermelli, S.; Ruggieri, A.; Marrero, J.A.; Ioannou, G.N.; Beretta, L. Circulating microRNAs in patients with chronic hepatitis C and non-alcoholic fatty liver disease. *PLoS ONE* **2011**, *6*, e23937. [CrossRef]

81. Miyaaki, H.; Ichikawa, T.; Kamo, Y.; Taura, N.; Honda, T.; Shibata, H.; Milazzo, M.; Fornari, F.; Gramantieri, L.; Bolondi, L.; et al. Significance of serum and hepatic microRNA-122 levels in patients with non-alcoholic fatty liver disease. *Liver Int.* **2014**, *34*, e302–e307. [CrossRef]

82. Yu, F.; Jiang, Z.; Chen, B.; Dong, P.; Zheng, J. NEAT1 accelerates the progression of liver fibrosis via regulation of microRNA-122 and Kruppel-like factor 6. *J. Mol. Med. (Berl.)* **2017**, *95*, 1191–1202. [CrossRef] [PubMed]

83. Hsu, S.H.; Wang, B.; Kota, J.; Yu, J.; Costinean, S.; Kutay, H.; Yu, L.; Bai, S.; La Perle, K.; Chivukula, R.R.; et al. Essential metabolic, anti-inflammatory, and anti-tumorigenic functions of miR-122 in liver. *J. Clin. Investig.* **2012**, *122*, 2871–2883. [CrossRef] [PubMed]

84. Salvoza, N.C.; Klinzing, D.C.; Gopez-Cervantes, J.; Baclig, M.O. Association of Circulating Serum miR-34a and miR-122 with Dyslipidemia among Patients with Non-Alcoholic Fatty Liver Disease. *PLoS ONE* **2016**, *11*, e0153497. [CrossRef] [PubMed]

85. Yamada, H.; Suzuki, K.; Ichino, N.; Ando, Y.; Sawada, A.; Osakabe, K.; Sugimoto, K.; Ohashi, K.; Teradaira, R.; Inoue, T.; et al. Associations between circulating microRNAs (miR-21, miR-34a, miR-122 and miR-451) and non-alcoholic fatty liver. *Clin. Chim. Acta* **2013**, *424*, 99–103. [CrossRef]

86. Chang, T.C.; Wentzel, E.A.; Kent, O.A.; Ramachandran, K.; Mullendore, M.; Lee, K.H.; Feldmann, G.; Yamakuchi, M.; Ferlito, M.; Lowenstein, C.J.; et al. Transactivation of miR-34a by p53 broadly influences gene expression and promotes apoptosis. *Mol. Cell* **2007**, *26*, 745–752. [CrossRef]

87. Ding, J.; Li, M.; Wan, X.; Jin, X.; Chen, S.; Yu, C.; Li, Y. Effect of miR-34a in regulating steatosis by targeting PPARalpha expression in nonalcoholic fatty liver disease. *Sci. Rep.* **2015**, *5*, 13729. [CrossRef]

88. Shan, W.; Gao, L.; Zeng, W.; Hu, Y.; Wang, G.; Li, M.; Zhou, J.; Ma, X.; Tian, X.; Yao, J. Activation of the SIRT1/p66shc antiapoptosis pathway via carnosic acid-induced inhibition of miR-34a protects rats against nonalcoholic fatty liver disease. *Cell Death Dis.* **2015**, *6*, e1833. [CrossRef]

89. Xu, Y.; Zalzala, M.; Xu, J.; Li, Y.; Yin, L.; Zhang, Y. A metabolic stress-inducible miR-34a-HNF4alpha pathway regulates lipid and lipoprotein metabolism. *Nat. Commun.* **2015**, *6*, 7466. [CrossRef]

90. Castro, R.E.; Ferreira, D.M.; Afonso, M.B.; Borralho, P.M.; Machado, M.V.; Cortez-Pinto, H.; Rodrigues, C.M. miR-34a/SIRT1/p53 is suppressed by ursodeoxycholic acid in the rat liver and activated by disease severity in human non-alcoholic fatty liver disease. *J. Hepatol.* **2013**, *58*, 119–125. [CrossRef]

91. Derdak, Z.; Villegas, K.A.; Harb, R.; Wu, A.M.; Sousa, A.; Wands, J.R. Inhibition of p53 attenuates steatosis and liver injury in a mouse model of non-alcoholic fatty liver disease. *J. Hepatol.* **2013**, *58*, 785–791. [CrossRef]

92. Braza-Boils, A.; Mari-Alexandre, J.; Molina, P.; Arnau, M.A.; Barcelo-Molina, M.; Domingo, D.; Girbes, J.; Giner, J.; Martinez-Dolz, L.; Zorio, E. Deregulated hepatic microRNAs underlie the association between non-alcoholic fatty liver disease and coronary artery disease. *Liver Int.* **2016**, *36*, 1221–1229. [CrossRef] [PubMed]

93. Roy, S.; Benz, F.; Luedde, T.; Roderburg, C. The role of miRNAs in the regulation of inflammatory processes during hepatofibrogenesis. *Hepatobiliary Surg. Nutr.* **2015**, *4*, 24–33. [PubMed]

94. Liu, M.X.; Gao, M.; Li, C.Z.; Yu, C.Z.; Yan, H.; Peng, C.; Li, Y.; Li, C.G.; Ma, Z.L.; Zhao, Y.; et al. Dicer1/miR-29/HMGCR axis contributes to hepatic free cholesterol accumulation in mouse non-alcoholic steatohepatitis. *Acta Pharmacol. Sin.* **2017**, *38*, 660–671. [CrossRef] [PubMed]

95. Mattis, A.N.; Song, G.; Hitchner, K.; Kim, R.Y.; Lee, A.Y.; Sharma, A.D.; Malato, Y.; McManus, M.T.; Esau, C.C.; Koller, E.; et al. A screen in mice uncovers repression of lipoprotein lipase by microRNA-29a as a mechanism for lipid distribution away from the liver. *Hepatology* **2015**, *61*, 141–152. [CrossRef] [PubMed]

96. Kurtz, C.L.; Fannin, E.E.; Toth, C.L.; Pearson, D.S.; Vickers, K.C.; Sethupathy, P. Inhibition of miR-29 has a significant lipid-lowering benefit through suppression of lipogenic programs in liver. *Sci. Rep.* **2015**, *5*, 12911. [CrossRef]

97. Bandyopadhyay, S.; Friedman, R.C.; Marquez, R.T.; Keck, K.; Kong, B.; Icardi, M.S.; Brown, K.E.; Burge, C.B.; Schmidt, W.N.; Wang, Y.; et al. Hepatitis C virus infection and hepatic stellate cell activation downregulate miR-29: miR-29 overexpression reduces hepatitis C viral abundance in culture. *J. Infect. Dis.* **2011**, *203*, 1753–1762. [CrossRef]

98. Pogribny, I.P.; Starlard-Davenport, A.; Tryndyak, V.P.; Han, T.; Ross, S.A.; Rusyn, I.; Beland, F.A. Difference in expression of hepatic microRNAs miR-29c, miR-34a, miR-155, and miR-200b is associated with strain-specific susceptibility to dietary nonalcoholic steatohepatitis in mice. *Lab. Investig.* **2010**, *90*, 1437–1446. [CrossRef]

99. Zhu, L.; Ren, T.; Zhu, Z.; Cheng, M.; Mou, Q.; Mu, M.; Liu, Y.; Yao, Y.; Cheng, Y.; Zhang, B.; et al. Thymosin-beta4 Mediates Hepatic Stellate Cell Activation by Interfering with CircRNA-0067835/miR-155/FoxO3 Signaling Pathway. *Cell. Physiol. Biochem.* **2018**, *51*, 1389–1398. [CrossRef]

100. Wang, L.; Zhang, N.; Wang, Z.; Ai, D.M.; Cao, Z.Y.; Pan, H.P. Decreased MiR-155 Level in the Peripheral Blood of Non-Alcoholic Fatty Liver Disease Patients may Serve as a Biomarker and may Influence LXR Activity. *Cell. Physiol. Biochem.* **2016**, *39*, 2239–2248. [CrossRef]

101. Csak, T.; Bala, S.; Lippai, D.; Satishchandran, A.; Catalano, D.; Kodys, K.; Szabo, G. microRNA-122 regulates hypoxia-inducible factor-1 and vimentin in hepatocytes and correlates with fibrosis in diet-induced steatohepatitis. *Liver Int.* **2015**, *35*, 532–541. [CrossRef]

102. Csak, T.; Bala, S.; Lippai, D.; Kodys, K.; Catalano, D.; Iracheta-Vellve, A.; Szabo, G. MicroRNA-155 Deficiency Attenuates Liver Steatosis and Fibrosis without Reducing Inflammation in a Mouse Model of Steatohepatitis. *PLoS ONE* **2015**, *10*, e0129251. [CrossRef] [PubMed]

103. Okamoto, K.; Koda, M.; Okamoto, T.; Onoyama, T.; Miyoshi, K.; Kishina, M.; Kato, J.; Tokunaga, S.; Sugihara, T.A.; Hara, Y.; et al. A Series of microRNA in the Chromosome 14q32.2 Maternally Imprinted Region Related to Progression of Non-Alcoholic Fatty Liver Disease in a Mouse Model. *PLoS ONE* **2016**, *11*, e0154676. [CrossRef] [PubMed]

104. Behrman, S.; Acosta-Alvear, D.; Walter, P. A CHOP-regulated microRNA controls rhodopsin expression. *J. Cell Biol.* **2011**, *192*, 919–927. [CrossRef] [PubMed]

105. Chitnis, N.S.; Pytel, D.; Bobrovnikova-Marjon, E.; Pant, D.; Zheng, H.; Maas, N.L.; Frederick, B.; Kushner, J.A.; Chodosh, L.A.; Koumenis, C.; et al. miR-211 is a prosurvival microRNA that regulates chop expression in a PERK-dependent manner. *Mol. Cell* **2012**, *48*, 353–364. [CrossRef] [PubMed]

106. Zhang, W.G.; Chen, L.; Dong, Q.; He, J.; Zhao, H.D.; Li, F.L.; Li, H. Mmu-miR-702 functions as an anti-apoptotic mirtron by mediating ATF6 inhibition in mice. *Gene* **2013**, *531*, 235–242. [CrossRef]

107. Belmont, P.J.; Chen, W.J.; Thuerauf, D.J.; Glembotski, C.C. Regulation of microRNA expression in the heart by the ATF6 branch of the ER stress response. *J. Mol. Cell Cardiol.* **2012**, *52*, 1176–1182. [CrossRef]

108. Malhi, H. Micrornas in er stress: Divergent roles in cell fate decisions. *Curr. Pathobiol. Rep.* **2014**, *2*, 117–122. [CrossRef]

109. WHO. Hepatitis C. Available online: https://www.who.int/news-room/fact-sheets/detail/hepatitis-c (accessed on 29 September 2019).

110. WHO. Hepatitis B. Available online: https://www.who.int/news-room/fact-sheets/detail/hepatitis-b (accessed on 29 September 2019).

111. Choi, J.; Ou, J.H. Mechanisms of liver injury. III. Oxidative stress in the pathogenesis of hepatitis C virus. *Am. J. Physiol. Gastrointest. Liver Physiol.* **2006**, *290*, G847–G851. [CrossRef]

112. Muriel, P. Role of free radicals in liver diseases. *Hepatol. Int.* **2009**, *3*, 526–536. [CrossRef]

113. Farinati, F.; Cardin, R.; De Maria, N.; Della Libera, G.; Marafin, C.; Lecis, E.; Burra, P.; Floreani, A.; Cecchetto, A.; Naccarato, R. Iron storage, lipid peroxidation and glutathione turnover in chronic anti-HCV positive hepatitis. *J. Hepatol.* **1995**, *22*, 449–456. [CrossRef]

114. Hou, W.; Tian, Q.; Zheng, J.; Bonkovsky, H.L. MicroRNA-196 represses Bach1 protein and hepatitis C virus gene expression in human hepatoma cells expressing hepatitis C viral proteins. *Hepatology* **2010**, *51*, 1494–1504. [CrossRef] [PubMed]

115. Jopling, C.L.; Yi, M.; Lancaster, A.M.; Lemon, S.M.; Sarnow, P. Modulation of hepatitis C virus RNA abundance by a liver-specific MicroRNA. *Science* **2005**, *309*, 1577–1581. [CrossRef]

116. Sarnow, P.; Sagan, S.M. Unraveling the Mysterious Interactions Between Hepatitis C Virus RNA and Liver-Specific MicroRNA-122. *Annu. Rev. Virol.* **2016**, *3*, 309–332. [CrossRef] [PubMed]

117. Lowey, B.; Hertz, L.; Chiu, S.; Valdez, K.; Li, Q.; Liang, T.J. Hepatitis C Virus Infection Induces Hepatic Expression of NF-kappaB-Inducing Kinase and Lipogenesis by Downregulating miR-122. *MBio* **2019**, *10*, e01617-19. [CrossRef]

118. Goh, G.Y.S.; Winter, J.J.; Bhanshali, F.; Doering, K.R.S.; Lai, R.; Lee, K.; Veal, E.A.; Taubert, S. NHR-49/HNF4 integrates regulation of fatty acid metabolism with a protective transcriptional response to oxidative stress and fasting. *Aging Cell* **2018**, *17*, e12743. [CrossRef] [PubMed]

119. Li, Q.; Pene, V.; Krishnamurthy, S.; Cha, H.; Liang, T.J. Hepatitis C virus infection activates an innate pathway involving IKK-alpha in lipogenesis and viral assembly. *Nat. Med.* **2013**, *19*, 722–729. [CrossRef]

120. Wang, S.; Qiu, L.; Yan, X.; Jin, W.; Wang, Y.; Chen, L.; Wu, E.; Ye, X.; Gao, G.F.; Wang, F.; et al. Loss of microRNA 122 expression in patients with hepatitis B enhances hepatitis B virus replication through cyclin G(1) -modulated P53 activity. *Hepatology* **2012**, *55*, 730–741. [CrossRef]

121. Wojcik, K.; Piekarska, A.; Szymanska, B.; Jablonowska, E. Hepatic expression of miR-122 and antioxidant genes in patients with chronic hepatitis B. *Acta Biochim. Pol.* **2016**, *63*, 527–531. [CrossRef]

122. Varnholt, H.; Drebber, U.; Schulze, F.; Wedemeyer, I.; Schirmacher, P.; Dienes, H.P.; Odenthal, M. MicroRNA gene expression profile of hepatitis C virus-associated hepatocellular carcinoma. *Hepatology* **2008**, *47*, 1223–1232. [CrossRef]

123. Bray, F.; Ferlay, J.; Soerjomataram, I.; Siegel, R.L.; Torre, L.A.; Jemal, A. Global cancer statistics 2018: GLOBOCAN estimates of incidence and mortality worldwide for 36 cancers in 185 countries. *CA Cancer J. Clin.* **2018**, *68*, 394–424. [CrossRef]

124. Petrick, J.L.; McGlynn, K.A. The changing epidemiology of primary liver cancer. *Curr. Epidemiol. Rep.* **2019**, *6*, 104–111. [CrossRef] [PubMed]

125. Anstee, Q.M.; Reeves, H.L.; Kotsiliti, E.; Govaere, O.; Heikenwalder, M. From NASH to HCC: Current concepts and future challenges. *Nat. Rev. Gastroenterol. Hepatol.* **2019**, *16*, 411–428. [CrossRef] [PubMed]

126. Neureiter, D.; Stintzing, S.; Kiesslich, T.; Ocker, M. Hepatocellular carcinoma: Therapeutic advances in signaling, epigenetic and immune targets. *World J. Gastroenterol.* **2019**, *25*, 3136–3150. [CrossRef] [PubMed]

127. Engedal, N.; Zerovnik, E.; Rudov, A.; Galli, F.; Olivieri, F.; Procopio, A.D.; Rippo, M.R.; Monsurro, V.; Betti, M.; Albertini, M.C. From Oxidative Stress Damage to Pathways, Networks, and Autophagy via MicroRNAs. *Oxid. Med. Cell. Longev.* **2018**, *2018*, 4968321. [CrossRef]

128. Mansouri, A.; Gattolliat, C.H.; Asselah, T. Mitochondrial Dysfunction and Signaling in Chronic Liver Diseases. *Gastroenterology* **2018**, *155*, 629–647. [CrossRef]

129. Masarone, M.; Rosato, V.; Dallio, M.; Gravina, A.G.; Aglitti, A.; Loguercio, C.; Federico, A.; Persico, M. Role of Oxidative Stress in Pathophysiology of Nonalcoholic Fatty Liver Disease. *Oxid. Med. Cell. Longev.* **2018**, *2018*, 9547613. [CrossRef]

130. He, Z.; Yu, Y.; Nong, Y.; Du, L.; Liu, C.; Cao, Y.; Bai, L.; Tang, H. Hepatitis B virus X protein promotes hepatocellular carcinoma invasion and metastasis via upregulating thioredoxin interacting protein. *Oncol. Lett.* **2017**, *14*, 1323–1332. [CrossRef]

131. Ivanov, A.V.; Bartosch, B.; Smirnova, O.A.; Isaguliants, M.G.; Kochetkov, S.N. HCV and oxidative stress in the liver. *Viruses* **2013**, *5*, 439–469. [CrossRef]

132. Medvedev, R.; Ploen, D.; Hildt, E. HCV and Oxidative Stress: Implications for HCV Life Cycle and HCV-Associated Pathogenesis. *Oxid. Med. Cell. Longev.* **2016**, *2016*, 9012580. [CrossRef]

133. Severi, T.; Vander Borght, S.; Libbrecht, L.; VanAelst, L.; Nevens, F.; Roskams, T.; Cassiman, D.; Fevery, J.; Verslype, C.; van Pelt, J.F. HBx or HCV core gene expression in HepG2 human liver cells results in a survival benefit against oxidative stress with possible implications for HCC development. *Chem. Biol. Interact.* **2007**, *168*, 128–134. [CrossRef]

134. Zhong, L.; Shu, W.; Dai, W.; Gao, B.; Xiong, S. Reactive Oxygen Species-Mediated c-Jun NH2-Terminal Kinase Activation Contributes to Hepatitis B Virus X Protein-Induced Autophagy via Regulation of the Beclin-1/Bcl-2 Interaction. *J. Virol.* **2017**, *91*, e00001-17. [CrossRef] [PubMed]

135. Marra, M.; Sordelli, I.M.; Lombardi, A.; Lamberti, M.; Tarantino, L.; Giudice, A.; Stiuso, P.; Abbruzzese, A.; Sperlongano, R.; Accardo, M.; et al. Molecular targets and oxidative stress biomarkers in hepatocellular carcinoma: An overview. *J. Transl. Med.* **2011**, *9*, 171. [CrossRef] [PubMed]

136. Tanaka, H.; Fujita, N.; Sugimoto, R.; Urawa, N.; Horiike, S.; Kobayashi, Y.; Iwasa, M.; Ma, N.; Kawanishi, S.; Watanabe, S.; et al. Hepatic oxidative DNA damage is associated with increased risk for hepatocellular carcinoma in chronic hepatitis C. *Br. J. Cancer* **2008**, *98*, 580–586. [CrossRef] [PubMed]

137. Li, S.; Wang, X.; Wu, Y.; Zhang, H.; Zhang, L.; Wang, C.; Zhang, R.; Guo, Z. 8-Hydroxy-2'-deoxyguanosine expression predicts hepatocellular carcinoma outcome. *Oncol. Lett.* **2012**, *3*, 338–342. [CrossRef] [PubMed]

138. Zhou, X.; Zhuang, Z.; Wang, W.; He, L.; Wu, H.; Cao, Y.; Pan, F.; Zhao, J.; Hu, Z.; Sekhar, C.; et al. OGG1 is essential in oxidative stress induced DNA demethylation. *Cell. Signal.* **2016**, *28*, 1163–1171. [CrossRef]

139. Sartorius, K.; Sartorius, B.; Winkler, C.; Chuturgoon, A.; Makarova, J. The biological and diagnostic role of miRNA's in hepatocellular carcinoma. *Front. Biosci.* **2018**, *23*, 1701–1720. [CrossRef]

140. Ji, J.; Shi, J.; Budhu, A.; Yu, Z.; Forgues, M.; Roessler, S.; Ambs, S.; Chen, Y.; Meltzer, P.S.; Croce, C.M.; et al. MicroRNA expression, survival, and response to interferon in liver cancer. *N. Engl. J. Med.* **2009**, *361*, 1437–1447. [CrossRef]

141. Wong, C.M.; Tsang, F.H.; Ng, I.O. Non-coding RNAs in hepatocellular carcinoma: Molecular functions and pathological implications. *Nat. Rev. Gastroenterol. Hepatol.* **2018**, *15*, 137–151. [CrossRef]

142. Deng, M.; Zhang, R.; He, Z.; Qiu, Q.; Lu, X.; Yin, J.; Liu, H.; Jia, X.; He, Z. TET-Mediated Sequestration of miR-26 Drives EZH2 Expression and Gastric Carcinogenesis. *Cancer Res.* **2017**, *77*, 6069–6082. [CrossRef]

143. Lu, J.; He, M.L.; Wang, L.; Chen, Y.; Liu, X.; Dong, Q.; Chen, Y.C.; Peng, Y.; Yao, K.T.; Kung, H.F.; et al. MiR-26a inhibits cell growth and tumorigenesis of nasopharyngeal carcinoma through repression of EZH2. *Cancer Res.* **2011**, *71*, 225–233. [CrossRef]

144. Lin, L.L.; Wang, W.; Hu, Z.; Wang, L.W.; Chang, J.; Qian, H. Erratum to: Negative feedback of miR-29 family TET1 involves in hepatocellular cancer. *Med. Oncol.* **2015**, *32*, 39. [CrossRef]

145. Chuang, K.H.; Whitney-Miller, C.L.; Chu, C.Y.; Zhou, Z.; Dokus, M.K.; Schmit, S.; Barry, C.T. MicroRNA-494 is a master epigenetic regulator of multiple invasion-suppressor microRNAs by targeting ten eleven translocation 1 in invasive human hepatocellular carcinoma tumors. *Hepatology* **2015**, *62*, 466–480. [CrossRef]

146. Zhang, W.; Lu, Z.; Gao, Y.; Ye, L.; Song, T.; Zhang, X. MiR-520b suppresses proliferation of hepatoma cells through targeting ten-eleven translocation 1 (TET1) mRNA. *Biochem. Biophys. Res. Commun.* **2015**, *460*, 793–798. [CrossRef]

147. Cardin, R.; Piciocchi, M.; Sinigaglia, A.; Lavezzo, E.; Bortolami, M.; Kotsafti, A.; Cillo, U.; Zanus, G.; Mescoli, C.; Rugge, M.; et al. Oxidative DNA damage correlates with cell immortalization and mir-92 expression in hepatocellular carcinoma. *BMC Cancer* **2012**, *12*, 177. [CrossRef]

148. Henrici, A.; Montalbano, R.; Neureiter, D.; Krause, M.; Stiewe, T.; Slater, E.P.; Quint, K.; Ocker, M.; Di Fazio, P. The pan-deacetylase inhibitor panobinostat suppresses the expression of oncogenic miRNAs in hepatocellular carcinoma cell lines. *Mol. Carcinog.* **2015**, *54*, 585–597. [CrossRef]

149. Wang, Z.; Liu, Y.; Han, N.; Chen, X.; Yu, W.; Zhang, W.; Zou, F. Profiles of oxidative stress-related microRNA and mRNA expression in auditory cells. *Brain Res.* **2010**, *1346*, 14–25. [CrossRef]

150. Lei, P.P.; Zhang, Z.J.; Shen, L.J.; Li, J.Y.; Zou, Q.; Zhang, H.X. Expression and hypermethylation of p27 kip1 in hepatocarcinogenesis. *World J. Gastroenterol.* **2005**, *11*, 4587–4591. [CrossRef]

151. Di Fazio, P.; Montalbano, R.; Neureiter, D.; Alinger, B.; Schmidt, A.; Merkel, A.L.; Quint, K.; Ocker, M. Downregulation of HMGA2 by the pan-deacetylase inhibitor panobinostat is dependent on hsa-let-7b expression in liver cancer cell lines. *Exp. Cell Res.* **2012**, *318*, 1832–1843. [CrossRef]

152. Pillon Barcelos, R.; Freire Royes, L.F.; Gonzalez-Gallego, J.; Bresciani, G. Oxidative stress and inflammation: Liver responses and adaptations to acute and regular exercise. *Free Radic. Res.* **2017**, *51*, 222–236. [CrossRef]

153. Farias, M.S.; Budni, P.; Ribeiro, C.M.; Parisotto, E.B.; Santos, C.E.; Dias, J.F.; Dalmarco, E.M.; Frode, T.S.; Pedrosa, R.C.; Wilhelm Filho, D. Antioxidant supplementation attenuates oxidative stress in chronic hepatitis C patients. *Gastroenterol. Hepatol.* **2012**, *35*, 386–394. [CrossRef]

154. Chakraborty, C.; Sharma, A.R.; Sharma, G.; Doss, C.G.P.; Lee, S.S. Therapeutic miRNA and siRNA: Moving from Bench to Clinic as Next Generation Medicine. *Mol. Ther. Nucleic Acids* **2017**, *8*, 132–143. [CrossRef]

155. Vienberg, S.; Geiger, J.; Madsen, S.; Dalgaard, L.T. MicroRNAs in metabolism. *Acta Physiol. (Oxf.)* **2017**, *219*, 346–361. [CrossRef] [PubMed]

156. Elmen, J.; Lindow, M.; Schutz, S.; Lawrence, M.; Petri, A.; Obad, S.; Lindholm, M.; Hedtjarn, M.; Hansen, H.F.; Berger, U.; et al. LNA-mediated microRNA silencing in non-human primates. *Nature* **2008**, *452*, 896–899. [CrossRef] [PubMed]

157. Petri, A.; Lindow, M.; Kauppinen, S. MicroRNA silencing in primates: Towards development of novel therapeutics. *Cancer Res.* **2009**, *69*, 393–395. [CrossRef] [PubMed]

158. Huang, Y. Preclinical and Clinical Advances of GalNAc-Decorated Nucleic Acid Therapeutics. *Mol. Ther. Nucleic Acids* **2017**, *6*, 116–132. [CrossRef]

159. Huang, X.; Le, Q.T.; Giaccia, A.J. MiR-210–micromanager of the hypoxia pathway. *Trends Mol. Med.* **2010**, *16*, 230–237. [CrossRef]

160. Tsuchiya, S.; Fujiwara, T.; Sato, F.; Shimada, Y.; Tanaka, E.; Sakai, Y.; Shimizu, K.; Tsujimoto, G. MicroRNA-210 regulates cancer cell proliferation through targeting fibroblast growth factor receptor-like 1 (FGFRL1). *J. Biol. Chem.* **2011**, *286*, 420–428. [CrossRef]

161. Puissegur, M.P.; Mazure, N.M.; Bertero, T.; Pradelli, L.; Grosso, S.; Robbe-Sermesant, K.; Maurin, T.; Lebrigand, K.; Cardinaud, B.; Hofman, V.; et al. miR-210 is overexpressed in late stages of lung cancer and mediates mitochondrial alterations associated with modulation of HIF-1 activity. *Cell Death Differ.* **2011**, *18*, 465–478. [CrossRef]

162. Devlin, C.; Greco, S.; Martelli, F.; Ivan, M. miR-210: More than a silent player in hypoxia. *IUBMB Life* **2011**, *63*, 94–100. [CrossRef]

163. Huang, X.; Ding, L.; Bennewith, K.L.; Tong, R.T.; Welford, S.M.; Ang, K.K.; Story, M.; Le, Q.T.; Giaccia, A.J. Hypoxia-inducible mir-210 regulates normoxic gene expression involved in tumor initiation. *Mol. Cell* **2009**, *35*, 856–867. [CrossRef]

164. Momen-Heravi, F.; Saha, B.; Kodys, K.; Catalano, D.; Satishchandran, A.; Szabo, G. Increased number of circulating exosomes and their microRNA cargos are potential novel biomarkers in alcoholic hepatitis. *J. Transl. Med.* **2015**, *13*, 261. [CrossRef]

165. Trajkovski, M.; Hausser, J.; Soutschek, J.; Bhat, B.; Akin, A.; Zavolan, M.; Heim, M.H.; Stoffel, M. MicroRNAs 103 and 107 regulate insulin sensitivity. *Nature* **2011**, *474*, 649–653. [CrossRef]

166. Bala, S.; Petrasek, J.; Mundkur, S.; Catalano, D.; Levin, I.; Ward, J.; Alao, H.; Kodys, K.; Szabo, G. Circulating microRNAs in exosomes indicate hepatocyte injury and inflammation in alcoholic, drug-induced, and inflammatory liver diseases. *Hepatology* **2012**, *56*, 1946–1957. [CrossRef]

167. Janssen, H.L.; Reesink, H.W.; Lawitz, E.J.; Zeuzem, S.; Rodriguez-Torres, M.; Patel, K.; van der Meer, A.J.; Patick, A.K.; Chen, A.; Zhou, Y.; et al. Treatment of HCV infection by targeting microRNA. *N. Engl. J. Med.* **2013**, *368*, 1685–1694. [CrossRef]

168. Gebert, L.F.; Rebhan, M.A.; Crivelli, S.E.; Denzler, R.; Stoffel, M.; Hall, J. Miravirsen (SPC3649) can inhibit the biogenesis of miR-122. *Nucleic Acids Res.* **2014**, *42*, 609–621. [CrossRef]

169. Streba, L.A.; Vere, C.C.; Rogoveanu, I.; Streba, C.T. Nonalcoholic fatty liver disease, metabolic risk factors, and hepatocellular carcinoma: An open question. *World J. Gastroenterol.* **2015**, *21*, 4103–4110. [CrossRef]

170. European Association for the Study of the Liver; European Association for the Study of Diabetes. EASL-EASD-EASO Clinical Practice Guidelines for the management of non-alcoholic fatty liver disease. *J. Hepatol.* **2016**, *64*, 1388–1402. [CrossRef]

171. Ardekani, A.M.; Naeini, M.M. The Role of MicroRNAs in Human Diseases. *Avicenna J. Med. Biotechnol.* **2010**, *2*, 161–179.

172. Liu, X.; Huang, K.; Niu, Z.; Mei, D.; Zhang, B. Protective effect of isochlorogenic acid B on liver fibrosis in non-alcoholic steatohepatitis of mice. *Basic Clin. Pharmacol. Toxicol.* **2019**, *124*, 144–153. [CrossRef]

173. Kumar, V.; Mondal, G.; Dutta, R.; Mahato, R.I. Co-delivery of small molecule hedgehog inhibitor and miRNA for treating liver fibrosis. *Biomaterials* **2016**, *76*, 144–156. [CrossRef]

174. Maurer, B.; Stanczyk, J.; Jungel, A.; Akhmetshina, A.; Trenkmann, M.; Brock, M.; Kowal-Bielecka, O.; Gay, R.E.; Michel, B.A.; Distler, J.H.; et al. MicroRNA-29, a key regulator of collagen expression in systemic sclerosis. *Arthritis Rheum.* **2010**, *62*, 1733–1743. [CrossRef] [PubMed]

175. Okada, H.; Honda, M.; Campbell, J.S.; Takegoshi, K.; Sakai, Y.; Yamashita, T.; Shirasaki, T.; Takabatake, R.; Nakamura, M.; Tanaka, T.; et al. Inhibition of microRNA-214 ameliorates hepatic fibrosis and tumor incidence in platelet-derived growth factor C transgenic mice. *Cancer Sci.* **2015**, *106*, 1143–1152. [CrossRef] [PubMed]

176. Dai, L.L.; Li, S.D.; Ma, Y.C.; Tang, J.R.; Lv, J.Y.; Zhang, Y.Q.; Miao, Y.L.; Ma, Y.Q.; Li, C.M.; Chu, Y.Y.; et al. MicroRNA-30b regulates insulin sensitivity by targeting SERCA2b in non-alcoholic fatty liver disease. *Liver Int.* **2019**, *39*, 1504–1513. [CrossRef] [PubMed]

177. Li, J.; Cheng, Z.J.; Liu, Y.; Yan, Z.L.; Wang, K.; Wu, D.; Wan, X.Y.; Xia, Y.; Lau, W.Y.; Wu, M.C.; et al. Serum thioredoxin is a diagnostic marker for hepatocellular carcinoma. *Oncotarget* **2015**, *6*, 9551–9563. [CrossRef]

178. Mollbrink, A.; Jawad, R.; Vlamis-Gardikas, A.; Edenvik, P.; Isaksson, B.; Danielsson, O.; Stal, P.; Fernandes, A.P. Expression of thioredoxins and glutaredoxins in human hepatocellular carcinoma: Correlation to cell proliferation, tumor size and metabolic syndrome. *Int. J. Immunopathol. Pharmacol.* **2014**, *27*, 169–183. [CrossRef]

179. Tamai, T.; Uto, H.; Takami, Y.; Oda, K.; Saishoji, A.; Hashiguchi, M.; Kumagai, K.; Kure, T.; Mawatari, S.; Moriuchi, A.; et al. Serum manganese superoxide dismutase and thioredoxin are potential prognostic markers for hepatitis C virus-related hepatocellular carcinoma. *World J. Gastroenterol.* **2011**, *17*, 4890–4898. [CrossRef]

180. Valavanidis, A.; Vlachogianni, T.; Fiotakis, K.; Loridas, S. Pulmonary oxidative stress, inflammation and cancer: Respirable particulate matter, fibrous dusts and ozone as major causes of lung carcinogenesis through reactive oxygen species mechanisms. *Int. J. Environ. Res. Public Health* **2013**, *10*, 3886–3907. [CrossRef]

181. Chuma, M.; Hige, S.; Nakanishi, M.; Ogawa, K.; Natsuizaka, M.; Yamamoto, Y.; Asaka, M. 8-Hydroxy-2′-deoxy-guanosine is a risk factor for development of hepatocellular carcinoma in patients with chronic hepatitis C virus infection. *J. Gastroenterol. Hepatol.* **2008**, *23*, 1431–1436. [CrossRef]

182. Ichiba, M.; Maeta, Y.; Mukoyama, T.; Saeki, T.; Yasui, S.; Kanbe, T.; Okano, J.; Tanabe, Y.; Hirooka, Y.; Yamada, S.; et al. Expression of 8-hydroxy-2′-deoxyguanosine in chronic liver disease and hepatocellular carcinoma. *Liver Int.* **2003**, *23*, 338–345. [CrossRef]

183. Ma-On, C.; Sanpavat, A.; Whongsiri, P.; Suwannasin, S.; Hirankarn, N.; Tangkijvanich, P.; Boonla, C. Oxidative stress indicated by elevated expression of Nrf2 and 8-OHdG promotes hepatocellular carcinoma progression. *Med. Oncol.* **2017**, *34*, 57. [CrossRef]

184. Tanaka, S.; Miyanishi, K.; Kobune, M.; Kawano, Y.; Hoki, T.; Kubo, T.; Hayashi, T.; Sato, T.; Sato, Y.; Takimoto, R.; et al. Increased hepatic oxidative DNA damage in patients with nonalcoholic steatohepatitis who develop hepatocellular carcinoma. *J. Gastroenterol.* **2013**, *48*, 1249–1258. [CrossRef] [PubMed]

185. Jopling, C. Liver-specific microRNA-122: Biogenesis and function. *RNA Biol.* **2012**, *9*, 137–142. [CrossRef] [PubMed]

186. Kilic, M.; Kasperczyk, H.; Fulda, S.; Debatin, K.M. Role of hypoxia inducible factor-1 alpha in modulation of apoptosis resistance. *Oncogene* **2007**, *26*, 2027–2038. [CrossRef] [PubMed]

187. Fu, H.; Zhang, X.; Wang, Q.; Sun, Y.; Liu, L.; Huang, L.; Ding, L.; Shen, M.; Zhang, L.; Duan, Y. Simple and rational design of a polymer nano-platform for high performance of HCV related miR-122 reduction in the liver. *Biomater Sci.* **2018**, *6*, 2667–2680. [CrossRef]

188. Fu, B.; Meng, W.; Zeng, X.; Zhao, H.; Liu, W.; Zhang, T. TXNRD1 Is an Unfavorable Prognostic Factor for Patients with Hepatocellular Carcinoma. *Biomed. Res. Int.* **2017**, *2017*, 4698167. [CrossRef]

189. Lee, D.; Xu, I.M.; Chiu, D.K.; Leibold, J.; Tse, A.P.; Bao, M.H.; Yuen, V.W.; Chan, C.Y.; Lai, R.K.; Chin, D.W.; et al. Induction of Oxidative Stress Through Inhibition of Thioredoxin Reductase 1 Is an Effective Therapeutic Approach for Hepatocellular Carcinoma. *Hepatology* **2019**, *69*, 1768–1786. [CrossRef]

190. Liu, L.; Cao, Y.; Chen, C.; Zhang, X.; McNabola, A.; Wilkie, D.; Wilhelm, S.; Lynch, M.; Carter, C. Sorafenib blocks the RAF/MEK/ERK pathway, inhibits tumor angiogenesis, and induces tumor cell apoptosis in hepatocellular carcinoma model PLC/PRF/5. *Cancer Res.* **2006**, *66*, 11851–11858. [CrossRef]

191. Wilhelm, S.M.; Carter, C.; Tang, L.; Wilkie, D.; McNabola, A.; Rong, H.; Chen, C.; Zhang, X.; Vincent, P.; McHugh, M.; et al. BAY 43-9006 exhibits broad spectrum oral antitumor activity and targets the RAF/MEK/ERK pathway and receptor tyrosine kinases involved in tumor progression and angiogenesis. *Cancer Res.* **2004**, *64*, 7099–7109. [CrossRef]

192. Coriat, R.; Nicco, C.; Chereau, C.; Mir, O.; Alexandre, J.; Ropert, S.; Weill, B.; Chaussade, S.; Goldwasser, F.; Batteux, F. Sorafenib-induced hepatocellular carcinoma cell death depends on reactive oxygen species production in vitro and in vivo. *Mol. Cancer Ther.* **2012**, *11*, 2284–2293. [CrossRef]

193. Paech, F.; Mingard, C.; Grunig, D.; Abegg, V.F.; Bouitbir, J.; Krahenbuhl, S. Mechanisms of mitochondrial toxicity of the kinase inhibitors ponatinib, regorafenib and sorafenib in human hepatic HepG2 cells. *Toxicology* **2018**, *395*, 34–44. [CrossRef]

194. Bull, V.H.; Rajalingam, K.; Thiede, B. Sorafenib-induced mitochondrial complex I inactivation and cell death in human neuroblastoma cells. *J. Proteome Res.* **2012**, *11*, 1609–1620. [CrossRef] [PubMed]

195. Awan, F.M.; Naz, A.; Obaid, A.; Ikram, A.; Ali, A.; Ahmad, J.; Naveed, A.K.; Janjua, H.A. MicroRNA pharmacogenomics based integrated model of miR-17-92 cluster in sorafenib resistant HCC cells reveals a strategy to forestall drug resistance. *Sci. Rep.* **2017**, *7*, 11448. [CrossRef] [PubMed]

International Journal of
Molecular Sciences

Review

Roles of Thyroid Hormone-Associated microRNAs Affecting Oxidative Stress in Human Hepatocellular Carcinoma

Po-Shuan Huang [1,2], Chia-Siu Wang [3], Chau-Ting Yeh [4] and Kwang-Huei Lin [1,2,4,5,*]

[1] Department of Biochemistry, College of Medicine, Chang-Gung University, Taoyuan 33302, Taiwan;
leo_6813@msn.com

[2] Department of Biomedical Sciences, College of Medicine, Chang-Gung University, Taoyuan 33302, Taiwan

[3] Department of General Surgery, Chang Gung Memorial Hospital, Chiayi 61363, Taiwan;
wangcs@cgmh.org.tw

[4] Liver Research Center, Chang Gung Memorial Hospital, Linkou, Taoyuan 33302, Taiwan;
chauting@adm.cgmh.org.tw

[5] Research Center for Chinese Herbal Medicine, College of Human Ecology, Chang Gung University of
Science and Technology, Taoyuan 33302, Taiwan

* Correspondence: khlin@mail.cgu.edu.tw; Tel.: +88-63-211-8263

Received: 26 September 2019; Accepted: 18 October 2019; Published: 21 October 2019

Abstract: Oxidative stress occurs as a result of imbalance between the generation of reactive oxygen species (ROS) and antioxidant genes in cells, causing damage to lipids, proteins, and DNA. Accumulating damage of cellular components can trigger various diseases, including metabolic syndrome and cancer. Over the past few years, the physiological significance of microRNAs (miRNA) in cancer has been a focus of comprehensive research. In view of the extensive level of miRNA interference in biological processes, the roles of miRNAs in oxidative stress and their relevance in physiological processes have recently become a subject of interest. In-depth research is underway to specifically address the direct or indirect relationships of oxidative stress-induced miRNAs in liver cancer and the potential involvement of the thyroid hormone in these processes. While studies on thyroid hormone in liver cancer are abundantly documented, no conclusive information on the potential relationships among thyroid hormone, specific miRNAs, and oxidative stress in liver cancer is available. In this review, we discuss the effects of thyroid hormone on oxidative stress-related miRNAs that potentially have a positive or negative impact on liver cancer. Additionally, supporting evidence from clinical and animal experiments is provided.

Keywords: oxidative stress; microRNA; thyroid hormone; liver cancer

---

1. Introduction

Hepatocellular carcinoma (HCC) is an inflammation-related cancer, with the majority of cases occurring in the context of hepatic injury and inflammation [1]. The risk factors correlated with HCC include chronic inflammation due to viral infection (such as hepatitis B virus (HBV) and hepatitis C virus (HCV)), excessive intake of alcohol, metabolic disease, non-alcoholic steatohepatitis (NASH), bacterial infection, type 2 diabetes, smoking, and chemical exposure [2]. Both HCC and the associated risk factors are significantly correlated with oxidative stress. Oxidative stress occurs when excessive production of reactive oxygen species (ROS) overpowers intrinsic antioxidant defense mechanisms. Accumulating levels of ROS can cause extensive damage to biological molecules, leading to cell injury, loss of function, development of cancer, and even death [3]. Therefore, elucidation of the relationship between oxidative stress and cancer is of clinical importance. Among the potential mechanisms involved in carcinogenesis, the pathways triggered by oxidative stress-induced microRNAs (miRNA) have been widely investigated. MiRNAs are small endogenous non-coding RNA molecules that regulate multiple gene expression at the post-transcriptional level. These molecules suppress messenger RNA through binding to stretches of complementary sequences [4,5]. The potential associations of miRNAs with human disease are widely documented. As crucial regulators of gene expression, miRNAs thus present promising candidates for biomarkers and treatment strategies.

Thyroid hormones play major roles in cell growth, development, and metabolism. Considerable research supports a relationship between the thyroid hormone and pathophysiology of various cancer types. Thyroid hormones exert their effects on cancer cells through either genomic or non-genomic pathways and their dysregulation has significant effects on cancer development and progression. Hypothyroidism is reported to contribute to liver carcinogenesis [6]. Notably, both hyperthyroidism and hypothyroidism appear to be associated with oxidative stress in animal and human diseases, indicating involvement of the thyroid hormone in disease progression [7]. Preliminary data from recent studies focusing on the potential relationship between miRNAs associated with oxidative stress and dysregulation of thyroid hormone in liver cancer progression are comprehensively summarized in the current review.

2. Effect of Thyroid Hormone on the Role of Oxidative Stress-Related microRNAs in Liver

2.1. Oxidative Stress Promotes HCC Progression

Liver cancer is the second leading cause of cancer-related deaths worldwide. Hepatocellular carcinoma (HCC), a type of inflammation-related cancer with >90% cases associated with hepatic injury and inflammation, is the most common primary malignant tumor type [8]. The incidence of HCC is highly correlated with inflammatory risk factors, such as hepatitis B virus (HBV), hepatitis C virus (HCV), liver disease (non-alcoholic fatty liver disease/non-alcoholic steatohepatitis), habitual drinking (high alcohol exposure), obesity, type 2 diabetes (T2D), and aflatoxin exposure [9–12].

Oxidative stress additionally plays an important role in HCC development. Excess ROS levels induce liver DNA injury, in turn leading to increased fatty liver, hepatitis B/C, liver cirrhosis, and consequently, HCC [13]. Oxidative stress is defined as an imbalance between production of reactive oxygen species (ROS) and antioxidant capacity of the cell, which causes damage to biomolecules, such as DNA, lipids, and proteins [14].

ROS simultaneously affect a series of signaling cascades and mediate the regulation of several transcription factors that control the expression of various genes involved in cell survival, proliferation, invasion, and metastasis [15–19]. Common ROS species include superoxide anion (O_2^-), hydrogen peroxide (H_2O_2), hydroxyl radical (OH^-), singlet oxygen ($1 O_2$), and ozone (O_3) [20,21]. Reactive species induce nicks in DNA and failure in mechanisms to repair DNA damage that lead to HCC.

ROS can react with cellular biomolecules, yielding oxidatively modified DNA products that eventually induce cell damage and death. For instance, protein carbonyl and 8-hydroxydeoxyguanosine (8-OHdG), the well-known oxidatively modified molecular products of proteins and DNA, are associated with poor survival in HCC patients [16].

The inflammation risk factors, HCV and HBV infection, cause malignant degeneration by induction of oxidative stress that is critical in HCC. Oxidative stress is present to a greater degree in HCV infection than other inflammatory liver diseases and proposed as a major mechanism of liver injury in patients with chronic hepatitis C [22]. The core protein of HCV, which induces excess ROS production through adjustment of mitochondrial electron transport and mitochondria, is a primary target of ROS. Therefore, damage to mitochondria via ROS induced by HCV presents a potential mechanism underlying the development of HCC [23].

In addition to HCV, HBV infection markedly increases the risk of development of HCC. Among the viral proteins, HBV encoding HBV X protein (HBx) appears to have the greatest oncogenic potential in HCC. Similar to HCV core protein, HBx is associated with mitochondria, leading to augmented ROS production and induction of oxidative stress in hepatocytes [24]. The key mechanisms used by HBx, such as inhibition of high-mobility group protein box1 (HMGB1) expression and generation of ROS via the NF-κB signaling pathway, are discussed in an earlier report [25].

Non-alcoholic fatty liver disease (NAFLD) is a complex disorder characterized by excessive lipid accumulation the in liver, controlled by multiple metabolic factors, that is often diagnosed in conjunction with obesity, type 2 diabetes (T2D), and hyperlipemia [26]. Among the numerous mechanisms underlying NAFLD pathogenesis, redox imbalance is suggested to be the most significantly correlated factor to HCC progression. In addition, conditions such as metabolic oxidative stress, cell autophagy, and inflammation induce more severe nonalcoholic steatohepatitis (NASH) progression [26]. In patients with NASH, the activities of mitochondrial respiratory chain complexes are decreased in liver tissue, resulting in reduced glutathione expression and consequent activation of the c-Jun N-terminal kinase (JNK)/c-Jun signaling pathway by oxidative stress that induces cell death in steatotic liver [27].

The issue of whether risk factors directly induce or are subject to oxidative stress to increase their effects remains to be established. However, the findings to date suggest that oxidative stress exerts harmful effects on liver cells through inducing lesions. Elucidation of the underlying mechanisms should facilitate the development of effective strategies to manage HCC.

2.2. Roles of microRNAs Correlated with Oxidative Stress in HCC

MicroRNAs (miRNAs) are small non-coding RNA molecules that regulate >70% human genes at the post-transcriptional level. The average miRNA length is ~21–23 nucleotides. DNA sequences are transcribed into primary miRNAs (pri-miRNA) and processed into precursor miRNAs (pre-miRNA) in the nucleus and mature miRNAs in the cytoplasm. In most cases, miRNAs interact with a specific sequence at the 3′ untranslated region (UTR) of target mRNAs to induce translational repression via post-transcriptional regulation of cleavage or simply suppressing translation [28,29].

Accumulating studies support the importance of a series of oxidative stress-induced miRNAs in progression of carcinogenesis (Table 1). For instance, using the robust rank aggregation (RRA) method, miR-34a-5p, miR-1915-3p, miR-638, and miR-150-3p were shown to be upregulated under conditions of H_2O_2 treatment as oxidative stress-responsive miRNAs in HCC cell lines [30]. The functions of these four miRNAs were further predicted using the TargetScan web tool and Gene Ontology (GO) pathway enrichment analysis. All four miRNAs were closely related to anti-apoptosis pathways and p53 signaling, clearly demonstrating a significant association between the p53 pathway and oxidative stress [31].

The importance of miRNAs in progression of chronic liver diseases to HCC is recognized. MiRNAs act as key mediators in the development of a number of cancer types owing to their involvement in inflammation and oncogenesis processes. Several miRNAs showing altered expression patterns in HCC and oxidative damage have been identified, including miRNA-92, miRNA-145, miRNA-199a, miRNA-199b, miRNA-195, and miRNA-122a [27].

In recent years, several miRNAs have been extensively investigated and their functions in association with oxidative stress determined. MiR-26a is reported to play a dual role in HCC. Considerable research has confirmed its activity as a tumor suppressor in HCC that inhibits proliferation, migration, and invasion by targeting F-box protein 11 (FBXO11), an E3 ubiquitin ligase, and type II methyltransferase [47,48]. DNA methyltransferase 3b (DNMT3B) is another direct target of miR-26a. Inhibition of DNMT3B associated with miR-26a upregulation led to a similar tumor suppressor effect in HCC cells [49]. In contrast, other studies suggest that miR-26a has potential oncogenic function in HCC. For instance, therapeutic miR-26a delivery suppresses tumorigenesis in an animal liver cancer model while other studies demonstrated that miR-26a expression promotes HCC cell migration and invasion in vivo. Another earlier in-depth study reported that miR-26a promotes cell migration and invasion by inhibiting the phosphatase prime time entertainment network (PTEN) [50,51]. Based on the metabolic perspective, increasing free fatty acid (FFA) supply into liver cells caused oxidative stress by ROS and lipid peroxidation generated during the metabolism of these accumulating fatty acids [52]. Recently, regulatory and protective roles of miR-26a on lipid metabolism and progression of NAFLD in human HepG2 cells loaded with FFA have been demonstrated. Upregulation of miR-26a resulted in the downregulation of triglyceride (TG), total cholesterol (TCL), and malondialdehyde (MDA) through modulation of mRNA levels of genes involved in lipid homeostasis, ER stress, inflammation, and fibrogenesis [36]. Additionally, miR-26a targets different metabolic relative genes involved in fatty acid and cholesterol metabolism and insulin signaling, such as ACSL3, ACSL4, PKCδ, PKCθ, GSK3β, and SERBF1, suggesting a crucial role in preventing development of metabolic disease [53]. Notably, these liver-related lipid metabolism abnormalities are strongly associated with oxidative stress in liver cells [54,55].

Table 1. Oxidation stress-related microRNAs.

microRNA	Correlative with Oxidative Stress	Ref.
miR-34a-5p miR-1915-3p miR-638 miR-150-3p	Associated with oxidative stress-related apoptosis	[30]
miR-92	Correlated positively with telomerase activity, 8-OHdG Target to anti-oxidative gene Sirt1	[32]
miR-199a/b	Prevents the liver cell oxidative stress induced by bile acid Target to Sirt1	[33]
miR-122	Correlative with HCV/ HBV infection Positive association with antioxidant enzyme NQO1	[34] [35]
miR-26a	Affecting liver lipid metabolism	[36]
miR-155	Affecting liver lipid metabolism	[37]
miR-214	Associated with oxidative stress-related apoptosis Target to ATF4 and EZH2	[38] [39]
miR-200	Target to p38α and repression anti-oxidative gene Nrf2	[40]
miR-181	Target to Sirt1 and impair insulin sensitivity	[41]
miR-128	Target to DJ-1 Target to Sirt1	[42] [43]
miR-29a/c	Controls the hepatic lipogenic process	[44]
miR-21	Leading to mitochondrial ROS accumulation	[45]
miR-196	Downregulates Bach1, and inhibition of HCV expression	[46]

* Potential functions of miRNAs related to oxidative stress.

MiR-155 acts as a multifunctional oncogenic miRNA in different human cancer types, including breast, pancreatic, and liver cancer [56–58]. The miRNA promotes proliferation, invasion, and migration in HCC by directly targeting and inhibiting PTEN. The negative correlation between miR-155 and PTEN is significantly associated with TNM stage in HCC [56]. MiR-155 additionally inhibits Forkhead box O3 (FoxO3a) expression to suppress downstream apoptotic gene B-cell lymphoma-2 (Bcl-2)-interacting mediator of cell death (BIM) and suppresses cleavage of caspase-3 and caspase-9, consequently inhibiting HCC cell apoptosis and facilitating proliferation [59]. Furthermore, high expression of miR-155 is associated with poor survival, and in combination with Alpha-fetoprotein (AFP) shows higher sensitivity and specificity as a biomarker panel for diagnosis of HCC, compared with a single marker [60]. However, conflicting results on the role of miR-155 in lipid metabolism have been reported to date. Suppression of miR-155 in peripheral blood may be utilized as a novel biomarker for NAFLD screening. The transcription factor, Liver X Receptor α (LXRα), that interacts with the promoter region of sterol regulatory element-binding protein (SREBP)-1c, has been identified as a direct target of miR-155 [37,61]. Other studies have highlighted a reduction in alcohol-induced fat accumulation in miR-155 knockout mice, associated with increased Peroxisome proliferator-activated receptor response element (PPRE) binding to the miR-155 target gene, Peroxisome proliferator-activated receptor (PPAR)α [62]. However, further studies are required to confirm the finding that miR-155 participates in lipid accumulation in liver, inducing generation of oxidative stress.

2.3. Role of Thyroid Hormone and Its Receptor in HCC

Thyroid hormone, 3,3′,5-tri-iodo-L-thyronine (T3), is a key mediator of multiple physiological processes, including cell development, differentiation, metabolism, and growth [38]. The pituitary gland secretes thyrotropin, which influences the thyroid gland to synthesize thyroid hormone mainly precursor T4. T4 moves across the cell membrane of responsive cells by specific transporters, including the monocarboxylate anion transporters 8 and 10 (MCT8 and MCT10), and is converted to the active T3 by type I 5′-deiodinase (DIO) 1 and 2, leading to increased levels of T3 [63]. T3 controls metabolic activities related to anabolism or catabolism, including carbohydrates, proteins, lipids, and damaged organelles in cells to maintain homeostasis under different physiological conditions [24]. To implement genomic effects, cytoplasmic T3 translocates to the nucleus and binds to specific high-affinity thyroid hormone receptors (TR) associated with thyroid hormone response elements (TRE) on DNA, thereby affecting transcriptional levels of downstream genes [14]. Typical TREs within promoter regions of downstream genes contain two half-site sequences (A/G)GGT(C/A/G)A in palindromic (Pal), direct repeat (DR), or inverted repeat arrangements (IP) recognized by TR. TRs bind to their respective TREs as monomers, homodimers, or heterodimers with retinoid X receptors (RXR). TRs usually form heterodimers with the RXR to interact with TREs within the promoter regions of target genes. Human TRs are encoded by two distinct genes, THRA (TRα) and THRB (TRβ), located on human chromosomes 17 and 3 [64]. Different TRs are composed of similar domains, including amino-terminal A/B domain to recruit regulatory proteins; central DNA-binding domain (DBD), or C region, which displays high affinity for DNA sequences of TREs; linker D region, which is necessary for nuclear translocation of the receptor; and carboxy-terminal ligand-binding domain (LBD), which interact with thyroid hormones [63,65]. In humans, TRβ/T3 regulates the metabolic activity of body and it is the major receptor isoform expressed in liver; in contrast, TRα is expressed mainly in the heart, skeletal muscle, adipose tissues, and specifically mediates adaptive thermogenesis [66].

Owing to its critical regulatory function in cellular homeostasis, imbalance of thyroid hormone in the body is highly associated with multiple chronic diseases including obesity, diabetes, cardiovascular, and liver disorders. The liver is the most important thyroid hormone target organ associated with cellular metabolic functions, such as hepatic fatty acid and cholesterol synthesis and metabolism. Hypothyroidism has been associated with increased serum expression of triglycerides and cholesterol as well as hypercholesterolemia or non-alcoholic fatty liver disease (NAFLD) [24,67]. Prevention of cardiovascular disease occurrence is important in patients with low-serum high-density lipoprotein cholesterol (HDL-C) due to thyroid dysfunction [68].

In addition to its effects on metabolism, thyroid hormone suppresses HCC development by protecting hepatocytes from HBx-induced damage through regulating mitochondrial quality control to suppress HBx protein stability. Mitochondrial quality maintenance by T3 prevents HBx-induced hepatocarcinogenesis and attenuates HCC progression [25,69]. In an earlier study, liver disease patients diagnosed with hepatic cirrhosis triggered by hepatitis B or C were screened for thyroid function status. The T3 levels of patients were lower than the normal range, suggesting that the serum T3 concentration is a good index of hepatic function, decreasing the severity of liver damage [70].

Several studies have demonstrated that treatment with T3 analogs can prevent hepatic steatosis and hepatitis. The thyroid hormone has potential therapeutic applications in hepatitis B and C, and T3 analogs may be effectively used as an alternative strategy to prevent HCC [71].

2.4. Thyroid Hormone Induces an Anti-Oxidative Stress Effect in Hepatocytes Mediated by microRNAs

Hypermetabolic effects of thyroid hormones as the major endocrine regulators of metabolic rate are well documented. Thyroid hormones have a profound impact on mitochondria, the organelles predominantly responsible for cellular energy metabolism, and are correlated with O_2 consumption and consequent ROS generation [72]. Effects of thyroid hormone on redox signaling to protect cellular function are documented. The pathways affected by thyroid hormone generally fall into two broad categories: Genomic and non-genomic. ROS production leads to activation of the redox-sensitive transcription factors nuclear factor-κB (NF-kB), signal transducer and activator of transcription 3 (STAT3), signal transducer and activator of transcription 1 (STAT1), and nuclear factor erythroid 2-related factor 2 (Nrf2), promoting cell protection and survival mechanisms. Functions of the thyroid hormone include enhancement of homeostatic potential, through induction of antioxidant, anti-apoptotic, and anti-inflammatory gene expression, and higher detoxification capabilities and energy supply through AMP-activated protein kinase (AMPK) upregulation [73]. Thyroid hormone additionally regulates miRNAs that promote antioxidant capacity in the liver to prevent HCC progression.

In a previous study, our group used qRT-PCR array to explore the expression patterns of different miRNAs regulated by thyroid hormone (Tables 2 and 3) [38,74–76]. The functions of potentially important thyroid hormone-regulated miRNAs in HCC and their correlation with oxidative stress are further discussed below.

Table 2. MicroRNAs positively associated with thyroid hormones in HepG2 liver cancer cell lines.

| | miRNAs Positively Affected by Thyroid Hormones * | | | | | | | |
| | HepG2-TRα1 | | | | HepG2-TRβ1 | | | |
TH/microRNAs	MicroRNA	Fold	HCC/ROS Ref.	TH	MicroRNA	Fold	HCC/ROS Ref.	TH
Three times repetitive experiments	miR-122	10.54	[35,77]	[78]	miR-29c	4.21	[44,79]	
	miR-152	3.42			miR-214	3.50	[38,80,81]	[38]
	miR-139-5p	10.38			miR-202	2.41		
	miR-128a	71.90	[43]					
	miR-139-3p	3.85						
	miR-548d-3p	3.09						
	miR-140-3p	2.77						
Two times repetitive experiments	miR-143	6.20			miR-193b	2.99		
	miR-210	5.22			miR-139-5p	3.12		
	miR-365	5.53			miR-210	2.52		
	miR-135b	4.38			miR-323-3p	4.12		
	miR-148a	5.16			miR-22	2.54		
	miR-193b	3.30			miR-29a	2.18	[44,82]	
	miR-125a-3p	2.92			miR-29b-1 *	3.30		
	miR-29a	3.15	[44,82]		miR-193a-3p	3.34		
	miR-24	2.40	[83]	[78]	miR-139-3p	2.22		
	miR-372	3.57			miR-510	2.32		[75,78]
	miR-372	5.07			miR-21 *	2.22	[45,84–86]	
	miR-188-3p	3.05						
	miR-100	4.11						
	miR-126	2.35						
	miR-21	3.30	[45,84–86]	[75,78]				

* HepG2 hepatoma cell lines overexpressing TRα1 or TRβ1 were treated with thyroid hormone (T3; 20 nM). After 24 h, qRT-PCR array analysis of microRNA (miRNA) expression was performed. The specified miRNAs were positively affected (>2-fold) upon thyroid hormone stimulation and selected candidates were identified from at least two times repetitive experiments. The references are to indicate oxidative stress (HCC/ROS) or thyroid hormone (TH) related miRNAs in liver cancer.

Table 3. MicroRNAs negativity associated with thyroid hormones in HepG2 liver cancer cell lines.

			miRNAs Negativity Affected by Thyroid Hormones *					
		HepG2-TRα1				HepG2-TRβ1		
TH/microRNAs	MicroRNA	Fold	HCC/ROS Ref.	TH	MicroRNA	Fold	HCC/ROS Ref.	TH
Three times repetitive experiments	miR-184	0.22			miR-455-3p	0.22		
	miR-455-3p	0.12			miR-148a	0.36		
	miR-499-3p	0.20			miR-425 *	0.24		
	miR-221	0.33			miR-187	0.27		
	miR-181b	0.30	[87]		miR-429	0.41		
	miR-130b	0.34		[76]				
	miR-149	0.35						
	miR-17	0.34	[85]	[74]				
Two times repetitive experiments	miR-425 *	0.22			miR-106a	0.23		
	miR-20a	0.31			miR-199a-5p	0.22	[33,87,88]	[38]
	miR-377	0.42			miR-548d-5p	0.24		
	miR-15b	0.43			miR-146a	0.31		
	miR-516a-5p	0.29			miR-221	0.27		
	miR-652	0.49			miR-30a *	0.35		
	miR-550	0.26			miR-499-3p	0.32		
	miR-18a	0.23			miR-888	0.27		
	miR-106a	0.28			miR-100	0.33		
	miR-628-3p	0.34			miR-339-3p	0.45		
	miR-146a	0.36			miR-18a	0.39		
	miR-181c	0.41	[87]		miR-18b	0.24		
	miR-92a	0.36	[32,89,90]		miR-10a	0.25		
	miR-106b	0.38			miR-421	0.30		
	miR-487b	0.35			miR-525-3p	0.41	[74,85]	
	miR-570	0.40			miR-17	0.37	[85,90]	[74]
	let-7d	0.44			miR-542-5p	0.33	[46,85]	
	miR-15b *	0.44			miR-196a *	0.42	[46]	
					miR-196b	0.46		
					miR-19a	0.46	[87]	
					miR-181d	0.32		
					miR-20b	0.40		

* HepG2 hepatoma cell lines overexpressing TRα1 or TRβ1 were treated with thyroid hormone (T3; 20 nM). After 24 h, qRT-PCR array analysis of microRNA (miRNA) expression was performed. The specified miRNAs were negatively affected (<0.5-fold) upon thyroid hormone stimulation and selected candidates were identified from at least two times repetitive experiments. The references are to indicate oxidative stress (HCC/ROS) or thyroid hormone (TH)-related miRNAs in liver cancer.

MiR-214 is dysregulated in many human cancer types including cervical, prostate, and ovarian cancer [91–93]. In HCC, miR-214 acts as a tumor suppressor and is used as a potential prognostic marker for overall survival [94,95]. Earlier studies indicate that miR-214 plays a tumor suppressor role by inhibiting proliferation and migration of HCC cells through targeting pyruvate dehydrogenase kinase 2 (PDK2) and plant homeodomain finger protein 6 (PHF6) [80]. Forkhead box protein M1 (FoxM1) is an important transcription factor in the progression of HCC. Direct targeting and downregulation of FoxM1 mRNA by miR-214 inhibits proliferation, migration, and invasion of HCC [81]. In the clinic, miR-214 downregulation is positively associated with higher tumor recurrence and poorer clinical outcomes. Ectopically expressed miR-214 inhibits xenograft tumor growth and microvascularity of tumors and their surrounding tissues via targeting and suppressing its downstream target gene, hepatoma-derived growth factor (HDGF) [94].

Several oncogenic long non-coding RNAs (lncRNA) are correlated with miR-214. Among these, myocardial infarction-associated transcript (MIAT) regulates proliferation and invasion of HCC cells via sponging miR-214 [96]. Plasmacytoma variant translocation 1 (PVT1) lncRNA is increased in HCC tissues and associated with tumor size, histological differentiation grade, and advanced TNM stage. PVT1 has been shown to promote proliferation and invasion of HCC via inhibition of miR-214 expression by interacting with enhancer of zeste homolog 2 (EZH2) [97].

MiR-214 is upregulated by the thyroid hormone through direct interactions with its receptor in the promoter region, leading to repression of the target oncogene, PIM-1, and in turn, suppression of HCC cell proliferation and inhibition of tumor formation [38]. Diethylnitrosamine (DEN) is a typical chemical carcinogen with the potential to cause tumors in multiple organs, such as liver, skin, gastrointestinal tract, and the respiratory system. This significant environmental carcinogen triggers ROS production, resulting in oxidative stress and cellular injury. DEN is considered a complete hepatocarcinogen [98–100]. As highlighted previously, thyroid hormone promotes selective autophagy via induction of the death-associated protein kinase 2-Sequestosome 1 (DAPK2-SQSTM1) pathway, thus protecting against DEN-induced carcinogenesis in hepatocytes [101]. Notably, thyroid hormone additionally plays a protective role against DEN-induced HCC through upregulation of miR-214 [38].

Thyroid hormone is a human hormone that mediates the cell differentiation and metabolism and acts as an anti-apoptosis factor upon challenge of thyroid hormone receptor expression in HCC cells with cancer therapy drugs, such as cisplatin, doxorubicin, and tumor necrosis factor-related apoptosis-inducing ligand (TRAIL). Doxorubicin (Dox), a DNA topoisomerase II inhibitor, belongs to the anthracycline anticancer drug family [102]. Dox is widely used to treat lymphoma breast, head-and-neck, prostate, and liver cancers [103–106]. Dox induces pathogenic mechanisms including apoptosis, oxidative stress, and inflammation, through formation of ROS, reduces anti-oxidative defense, and stabilizes mitochondrial damage [107–109]. Thyroid hormone and its receptor signaling pathway promote chemotherapeutic resistance through negatively regulating the pro-apoptotic protein, BCL2-like 11 (BCL2L11/Bim), resulting in Dox-induced metastasis of chemotherapy-resistant HCC cells [110].

In addition, HCV infection promotes mitochondrion-mediated apoptosis through stimulating the upstream ROS/JNK signaling pathway to affect Bax-triggered mechanisms. In brief, HCV-induced ROS/JNK signaling transcriptionally activates Bim expression, which leads to Bax activation and apoptosis induction [111]. Bim is a direct target gene of miR-214 in nasopharyngeal carcinoma (NPC) and other tissues [112–114]. One possibility is that the thyroid hormone induces miR-214 to suppress Bim expression through negatively regulating the transcription factor Forkhead box protein O1 (FoxO1) to avoid liver cell apoptosis and ROS-induced stress [110]. In addition to miR-214, there are many miRNAs that have the potential to affect the apoptosis of liver cancer cells, such as miR-155, miR-4417, miR-199a, and miR-122 [59,77,115,116]. Among them, the expression levels of miR-199a and miR-122 are associated with thyroid hormone and oxidative stress (Tables 2 and 3) [27]. Previous studies have indicated that Dauricine (Dau) is a natural alkaloid, which promoted apoptosis of HCC cells induced by chemotherapeutic reagents. Dau stimulates the expression of miR-199a and results in inhibition of the target gene hexokinase 2 (HK2) and pyruvate kinase M2 (PKM2), resulting in sensitivity to chemotherapeutic reagents, including Cisplatin, Sorafenib, and Isoliensinine in HCC cells [116]. However, thyroid hormones, which are inversely related to miR-199a, may also involve in this action of apoptosis. Interestingly, in addition to the target gene of miR-199a, PKM2, which is strongly associated with apoptosis, is also affected by miR-4417 and miR-122 in liver cancer cells [77,115].

Gemcitabine (GEM) is a commonly used chemotherapeutic agent for HCC that uses oxidative stress induction as a common effector pathway. Overexpression of mitochondrial uncoupling protein 2 (UCP2) causes resistance to GEM. GEM administered alone or in combination with oxaliplatin renders minimal survival benefits to HCC patients. The tumor suppressor activity of miR-214 is activated through targeting UCP2, which may solve the problem of GEM efficacy [117,118]. Combined usage of thyroid hormone combined with GEM could provide new insights into strategies to treat liver cancer based on this novel mechanism of action.

MiR-214 protects red blood cells against oxidative stress by targeting activating transcription factor 4 (ATF4) and enhancer of zeste homolog 2 (EZH2) [39]. Direct targeting of the transcriptional factor, ATF4, by miR-214, attenuates stress responses. Suppression of miR-214 leads to enhanced ATF4 translation and consequently, upregulation of ATF4 protein. Additionally, miRNA-214 is reported to reduce oxidative stress in diabetic nephropathy mediated the ROS/Akt/mTOR signaling pathway [119]. An earlier study by Liu et al. [120] showed that miR-424 inhibits oxidative stress and protects against transient cerebral ischemia injury.

MiR-122 plays a complex role in HBV and HCV infection [34,121,122]. MiR-122 is a liver-specific miRNA that acts as a host factor to increase the abundance of HCV RNA by stabilizing the positive strand of HCV RNA genome and promotes HCV synthesis by binding two sites near the HCV 5′ end and associating with Ago2 [34,123–125]. In contrast to its role in HCV infection in HCC, miRNA-122 is significantly downregulated in patients with HBV infection [126,127]. Adenosine deaminases act on RNA-1 (ADAR1), an important gene involved in adenosine to inosine RNA editing and miRNA processing. ADAR1 also plays an anti-viral role against HBV infection by increasing the miRNA-122 level in hepatocytes [128]. In terms of the role of miR-122 in carcinogenesis, this liver-specific miRNA is reported to be dramatically downregulated in most HCCs. The tumor suppressor role of miR-122 in HCC is exerted by targeting the genes involved in cell proliferation, differentiation, apoptosis, and angiogenesis, and its expression is inversely associated with poor prognosis and metastasis [129]. Many studies have demonstrated that miR-122 acts as an important tumor suppressor through regulating different target genes including WNT1, Cyclin G1, MDR, ADAM17, CUTL1, and AKT3 in HCC [130,131]. Associations of miR-122 expressed in liver and anti-oxidant genes, such as heme oxygenase 1 (HMOX-1), NAD(P)H, quinone oxidoreductase-1 (NQO1), and growth factor erv1-like (GFER1) in liver tissue specimens obtained from patients with chronic hepatitis B, have been uncovered. A significant positive association between expression of NQO1 and miR-122 was determined [35]. NQO1 is a multifunctional antioxidant enzyme and exceptionally versatile cytoprotective agent that regulates the proteasomal degradation of specific antioxidant proteins, such as nuclear factor erythroid 2-like 2 (Nrf2) [132], one of the major mediators of inflammation and a transcription factor. Nrf2 promotes the expression of antioxidant as well as cytoprotective genes, resulting in anti-inflammatory effects [133].

Perfluorooctanesulfonate (PFOS) has been widely used in commercial applications as a surfactant and stain repellent. PFOS has been shown to cause liver damage (including liver tumors) in experimental animals through interactions with peroxisome proliferator-activated receptor α (PPARα) and constitutive androstane receptor (CAR)/pregnane X receptor (PXR). Further studies have highlighted the ability of PFOS to disrupt thyroid function and induce thyroid hormone alterations, leading to hypothyroxinemia [78,134]. Assessment of changes in miRNA levels in rats with PFOS-induced hypothyroxinemia revealed that three members of the miR-200 family were the most significantly increased while miR-122 and miR-21 showed the greatest decrease in expression. Moreover, expression of the miR-23b/27b/24 cluster was decreased in PFOS-treated animals [78]. Consistently, experiments by our group demonstrated upregulation of miR-122, miR-21, and miR-24 by thyroid hormone treatment in HCC cells (Table 2).

Among the miRNAs affected by thyroid hormone, members of the miR-200 family were markedly enhanced in hepatic cells following hydrogen peroxide (H_2O_2) treatment. Among these, miR-200-3p modulates the H_2O_2-mediated oxidative stress response by targeting mitogen-activated protein kinase 14 (p38α). p38α acts as a stress-activated protein kinase that negatively regulates tumorigenesis by acting on cell apoptosis, survival, and stress response. p38α inhibition leads to increased ROS levels in liver cells through repression of Nrf2, a master regulator of antioxidant and detoxifying genes [40]. These results support a hepatoprotective role of thyroid hormone through effects on the pathway of oxidative stress-induced miR-200 to repress p38α and Nrf2.

MiR-92a is highly expressed and specifically altered in HBV/HCV-related HCC [135,136]. This miRNA plays a critical role in HCC proliferation and invasion and could serve as a novel therapeutic target via repression of Forkhead Box A2 (FOXA2) [137,138]. Clinical association analysis revealed a correlation of high expression of miR-92a with poor prognostic characteristics of HCC. Diagnostic efficacy of a combination of miR-92a and AFP was powerful for HCC, in particular screening of early tumor and low-level AFP patients [139]. A combination of the tumor suppressor gene phosphatase and tensin homolog (PTEN) and miR-92a also provided significant clinical value for early diagnosis and prognosis of HCC based on their significant negative correlation in HCC and para-cancerous tissue [140]. MiR-92a has been shown to promote tumor growth of HCC by targeting F-box and WD repeat domain-containing 7 (FBXW7) and may serve as a novel prognostic biomarker and therapeutic target [141]. ROS trigger DNA oxidation leading to multiple modifications in DNA bases, among which 8-OHdG is the most frequent [142]. 8-OHdG induces point mutations in DNA strands and accumulates in DNA to cause mispairing, resulting in mutagenic and potentially carcinogenic activity. HCC tissues are frequently characterized by increased oxidative damage, which contributes to acceleration of telomere shortening and telomerase activation in cancer cells. The telomere acts as a protective cap at the ends of chromosomes and telomere shortening promotes chromosomal instability [143,144]. Oncogenic miR-92a expression is significantly correlated with telomerase activity and 8-OHdG levels in HCC tissues, indicating a link with ROS-mediated oxidative DNA damage [32]. The pre-mRNA-splicing factor, SLU7, is essential for HCC cell viability. SLU7 expression is reduced in HCC cells, and its depletion triggers autophagy-related cellular apoptosis in association with generation of ROS. Low expression of SLU7 leads to altered splicing of the C13orf25 primary transcript and reduced expression of its miR-17-92a constituents, leading to upregulation of its target genes, CDKN1A (P21) and BCL2L11 (BIM), and mediators of pro-survival and tumorigenic activities [90]. Previous studies have shown that miR-92a and its cluster miR-17, miR-18a, and paralog, miR-20b, are downregulated by the thyroid hormone in HCC cells (Table 3). MiR-92 also plays a key regulatory role in neovascularization and is predicted to target Sirtuin-1 (Sirt1) [89], a NAD$^+$-dependent deacetylase with potential anti-oxidative stress activity in vascular endothelial cells. The mechanisms underlying the protective effects involve Sirt1/FOXOs, Sirt1/NF-κB, Sirt1/NOX, Sirt1/SOD, and Sirt1/eNOS pathways [145]. Several other miRNAs, such as miR-181, miR-138, and miR-199, suppress Sirt1 in different cells/tissue types. Among these, miRNA-181 is upregulated under conditions of a high-fat diet and is reported to suppress Sirt1 and impair insulin sensitivity in liver [41,87]. Similar results have been reported for miR-200 and miR-199 in the DEN model [88]. Data from our qRT-PCR array disclosed downregulation of miR-181 and miR-199 by thyroid hormone (Table 3) [38]. Notably, miR-181 is inversely correlated with TRβ1 in human cirrhotic peritumoral tissue, compared to normal liver [146]. These findings support the theory that thyroid hormone decreases oxidative stress via repression of miR-181 and miR-199 to increase the target gene Sirt1 expression in liver.

MiR-206 is downregulated during tumorigenesis and plays an important role in modulating the growth of multiple HCC cells via targeting cyclin-dependent kinase 9 (CDK9), which stimulates the production of abundant prosurvival proteins, leading to impaired cancer cell apoptosis [147]. Overexpression of miR-206 has been shown to inhibit proliferation, invasion, and migration of the HCC cell lines HepG2 and Huh7. Conversely, inhibition of miR-206 inhibition enhances expression of protein tyrosine phosphatase 1B (PTP1B) that plays an oncogenic role in HCC, in HepG2, and Huh7 cells [148]. MiR-206 also directly targets the c-Met gene for silencing and restoration of c-Met expression reverses the inhibitory effect of miR-206 on HCC [149]. Nrf2, involved in cellular antioxidant defense systems, protects against excessive ROS damage to macromolecules and consequent senescence and apoptosis. Upregulation of Nrf2-dependent antioxidant and metabolic genes and significantly reduced miR-1 and miR-206 expression in lung tumors are associated with reduced survival in patients with lung adenocarcinoma [150]. MiR-206 is involved in thyroid hormone-mediated regulation of lipid metabolism in HepG2 cells, and its expression is suppressed in patients with hyperthyroidism, indicating a role in thyroid hormone-induced disorders of lipid metabolism in the liver [151].

2.5. Thyroid Hormone Promotes Oxidative Stress in Hepatocytes by Regulating microRNAs

Thyroid hormone not only regulates miRNAs to prevent oxidative stress in hepatocytes, but also exerts effects on miRNAs that result in increased oxidative stress-induced damage to liver. Below, we have discussed a few examples of miRNAs positively correlated with oxidative stress and regulated by thyroid hormone.

MiR-128 is downregulated in HCC and suppresses cell proliferation through inducing G1 phase cell arrest via regulating phosphoinositide 3 kinase regulatory subunit 1 (PIK3R1) expression, which inhibits the phosphatidylinositol 3-kinase (PI3K)/AKT signal pathway [152]. In addition, miR-128 significantly inhibits HCC cell metastasis and stem-cell like properties through direct targeting of integrin alpha 2 (ITGA2) and integrin alpha 5 (ITGA5) [153]. Parkinson disease protein 7 (PARK7/ DJ-1) expression is elevated in various tumors and related to the survival of tumor cells under adverse stimuli, including oxidative stress. DJ-1, also known as Parkinson's disease-associated protein (PDAP), performs multiple functions, including cysteine protease, anti-oxidative stress reaction, and tumorigenesis activities [154,155]. MiR-128 is downregulated and negatively correlated with DJ-1, which is a direct target of the miRNA, in HCC cells [42]. Dox also markedly upregulates miR-128 and downregulates Sirt1 expression by direct targeting and affecting the expression of other antioxidant proteins, such as Nrf2, Keap1, Sirt3, NQO1, and HO-1, leading to excessive oxidative stress in liver [43]. In our qRT-PCR array, the tumor suppressor, miR-128, was upregulated by thyroid hormone, suggesting a correlation between thyroid hormone and miR-128-affected antioxidant genes, such as Sirt1 (Table 2).

According to our qRT-PCR data, thyroid hormone enhances miR-128 to suppress Sirt1 expression in liver. MiR-29a and miR-29c are also upregulated by the thyroid hormone and associated with Sirt1 expression (Table 2). Previous experiments have shown that miR-29 controls the hepatic lipogenic process through regulation of anti-lipogenic transcription factor aryl hydrocarbon receptor (AHR) and Sirt1 in liver [44]. MiR-29a suppresses cell proliferation through direct targeting of Sirt1 in HCC [82]. PU box binding protein (PU.1) is a critical transcription factor involved in many pathological processes. In PU.1-deficient mice, miR-34a and miR-29c are highly expressed and regulate Sirt1 expression in hepatic stellate cells to resistant hepatic fibrosis [79]. The data suggest that the thyroid hormone may suppresses the anti-oxidative stress reaction by miR-128, miR-29, miR-29a, and miR-29c to direct targeting Sirt1 and indirect effect other anti-oxidative stress genes expression.

In HCC cells and tissues, miR-21 is upregulated and positively associated with cell migration and invasion abilities. Krueppel-like factor 5 (KLF5) acts as a tumor inhibitor in some cancer types. In an earlier study, KLF5 expression was inhibited through direct targeting by miR-21, leading to the induction of migratory and invasive abilities in HCC [156]. Betulinic acid (BA) is a pentacyclic triterpene that possesses potential pro-apoptotic activities through increasing mitochondrial ROS generation. Mitochondrial dysfunction activates the molecular apoptotic events leading to cell death in HCC. BA suppresses superoxide dismutase 2 (Sod2) expression through upregulation of miR-21, leading to mitochondrial ROS accumulation and apoptosis in HCC [45]. MiR-21 is reported to be activated via thyroid hormone-receptor interactions at the native TRE site in the promoter region [75]. The thyroid hormone may thus have a similar function as BA in increasing mitochondrial ROS generation and mitochondrial dysfunction through miR-21 expression in HCC.

MiR-196 is readily released in body fluids and blood during HBV/HCV-associated hepatitis as well as metabolic, alcohol-associated, drug-induced, and autoimmune hepatitis. Liver-specific miR-196 is a potential indicator of liver injury (mainly apoptosis, necrosis, and necroptosis) or hepatitis, showing variable expression during acute/fulminant, chronic, liver fibrosis/cirrhosis, and HCC [157]. Bach1, a basic leucine-zipper mammalian transcriptional repressor, negatively regulates HMOX1, a key cytoprotective enzyme with antioxidant and anti-inflammatory activities. MiR-196 significantly downregulates Bach1, leading to upregulation of HMOX1 gene expression and inhibition of HCV expression, further affecting oxidative stress and liver injury induced by HCV [46]. Data from our qRT-PCR experiments suggest that miR-196 is a potential oncogenic miRNA downregulated by thyroid hormone (Table 3).

MiR-199 family members (miR-199a/b-5p) are downregulated in HCC. Notably, the lower expression of miR-199a is also associated with poorer overall survival of HCC patients. MiR-199a overexpression in HCC cell lines is reported to inhibit cell proliferation, migration, and invasion. The miR199a family suppresses Rho-associated coiled-coil kinase 1 (ROCK1) post-transcriptionally to inhibit PI3K/Akt signaling, which is necessary for HCC proliferation and metastasis [158]. Moreover, miR-199 targets and negatively regulates X-box binding protein 1 (XBP1) and affects cyclin D, which is associated with cell cycle regulation in HCC cells [159]. Bile salts retained within the liver play a major role in liver injury during cholestasis and trigger cellular stress events, including protein misfolding, DNA damage, endoplasmic reticulum (ER) stress, and oxidative stress, that may result in cell death and pathogenesis of several liver diseases [160]. Another study reported elevated miR-199a-5p levels in bile acid-stimulated cultured hepatocytes of liver from bile duct-ligated mice. Elevated miR-199-5p disrupted sustained ER stress and prevented hepatocytes from undergoing bile acid-induced cell death, supporting the potential of this miRNA as a target for clinical approaches aiming to protect against liver toxicity from bile salts in hepatocytes [33]. Analysis of the association between thyroid hormone and miR-199 unexpectedly revealed a negative correlation between the two molecules. MiR-199/miR-214 are clustered and located on opposite strands of the Dynamin3 gene (DNM3). Under most conditions, while clusters have the same performance, but the thyroid hormone exerts differential effects on the two molecules, which seems to be due to the existence between miR-199 and miR-214 additional positivity TRE affecting miR-214 [38].

3. Discussion

The specific roles of thyroid hormone in different human cancer types are controversial. A number of investigators have reported that thyroid hormone promotes development of various cancers, whereas others suggest a tumor suppressor role [161–164]. Accumulating evidence from animal models and epidemiologic studies indicate an association between higher thyroid hormone levels and prevention of liver diseases, supporting the suppressor role of thyroid hormone and its receptor in HCC [25,38,164–166]. Moreover, clinical findings support a positive correlation between hypothyroidism and HCC development [167–169]. Oxidative stress-induced liver inflammation is the most important factor for HCC progression [1,170]. Oxidative stress is also related to thyroid hormone derangement, with the hormone reported to influence the antioxidant level or generation of ROS. Hyperthyroidism and hypothyroidism have been shown to be associated with oxidative stress in acute and chronic nonthyroidal illness syndrome (NTIS) [171,172].

In this report, we have discussed the involvement of a range of miRNAs in correlation with thyroid hormone and oxidative stress in HCC. Inconsistent results have been obtained from multiple studies on the role of the thyroid hormone in multiple cancer types. Based on the collective findings, thyroid hormone clearly regulates the expression of different miRNAs either directly or indirectly to affect oxidative stress (HCV/HBV-induced or DEN, Dox-induced) in liver. As shown in Figure 1, thyroid hormones can influence oxidative stress-induced hepatocarcinogenesis mediated by miRNAs, specifically, via upregulating miR-214, miR-122, and miR-206 in HCC. Several other studies indicate these miRNAs act as tumor suppressors in HCC. Simultaneously, these miRNAs regulate different oxidative stress-related genes that participate in liver cell antioxidant capacity, including Bim, NQO1, and Nrf2. A number of oncogenic miRNAs have been shown to be downregulated by thyroid hormone in HCC, including miR-200, miR-92a, and miR-181. Their target genes, p38α, 8-OHdG, and Sirt1, participate in oxidative stress. Multiple studies have shown that thyroid hormones participate in the regulation of miRNA expression to prevent excessive generation of ROS (miR-122, miR-200, miR-206) and reduce DNA damage (miR-214, miR-92a, miR-181) in hepatocytes.

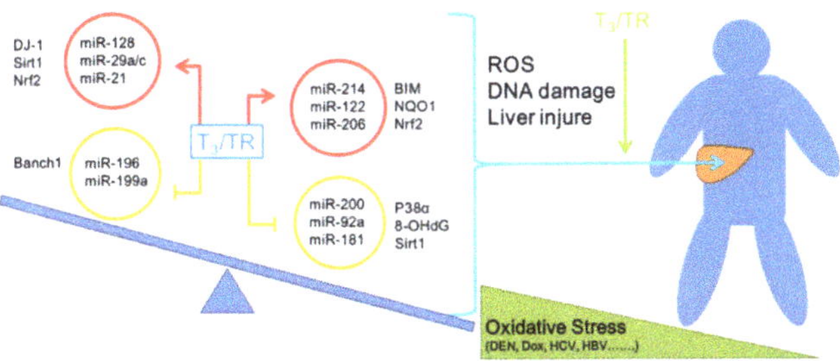

Figure 1. Thyroid hormones affect oxidative stress-induced hepatocarcinogenesis through effects on miRNAs. Oxidative stress is a risk factor associated with liver cancer. Among them, miRNAs strongly related to physiological significance are also involved. We analyzed the associations of thyroid hormones with oxidative stress and miRNAs in liver cells. While miRNAs related to promotion of resistance to oxidative stress were also affected by thyroid hormones, based on empirical evidence from other experimental animal and clinical studies, we believe that thyroid hormone plays a largely hepatoprotective role under conditions of oxidative stress. (The red arrow is a positive association, the yellow T bar is a negative association, blue arrow indicates the miRNAs affected liver genes, and green arrow is thyroid hormone associated-miRNAs affected liver genes)

Conversely, an oncogenic role of thyroid hormone in HCC has been reported by other investigators [14,173]. The thyroid hormone is reported to induce oxidative stress through enhancing the expression of different miRNAs in liver, such as miR-128, miR-29a/c, and miR-21, which directly regulate anti-oxidant genes, such as DJ-1, Sirt1, and Bach1, or affect related antioxidant genes, including Nrf2, Keap1, Sirt3, NQO1, and HO-1.

The role of thyroid hormone in various human organs is complex, especially under conditions of oxidative stress. In patients with hypothyroidism, high plasma levels of NO and malondialdehyde (MDA), a marker of oxidative stress, were measured in hepatic vein, along with lower activity of paraoxonase-1 (PON-1), a liver enzyme with antioxidant features. Data from this study further indicated that increased oxidative stress in hypothyroidism is primarily attributed not only to a decrease in antioxidant levels, but also effects on lipid metabolism [174,175]. The collective evidence suggests that thyroid hormone has potential antioxidant activity, at least in the liver [7,73]. The same findings were reported in experimental animals. Thyroid hormone signaling was altered following stimulation with various stress signals and played a crucial role in response to stress, post-stress recovery, and tissue repair by reducing the inflammatory response associated with NF-κB and STAT3 activation as well as the acute phase response [176–178]. Hepatic autophagy regulates lipid metabolism through elimination of triglyceride accumulation in liver, prevents the development of steatosis, and reduces oxidative stress [179]. Mitochondrial balance is a precisely regulated process that influences cellular homeostasis. Since activation of hepatic mitophagy eliminates the lipid content and oxidative stress, its dysregulation is implicated in progression of NAFLD. Dysregulation of autophagy and defective mitochondrial homeostasis contribute to hepatocyte injury and liver-related diseases [180]. In a murine model, disruption of thyroid hormone production led to a marked increase in progression of DEN-induced HCC. DEN triggers generation of ROS, resulting in oxidative stress and cellular injury, and consequently, progression of HCC. Data from this model indicate that thyroid hormone promotes autophagy via induction of the DAPK2-SQSTM1 cascade, thus protecting hepatocytes from DEN-induced hepatotoxicity or carcinogenesis. Furthermore, thyroid hormone is reported to participate in regulation of lipid metabolism through a chromosome 19 open reading frame 80 (C19orf80)-activated autophagic process in HCC [99,101,181].

Although the role of thyroid hormone in relation to oxidative stress in HCC remains a controversial issue, abundant reports support antioxidant over pro-oxidant activity, which prevents induction of liver cancer progression by oxidative stress in hepatocytes. This situation is similar to that of the antioxidant genes Sirt1 and Nrf2. Sirt1 and Nrf2 are oppositely regulated by thyroid hormone in association with different miRNAs. However, the majority of reports indicate that the thyroid hormone activates Sirt1 [182,183]. Moreover, the thyroid hormone positively affects FOXO1 in addition to Sirt1 to stimulate genes that enhance the autophagic process [184–187]. Several animal studies have provided evidence to support a positive association of thyroid hormone with Nrf2, a key redox-sensitive transcription factor of antioxidant genes, in the liver. Thyroid hormone not only upregulates Nrf2, but also promotes antioxidant gene expression, since Nrf2 translocates from the cytosol to nucleus, mediating hepatic cytoprotection [188–191].

In this review, we have discussed several miRNAs associated with oxidative stress in HCC. Importantly, the relationships among miRNAs, thyroid hormone, and oxidative stress have been comprehensively explored. Despite conflicting results in the literature, thyroid hormone is considered a protective factor overall in hepatocytes. Thyroid hormone may aid in maintaining the normal environment of hepatocytes through effects on lipid metabolism and mitochondrial activity. Moreover, thyroid hormone protects against liver injury by reducing oxidative stress induced by harmful chemicals or HBV/HCV. Further studies should focus on the development of thyroid hormone analogs beneficial for human health.

Author Contributions: Conceptualization, P.-S.H. and K.-H.L.; investigation, P.-S.H.; writing—original draft preparation, P.-S.H.; writing—review and editing, K.-H.L.; visualization, supervision, C.-T.Y.; K.-H.L.; funding acquisition, C.-S.W. and K.-H.L.

Funding: This research was funded by grants from Chang Gung Memorial Hospital, Taoyuan, Taiwan (CMRPD1J0051, CMRPD1H0631 to K. H. Lin; CMRPG6F0621, CMRPG6F0622, CMRPG6F0623 to C. S. Wang) and from the Ministry of Science and Technology of the Republic of China (MOST 106-2320-B-182-031-MY3 and 106-2320-B-182-032-MY3 to K.H. Lin). And The APC was funded by Chang Gung Memorial Hospital, Taoyuan, Taiwan.

Conflicts of Interest: The authors declare no conflict of interest. The funders had no role in the design of the study; in the collection, analyses, or interpretation of data; in the writing of the manuscript, or in the decision to publish the results.

References

1. Bishayee, A. The role of inflammation and liver cancer. *Adv. Exp. Med. Biol.* **2014**, *816*, 401–435. [PubMed]
2. Balogh, J.; Victor, D., 3rd; Asham, E.H.; Burroughs, S.G.; Boktour, M.; Saharia, A.; Li, X.; Ghobrial, R.M.; Monsour, H.P., Jr. Hepatocellular carcinoma: A review. *J. Hepatocell. Carcinoma* **2016**, *3*, 41–53. [CrossRef] [PubMed]
3. Burton, G.J.; Jauniaux, E. Oxidative stress. *Best Pract. Research. Clin. Obstet. Gynaecol.* **2011**, *25*, 287–299. [CrossRef]
4. Lu, T.X.; Rothenberg, M.E. MicroRNA. *J. Allergy Clin. Immunol.* **2018**, *141*, 1202–1207. [CrossRef] [PubMed]
5. Mohr, A.M.; Mott, J.L. Overview of microRNA biology. *Semin. Liver Dis.* **2015**, *35*, 3–11. [CrossRef] [PubMed]
6. Krashin, E.; Piekielko-Witkowska, A.; Ellis, M.; Ashur-Fabian, O. Thyroid Hormones and Cancer: A Comprehensive Review of Preclinical and Clinical Studies. *Front. Endocrinol.* **2019**, *10*, 59. [CrossRef]
7. Mancini, A.; Di Segni, C.; Raimondo, S.; Olivieri, G.; Silvestrini, A.; Meucci, E.; Curro, D. Thyroid Hormones, Oxidative Stress, and Inflammation. *Mediat. Inflamm.* **2016**, *2016*, 6757154. [CrossRef]
8. Kim, B.Y.; Cho, M.H.; Kim, K.J.; Cho, K.J.; Kim, S.W.; Kim, H.S.; Jung, W.W.; Lee, B.H.; Lee, B.H.; Lee, S.G. Effects of PRELI in Oxidative-Stressed HepG2 Cells. *Ann. Clin. Lab. Sci.* **2015**, *45*, 419–425.
9. Huang, F.Y.; Wong, D.K.; Tsui, V.W.; Seto, W.K.; Mak, L.Y.; Cheung, T.T.; Lai, K.K.; Yuen, M.F. Targeted genomic profiling identifies frequent deleterious mutations in FAT4 and TP53 genes in HBV-associated hepatocellular carcinoma. *BMC Cancer* **2019**, *19*, 789. [CrossRef]
10. Feld, J. Update on the Risk of Primary and Recurrent HCC With the Use of DAA Therapy for HCV Infection. *Gastroenterol. Hepatol.* **2019**, *15*, 303–306.
11. Vandenbulcke, H.; Moreno, C.; Colle, I.; Knebel, J.F.; Francque, S.; Serste, T.; George, C.; de Galocsy, C.; Laleman, W.; Delwaide, J.; et al. Alcohol intake increases the risk of HCC in hepatitis C virus-related compensated cirrhosis: A prospective study. *J. Hepatol.* **2016**, *65*, 543–551. [CrossRef] [PubMed]

12. Yang, J.D.; Ahmed, F.; Mara, K.C.; Addissie, B.D.; Allen, A.M.; Gores, G.J.; Roberts, L.R. Diabetes is Associated with Increased Risk of Hepatocellular Carcinoma in Cirrhosis Patients with Nonalcoholic Fatty Liver Disease. *Hepatology* **2019**. [CrossRef] [PubMed]

13. Bartsch, H.; Nair, J. Chronic inflammation and oxidative stress in the genesis and perpetuation of cancer: Role of lipid peroxidation, DNA damage, and repair. *Langenbeck's Arch. Surg.* **2006**, *391*, 499–510. [CrossRef] [PubMed]

14. Chen, C.Y.; Wu, S.M.; Lin, Y.H.; Chi, H.C.; Lin, S.L.; Yeh, C.T.; Chuang, W.Y.; Lin, K.H. Induction of nuclear protein-1 by thyroid hormone enhances platelet-derived growth factor A mediated angiogenesis in liver cancer. *Theranostics* **2019**, *9*, 2361–2379. [CrossRef] [PubMed]

15. Fu, Y.; Chung, F.L. Oxidative stress and hepatocarcinogenesis. *Hepatoma Res.* **2018**, *4*, 39. [CrossRef] [PubMed]

16. Ma-On, C.; Sanpavat, A.; Whongsiri, P.; Suwannasin, S.; Hirankarn, N.; Tangkijvanich, P.; Boonla, C. Oxidative stress indicated by elevated expression of Nrf2 and 8-OHdG promotes hepatocellular carcinoma progression. *Med. Oncol.* **2017**, *34*, 57. [CrossRef]

17. Bisht, S.; Dada, R. Oxidative stress: Major executioner in disease pathology, role in sperm DNA damage and preventive strategies. *Front. Biosci.* **2017**, *9*, 420–447.

18. Veskoukis, A.S.; Tsatsakis, A.M.; Kouretas, D. Dietary oxidative stress and antioxidant defense with an emphasis on plant extract administration. *Cell Stress Chaperones* **2012**, *17*, 11–21. [CrossRef]

19. Fu, N.; Yao, H.; Nan, Y.; Qiao, L. Role of Oxidative Stress in Hepatitis C Virus Induced Hepatocellular Carcinoma. *Curr. Cancer Drug Targets* **2017**, *17*, 498–504. [CrossRef]

20. Marra, M.; Sordelli, I.M.; Lombardi, A.; Lamberti, M.; Tarantino, L.; Giudice, A.; Stiuso, P.; Abbruzzese, A.; Sperlongano, R.; Accardo, M.; et al. Molecular targets and oxidative stress biomarkers in hepatocellular carcinoma: An overview. *J. Transl. Med.* **2011**, *9*, 171. [CrossRef]

21. Simic, M.G.; Bergtold, D.S.; Karam, L.R. Generation of oxy radicals in biosystems. *Mutat. Res.* **1989**, *214*, 3–12. [CrossRef]

22. Fujita, N.; Sugimoto, R.; Ma, N.; Tanaka, H.; Iwasa, M.; Kobayashi, Y.; Kawanishi, S.; Watanabe, S.; Kaito, M.; Takei, Y. Comparison of hepatic oxidative DNA damage in patients with chronic hepatitis B and C. *J. Viral Hepat.* **2008**, *15*, 498–507. [CrossRef] [PubMed]

23. Hino, K.; Nishina, S.; Sasaki, K.; Hara, Y. Mitochondrial damage and iron metabolic dysregulation in hepatitis C virus infection. *Free Radic. Biol. Med.* **2019**, *133*, 193–199. [CrossRef] [PubMed]

24. Sinha, R.A.; Singh, B.K.; Yen, P.M. Direct effects of thyroid hormones on hepatic lipid metabolism. Nature reviews. *Endocrinology* **2018**, *14*, 259–269.

25. Chi, H.C.; Chen, S.L.; Lin, S.L.; Tsai, C.Y.; Chuang, W.Y.; Lin, Y.H.; Huang, Y.H.; Tsai, M.M.; Yeh, C.T.; Lin, K.H. Thyroid hormone protects hepatocytes from HBx-induced carcinogenesis by enhancing mitochondrial turnover. *Oncogene* **2017**, *36*, 5274–5284. [CrossRef]

26. Spahis, S.; Delvin, E.; Borys, J.M.; Levy, E. Oxidative Stress as a Critical Factor in Nonalcoholic Fatty Liver Disease Pathogenesis. *Antioxid. Redox Signal.* **2017**, *26*, 519–541. [CrossRef]

27. Cardin, R.; Piciocchi, M.; Bortolami, M.; Kotsafti, A.; Barzon, L.; Lavezzo, E.; Sinigaglia, A.; Rodriguez-Castro, K.I.; Rugge, M.; Farinati, F. Oxidative damage in the progression of chronic liver disease to hepatocellular carcinoma: An intricate pathway. *World J. Gastroenterol.* **2014**, *20*, 3078–3086. [CrossRef]

28. Cai, Y.; Yu, X.; Hu, S.; Yu, J. A brief review on the mechanisms of miRNA regulation. *Genom. Proteom. Bioinform.* **2009**, *7*, 147–154. [CrossRef]

29. O'Brien, J.; Hayder, H.; Zayed, Y.; Peng, C. Overview of MicroRNA Biogenesis, Mechanisms of Actions, and Circulation. *Front. Endocrinol.* **2018**, *9*, 402. [CrossRef]

30. Wan, Y.; Cui, R.; Gu, J.; Zhang, X.; Xiang, X.; Liu, C.; Qu, K.; Lin, T. Identification of Four Oxidative Stress-Responsive MicroRNAs, miR-34a-5p, miR-1915-3p, miR-638, and miR-150-3p, in Hepatocellular Carcinoma. *Oxidative Med. Cell. Longev.* **2017**, *2017*, 5189138. [CrossRef]

31. Tian, Y.; Kuo, C.F.; Sir, D.; Wang, L.; Govindarajan, S.; Petrovic, L.M.; Ou, J.H. Autophagy inhibits oxidative stress and tumor suppressors to exert its dual effect on hepatocarcinogenesis. *Cell Death Differ.* **2015**, *22*, 1025–1034. [CrossRef]

32. Cardin, R.; Piciocchi, M.; Sinigaglia, A.; Lavezzo, E.; Bortolami, M.; Kotsafti, A.; Cillo, U.; Zanus, G.; Mescoli, C.; Rugge, M.; et al. Oxidative DNA damage correlates with cell immortalization and mir-92 expression in hepatocellular carcinoma. *BMC Cancer* **2012**, *12*, 177. [CrossRef]

33. Dai, B.H.; Geng, L.; Wang, Y.; Sui, C.J.; Xie, F.; Shen, R.X.; Shen, W.F.; Yang, J.M. microRNA-199a-5p protects hepatocytes from bile acid-induced sustained endoplasmic reticulum stress. *Cell Death Dis.* **2013**, *4*, e604. [CrossRef] [PubMed]

34. Sendi, H. Dual Role of miR-122 in Molecular Pathogenesis of Viral Hepatitis. *Hepat. Mon.* **2012**, *12*, 312–314. [CrossRef]

35. Wojcik, K.; Piekarska, A.; Szymanska, B.; Jablonowska, E. Hepatic expression of miR-122 and antioxidant genes in patients with chronic hepatitis B. *Acta Biochim. Pol.* **2016**, *63*, 527–531. [CrossRef] [PubMed]

36. Ali, O.; Darwish, H.A.; Eldeib, K.M.; Abdel Azim, S.A. miR-26a Potentially Contributes to the Regulation of Fatty Acid and Sterol Metabolism In Vitro Human HepG2 Cell Model of Nonalcoholic Fatty Liver Disease. *Oxidative Med. Cell. Longev.* **2018**, *2018*, 8515343. [CrossRef] [PubMed]

37. Wang, L.; Zhang, N.; Wang, Z.; Ai, D.M.; Cao, Z.Y.; Pan, H.P. Decreased MiR-155 Level in the Peripheral Blood of Non-Alcoholic Fatty Liver Disease Patients may Serve as a Biomarker and may Influence LXR Activity. *Cell. Physiol. Biochem.* **2016**, *39*, 2239–2248. [CrossRef]

38. Huang, P.S.; Lin, Y.H.; Chi, H.C.; Chen, P.Y.; Huang, Y.H.; Yeh, C.T.; Wang, C.S.; Lin, K.H. Thyroid hormone inhibits growth of hepatoma cells through induction of miR-214. *Sci. Rep.* **2017**, *7*, 14868. [CrossRef]

39. Gao, M.; Liu, Y.; Chen, Y.; Yin, C.; Chen, J.J.; Liu, S. miR-214 protects erythroid cells against oxidative stress by targeting ATF4 and EZH2. *Free Radic. Biol. Med.* **2016**, *92*, 39–49. [CrossRef]

40. Xiao, Y.; Yan, W.; Lu, L.; Wang, Y.; Lu, W.; Cao, Y.; Cai, W. p38/p53/miR-200a-3p feedback loop promotes oxidative stress-mediated liver cell death. *Cell Cycle* **2015**, *14*, 1548–1558. [CrossRef]

41. Zhou, B.; Li, C.; Qi, W.; Zhang, Y.; Zhang, F.; Wu, J.X.; Hu, Y.N.; Wu, D.M.; Liu, Y.; Yan, T.T.; et al. Downregulation of miR-181a upregulates sirtuin-1 (SIRT1) and improves hepatic insulin sensitivity. *Diabetologia* **2012**, *55*, 2032–2043. [CrossRef] [PubMed]

42. Guo, X.L.; Wang, H.B.; Yong, J.K.; Zhong, J.; Li, Q.H. MiR-128-3p overexpression sensitizes hepatocellular carcinoma cells to sorafenib induced apoptosis through regulating DJ-1. *Eur. Rev. Med Pharmacol. Sci.* **2018**, *22*, 6667–6677. [PubMed]

43. Zhao, X.; Jin, Y.; Li, L.; Xu, L.; Tang, Z.; Qi, Y.; Yin, L.; Peng, J. MicroRNA-128-3p aggravates doxorubicin-induced liver injury by promoting oxidative stress via targeting Sirtuin-1. *Pharmacol. Res.* **2019**, *146*, 104276. [CrossRef] [PubMed]

44. Kurtz, C.L.; Fannin, E.E.; Toth, C.L.; Pearson, D.S.; Vickers, K.C.; Sethupathy, P. Inhibition of miR-29 has a significant lipid-lowering benefit through suppression of lipogenic programs in liver. *Sci. Rep.* **2015**, *5*, 12911. [CrossRef] [PubMed]

45. Yang, J.; Qiu, B.; Li, X.; Zhang, H.; Liu, W. p53-p66(shc)/miR-21-Sod2 signaling is critical for the inhibitory effect of betulinic acid on hepatocellular carcinoma. *Toxicol. Lett.* **2015**, *238*, 1–10. [CrossRef] [PubMed]

46. Hou, W.; Tian, Q.; Zheng, J.; Bonkovsky, H.L. MicroRNA-196 represses Bach1 protein and hepatitis C virus gene expression in human hepatoma cells expressing hepatitis C viral proteins. *Hepatology* **2010**, *51*, 1494–1504. [CrossRef] [PubMed]

47. Ma, Y.; Deng, F.; Li, P.; Chen, G.; Tao, Y.; Wang, H. The tumor suppressive miR-26a regulation of FBXO11 inhibits proliferation, migration and invasion of hepatocellular carcinoma cells. *Biomed. Pharmacother.* **2018**, *101*, 648–655.

48. Liang, L.; Zeng, J.H.; Wang, J.Y.; He, R.Q.; Ma, J.; Chen, G.; Cai, X.Y.; Hu, X.H. Down-regulation of miR-26a-5p in hepatocellular carcinoma: A qRT-PCR and bioinformatics study. *Pathol. Res. Pract.* **2017**, *213*, 1494–1509. [CrossRef]

49. Li, Y.; Ren, M.; Zhao, Y.; Lu, X.; Wang, M.; Hu, J.; Lu, G.; He, S. MicroRNA-26a inhibits proliferation and metastasis of human hepatocellular carcinoma by regulating DNMT3B-MEG3 axis. *Oncol. Rep.* **2017**, *37*, 3527–3535. [CrossRef]

50. Zhao, W.T.; Lin, X.L.; Liu, Y.; Han, L.X.; Li, J.; Lin, T.Y.; Shi, J.W.; Wang, S.C.; Lian, M.; Chen, H.W.; et al. miR-26a promotes hepatocellular carcinoma invasion and metastasis by inhibiting PTEN and inhibits cell growth by repressing EZH2. *Lab. Investig.* **2019**, *99*, 1484–1500. [CrossRef]

51. Zhang, L.; Hu, J.; Hao, M.; Bu, L. Long noncoding RNA Linc01296 promotes hepatocellular carcinoma development through regulation of the miR-26a/PTEN axis. *Biol. Chem.* **2019**. [CrossRef] [PubMed]

52. Koek, G.H.; Liedorp, P.R.; Bast, A. The role of oxidative stress in non-alcoholic steatohepatitis. *Clin. Chim. Acta* **2011**, *412*, 1297–1305. [CrossRef] [PubMed]

53. Fu, X.; Dong, B.; Tian, Y.; Lefebvre, P.; Meng, Z.; Wang, X.; Pattou, F.; Han, W.; Wang, X.; Lou, F.; et al. MicroRNA-26a regulates insulin sensitivity and metabolism of glucose and lipids. *J. Clin. Investig.* **2015**, *125*, 2497–2509. [CrossRef]

54. Greene, M.W.; Burrington, C.M.; Lynch, D.T.; Davenport, S.K.; Johnson, A.K.; Horsman, M.J.; Chowdhry, S.; Zhang, J.; Sparks, J.D.; Tirrell, P.C. Lipid metabolism, oxidative stress and cell death are regulated by PKC delta in a dietary model of nonalcoholic steatohepatitis. *PLoS ONE* **2014**, *9*, e85848. [CrossRef]

55. Chen, Q.; Tang, L.; Xin, G.; Li, S.; Ma, L.; Xu, Y.; Zhuang, M.; Xiong, Q.; Wei, Z.; Xing, Z.; et al. Oxidative stress mediated by lipid metabolism contributes to high glucose-induced senescence in retinal pigment epithelium. *Free Radic. Biol. Med.* **2019**, *130*, 48–58. [CrossRef] [PubMed]

56. Fu, X.; Wen, H.; Jing, L.; Yang, Y.; Wang, W.; Liang, X.; Nan, K.; Yao, Y.; Tian, T. MicroRNA-155-5p promotes hepatocellular carcinoma progression by suppressing PTEN through the PI3K/Akt pathway. *Cancer Sci.* **2017**, *108*, 620–631. [CrossRef]

57. Liu, J.H.; Yang, Y.; Song, Q.; Li, J.B. MicroRNA-155regulates the proliferation and metastasis of human breast cancers by targeting MAPK7. *J. B. U.* **2019**, *24*, 1075–1080.

58. Wang, J.; Guo, J.; Fan, H. MiR-155 regulates the proliferation and apoptosis of pancreatic cancer cells through targeting SOCS3. *Eur. Rev. Med. Pharmacol. Sci.* **2019**, *23*, 5168–5175.

59. Liao, W.W.; Zhang, C.; Liu, F.R.; Wang, W.J. Effects of miR-155 on proliferation and apoptosis by regulating FoxO3a/BIM in liver cancer cell line HCCLM3. *Eur. Rev. Med. Pharmacol. Sci.* **2018**, *22*, 1277–1285.

60. Ning, S.; Liu, H.; Gao, B.; Wei, W.; Yang, A.; Li, J.; Zhang, L. miR-155, miR-96 and miR-99a as potential diagnostic and prognostic tools for the clinical management of hepatocellular carcinoma. *Oncol. Lett.* **2019**, *18*, 3381–3387. [CrossRef]

61. Csak, T.; Bala, S.; Lippai, D.; Kodys, K.; Catalano, D.; Iracheta-Vellve, A.; Szabo, G. MicroRNA-155 Deficiency Attenuates Liver Steatosis and Fibrosis without Reducing Inflammation in a Mouse Model of Steatohepatitis. *PLoS ONE* **2015**, *10*, e0129251. [CrossRef]

62. Bala, S.; Csak, T.; Saha, B.; Zatsiorsky, J.; Kodys, K.; Catalano, D.; Satishchandran, A.; Szabo, G. The pro-inflammatory effects of miR-155 promote liver fibrosis and alcohol-induced steatohepatitis. *J. Hepatol.* **2016**, *64*, 1378–1387. [CrossRef] [PubMed]

63. Perra, A.; Plateroti, M.; Columbano, A. T3/TRs axis in hepatocellular carcinoma: New concepts for an old pair. *Endocr. Relat. Cancer* **2016**, *23*, R353–R369. [CrossRef] [PubMed]

64. Wu, S.M.; Cheng, W.L.; Lin, C.D.; Lin, K.H. Thyroid hormone actions in liver cancer. *Cell. Mol. Life Sci.* **2013**, *70*, 1915–1936. [CrossRef] [PubMed]

65. Liu, Y.C.; Yeh, C.T.; Lin, K.H. Molecular Functions of Thyroid Hormone Signaling in Regulation of Cancer Progression and Anti-Apoptosis. *Int. J. Mol. Sci.* **2019**, *20*, 4986. [CrossRef]

66. Manka, P.; Coombes, J.D.; Boosman, R.; Gauthier, K.; Papa, S.; Syn, W.K. Thyroid hormone in the regulation of hepatocellular carcinoma and its microenvironment. *Cancer Lett.* **2018**, *419*, 175–186. [CrossRef]

67. Chi, H.C.; Tsai, C.Y.; Tsai, M.M.; Yeh, C.T.; Lin, K.H. Molecular functions and clinical impact of thyroid hormone-triggered autophagy in liver-related diseases. *J. Biomed. Sci.* **2019**, *26*, 24. [CrossRef]

68. Huang, F.; Wu, L.; Qiu, Y.; Bu, K.; Huang, H.; Li, B. The role of free triiodothyronine in high-density lipoprotein cholesterol metabolism. *Medicine* **2019**, *98*, e17016. [CrossRef]

69. Orsagova, I.; RoZnovsky, L.; Petrousova, L.; Konecna, M.; Kabieszova, L.; Safarcik, K.; Kloudova, A. Thyroid dysfunction during interferon alpha therapy for chronic hepatitis B and C—Twenty years of experience. *Klin. Mikrobiol. A Infekcni Lek.* **2014**, *20*, 92–97.

70. Mansour-Ghanaei, F.; Mehrdad, M.; Mortazavi, S.; Joukar, F.; Khak, M.; Atrkar-Roushan, Z. Decreased serum total T3 level in hepatitis B and C related cirrhosis by severity of liver damage. *Ann. Hepatol.* **2012**, *11*, 667–671. [CrossRef]

71. Chi, H.C.; Chen, C.Y.; Tsai, M.M.; Tsai, C.Y.; Lin, K.H. Molecular functions of thyroid hormones and their clinical significance in liver-related diseases. *Biomed Res. Int.* **2013**, *2013*, 601361. [CrossRef] [PubMed]

72. Lanni, A.; Moreno, M.; Goglia, F. Mitochondrial Actions of Thyroid Hormone. *Compr. Physiol.* **2016**, *6*, 1591–1607. [PubMed]

73. Videla, L.A.; Fernandez, V.; Cornejo, P.; Vargas, R.; Castillo, I. Thyroid hormone in the frontier of cell protection, survival and functional recovery. *Expert Rev. Mol. Med.* **2015**, *17*, e10. [CrossRef] [PubMed]

74. Lin, Y.H.; Liao, C.J.; Huang, Y.H.; Wu, M.H.; Chi, H.C.; Wu, S.M.; Chen, C.Y.; Tseng, Y.H.; Tsai, C.Y.; Chung, I.H.; et al. Thyroid hormone receptor represses miR-17 expression to enhance tumor metastasis in human hepatoma cells. *Oncogene* **2013**, *32*, 4509–4518. [CrossRef]

75. Huang, Y.H.; Lin, Y.H.; Chi, H.C.; Liao, C.H.; Liao, C.J.; Wu, S.M.; Chen, C.Y.; Tseng, Y.H.; Tsai, C.Y.; Lin, S.Y.; et al. Thyroid hormone regulation of miR-21 enhances migration and invasion of hepatoma. *Cancer Res.* **2013**, *73*, 2505–2517. [CrossRef]

76. Lin, Y.H.; Wu, M.H.; Liao, C.J.; Huang, Y.H.; Chi, H.C.; Wu, S.M.; Chen, C.Y.; Tseng, Y.H.; Tsai, C.Y.; Chung, I.H.; et al. Repression of microRNA-130b by thyroid hormone enhances cell motility. *J. Hepatol.* **2015**, *62*, 1328–1340. [CrossRef]

77. Xu, Q.; Zhang, M.; Tu, J.; Pang, L.; Cai, W.; Liu, X. MicroRNA-122 affects cell aggressiveness and apoptosis by targeting PKM2 in human hepatocellular carcinoma. *Oncol. Rep.* **2015**, *34*, 2054–2064. [CrossRef]

78. Dong, H.; Curran, I.; Williams, A.; Bondy, G.; Yauk, C.L.; Wade, M.G. Hepatic miRNA profiles and thyroid hormone homeostasis in rats exposed to dietary potassium perfluorooctanesulfonate (PFOS). *Environ. Toxicol. Pharmacol.* **2016**, *41*, 201–210. [CrossRef]

79. Liu, Q.; Zhang, Y.; Yang, S.; Wu, Y.; Wang, J.; Yu, W.; Liu, Y. PU.1-deficient mice are resistant to thioacetamide-induced hepatic fibrosis: PU.1 finely regulates Sirt1 expression via transcriptional promotion of miR-34a and miR-29c in hepatic stellate cells. *Biosci. Rep.* **2017**, *37*. [CrossRef]

80. Yu, Q.; Zhou, J.; Jian, Y.; Xiu, Z.; Xiang, L.; Yang, D.; Zeng, W. MicroRNA-214 suppresses cell proliferation and migration and cell metabolism by targeting PDK2 and PHF6 in hepatocellular carcinoma. *Cell Biol. Int.* **2019**. [CrossRef]

81. Tian, C.; Wu, H.; Li, C.; Tian, X.; Sun, Y.; Liu, E.; Liao, X.; Song, W. Downreguation of FoxM1 by miR-214 inhibits proliferation and migration in hepatocellular carcinoma. *Gene Ther.* **2018**, *25*, 312–319. [CrossRef] [PubMed]

82. Zhang, Y.; Yang, L.; Wang, S.; Liu, Z.; Xiu, M. MiR-29a suppresses cell proliferation by targeting SIRT1 in hepatocellular carcinoma. *Cancer Biomark.* **2018**, *22*, 151–159. [CrossRef] [PubMed]

83. Fan, J.C.; Zeng, F.; Le, Y.G.; Xin, L. LncRNA CASC2 inhibited the viability and induced the apoptosis of hepatocellular carcinoma cells through regulating miR-24-3p. *J. Cell. Biochem.* **2018**, *119*, 6391–6397. [CrossRef] [PubMed]

84. Jiang, J.; Yang, P.; Guo, Z.; Yang, R.; Yang, H.; Yang, F.; Li, L.; Xiang, B. Overexpression of microRNA-21 strengthens stem cell-like characteristics in a hepatocellular carcinoma cell line. *World J. Surg. Oncol.* **2016**, *14*, 278. [CrossRef] [PubMed]

85. Zhang, J.; Wang, Y.; Zhen, P.; Luo, X.; Zhang, C.; Zhou, L.; Lu, Y.; Yang, Y.; Zhang, W.; Wan, J. Genome-wide analysis of miRNA signature differentially expressed in doxorubicin-resistant and parental human hepatocellular carcinoma cell lines. *PLoS ONE* **2013**, *8*, e54111. [CrossRef] [PubMed]

86. Rodrigues, P.M.; Rodrigues, C.M.P.; Castro, R.E. Modulation of liver steatosis by miR-21/PPARalpha. *Cell Death Discov.* **2018**, *4*, 9. [CrossRef] [PubMed]

87. Chen, Z.; Shentu, T.P.; Wen, L.; Johnson, D.A.; Shyy, J.Y. Regulation of SIRT1 by oxidative stress-responsive miRNAs and a systematic approach to identify its role in the endothelium. *Antioxid. Redox Signal.* **2013**, *19*, 1522–1538. [CrossRef]

88. Tang, Q.; Wang, Q.; Zhang, Q.; Lin, S.Y.; Zhu, Y.; Yang, X.; Guo, A.Y. Gene expression, regulation of DEN and HBx induced HCC mice models and comparisons of tumor, para-tumor and normal tissues. *BMC Cancer* **2017**, *17*, 862. [CrossRef]

89. Bonauer, A.; Carmona, G.; Iwasaki, M.; Mione, M.; Koyanagi, M.; Fischer, A.; Burchfield, J.; Fox, H.; Doebele, C.; Ohtani, K.; et al. MicroRNA-92a controls angiogenesis and functional recovery of ischemic tissues in mice. *Science* **2009**, *324*, 1710–1713. [CrossRef]

90. Urtasun, R.; Elizalde, M.; Azkona, M.; Latasa, M.U.; Garcia-Irigoyen, O.; Uriarte, I.; Fernandez-Barrena, M.G.; Vicent, S.; Alonso, M.M.; Muntane, J.; et al. Splicing regulator SLU7 preserves survival of hepatocellular carcinoma cells and other solid tumors via oncogenic miR-17-92 cluster expression. *Oncogene* **2016**, *35*, 4719–4729. [CrossRef]

91. Peng, R.; Men, J.; Ma, R.; Wang, Q.; Wang, Y.; Sun, Y.; Ren, J. miR-214 down-regulates ARL2 and suppresses growth and invasion of cervical cancer cells. *Biochem. Biophys. Res. Commun.* **2017**, *484*, 623–630. [CrossRef] [PubMed]

92. Zheng, C.; Guo, K.; Chen, B.; Wen, Y.; Xu, Y. miR-214-5p inhibits human prostate cancer proliferation and migration through regulating CRMP5. *Cancer Biomark.* **2019**. [CrossRef] [PubMed]

93. Liu, Y.; Zhou, H.; Ma, L.; Hou, Y.; Pan, J.; Sun, C.; Yang, Y.; Zhang, J. MiR-214 suppressed ovarian cancer and negatively regulated semaphorin 4D. *Tumour Biol.* **2016**, *37*, 8239–8248. [CrossRef] [PubMed]

94. Shih, T.C.; Tien, Y.J.; Wen, C.J.; Yeh, T.S.; Yu, M.C.; Huang, C.H.; Lee, Y.S.; Yen, T.C.; Hsieh, S.Y. MicroRNA-214 downregulation contributes to tumor angiogenesis by inducing secretion of the hepatoma-derived growth factor in human hepatoma. *J. Hepatol.* **2012**, *57*, 584–591. [CrossRef] [PubMed]

95. Jin, Y.; Wong, Y.S.; Goh, B.K.P.; Chan, C.Y.; Cheow, P.C.; Chow, P.K.H.; Lim, T.K.H.; Goh, G.B.B.; Krishnamoorthy, T.L.; Kumar, R.; et al. Circulating microRNAs as Potential Diagnostic and Prognostic Biomarkers in Hepatocellular Carcinoma. *Sci. Rep.* **2019**, *9*, 10464. [CrossRef] [PubMed]

96. Huang, X.; Gao, Y.; Qin, J.; Lu, S. lncRNA MIAT promotes proliferation and invasion of HCC cells via sponging miR-214. *Am. J. Physiol. Gastrointest. Liver Physiol.* **2018**, *314*, G559–G565. [CrossRef]

97. Gou, X.; Zhao, X.; Wang, Z. Long noncoding RNA PVT1 promotes hepatocellular carcinoma progression through regulating miR-214. *Cancer Biomark.* **2017**, *20*, 511–519. [CrossRef]

98. Park, D.H.; Shin, J.W.; Park, S.K.; Seo, J.N.; Li, L.; Jang, J.J.; Lee, M.J. Diethylnitrosamine (DEN) induces irreversible hepatocellular carcinogenesis through overexpression of G1/S-phase regulatory proteins in rat. *Toxicol. Lett.* **2009**, *191*, 321–326. [CrossRef]

99. Zhou, F.; Shen, T.; Duan, T.; Xu, Y.Y.; Khor, S.C.; Li, J.; Ge, J.; Zheng, Y.F.; Hsu, S.; De Stefano, J.; et al. Antioxidant effects of lipophilic tea polyphenols on diethylnitrosamine/phenobarbital-induced hepatocarcinogenesis in rats. *In Vivo* **2014**, *28*, 495–503.

100. Unsal, V.; Belge-Kurutas, E. Experimental Hepatic Carcinogenesis: Oxidative Stress and Natural Antioxidants. *Maced. J. Med. Sci.* **2017**, *5*, 686–691. [CrossRef]

101. Chi, H.C.; Chen, S.L.; Tsai, C.Y.; Chuang, W.Y.; Huang, Y.H.; Tsai, M.M.; Wu, S.M.; Sun, C.P.; Yeh, C.T.; Lin, K.H. Thyroid hormone suppresses hepatocarcinogenesis via DAPK2 and SQSTM1-dependent selective autophagy. *Autophagy* **2016**, *12*, 2271–2285. [CrossRef] [PubMed]

102. Chen, C.; Lu, L.; Yan, S.; Yi, H.; Yao, H.; Wu, D.; He, G.; Tao, X.; Deng, X. Autophagy and doxorubicin resistance in cancer. *Anti-Cancer Drugs* **2018**, *29*, 1–9. [CrossRef] [PubMed]

103. Ozlu, B.; Kabay, G.; Bocek, I.; Yilmaz, M.; Piskin, A.K.; Shim, B.S.; Mutlu, M. Controlled release of doxorubicin from polyethylene glycol functionalized melanin nanoparticles for breast cancer therapy: Part I. Production and drug release performance of the melanin nanoparticles. *Int. J. Pharm.* **2019**, *570*, 118613. [CrossRef] [PubMed]

104. Damiani, V.; Falvo, E.; Fracasso, G.; Federici, L.; Pitea, M.; De Laurenzi, V.; Sala, G.; Ceci, P. Therapeutic Efficacy of the Novel Stimuli-Sensitive Nano-Ferritins Containing Doxorubicin in a Head and Neck Cancer Model. *Int. J. Mol. Sci.* **2017**, *18*, 1555. [CrossRef] [PubMed]

105. Zheng, F.M.; Chen, W.B.; Qin, T.; Lv, L.N.; Feng, B.; Lu, Y.L.; Li, Z.Q.; Wang, X.C.; Tao, L.J.; Li, H.W.; et al. ACOX1 destabilizes p73 to suppress intrinsic apoptosis pathway and regulates sensitivity to doxorubicin in lymphoma cells. *BMB Rep.* **2019**, *52*, 566–571. [CrossRef] [PubMed]

106. Fang, Y.; Yang, W.; Cheng, L.; Meng, F.; Zhang, J.; Zhong, Z. EGFR-targeted multifunctional polymersomal doxorubicin induces selective and potent suppression of orthotopic human liver cancer in vivo. *Acta Biomater.* **2017**, *64*, 323–333. [CrossRef]

107. Akolkar, G.; da Silva Dias, D.; Ayyappan, P.; Bagchi, A.K.; Jassal, D.S.; Salemi, V.M.C.; Irigoyen, M.C.; De Angelis, K.; Singal, P.K. Vitamin C mitigates oxidative/nitrosative stress and inflammation in doxorubicin-induced cardiomyopathy. *Am. J. Physiol. Heart Circ. Physiol.* **2017**, *313*, H795–H809. [CrossRef]

108. Abdel-Daim, M.M.; Kilany, O.E.; Khalifa, H.A.; Ahmed, A.A.M. Allicin ameliorates doxorubicin-induced cardiotoxicity in rats via suppression of oxidative stress, inflammation and apoptosis. *Cancer Chemother. Pharmacol.* **2017**, *80*, 745–753. [CrossRef]

109. Govender, J.; Loos, B.; Marais, E.; Engelbrecht, A.M. Mitochondrial catastrophe during doxorubicin-induced cardiotoxicity: A review of the protective role of melatonin. *J. Pineal Res.* **2014**, *57*, 367–380. [CrossRef]

110. Chi, H.C.; Chen, S.L.; Cheng, Y.H.; Lin, T.K.; Tsai, C.Y.; Tsai, M.M.; Lin, Y.H.; Huang, Y.H.; Lin, K.H. Chemotherapy resistance and metastasis-promoting effects of thyroid hormone in hepatocarcinoma cells are mediated by suppression of FoxO1 and Bim pathway. *Cell Death Dis.* **2016**, *7*, e2324. [CrossRef]

111. Deng, L.; Chen, M.; Tanaka, M.; Ku, Y.; Itoh, T.; Shoji, I.; Hotta, H. HCV upregulates Bim through the ROS/JNK signalling pathway, leading to Bax-mediated apoptosis. *J. Gen. Virol.* **2015**, *96*, 2670–2683. [CrossRef] [PubMed]

112. Zhang, Z.C.; Li, Y.Y.; Wang, H.Y.; Fu, S.; Wang, X.P.; Zeng, M.S.; Zeng, Y.X.; Shao, J.Y. Knockdown of miR-214 promotes apoptosis and inhibits cell proliferation in nasopharyngeal carcinoma. *PLoS ONE* **2014**, *9*, e86149. [CrossRef] [PubMed]

113. Aurora, A.B.; Mahmoud, A.I.; Luo, X.; Johnson, B.A.; van Rooij, E.; Matsuzaki, S.; Humphries, K.M.; Hill, J.A.; Bassel-Duby, R.; Sadek, H.A.; et al. MicroRNA-214 protects the mouse heart from ischemic injury by controlling Ca(2)(+) overload and cell death. *J. Clin. Investig.* **2012**, *122*, 1222–1232. [CrossRef] [PubMed]

114. Wang, X.; Ha, T.; Hu, Y.; Lu, C.; Liu, L.; Zhang, X.; Kao, R.; Kalbfleisch, J.; Williams, D.; Li, C. MicroRNA-214 protects against hypoxia/reoxygenation induced cell damage and myocardial ischemia/reperfusion injury via suppression of PTEN and Bim1 expression. *Oncotarget* **2016**, *7*, 86926–86936. [CrossRef] [PubMed]

115. Song, L.; Zhang, W.; Chang, Z.; Pan, Y.; Zong, H.; Fan, Q.; Wang, L. miR-4417 Targets Tripartite Motif-Containing 35 (TRIM35) and Regulates Pyruvate Kinase Muscle 2 (PKM2) Phosphorylation to Promote Proliferation and Suppress Apoptosis in Hepatocellular Carcinoma Cells. *Med. Sci. Monit.* **2017**, *23*, 1741–1750. [CrossRef]

116. Li, W.; Qiu, Y.; Hao, J.; Zhao, C.; Deng, X.; Shu, G. Dauricine upregulates the chemosensitivity of hepatocellular carcinoma cells: Role of repressing glycolysis via miR-199a:HK2/PKM2 modulation. *Food Chem. Toxicol.* **2018**, *121*, 156–165. [CrossRef]

117. Yu, G.; Wang, J.; Xu, K.; Dong, J. Dynamic regulation of uncoupling protein 2 expression by microRNA-214 in hepatocellular carcinoma. *Biosci. Rep.* **2016**, *36*, e0035. [CrossRef]

118. Yokoyama, C.; Sueyoshi, Y.; Ema, M.; Mori, Y.; Takaishi, K.; Hisatomi, H. Induction of oxidative stress by anticancer drugs in the presence and absence of cells. *Oncol. Lett.* **2017**, *14*, 6066–6070. [CrossRef]

119. Yang, S.; Fei, X.; Lu, Y.; Xu, B.; Ma, Y.; Wan, H. miRNA-214 suppresses oxidative stress in diabetic nephropathy via the ROS/Akt/mTOR signaling pathway and uncoupling protein 2. *Exp. Ther. Med.* **2019**, *17*, 3530–3538. [CrossRef]

120. Liu, P.; Zhao, H.; Wang, R.; Wang, P.; Tao, Z.; Gao, L.; Yan, F.; Liu, X.; Yu, S.; Ji, X.; et al. MicroRNA-424 protects against focal cerebral ischemia and reperfusion injury in mice by suppressing oxidative stress. *Stroke* **2015**, *46*, 513–519. [CrossRef]

121. Henke, J.I.; Goergen, D.; Zheng, J.; Song, Y.; Schuttler, C.G.; Fehr, C.; Junemann, C.; Niepmann, M. microRNA-122 stimulates translation of hepatitis C virus RNA. *EMBO J.* **2008**, *27*, 3300–3310. [CrossRef] [PubMed]

122. Nabih, H.K. The Significance of HCV Viral Load in the Incidence of HCC: A Correlation Between Mir-122 and CCL2. *J. Gastrointest. Cancer* **2019**. [CrossRef] [PubMed]

123. Jopling, C.L. Regulation of hepatitis C virus by microRNA-122. *Biochem. Soc. Trans.* **2008**, *36*, 1220–1223. [CrossRef] [PubMed]

124. Li, Y.; Wang, L.; Rivera-Serrano, E.E.; Chen, X.; Lemon, S.M. TNRC6 proteins modulate hepatitis C virus replication by spatially regulating the binding of miR-122/Ago2 complexes to viral RNA. *Nucleic Acids Res.* **2019**, *47*, 6411–6424. [CrossRef] [PubMed]

125. Chahal, J.; Gebert, L.F.R.; Gan, H.H.; Camacho, E.; Gunsalus, K.C.; MacRae, I.J.; Sagan, S.M. miR-122 and Ago interactions with the HCV genome alter the structure of the viral 5' terminus. *Nucleic Acids Res.* **2019**, *47*, 5307–5324. [CrossRef]

126. Mahmoudian-Sani, M.R.; Asgharzade, S.; Alghasi, A.; Saeedi-Boroujeni, A.; Adnani Sadati, S.J.; Moradi, M.T. MicroRNA-122 in patients with hepatitis B and hepatitis B virus-associated hepatocellular carcinoma. *J. Gastrointest. Oncol.* **2019**, *10*, 789–796. [CrossRef]

127. Rana, M.A.; Ijaz, B.; Daud, M.; Tariq, S.; Nadeem, T.; Husnain, T. Interplay of Wnt beta-catenin pathway and miRNAs in HBV pathogenesis leading to HCC. *Clin. Res. Hepatol. Gastroenterol.* **2019**, *43*, 373–386. [CrossRef]

128. Liu, G.; Ma, X.; Wang, Z.; Wakae, K.; Yuan, Y.; He, Z.; Yoshiyama, H.; Iizasa, H.; Zhang, H.; Matsuda, M.; et al. Adenosine deaminase acting on RNA-1 (ADAR1) inhibits HBV replication by enhancing microRNA-122 processing. *J. Biol. Chem.* **2019**, *294*, 14043–14054. [CrossRef]

129. He, J.; Zhao, K.; Zheng, L.; Xu, Z.; Gong, W.; Chen, S.; Shen, X.; Huang, G.; Gao, M.; Zeng, Y.; et al. Upregulation of microRNA-122 by farnesoid X receptor suppresses the growth of hepatocellular carcinoma cells. *Mol. Cancer* **2015**, *14*, 163. [CrossRef]

130. Nassirpour, R.; Mehta, P.P.; Yin, M.J. miR-122 regulates tumorigenesis in hepatocellular carcinoma by targeting AKT3. *PLoS ONE* **2013**, *8*, e79655. [CrossRef]

131. Ahsani, Z.; Mohammadi-Yeganeh, S.; Kia, V.; Karimkhanloo, H.; Zarghami, N.; Paryan, M. WNT1 Gene from WNT Signaling Pathway Is a Direct Target of miR-122 in Hepatocellular Carcinoma. *Appl. Biochem. Biotechnol.* **2017**, *181*, 884–897. [CrossRef]

132. Dinkova-Kostova, A.T.; Talalay, P. NAD(P)H:quinone acceptor oxidoreductase 1 (NQO1), a multifunctional antioxidant enzyme and exceptionally versatile cytoprotector. *Arch. Biochem. Biophys.* **2010**, *501*, 116–123. [CrossRef]

133. Vomund, S.; Schafer, A.; Parnham, M.J.; Brune, B.; von Knethen, A. Nrf2, the Master Regulator of Anti-Oxidative Responses. *Int. J. Mol. Sci.* **2017**, *18*, 2772. [CrossRef]

134. Yu, W.G.; Liu, W.; Jin, Y.H. Effects of perfluorooctane sulfonate on rat thyroid hormone biosynthesis and metabolism. *Environ. Toxicol. Chem.* **2009**, *28*, 990–996. [CrossRef]

135. Wang, G.; Dong, F.; Xu, Z.; Sharma, S.; Hu, X.; Chen, D.; Zhang, L.; Zhang, J.; Dong, Q. MicroRNA profile in HBV-induced infection and hepatocellular carcinoma. *BMC Cancer* **2017**, *17*, 805. [CrossRef]

136. Sadri Nahand, J.; Bokharaei-Salim, F.; Salmaninejad, A.; Nesaei, A.; Mohajeri, F.; Moshtzan, A.; Tabibzadeh, A.; Karimzadeh, M.; Moghoofei, M.; Marjani, A.; et al. microRNAs: Key players in virus-associated hepatocellular carcinoma. *J. Cell. Physiol.* **2019**, *234*, 12188–12225. [CrossRef]

137. Shigoka, M.; Tsuchida, A.; Matsudo, T.; Nagakawa, Y.; Saito, H.; Suzuki, Y.; Aoki, T.; Murakami, Y.; Toyoda, H.; Kumada, T.; et al. Deregulation of miR-92a expression is implicated in hepatocellular carcinoma development. *Pathol. Int.* **2010**, *60*, 351–357. [CrossRef]

138. Wang, L.; Wu, J.; Xie, C. miR-92a promotes hepatocellular carcinoma cells proliferation and invasion by FOXA2 targeting. *Iran. J. Basic Med. Sci.* **2017**, *20*, 783–790.

139. Zhang, Y.; Li, T.; Qiu, Y.; Zhang, T.; Guo, P.; Ma, X.; Wei, Q.; Han, L. Serum microRNA panel for early diagnosis of the onset of hepatocellular carcinoma. *Medicine* **2017**, *96*, e5642. [CrossRef]

140. Zhao, B.; Zhu, Y.; Cui, K.; Gao, J.; Yu, F.; Chen, L.; Li, S. Expression and significance of PTEN and miR-92 in hepatocellular carcinoma. *Mol. Med. Rep.* **2013**, *7*, 1413–1416. [CrossRef]

141. Yang, W.; Dou, C.; Wang, Y.; Jia, Y.; Li, C.; Zheng, X.; Tu, K. MicroRNA-92a contributes to tumor growth of human hepatocellular carcinoma by targeting FBXW7. *Oncol. Rep.* **2015**, *34*, 2576–2584. [CrossRef]

142. Ock, C.Y.; Kim, E.H.; Choi, D.J.; Lee, H.J.; Hahm, K.B.; Chung, M.H. 8-Hydroxydeoxyguanosine: Not mere biomarker for oxidative stress, but remedy for oxidative stress-implicated gastrointestinal diseases. *World J. Gastroenterol.* **2012**, *18*, 302–308. [CrossRef]

143. Liu, D.Y.; Peng, Z.H.; Qiu, G.Q.; Zhou, C.Z. Expression of telomerase activity and oxidative stress in human hepatocellular carcinoma with cirrhosis. *World J. Gastroenterol.* **2003**, *9*, 1859–1862. [CrossRef]

144. Saini, N.; Srinivasan, R.; Chawla, Y.; Sharma, S.; Chakraborti, A.; Rajwanshi, A. Telomerase activity, telomere length and human telomerase reverse transcriptase expression in hepatocellular carcinoma is independent of hepatitis virus status. *Liver Int.* **2009**, *29*, 1162–1170. [CrossRef]

145. Zhang, W.; Huang, Q.; Zeng, Z.; Wu, J.; Zhang, Y.; Chen, Z. Sirt1 Inhibits Oxidative Stress in Vascular Endothelial Cells. *Oxidative Med. Cell. Longev.* **2017**, *2017*, 7543973. [CrossRef]

146. Frau, C.; Loi, R.; Petrelli, A.; Perra, A.; Menegon, S.; Kowalik, M.A.; Pinna, S.; Leoni, V.P.; Fornari, F.; Gramantieri, L.; et al. Local hypothyroidism favors the progression of preneoplastic lesions to hepatocellular carcinoma in rats. *Hepatology* **2015**, *61*, 249–259. [CrossRef]

147. Pang, C.; Huang, G.; Luo, K.; Dong, Y.; He, F.; Du, G.; Xiao, M.; Cai, W. miR-206 inhibits the growth of hepatocellular carcinoma cells via targeting CDK9. *Cancer Med.* **2017**, *6*, 2398–2409. [CrossRef]

148. Yang, Q.; Zhang, L.; Zhong, Y.; Lai, L.; Li, X. miR-206 inhibits cell proliferation, invasion, and migration by down-regulating PTP1B in hepatocellular carcinoma. *Biosci. Rep.* **2019**, *39*. [CrossRef]

149. Wang, Y.; Tai, Q.; Zhang, J.; Kang, J.; Gao, F.; Zhong, F.; Cai, L.; Fang, F.; Gao, Y. MiRNA-206 inhibits hepatocellular carcinoma cell proliferation and migration but promotes apoptosis by modulating cMET expression. *Acta Biochim. Et Biophys. Sin.* **2019**, *51*, 243–253. [CrossRef]

150. Singh, A.; Happel, C.; Manna, S.K.; Acquaah-Mensah, G.; Carrerero, J.; Kumar, S.; Nasipuri, P.; Krausz, K.W.; Wakabayashi, N.; Dewi, R.; et al. Transcription factor NRF2 regulates miR-1 and miR-206 to drive tumorigenesis. *J. Clin. Investig.* **2013**, *123*, 2921–2934. [CrossRef]

151. Zheng, Y.; Zhao, C.; Zhang, N.; Kang, W.; Lu, R.; Wu, H.; Geng, Y.; Zhao, Y.; Xu, X. Serum microRNA miR-206 is decreased in hyperthyroidism and mediates thyroid hormone regulation of lipid metabolism in HepG2 human hepatoblastoma cells. *Mol. Med. Rep.* **2018**, *17*, 5635–5641. [CrossRef]

152. Huang, C.Y.; Huang, X.P.; Zhu, J.Y.; Chen, Z.G.; Li, X.J.; Zhang, X.H.; Huang, S.; He, J.B.; Lian, F.; Zhao, Y.N.; et al. miR-128-3p suppresses hepatocellular carcinoma proliferation by regulating PIK3R1 and is correlated with the prognosis of HCC patients. *Oncol. Rep.* **2015**, *33*, 2889–2898. [CrossRef]

153. Zhao, X.; Wu, Y.; Lv, Z. miR-128 modulates hepatocellular carcinoma by inhibition of ITGA2 and ITGA5 expression. *Am. J. Transl. Res.* **2015**, *7*, 1564–1573.

154. Lee, D.Y.; Kim, H.S.; Won, K.J.; Lee, K.P.; Jung, S.H.; Park, E.S.; Choi, W.S.; Lee, H.M.; Kim, B. DJ-1 regulates the expression of renal (pro)renin receptor via reactive oxygen species-mediated epigenetic modification. *Biochim. Et Biophys. Acta* **2015**, *1850*, 426–434. [CrossRef]

155. Bonilha, V.L.; Bell, B.A.; Rayborn, M.E.; Yang, X.; Kaul, C.; Grossman, G.H.; Samuels, I.S.; Hollyfield, J.G.; Xie, C.; Cai, H.; et al. Loss of DJ-1 elicits retinal abnormalities, visual dysfunction, and increased oxidative stress in mice. *Exp. Eye Res.* **2015**, *139*, 22–36. [CrossRef]

156. Wang, J.; Chu, Y.; Xu, M.; Zhang, X.; Zhou, Y.; Xu, M. miR-21 promotes cell migration and invasion of hepatocellular carcinoma by targeting KLF5. *Oncol. Lett.* **2019**, *17*, 2221–2227. [CrossRef]

157. Musaddaq, G.; Shahzad, N.; Ashraf, M.A.; Arshad, M.I. Circulating liver-specific microRNAs as noninvasive diagnostic biomarkers of hepatic diseases in human. *Biomarkers* **2019**, *24*, 103–109. [CrossRef]

158. Zhan, Y.; Zheng, N.; Teng, F.; Bao, L.; Liu, F.; Zhang, M.; Guo, M.; Guo, W.; Ding, G.; Wang, Q. MiR-199a/b-5p inhibits hepatocellular carcinoma progression by post-transcriptionally suppressing ROCK1. *Oncotarget* **2017**, *8*, 67169–67180. [CrossRef]

159. Lou, Z.; Gong, Y.Q.; Zhou, X.; Hu, G.H. Low expression of miR-199 in hepatocellular carcinoma contributes to tumor cell hyper-proliferation by negatively suppressing XBP1. *Oncol. Lett.* **2018**, *16*, 6531–6539. [CrossRef]

160. Bernstein, H.; Payne, C.M.; Bernstein, C.; Schneider, J.; Beard, S.E.; Crowley, C.L. Activation of the promoters of genes associated with DNA damage, oxidative stress, ER stress and protein malfolding by the bile salt, deoxycholate. *Toxicol. Lett.* **1999**, *108*, 37–46. [CrossRef]

161. Silva, T.M.; Moretto, F.C.F.; Sibio, M.T.; Goncalves, B.M.; Oliveira, M.; Olimpio, R.M.C.; Oliveira, D.A.M.; Costa, S.M.B.; Depra, I.C.; Namba, V.; et al. Triiodothyronine (T3) upregulates the expression of proto-oncogene TGFA independent of MAPK/ERK pathway activation in the human breast adenocarcinoma cell line, MCF7. *Arch. Endocrinol. Metab.* **2019**, *63*, 142–147. [CrossRef]

162. Cicatiello, A.G.; Ambrosio, R.; Dentice, M. Thyroid hormone promotes differentiation of colon cancer stem cells. *Mol. Cell. Endocrinol.* **2017**, *459*, 84–89. [CrossRef]

163. Latteyer, S.; Christoph, S.; Theurer, S.; Hones, G.S.; Schmid, K.W.; Fuehrer, D.; Moeller, L.C. Thyroxine promotes lung cancer growth in an orthotopic mouse model. *Endocr. Relat. Cancer* **2019**. [CrossRef]

164. Lin, Y.H.; Wu, M.H.; Huang, Y.H.; Yeh, C.T.; Chi, H.C.; Tsai, C.Y.; Chuang, W.Y.; Yu, C.J.; Chung, I.H.; Chen, C.Y.; et al. Thyroid hormone negatively regulates tumorigenesis through suppression of BC200. *Endocr. Relat. Cancer* **2018**, *25*, 967–979. [CrossRef]

165. Sap, J.; Munoz, A.; Damm, K.; Goldberg, Y.; Ghysdael, J.; Leutz, A.; Beug, H.; Vennstrom, B. The c-erb-A protein is a high-affinity receptor for thyroid hormone. *Nature* **1986**, *324*, 635–640. [CrossRef]

166. Barlow, C.; Meister, B.; Lardelli, M.; Lendahl, U.; Vennstrom, B. Thyroid abnormalities and hepatocellular carcinoma in mice transgenic for v-erbA. *EMBO J.* **1994**, *13*, 4241–4250. [CrossRef]

167. Kabir, A. Decreased serum total T3 level in hepatitis B and C related cirrhosis by severity of liver damage. *Ann. Hepatol.* **2013**, *12*, 506–507. [CrossRef]

168. Hassan, M.M.; Kaseb, A.; Li, D.; Patt, Y.Z.; Vauthey, J.N.; Thomas, M.B.; Curley, S.A.; Spitz, M.R.; Sherman, S.I.; Abdalla, E.K.; et al. Association between hypothyroidism and hepatocellular carcinoma: A case-control study in the United States. *Hepatology* **2009**, *49*, 1563–1570. [CrossRef]

169. Reddy, A.; Dash, C.; Leerapun, A.; Mettler, T.A.; Stadheim, L.M.; Lazaridis, K.N.; Roberts, R.O.; Roberts, L.R. Hypothyroidism: A possible risk factor for liver cancer in patients with no known underlying cause of liver disease. *Clin. Gastroenterol. Hepatol.* **2007**, *5*, 118–123. [CrossRef]

170. Yu, L.X.; Ling, Y.; Wang, H.Y. Role of nonresolving inflammation in hepatocellular carcinoma development and progression. *Npj Precis. Oncol.* **2018**, *2*, 6. [CrossRef]

171. Adler, S.M.; Wartofsky, L. The nonthyroidal illness syndrome. *Endocrinol. Metab. Clin. North Am.* **2007**, *36*, 657–672. [CrossRef]

172. De Groot, L.J. Non-thyroidal illness syndrome is a manifestation of hypothalamic-pituitary dysfunction, and in view of current evidence, should be treated with appropriate replacement therapies. *Crit. Care Clin.* **2006**, *22*, 57–86. [CrossRef]

173. Lin, S.L.; Wu, S.M.; Chung, I.H.; Lin, Y.H.; Chen, C.Y.; Chi, H.C.; Lin, T.K.; Yeh, C.T.; Lin, K.H. Stimulation of Interferon-Stimulated Gene 20 by Thyroid Hormone Enhances Angiogenesis in Liver Cancer. *Neoplasia* **2018**, *20*, 57–68. [CrossRef]

174. Baskol, G.; Atmaca, H.; Tanriverdi, F.; Baskol, M.; Kocer, D.; Bayram, F. Oxidative stress and enzymatic antioxidant status in patients with hypothyroidism before and after treatment. *Exp. Clin. Endocrinol. Diabetes* **2007**, *115*, 522–526. [CrossRef]

175. Haribabu, A.; Reddy, V.S.; Pallavi, C.; Bitla, A.R.; Sachan, A.; Pullaiah, P.; Suresh, V.; Rao, P.V.; Suchitra, M.M. Evaluation of protein oxidation and its association with lipid peroxidation and thyrotropin levels in overt and subclinical hypothyroidism. *Endocrine* **2013**, *44*, 152–157. [CrossRef]

176. Mourouzis, I.; Politi, E.; Pantos, C. Thyroid hormone and tissue repair: New tricks for an old hormone? *J. Thyroid Res.* **2013**, *2013*, 312104. [CrossRef]

177. D'Espessailles, A.; Dossi, C.; Intriago, G.; Leiva, P.; Romanque, P. Hormonal pretreatment preserves liver regenerative capacity and minimizes inflammation after partial hepatectomy. *Ann. Hepatol.* **2013**, *12*, 881–891. [CrossRef]

178. Fernandez, V.; Castillo, I.; Tapia, G.; Romanque, P.; Uribe-Echevarria, S.; Uribe, M.; Cartier-Ugarte, D.; Santander, G.; Vial, M.T.; Videla, L.A. Thyroid hormone preconditioning: Protection against ischemia-reperfusion liver injury in the rat. *Hepatology* **2007**, *45*, 170–177. [CrossRef]

179. Singh, R.; Kaushik, S.; Wang, Y.; Xiang, Y.; Novak, I.; Komatsu, M.; Tanaka, K.; Cuervo, A.M.; Czaja, M.J. Autophagy regulates lipid metabolism. *Nature* **2009**, *458*, 1131–1135. [CrossRef]

180. Ueno, T.; Komatsu, M. Autophagy in the liver: Functions in health and disease. *Nat. Reviews. Gastroenterol. Hepatol.* **2017**, *14*, 170–184. [CrossRef]

181. Tseng, Y.H.; Ke, P.Y.; Liao, C.J.; Wu, S.M.; Chi, H.C.; Tsai, C.Y.; Chen, C.Y.; Lin, Y.H.; Lin, K.H. Chromosome 19 open reading frame 80 is upregulated by thyroid hormone and modulates autophagy and lipid metabolism. *Autophagy* **2014**, *10*, 20–31. [CrossRef]

182. Vargas, R.; Riquelme, B.; Fernandez, J.; Alvarez, D.; Perez, I.F.; Cornejo, P.; Fernandez, V.; Videla, L.A. Docosahexaenoic acid-thyroid hormone combined protocol as a novel approach to metabolic stress disorders: Relation to mitochondrial adaptation via liver PGC-1alpha and sirtuin1 activation. *BioFactors* **2019**, *45*, 271–278. [CrossRef]

183. El Agaty, S.M. Triiodothyronine improves age-induced glucose intolerance and increases the expression of sirtuin-1 and glucose transporter-4 in skeletal muscle of aged rats. *Gen. Physiol. Biophys.* **2018**, *37*, 677–686.

184. Singh, B.K.; Sinha, R.A.; Ohba, K.; Yen, P.M. Role of thyroid hormone in hepatic gene regulation, chromatin remodeling, and autophagy. *Mol. Cell. Endocrinol.* **2017**, *458*, 160–168. [CrossRef]

185. Singh, B.K.; Sinha, R.A.; Zhou, J.; Tripathi, M.; Ohba, K.; Wang, M.E.; Astapova, I.; Ghosh, S.; Hollenberg, A.N.; Gauthier, K.; et al. Hepatic FOXO1 Target Genes Are Co-regulated by Thyroid Hormone via RICTOR Protein Deacetylation and MTORC2-AKT Protein Inhibition. *J. Biol. Chem.* **2016**, *291*, 198–214. [CrossRef]

186. Sinha, R.A.; Singh, B.K.; Zhou, J.; Wu, Y.; Farah, B.L.; Ohba, K.; Lesmana, R.; Gooding, J.; Bay, B.H.; Yen, P.M. Thyroid hormone induction of mitochondrial activity is coupled to mitophagy via ROS-AMPK-ULK1 signaling. *Autophagy* **2015**, *11*, 1341–1357. [CrossRef]

187. Webb, A.E.; Brunet, A. FOXO transcription factors: Key regulators of cellular quality control. *Trends Biochem. Sci.* **2014**, *39*, 159–169. [CrossRef]

188. Videla, L.A.; Cornejo, P.; Romanque, P.; Santibanez, C.; Castillo, I.; Vargas, R. Thyroid hormone-induced cytosol-to-nuclear translocation of rat liver Nrf2 is dependent on Kupffer cell functioning. *Sci. World J.* **2012**, *2012*, 301494. [CrossRef]

189. Cornejo, P.; Vargas, R.; Videla, L.A. Nrf2-regulated phase-II detoxification enzymes and phase-III transporters are induced by thyroid hormone in rat liver. *BioFactors* **2013**, *39*, 514–521. [CrossRef]

Int. J. Mol. Sci. **2019**, *20*, 5220

190. Romanque, P.; Cornejo, P.; Valdes, S.; Videla, L.A. Thyroid hormone administration induces rat liver Nrf2 activation: Suppression by N-acetylcysteine pretreatment. *Thyroid* **2011**, *21*, 655–662. [CrossRef]
191. Li, N.; Alam, J.; Venkatesan, M.I.; Eiguren-Fernandez, A.; Schmitz, D.; Di Stefano, E.; Slaughter, N.; Killeen, E.; Wang, X.; Huang, A.; et al. Nrf2 is a key transcription factor that regulates antioxidant defense in macrophages and epithelial cells: Protecting against the proinflammatory and oxidizing effects of diesel exhaust chemicals. *J. Immunol.* **2004**, *173*, 3467–3481. [CrossRef]

International Journal of
Molecular Sciences

Review

Cross-Talk between Mitochondrial Dysfunction-Provoked Oxidative Stress and Aberrant Noncoding RNA Expression in the Pathogenesis and Pathophysiology of SLE

Chang-Youh Tsai [1,*,†], Song-Chou Hsieh [2,†], Cheng-Shiun Lu [2,3], Tsai-Hung Wu [4], Hsien-Tzung Liao [1], Cheng-Han Wu [2,3], Ko-Jen Li [2], Yu-Min Kuo [2,3], Hui-Ting Lee [5], Chieh-Yu Shen [2,3] and Chia-Li Yu [2,*]

[1] Division of Allergy, Immunology & Rheumatology, Taipei Veterans General Hospital & National Yang-Ming University, #201 Sec.2, Shih-Pai Road, Taipei 11217, Taiwan; darryliao@yahoo.com.tw

[2] Department of Internal Medicine, National Taiwan University Hospital, #7 Chung-Shan South Road, Taipei 10002, Taiwan; hsiehsc@ntu.edu.tw (S.-C.H.); b89401085@ntu.edu.tw (C.-S.L.); chenghanwu@ntu.edu.tw (C.-H.W.); dtmed170@yahoo.com.tw (K.-J.L.); 543goole@gmail.com (Y.-M.K.); tsichhl@gmail.com (C.-Y.S.)

[3] Institute of Clinical Medicine, National Taiwan University College of Medicine, #7 Chung-Shan South Road, Taipei 10002, Taiwan

[4] Division of Nephrology, Taipei Veterans General Hospital & National Yang-Ming University, #201 Sec. 2, Shih-Pai Road, Taipei 11217, Taiwan; thwu@vghtpe.gov.tw

[5] Section of Allergy, Immunology & Rheumatology, Mackay Memorial Hospital, #92 Sec. 2, Chung-Shan North Road, Taipei 10449, Taiwan; htlee1228@gmail.com

* Correspondence: cytsai@vghtpe.gov.tw (C.-Y.T.); chialiyu0717@gmail.com (C.-L.Y.)

† These authors contributed equally to this work.

Received: 10 September 2019; Accepted: 14 October 2019; Published: 19 October 2019

Abstract: Systemic lupus erythematosus (SLE) is a prototype of systemic autoimmune disease involving almost every organ. Polygenic predisposition and complicated epigenetic regulations are the upstream factors to elicit its development. Mitochondrial dysfunction-provoked oxidative stress may also play a crucial role in it. Classical epigenetic regulations of gene expression may include DNA methylation/acetylation and histone modification. Recent investigations have revealed that intracellular and extracellular (exosomal) noncoding RNAs (ncRNAs), including microRNAs (miRs), and long noncoding RNAs (lncRNAs), are the key molecules for post-transcriptional regulation of messenger (m)RNA expression. Oxidative and nitrosative stresses originating from mitochondrial dysfunctions could become the pathological biosignatures for increased cell apoptosis/necrosis, nonhyperglycemic metabolic syndrome, multiple neoantigen formation, and immune dysregulation in patients with SLE. Recently, many authors noted that the cross-talk between oxidative stress and ncRNAs can trigger and perpetuate autoimmune reactions in patients with SLE. Intracellular interactions between miR and lncRNAs as well as extracellular exosomal ncRNA communication to and fro between remote cells/tissues via plasma or other body fluids also occur in the body. The urinary exosomal ncRNAs can now represent biosignatures for lupus nephritis. Herein, we'll briefly review and discuss the cross-talk between excessive oxidative/nitrosative stress induced by mitochondrial dysfunction in tissues/cells and ncRNAs, as well as the prospect of antioxidant therapy in patients with SLE.

Keywords: noncoding RNA; microRNA; long noncoding RNA; mitochondrial dysfunction; oxidative stress; nitrosative stress. exosome; cross-talk; systemic lupus erythematosus

---

1. Introduction

Systemic lupus erythematosus (SLE) is a highly heterogeneous disorder with chronic inflammatory and autoimmune reactions all over the body. It is characterized by the production of diverse autoantibodies [1,2] and chronic tissue inflammation [3–6]. There are multiple factors associated with lupus pathogenesis, including genetic predisposition [7–15], epigenetic dysregulation of gene transcription [16–21] and aberrant post-transcriptional events by noncoding (nc)RNAs [19,22–25], sex hormonal imbalance [26–29], environmental stimulation [30,31], mental/psychological stresses [28], dietary/nutritional influence [32–35], mitochondrial dysfunctions [36–39], and other yet-undefined factors [40]. Figure 1 shows the factors contributing to the pathogenesis of SLE, in which environmental factors such as infections, chemicals, heavy metals, medications, exogenous estrogens, and phthalate trigger its development in susceptible individuals. The genome-wide association study (GWAS) has identified over 100 risk loci for SLE susceptibility across populations [13]. However, functional studies have revealed that many of them fall in the category of noncoding regions of genomes, suggesting that they probably play a regulatory role. Many loci exhibit protean environmental interactions, epigenetic modifications, or association with genetic variants [10]. Nevertheless, the expression of IFN-α in tissues and circulation has been consistently found at a hereditary risk locus in patients with SLE [14]. The genetic predisposition for lupus pathogenesis is summarized in Table 1.

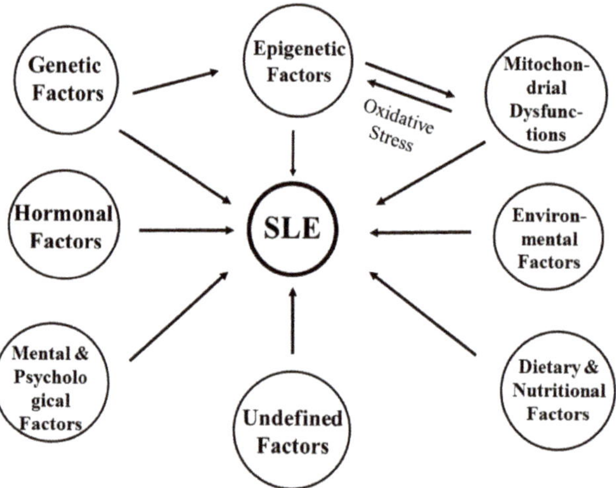

Figure 1. Factors contributing to the development of systemic lupus erythematosus. It is worthy to note that cross-talk between mitochondrial dysfunction and aberrant epigenetic regulation is mediated via excessive oxidative stress.

Recent investigations revealed that increased oxidative and/or nitrosative stress could induce structural and functional changes in different biomolecules, including proteins, lipids, nucleic acids, and glycoproteins [41,42]. The oxidative stress may also modulate proinflammatory cytokine gene expression [43–46] and cell senescence/apoptosis [47,48]. Antioxidants have been tried in the treatment of SLE with effectiveness [49–53]. Accordingly, the presence of oxidative stresses and their associated biomarkers are definitely playing a decisive role in the pathogenesis of SLE [54].

Epigenetics is an investigation of the changes in phenotypic presentation (or gene expression) that are caused by mechanisms other than the polymorphism of genome per se. It is conceivable that more than 97% of cellular RNAs are not transcribed for protein coding in nature. These ncRNAs, including microRNAs (miRs, 20–24 bp in length) and long noncoding (lnc) RNAs, which are >200 bp in length are the major molecules for post-transcriptional modifications of messenger (m)RNAs [55,56]. Interestingly, many reports have demonstrated that oxidative stress can modulate ncRNA expression in different diseases [57,58]. Conversely, ncRNAs have also been found to be regulators of oxidative stresses in different pathological conditions [59]. Furthermore, the cross-talk between miRs and lncRNAs has also been found [60,61]. Based on these facts, we hereby review and discuss briefly the molecular basis of epigenetic regulations, the underlying mechanism of mitochondrial dysfunctions, and the cross-talk between mitochondrial dysfunction-provoked oxidative stress and abnormal expression of ncRNAs during the pathologic development of SLE. At the end, a potential use of antioxidants as the therapy for SLE will also be concisely overviewed.

Table 1. Some of the genetic loci involved in the risk for SLE.

- **MHC association** [7–9]

    - MHC class II: DR_2, DR_3
    - MHC class III: C_4 null, TNF-α

- **Immune complex processing and phagocytosis** [7–15]

    - $C1_{q/r/s}$, $C_{4A/B}$, CFB
    - *FCGR2A/B*, CR2, CR3
    - CRP
    - *ICHMs* (intercellular adhesion molecules)
    - *ITGAM* (integrin subunit alpha M)

- **TLR and type I IFN signaling** [7–15]:

    - *TLR7* (toll-like receptor 7)
    - *TREX1* (three prime repair exonuclease 1)
    - *DNASE1* (DNA degrading enzyme 1)
    - *IRAK1/MECP2* (interleukin-receptor-associated kinase 1)
    - *IRF5/7/8* (interferon regulatory factor 5, 7, 8)
    - *STAT1* (signal transducer and activator of transcription 1)
    - *STAT4* (signal transducer and activator of transcription 4)

- **B and T cell function and signal genes** [7–15]

    - *IL10* (interleukin 10)
    - *STAT4* (signal transducer and activator of transcription 4)
    - *PTPN22* (protein tyrosine phosphatase non-receptor type 22)
    - *PDCD1* (programmed cell death 1)
    - *TNFSF4* (TNF superfamily member 4)
    - *BLK* (B lymphoid tyrosine kinase)
    - *BANK1* (B cell scaffold protein with ankyrin repeats 1)

- **Others**

    - *PXK/ABHD6* (PX domain containing serine/threonine kinase likes)
    - *XKR6* (XK related 6)
    - *UPF1/SMG7* (RNA helicase and ATPase)
    - *NMNAT2* (nicotinamide nucleotide adenyltransferase 2)
    - *UHRF1BP1* (ubiquitin like with PHD and ring finger domains 1 binding protein 1)

2. Epigenetic Regulations of Gene Expression/Silencing in Physiological Conditions

Epigenetic variation is a reversible but heritable change in gene expression without alterations in genetic code. It may include DNA methylation, histone modification, and post-transcriptional mRNA modification by ncRNAs [16]. DNA methylation is a biochemical process that involves a methyl group being added to a cytosine or adenine residue at the position of a repeated CpG dinucleotide (CpG island) in the promoter region to repress gene expression by DNA methyl- transferase (DNMT) 1, 3a, and 3b. In contrast, reactivation of DNA by demethylation to restore gene transcription can be achieved by ten-eleven translocation (TET) enzymes TET1, TET2, and TET3.

2.1. Abnormal DNA Methylation/Demethylation in SLE

DNA methylation is catalyzed by DNMT1 for gene silencing. A status of DNA hypomethylation to enhance gene expression can be found in CD4[+]T cells of SLE patients as a result of decreased expression of DNMT1 originating from a deficient *ras-MAPK* signature [62,63]. In addition, DNA methylation acts as a housekeeping mechanism for physiological inactivation of X-chromosomes in female [26,27,64]. Recent studies have suggested that *CD40L* demethylation is responsible for CD40L overexpression in T cells of women with SLE [64].

2.2. Abnormal Histone Modification in SLE

The degree of chromatin tightness is regulated via complex mechanisms, including structural changes in histones. Usually, double helix-chromatin coils around a protein core composed of histone octamers (H2A, H2B, H3, and H4 with two copies of each). The biochemical processes to change the 3D structure of histones include ubiquitination, phosphorylation, SUMOylation, methylation, and acetylation. The methylation and acetylation of histones are the most extensively studied [17]. These two biochemical changes are controlled by two major enzymes, histone acetyl transferase (HATs) and histone deacetylase (HDACs), that catalyze the addition/removal of an acetyl group on the lysine residues of histones. Acetylation relaxes the chromatin structures by diminishing the electric charge between histone and DNA as a result of offering an acetyl group. Conversely, deacetylation tightens the chromatin structure to silence gene expression.

The participation of histone modifications in lupus pathogenesis has been well documented. Hu et al. [65] demonstrated a global hyperacetylation of histones H3 and H4 in lupus CD4[+]T cells. Zhou et al. [66] reported that abnormal histone modifications within TNFSF7 promotor caused CD70 (a ligand for CD27) overexpression in SLE-T cells. Furthermore, Hedrich et al. [67] demonstrated that CREM, a transcription factor, participated in histone deacetylation in active T cells of SLE patients by way of silencing IL-2 expression, which normally recruits HDAC to cis-regulatory element (Cre) sites in IL-2 promotors. Dai et al. [68] showed in GWAS an alteration in histone H3 lysine K4 trimethylation (H3K4me3) by chromatin immunoprecipitation linked to microarray in peripheral blood mononuclear cells of some SLE patients. In addition, Zhang et al. [69] have found global H4 acetylation occurs in monocytes/macrophages in SLE subjects, which is regulated by IFN regulatory factors. The release of SLE-related cytokines such as IL-17, IL-10, and TNF-α was also abnormally increased in H3 acetylation by *stat*3 [70–72]. In lupus-prone MRL/*lpr* mice, a histone deacetylation gene, *sirtuin-1* (*Sirt*-1), was found overexpressed [73], indicating a compensatory repression of gene over-reactivation. Hu et al. [73] further noted downregulation of *Sirt*-1 would transiently enhance H3 and H4 acetylation and subsequently mitigate serum levels of anti-dsDNA, as well as kidney damage in lupus mice. Javierre et al. [74] reported a global decrease in the 5-methylcytosine content in parallel with DNA hypomethylation and high expression levels of ribosomal RNA genes relevant to SLE pathogenesis. In short, abnormal histone modifications are implicated in lupus pathogenesis and immunopathological changes in these patients.

2.3. Physiological Functions of ncRNAs

Besides DNA methylation/acetylation and histone modification, the most recently discovered epigenetic mechanisms for gene expression are dependent on the class of ncRNAs that are not translated into proteins. These molecules include both housekeeping ncRNAs and regulatory ncRNA [55]. In total 50% of mRNAs are located in chromosomal regions with liability to undergo structural changes [75]. On the other hand, lncRNA can regulate gene expression by different ways, including epigenetic, transcriptional, post-transcriptional, translational, and peptide localization modifications [56]. Interestingly, the interactions between lncRNAs and miRs, as well as their pathophysiological significance, have recently been reported [60,61]. It is believed that lncRNAs mediate "sponge-like" effects on various miRs and subsequently inhibit miR-mediated functions [60,61].

The regulatory effects of intracellular and extracellular (exosomal) ncRNA on cell functions are illustrated in Figure 2.

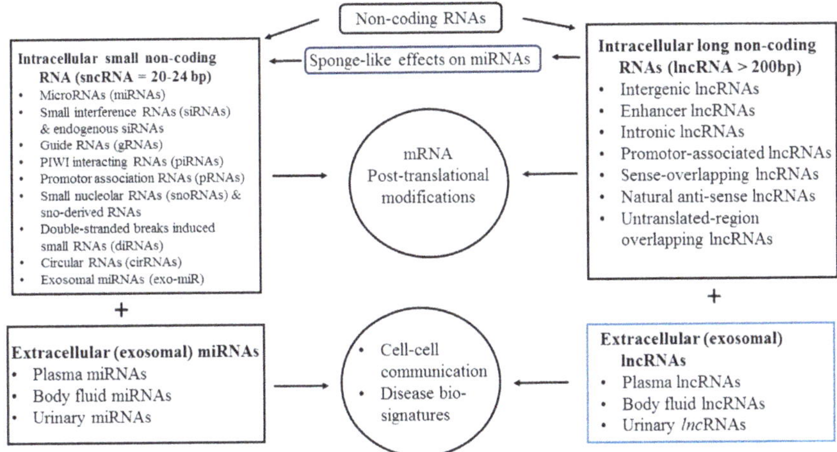

Figure 2. Different kinds of noncoding RNAs, including groups of small noncoding and long noncoding RNA, distributed in the intracellular and extracellular compartments, such as plasma, urine, and other body fluids, for regulation of messenger RNA translation and remote cell–cell communications in the body.

2.4. Aberrant Intracellular and Extracellular Exosomal ncRNA Expression in Association with Pathological Changes in Patients with SLE

It is not surprising that miRs play important roles in the regulation of innate and adaptive immunity, and the aberrantly expressed miRs are associated with autoimmune diseases [22,76–80]. Lu et al. [23,81–83] and Su et al. [84] have found various aberrantly expressed intracellular miRs implicated in the cell signaling abnormalities, deranged cytokine and chemokine release, and Th17/Treg ratio alterations in patients with SLE. Different from miRs, lncRNAs are expressed at lower levels in cells and tissues, more specifically [85–87]. These lncRNA are obviously modulating innate immunity [88] and inflammatory responses [89]. Luo et al. [90], Zhao et al. [91], and Wang et al. [92] reviewed the literature and found that lncRNA expression profiles in SLE were remarkably different from the normal.

The regulatory functions of miRNAs can be validated by transfecting miRNA mimics or antagonists using electroporator. Lu et al. [81] found increased miR-224 could target apoptosis inhibitory protein 5 (API5) and enhance T cell activation, and then activate induced cell apoptosis. Besides, the same group found decreased miR-31 in SLE T cells targeted the *Ras* homologue gene family member A (*Rho*A), which led to a decreased nuclear factor of activated T cells (NFAT) and cell apoptosis [23]. In addition, decreased miR-146a may result in upregulation of interferon regulatory factor 5 (IRF-5) and then enhanced production of IFN-α, STAT-1, IL-1 receptor associated kinase-1 (IRAK1), and TRAF6, which then increase innate immune responses, lupus disease activity, and lupus nephritis [23]. Furthermore, increased miR-524-5p that targets Jagged-1 and Hes-1mRNA may enhance IFN-γ production and then increase disease activity of SLE [82]. Su et al. [84] demonstrated that increased expression of miR-199-3p promoted ERK-mediated IL-10 production by targeting poly-(ADP-ribose) polymerase-1 (PARP-1) in SLE.

While their major functions are executed intracellularly, many miRs can be detected extracellularly in plasma/serum and urine. This extracellular form of ncRNA is protected from degradation by conjugation with carrier proteins or by being enclosed in subcellular vesicles by lipid bilayer exosomes [85]. With characteristics of the tissue- and disease-specific expression, these extracellular ncRNAs can carry out intercellular communication, signal transduction, transport of genetic information, immunomodulation, and can be taken as diagnostic biosignatures or as research tools for understanding the pathophysiology of autoimmune diseases [85–92]. Plasma circulating microRNAs exist in a rather stable form and are incorporated into distant cells to regulate protein translation and synthesis there. Carlsen et al. [87] have found plasma exosomal miR-142-3p, which targets IL-1β, and miR-181a, which targets FoxO1, are increased in active SLE patients. Kim et al. [88] demonstrated that increased plasma circulatory hsa-miR-30e-5p, hsa-miR-92a-3p, and hsa-miR-223-3p could become novel biosignatures in patients with SLE. The exosomal miRs can be found in other body fluids including breast milk, saliva, and urine, in addition to plasma [89]. Hsieh et al. [93] and Tsai et al. [94] concluded that urinary exosomal miRs could be used as biomarkers/biosignatures in lupus nephritis. Tsai et al. [94] have also noted aberrant miRNA expression in the immune-related cells could become biosignatures in correlation with pathological processes in different autoimmune and inflammatory rheumatic diseases. In addition, Perez-Hernandez et al. [95] and Xu et al. [96] have suggested the potential therapeutic application of exosomal ncRNA in different autoimmune diseases. Not only exosomal miRs, extracellularly expressed lncRNA profiles could also become potential biomarkers for human diseases [97,98]. lncRNAs are another regulatory noncoding RNA, capable of modulating many biological functions more specifically than miRs [99–102]. Aberrant expression of lncRNAs obviously induces different disease entities [99–106]. Table 2 summarizes the aberrant intracellular and circulating plasma exosomal lncRNA expression, their target mRNA, and related pathological processes in patients with SLE. Wang et al. [103] found that increased lncRNA ENST00000604411.1 expression in macrophages/dendritic cells, through targeting the X inactive specific transcript (XIST) that is normally implicated in keeping the active X chromosome in an activated state by protecting it from ectopic silencing after commencement of the silencing process of the haplotype X chromosome, could induce lupus development. Another lncRNA ENST 00000501122.2 (also known as NEAT1) overexpressed in SLE monocytes may activate CXCL-10 and IL-6 expression. Furthermore, Wu et al. [98] reported that elevated expression of plasma GAS-5, linc 0640, and linc 5150 may activate MAPK signaling pathway. The five lncRNA panels, including GAS-5, linc7074, linc 0597, linc 0640, and linc 5150 in plasma, could be regarded as biosignatures in SLE. The biochemical properties of extracellular ncRNAs and the pathophysiological roles of these aberrant exosomal ncRNAs in SLE are further discussed in the following paragraph.

Table 2. Aberrant expression of long none-coding RNAs, their target mRNAs, and related pathological processes in patients with systemic lupus erythematosus.

SLE	lnc RNA Expression	Target mRNA	Pathological Processes
	Intracellular [103–106]		
	NEAT$_1$↑*	IL-6↑, IFN↑, CXCL10↑	DNA hypomethylation
	MALAT$_1$↑	IL-21↑, SIRT$_1$↑	SLEDAI-2K↑
	Linc0597↑	TNF-α↑, IL-6↑	ESR↑, CRP↑, C3 ↓,
	Linc DC↑	STAT3↑	Th1↑
	ENST00000604411.1↑	XIST	SLEDAI score↑
	ENST000005011222↑	NEAT$_1$	
	Linc 0949↓	TNF-α↑, IL-6↑	Inflammation↑
	Linc-HSFY2-3:3↓	-	SLEDAI score↑
	Linc-SERPIN139-1:2↓	-	
	Gas 5↓	Apoptotic gene↓	T cell apoptosis↓
	Circulating plasma exosomal [98]		
	Linc0597↑	TNF-α↑, IL-6↑	MAPK signaling↑
	Lnc0640↑	Phosphatase 4 (DUSP4)↑	Lupus pathogenesis
	Lnc5150↑	Arrestin β2 (ARRB$_2$)↑	
		Ribosomal protein S$_6$ kinase A$_5$ (RPS6KA5)↑	
	Gas 5↓	Apoptotic gene↓	T cell apoptosis↓
	Lnc 7074↓		

↑: increased expression or production; ↓: decreased expression or production; *: Oxidative stress-induced [107].

3. Increased Oxidative Stress in Patients with SLE

3.1. Causes of Excessive Oxidative Stress in SLE

Li et al. [108] have compared the reduction–oxidation (redox) capacity between normal and SLE immune cells. They found decreased plasma and intracellular glutathione (GSH) levels, and decreased intracellular GSH-peroxidase and gamma-glutamyl-transpeptidase activity in patients with SLE. Besides, the defective expression of facilitative glucose transporter (GLUT) 3 and 6 led to increased intracellular basal lactate levels, as well as decreased ATP production in SLE T cells and polymorphonuclear leukocytes. These results may indicate deranged cellular bioenergetics and defective redox capacity in immune cells that would increase oxidative stress in SLE. Lee et al. [36–39] demonstrated that mitochondrial dysfunctions in SLE patients included decreased mitochondrial DNA (mtDNA) copy number, increased mtDNA D-310 (4977 bp) heteroplasmy, and variants, as well as polymorphism of $C_{1245}G$ in *hOGG1* gene in leukocytes. Leishangthem et al. [41] found a significant decrease in enzyme activity of complex I, IV, and V in mitochondria of patients with SLE. Lee et al. [109] have extensively investigated the cause of excessive stress in patients with SLE. They reported a number of antioxidant enzyme deficiencies in SLE leukocytes, including copper/zinc superoxide dismutase (Cu/ZnSOD), catalase, glutathione peroxidase 4 (GPx-4), glutathione reductase (GR), and glutathione synthetase (GS). In addition, the mitochondrial biogenesis-related proteins, such as mtDNA-encoded ND1 peptide (ND1), ND6, nuclear respiratory factor 1(NRF-1), and pyruvate dehydrogenase E1 component alpha subunit (PDHA1), and glycolytic enzymes, including hexokinase II (HK-II), glucose 6-phosphatate isomerase (GPI), phosphofructokinase (PFK), and glyceraldehyde 3-phosphate dehydrogenase (GAPDH), are also reduced in SLE immune cells. These mitochondrial functional abnormalities may further increase oxidative stress and cell apoptosis in patients with SLE, in addition to the defective bioenergetics. Yang et al. [110] and Tsai et al. [111] concluded that enhanced oxidative stress could facilitate mitophagy, inflammatory reactions, cell senescence/apoptosis, neoantigen formation, and NETosis in SLE. The causes of mitochondrial dysfunction to induce excessive oxidative stresses and their effects on the lupus pathogenesis and pathological processes are illustrated in Figure 3.

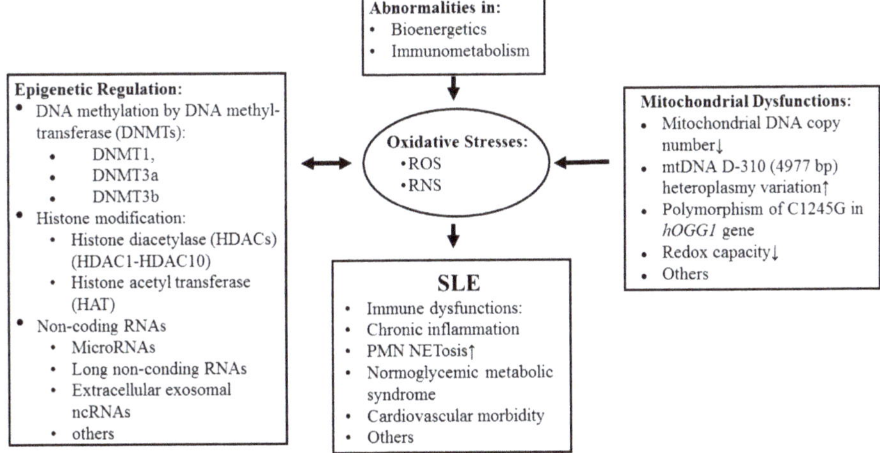

Figure 3. The origins of excessive oxidative stresses and their roles in abnormal epigenetic regulation and pathological processes in patients with SLE.

3.2. Effects of Excessive Oxidative Stress on the Pathogenesis and Pathophysiology in SLE Patients

The modifications of intra- and extracellular biomolecules by oxidative stress result in glycation and nitrosation of proteins [112], lipid peroxidation [42], as well as mitochondrial [113] and nuclear DNA strand breaks [114]. These biochemical and structural modifications of intracellular biomolecules would induce histone modification, nuclear and mitochondrial DNA damage, and aberrant ncRNA expression. As a consequence, the resulting sensitivity to environmental stress and sex hormone dysregulation [26–31] may further trigger the occurrence of lupus flare-ups. In addition, cardiovascular morbidities are enhanced due to increased glycation end products in patients with SLE [111,112,115]. The molecular basis and adverse effects of excessive oxidative stress in lupus pathogenesis and pathology are summarized in Figure 4.

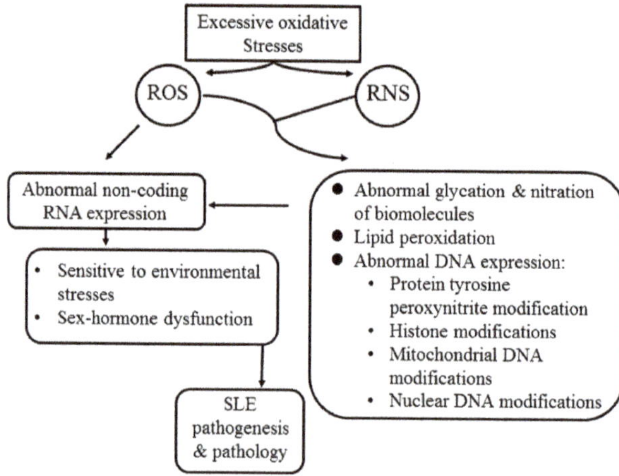

Figure 4. The molecular basis of excessive oxidative stress in the pathogenesis and pathological changes in patients with SLE.

4. Cross-Talk Between Oxidative Stress and ncRNAs in Physiological Condition

Recently, ever-increasing studies have emphasized the significance of the interactions between redox signaling and expression of ncRNAs in normal physiological conditions, as well as in disease status [44–46,57–59]. Sustained high levels of oxidative stress can cause cell senescence and even cell death, while optimal oxygen radicals are important for cell signaling. Dandekar et al. [44] and Lin et al. [116] have found mutual cross-talk among endoplasmic reticulum stress, oxidative stress, inflammatory response, and autophagy.

4.1. Excessive Oxidative Stress May Influence ncRNA Expression in Various Diseases

Many authors have demonstrated that redox-dependent signaling is essential for host's cellular decisions on differentiation, senescence, or death to maintain homeostasis of the body [117–119]. Figure 5 summarizes the aberrant miR expression resulting from excessive oxidative stress in different diseases, which include Alzheimer's disease [120], Parkinson's disease [121], hearing disorders [122], aging [123], osteoarthritis [124], cardiomyopathy in diabetes [125], and cancers [126]. However, despite the association of aberrant ncRNA expression with various pathological changes in SLE, as listed in Tables 2 and 3, there has been no literature demonstrating direct evidence for specific oxidative-induced ncRNA in patients with SLE. The combination of Table 3 and Figure 5 leads us to speculate that miR-21, miR-29b, miR-146a, and miR-126b may be induced by excessive oxidative stress in SLE as asterisked in Table 3 and its footnote.

Table 3. Aberrant expression of microRNAs, their target mRNAs, and pathological effects in patients with SLE.

SLE	miRNA	Target mRNA	Pathological Process
Intracellular [82–86]	● Increase in:		
	miR-21*	Arylamide small nucleotide inhibiors	DNA hypomethylation↑
	miR-524-5p	Jagged-1, Hes-1	IFN-γ↑, SLEDAI↑
	miR-126	KRAS	
	miR-148a	PTEN	
	● Decrease in:		
	miR-142-3p	HMGB-1	T and B activation↑
	miR-142-5p	PD-L1	
	miR-146a*	IRF-5, STAF-1	Innate immune response↑, lupus nephritis↑
	miR-224↑	API5	Type 1, IFN↑
	miR199-3p↑	PARP-1	IL-10↑
	● Decrease in:		
	miR-31	RhoA	Cell apoptosis↑
	miR-142-3p	HMGB-1	
	miR410	STAT3	
	miR-125a	STAT3, hexokinase 2, NEDDG	IL-10↑
	miR-125b*	Claudin 2, cingulin, SYVN1	
	mi-1273e		Th17/Treg ratio↑
	miR-3201		
Circulating plasma [87–94]	● Increase in:		
	miR-142-3p	IL-1β	
	miR-181a	FoxO1	
	hsa-miR-30e-5p		Oral ulcer and lupus anticoagulant
	hsa-miR-92a-3p		
	hsa-miR-223-3p		
	miR-16-5p	p38MAPK, NF-κB	
	miR-223-3p	Voltage-gated K$^+$ channel K$_{V4.2}$	
	miR-451	LKB1/AMPK	
	● Decrease in:		
	miR-106a	THBS$_2$	
	miR-17	JAB1/CSN5	
	miR-20a	IkBβ	
	miR-203	ZEB1	
	miR-92a	p63	
	miR-146a	JAK2/STAT3	
	miR-1202	cyclin dependent kinase 14	
Urinary exosomal (lupus Nephritis) [95,96]	● Increase in:		
	miR-125a	STAT3, hexokinase 2, NEDDG	Glomerulonephritis
	miR-146*	NF-κB	
	miR-150	Akt3	
	miR-155	PTEN, Wnt/β-catenin	
	● Decrease in:		
	miR-141	Tram1, GL/2, TGF-β	Glomerulonephritis
	miR-192	nin one binding protein	
	miR-200a	HMGB1/RAGE	
	miR-200c	ZEB1, Notch 1	
	miR-221	BIM-Bax/Bak, TIMP3	
	miR-222	PPP2R2A/Akt/mTOR, PCSK9	
	miR-429	TRAF6, DLC-1, HIF-1α	
	● Decrease in:		
	miR-3201		Endocapillary glomerular inflammation
	miR-1273e		

↑: increased expression or function; *: oxidative stress-induced microRNAs.

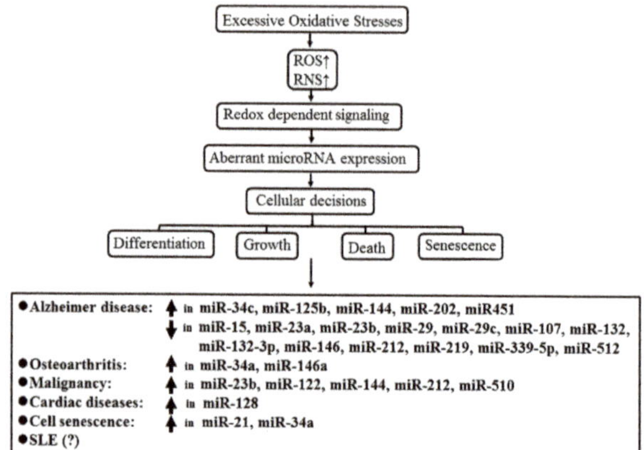

Figure 5. The effect of excessive oxidative stress on aberrant microRNA expression in various degenerative, malignant, cardiovascular, and autoimmune diseases. (?): increased miR-21, miR-29, miR-126b, and miR-146a expression induced by excessive oxidative stress is suspected in SLE patients, but no direct evidence has been published in the literature.

4.2. Aberrant ncRNA Expression Induces Oxidant/Antioxidant Imbalance in Different Pathological Processes

It has been demonstrated that excessive oxidative stress can affect ncRNA expression in Section 4.1. However, it is quite interesting that aberrant expression of ncRNAs conversely regulates redox balance in some pathological conditions. Esposti et al. [127] found miR-500a-5p could modulate oxidative stress-responsive genes in breast cancer and predict breast cancer progression as well as survival. Sangokoya et al. [128] have demonstrated that miR-144 modulates oxidative stress tolerance and, thus, is associated with changes in anemia severity in sickle cell disease. Kim et al. [129] found the roles of lncRNA and RNA-binding proteins in oxidative stress, cellular senescence, and age-related diseases. Tehrani et al. [130] further demonstrated multiple functions of lncRNAs in regulating oxidative stress, DNA damage response, and cancer progression. Mechanistically, ncRNAs can regulate enzymatic activity of different glutathione S-transferases (GSTs) to affect redox homeostasis [58]. These GSTs include microsomal GST, GST zeta 1, GST mu1, GST theca 1, and sirtuin 1, superoxide dismutase 2 and thioredoxin reductase 2. In addition, the cellular oxidant/antioxidant balance can also be regulated by lncRNAs [59]. The abnormal ncRNA expression to affect the oxidant/antioxidant system is summarized in Figure 6.

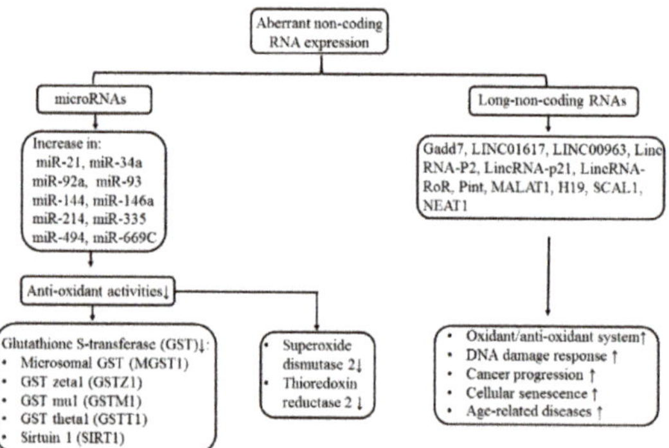

Figure 6. The effects of aberrant noncoding RNA expression on redox capacity and the induction of various age-related and malignant diseases.

5. Antioxidant Therapy and Manipulation of Epigenetic Expression to Treat Patients with SLE

In addition to increased oxygen free radicals in the plasma of SLE patients, there are other novel findings regarding the pro-oxidant/antioxidant balance in SLE. Mohan et al. [131] firstly confirmed that plasma concentrations of lipid peroxidase and nitric oxide were increased, whereas antioxidant molecules such as catalase, superoxide dismutase (SOD), GSH peroxidase, and vitamin E were decreased. Obviously, the pro-oxidant/antioxidant balance in SLE is disturbed [53]. Antioxidant therapy has been advocated for ameliorating tissue damage caused by excessive pro-oxidant radicals. Supplemented with GSH precursor, N-acetyl-cysteine (NAC) can improve disease activity in lupus-prone mice [50]. Delivering the *oxidation resistance*-1 (*OXR*1) gene to mouse kidneys by genetic manipulation can protect the kidney from damage induced by serum nephrotoxic agents, and prevent the animal from developing lupus nephritis [52]. Many authors, by administering NAC, have found remedies to ameliorate lupus activities in human SLE. Kudaravalli et al. [132] reported the improvement of endothelial dysfunction in patients with SLE by NAC and atorvastatin. Lai et al. [133] reported that NAC reduced disease activity by blocking mammalian targets of rapamycin (mTOR) in T cells of SLE patients. Tzang et al. [134] found cystamine attenuated lupus-associated apoptosis in ventricular tissue by suppressing both intrinsic and extrinsic apoptotic pathways. Nevertheless, much more clinical data are necessary to validate the efficacy of antioxidant therapy in managing patients with SLE.

Since there are so many intricate interactions among oxidative/nitrosative stress, epigenetic regulations, and gene expression in SLE, as discussed in the above sections, interference with epigenetic mechanisms such as modifying the activity of histone acetylase and/or DNA methylation, or inducing up- or downregulation of ncRNA expression may be helpful and can also be advocated to detour lupus pathogenesis and to diminish SLE disease activity in the future [135,136].

6. Conclusions

Mitochondrial dysfunction-provoked excessive oxidative stress is a crucial downstream contributory factor for lupus pathogenesis in addition to the dysregulation of upstream genetic/epigenetic functions. Recent studies have revealed that mutual interactions between oxidative stress and epigenetic regulation can perpetuate pathogenesis and pathological processes in SLE and other autoimmune diseases, as well as ageing-related diseases. In the ncRNA regulatory system, cross-talk between lncRNAs and miRs can occur for fine tuning of gene expression. Excessive oxidative stress-derived ROS and RNS may trigger autoimmune reaction and increase cell senescence/cell death in lupus-susceptible individuals. Antioxidant therapy and epigenetic modulators might become novel therapeutic strategies to treat SLE in the future.

Author Contributions: C.-L.Y. and C.-Y.T. supervised the writing project of the manuscript; C.-Y.T. and S.-C.H. prepared the manuscript and wrote the draft together; H.-T.L. prepared the figures; C.-S.L., C.-H.W., K.-J.L., Y.-M.K., H.-T.L., T.-H.W. and C.-Y.S. actively participated in the discussion and suggestions for the manuscript.

Funding: This research was funded by Ministry of Science & Technology, Executive Yuan, (MOST107-2314-B075-051-MY3, and MOST106-2634-F-075-001) and Taipei Veterans General Hospital (V107D37-002-MY3), Taiwan.

Acknowledgments: The authors thanks all of the individuals participating in the investigations.

Conflicts of Interest: The authors declare no conflict of interest.

Abbreviations

C.V	cardiovascular
DNA	deoxyribonucleic acid
DNMT	DNA methyltransfersase
FcγR	Immunoglobulin G Fragment C-gamma receptor
GLUT	glucose transporter
GSH	reduced form glutathione
GPx	glutathione peroxidase
GST	glutathione S-transferase

HAT	histone acetyltransferase
HDAC	histone deacetylase
IFN	interferon
IL	interleukin
LN	lupus nephritis
lncRNA	long noncoding ribonucleic acid
MAPK	mitogen-activated protein kinase
MHC	major histocompatibility complex
miR	microRNA
mtDNA	mitochondrial DNA
mTOR	mammalian target of rapamycin
NAC	N-acetylcysteine
ncRNA	non-coding RNA
NET	neutrophil extracellular trap
Ras	rat sarcoma protein, a superfamily of small GTPase
RNS	reactive nitrogen species
ROS	reactive oxygen species
SIRT1	sirtuin 1
SLE	systemic lupus erythematosus
SLEDAI	SLE disease activity index
SLEDAI-2K	SLEDAI in 2000 year
SOD	superoxide dismutase
TET	ten-eleven translocation DNA dioxygenase
Th	helper T cell
Treg	regulatory T cell

References

1. Wang, L.; Mohan, C.; Li, Q.Z. Arraying autoantibodies in SLE-lessons learned. *Curr. Mol. Med.* **2015**, *15*, 456–461. [CrossRef]

2. Yaniv, G.; Twig, G.; Shor, D.B.; Furer, A.; Sherer, Y.; Mozes, O.; Kimisar, O.; Slonimsky, E.; Klang, E.; Lotan, E.; et al. A volcano explosion of autoantibodies in systemic lupus erythematosus: A diversity of 180 different antibodies found in SLE patients. *Autoimmun. Rev.* **2015**, *14*, 75–79. [CrossRef]

3. Kahlenberg, J.M.; Kaplan, M.J. The inflammasome and lupus—Another innate immune mechanism contributing to disease pathogenesis? *Curr. Opin. Rheumatol.* **2014**, *26*, 475–481. [CrossRef]

4. Weidenbusch, M.; Kulkarni, O.P.; Anders, H.J. The innate immune system in human systemic lupus erythematosus. *Clin. Sci.* **2017**, *131*, 625–634. [CrossRef]

5. Tsai, C.Y.; Li, K.J.; Hsieh, S.C.; Liao, H.T.; Yu, C.L. What's wrong with neutrophils in lupus? *Clin. Exp. Rheumatol.* **2019**, *37*, 684–693. [PubMed]

6. Zharkova, O.; Celhar, T.; Crarens, P.D.; Satherthwaite, A.B.; Fairhurst, A.W.; Davis, L.S. Pathways leading to an immunological diseases: Systemic lupus erythematosus. *Rheumatology* **2017**, *56*, i55–i66. [CrossRef] [PubMed]

7. Harley, I.T.W.; Kaufman, K.M.; Langefeld, C.D.; Haraey, J.B.; Kelly, J.A. Genetic susceptibility to SLE: New insights from fine mapping and genome-wide association studies. *Nat. Rev. Genet.* **2009**, *10*, 285–290. [CrossRef] [PubMed]

8. Liu, Z.; Davidson, A. Taming lupus—A new understanding of pathogenesis is leading to clinical advances. *Nat. Med.* **2012**, *18*, 870–882. [CrossRef]

9. Ghodke-Puranick, Y.; Niewold, T.T. Immunogenetics of systemic lupus erythematosus: A comprehensive review. *J. Autoimmun.* **2015**, *64*, 125–136. [CrossRef]

10. Teruel, M.; Alacon-Riguelme, M.E. The genetic basis of systemic lupus erythematosus: What are the risk factors and what have we learned. *J. Autoimmun.* **2016**, *74*, 161–175. [CrossRef]

11. Iwamoto, T.; Niewold, T.B. Genetics of human lupus nephritis. *Clin. Immunol.* **2017**, *185*, 32–39. [CrossRef] [PubMed]

12. Hiraki, L.T.; Silverman, E.D. Genomics of systemic lupus erythematosus: Insights gained by studying monogenic young-onset systemic lupus erythematosus. *Rheum. Dis. N. Am.* **2017**, *43*, 415–434. [CrossRef] [PubMed]

13. Saeed, M. Lupus pathology based on genomics. *Immunogenetics* **2017**, *69*, 1–12. [CrossRef] [PubMed]

14. Goulielmos, G.N.; Zervou, M.I.; Vazgiourakis, V.M.; Ghodke-Puranik, Y.; Garyballos, A.; Niewold, T.B. The genetics and molecular pathogenesis of systemic lupus erythematosus (SLE) in populations of different ancestry. *Gene* **2018**, *668*, 59–72. [CrossRef] [PubMed]

15. Javinani, A.; Ashraf-Ganjouei, A.; Asloni, S.; Janshidi, A.; Mahmoudi, M. Exploring the etiopathogenesis of systemic lupus erythematosus: A genetic perspective. *Immunogenetics* **2019**, *71*, 283–297. [CrossRef] [PubMed]

16. Wu, H.; Zhao, M.; Chang, C.; Lu, Q. The real culprit in systemic lupus erythematosus: Abnormal epigenetic regulation. *Int. J. Mol. Sci.* **2015**, *16*, 11013–11033. [CrossRef] [PubMed]

17. Miceli-Richard, C. Epigenetics and lupus. *Jt. Bone Spine* **2015**, *82*, 90–93. [CrossRef]

18. Hedrick, C.M.; Mabert, K.; Rouen, T.; Tsokos, G.C. DNA methylation in systemic lupus erythematosus. *Epigenomics* **2017**, *9*, 505–525. [CrossRef]

19. Zhan, Y.; Guo, Y.; Lu, Q. Aberrant epigenetic regulation in the pathogenesis of systemic lupus erythematosus and its implications in precision medicine. *Cytogent. Genome Res.* **2016**, *149*, 141–155. [CrossRef]

20. Wang, Z.; Chang, C.; Peng, M.; Lu, Q. Translating epigenetics into clinic: Focus on lupus. *Clin. Epigenet.* **2017**, *9*, 78. [CrossRef]

21. Ren, J.; Panther, E.; Liao, X.; Grammer, A.C.; Lipsky, P.E.; Reilly, C.M. The impact of protein acetylation/deacetylation on systemic lupus erythematosus. *Int. J. Mol. Sci.* **2018**, *19*, 4007. [CrossRef] [PubMed]

22. Long, H.; Yin, H.; Wang, L.; Gershwin, M.E.; Lu, Q. The critical role of epigenetics in systemic lupus erythematosus and autoimmunity. *J. Autoimmun.* **2016**, *74*, 118–138. [CrossRef] [PubMed]

23. Lai, N.-S.; Koo, M.; Yu, C.-L.; Lu, M.C. Immunopathogenesis of systemic lupus erythematosus and rheumatoid arthritis: The role of aberrant expression of non-coding RNAs in T cells. *Clin. Exp. Immunol.* **2017**, *187*, 327–336. [CrossRef] [PubMed]

24. Zununi Vahed, S.; Nakhjarvani, M.; Etemadi, J.; Jamashidi, N.; Pourlak, T.; Abedizar, S. Altered levels of immune-regulatory microRNAs in plasma samples of patients with lupus nephritis. *Bioimpacts* **2018**, *8*, 177–183. [CrossRef] [PubMed]

25. Honarpisheh, M.; Kohler, P.; von Rauchhaupt, E.; Lech, M. The involvement of microRNAs in modulation of innate and adaptive immunity in systemic lupus erythematosus and lupus nephritis. *J. Immunol. Res.* **2018**, 4126106. [CrossRef] [PubMed]

26. McMurray, R.W. Sex-hormones in the pathogenesis in systemic lupus erythematosus. *Front Biosci.* **2001**, *6*, E193–E206. [CrossRef] [PubMed]

27. Khan, D.; Dai, R.; Ahmed, S.A. Sex differences and estrogen regulation of miRNA in lupus, a prototypical autoimmune disease. *Cell Immunol.* **2015**, *294*, 70–79. [CrossRef]

28. Assad, S.; Khan, H.H.; Ghazanfar, H.; Khan, Z.-H.; Mansor, S.; Rahman, M.A.; Khan, G.H.; Zafar, B.; Tariz, U.; Malik, S.A. Role of sex-hormone levels and psychological stress in the pathogenesis of autoimmune diseases. *Cureus* **2017**, *9*, E1315. [CrossRef]

29. Christou, E.A.A.; Banos, A.; Kosmara, D.; Bertsias, G.K.; Boumpas, D.T. Sexual dimorphism in SLE; above and beyond sex hormones. *Lupus* **2019**, *28*, 3–10. [CrossRef]

30. Sari-Puttini, P.; Atzeni, F.; Laccarino, L.; Doria, A. Environment and systemic lupus erythematosus: An overview. *Autoimmunity* **2005**, *38*, 465–472. [CrossRef]

31. Parks, C.G.; de Souza Espinodola Santos, A.; Barbhaiya, M.; Costenbader, K.H. Understanding the role of environmental factors in the development of systemic lupus erythematosus. *Best Pract. Res. Clin. Rheumatol.* **2017**, *31*, 306–320. [CrossRef] [PubMed]

32. Brown, A.C. Lupus erythematosus and nutrition: A review of the literature. *J. Ren. Nutr.* **2000**, *10*, 170–183. [CrossRef] [PubMed]

33. Minami, Y.; Sasaki, T.; Arai, Y.; Kurisu, Y.; Hisamichi, S. Diet and systemic lupus erythematosus: A 4 year prospective study of Japanese patients. *J. Rheumatol.* **2003**, *30*, 747–754. [PubMed]

34. Hsieh, C.-C.; Lin, B.-F. Dietary factors regulate cytokines in murine models of systemic lupus erythematosus. *Autoimmun. Rev.* **2011**, *11*, 22–27. [CrossRef]

35. Klack, K.; Bonfa, E.; Borba Neto, E.F. Diet and nutritional aspects in systemic lupus erythematosus. *Rev. Bras. Rheumatol.* **2012**, *52*, 384–408.

36. Lee, H.-T.; Lin, C.-S.; Chen, W.-S.; Liao, H.-T.; Tsai, C.-Y.; Wei, Y.-H. Leukocyte mitochondrial DNA alteration in systemic lupus erythematosus and its relevance to the susceptibility to lupus nephritis. *Int. J. Mol. Sci.* **2012**, *13*, 8853–8868. [CrossRef]

37. Lee, H.-T.; Wu, T.-H.; Lin, C.-S.; Lee, C.-S.; Wei, Y.-H.; Tsai, C.-Y.; Chang, D.-M. The pathogenesis of systemic lupus erythematosus—From the viewpoint of oxidative stress and mitochondrial dysfunction. *Mitochondrion* **2016**, *30*, 1–7. [CrossRef]

38. Lee, H.-T.; Wu, T.-H.; Lin, C.-S.; Lee, C.-S.; Pan, S.-C.; Chang, D.-M.; Wei, Y.-H.; Tsai, C.-Y. Oxidative DNA and mitochondrial DNA change in patients with SLE. *Front Biosci. Landmark.* **2017**, *22*, 493–503.

39. Lee, H.-T.; Lin, C.-S.; Pan, S.-C.; Wu, T.-H.; Lee, C.-S.; Chang, D.-M.; Tsai, C.-Y.; Wei, Y.-H. Alterations of oxygen consumption and extracellular acidification rates by glutamine in PBMCs of SLE patients. *Mitochondrion* **2019**, *44*, 65–74. [CrossRef]

40. Marion, T.N.; Postlethwaite, A.E. Chance, genetics, and the heterogeneity of disease and pathogenesis in systemic lupus erythematosus. *Sem. Immunopathol.* **2014**, *36*, 495–517. [CrossRef]

41. Leishangthem, B.D.; Sharma, A.; Bhatnagar, A. Role of altered mitochondria functions in the pathogenesis of systemic lupus erythematosus. *Lupus* **2016**, *25*, 272–281. [CrossRef] [PubMed]

42. Kuren, B.T.; Scofield, R.H. Lipid peroxidation in systemic lupus erythematosus. *Indian J. Exp. Biol.* **2006**, *44*, 349–356.

43. Das, U.N. Oxidative, anti-oxidants, essential fatty acids, eicosanoids, cytokines, gene/oncogene expression and apoptosis in systemic lupus erythematosus. *J. Assoc. Physicians India* **1998**, *46*, 630–634.

44. Dandekar, A.; Mendez, R.; Zhang, K. Cross talk between ER stress, oxidative stress, and inflammation in health and disease. *Methods Mol. Biol.* **2015**, *1292*, 205–214.

45. Hussain, T.; Tan, B.; Yin, Y.; Blachier, F.; Tossou, M.C.; Rahu, N. Oxidative stress and inflammation: What polyphenols can do for us? *Oxidative Med. Cell Longev.* **2016**, *2016*, 7432797. [CrossRef]

46. Guzik, T.J.; Touyz, R.M. Oxidative stress, inflammation, and vascular aging in hypertension. *Hypertension* **2017**, *70*, 660–667. [CrossRef]

47. Plotnikov, E.; Losenkov, I.; Epimakhova, E.; Bohan, N. Protective effects of pyruvic acid salt against lithium toxicity and oxidative damage in human blood mononuclear cells. *Adv. Pharm. Bull.* **2019**, *9*, 302–306. [CrossRef]

48. Lee, D.; Lee, S.H.; Noh, I.; Oh, E.; Ryu, H.; Ha, J.; Jeong, S.; Yoo, J.; Jeon, T.J.; Yun, C.O.; et al. A helical polypeptide-based potassium ionophore induces endoplasmic reticulum stress-mediated apoptosis by perturbing ion homeostasis. *Adv. Sci. (Weinheim)* **2019**, *6*, 1801995. [CrossRef]

49. Das, U.N. Current and emerging strategies for the treatment and management of systemic lupus erythematosus based on molecular signatures of acute and chronic inflammation. *J. Inflamm. Res.* **2010**, *3*, 143–170. [CrossRef]

50. Perl, A. Oxidative stress in the pathology and treatment of systemic lupus erythematosus. *Nat. Rev. Rheumatol.* **2013**, *9*, 674–686. [CrossRef]

51. Su, Y.J.; Cheng, T.T.; Chen, C.J.; Chiu, W.C.; Chang, W.N.; Tsai, N.W.; Kung, C.T.; Lin, W.C.; Huang, C.C.; Chang, Y.T.; et al. The association among antioxidant enzymes, autoantibodies, and disease severity score in systemic lupus erythematosus: Comparison of neuropsychiatric and nonneuropsychiatric groups. *BioMed Res. Int.* **2014**, *2014*, 137231. [CrossRef] [PubMed]

52. Li, Y.; Li, W.; Liu, C.; Yan, M.; Raman, I.; Du, Y.; Fang, X.; Zhou, X.J.; Mohan, C.; Li, Q.Z. Delivering oxidation resistance-1 (OXR1) to mouse kidney by genetic modified mesenchymal stem cells exhibited enhanced protection against nephrotoxic serum induced renal injury and lupus nephritis. *J. Stem Cell Res. Ther.* **2014**, *4*, 231. [PubMed]

53. Jafari, S.M.; Salimi, S.; Nakhaee, A.; Kalani, H.; Tavallaie, S.; Farajian-Mashhkadi, F.; Zakerj, Z.; Sandoughi, M. Prooxidant-antioxidant balance in patients with systemic lupus erythematosus and its relationship with clinical and laboratory findings. *Autoimmune Dis.* **2016**, *2016*, 4343514. [CrossRef] [PubMed]

54. Shah, D.; Mahajan, N.; Sah, S.; Nath, S.K.; Paudyal, B. Oxidative stress and its biomarkers in systemic lupus erythematosus. *J. Biomed. Sci.* **2014**, *21*, 23. [CrossRef] [PubMed]

55. Wei, J.-W.; Huang, K.; Yang, C.; Kang, C.-S. Non-coding RNAs as regulators in epigenetics. *Oncol. Rep.* **2017**, *37*, 3–9. [CrossRef] [PubMed]

56. Li, J.; Liu, C. Coding or noncoding, the converging concepts of RNAs. *Front. Genet.* **2019**, *10*, 496. [CrossRef]

57. Banerjee, J.; Khanna, S.; Bhattacharya, A. MicroRNA regulation of oxidative stress. *Oxid. Med. Cell Longev.* **2017**, 2872156. [CrossRef]

58. Bu, H.; Wedel, S.; Cavinato, M.; Jansen-Dürr, P. MicroRNA regulation of oxidative stress-induced cellular senescence. *Oxid. Med. Cell Longev.* **2017**, *2017*, 2398696. [CrossRef]

59. Wang, X.; Shen, C.; Zhu, J.; Shen, G.; Li, Z.; Dong, J. Long non-coding RNAs in the regulation of oxidative stress. *Oxid. Med. Cell Longev.* **2019**, 1318795. [CrossRef]

60. Bayoumi, A.S.; Sayed, A.; Broskova, Z.; Teoh, J.-P.; Wilson, J.; Su, H.; Tang, Y.-L.; Kim, I. Crosstalk between long noncoding RNAs and microRNAs in health and disease. *Int. J. Mol. Sci.* **2016**, *17*, 356. [CrossRef]

61. Yamamura, S.; Imai-Sumida, M.; Tanaka, Y.; Dahiya, R. Interaction and cross-talk between non-coding RNAs. *Cell Mol. Life Sci.* **2018**, *75*, 467–484. [CrossRef] [PubMed]

62. Deng, C.; Yang, J.; Scott, J.; Hanash, S.; Richardson, B.C. Role of the ras-MAPK signaling pathway in the DNA methyltransferase response to DNA hypomethylation. *Biol. Chem.* **1998**, *379*, 1113–1120. [CrossRef] [PubMed]

63. Sawalha, A.H.; Jeffries, M.; Webb, R.; Lu, Q.; Gorelik, G.; Ray, D.; Osban, J.; Knowlion, N.; Johnson, K.; Richardsonm, B. Defective T cell ERK signaling induces interferon-regulated gene expression and overexpression of methylation sensitive genes similar to lupus patients. *Genes Immun.* **2008**, *9*, 368–378. [CrossRef] [PubMed]

64. Lu, Q.; Wu, A.; Tesmer, L.; Ray, D.; Yousif, N.; Richardson, B. Demethylation of CD40LG on the inactive X in T cells from women with lupus. *J. Immunol.* **2007**, *179*, 6352–6358. [CrossRef] [PubMed]

65. Hu, N.; Qiu, X.; Luo, Y.; Yuan, J.; Li, Y.; Lei, W.; Zhang, G.; Zhou, Y.; Su, Y.; Lu, Q. Abnormal histone modification patterns in lupus CD4+T cells. *J. Rheumatol.* **2008**, *35*, 804–810. [PubMed]

66. Zhou, Y.; Qiu, X.; Luo, Y.; Yuan, J.; Li, Y.; Zhong, Q.; Zhao, M.; Lu, Q. Histone modifications and methyl-CpG binding domain protein levels at the TNFSF7 (CD70) promoter in SLE CD4+ T cells. *Lupus* **2011**, *20*, 1365–1371. [CrossRef] [PubMed]

67. Hedrich, C.M.; Tsokos, G.C. Epigenetic mechanisms in systemic lupus erythematosus and other autoimmune diseases. *Trends Mol. Med.* **2011**, *17*, 714–724. [CrossRef]

68. Dai, Y.; Zhang, L.; Hu, C.; Zhang, Y. Genome-wide analysis of histone H3 lysine 4 trimethylation by CHIP-chip in peripheral blood mononuclear cells of systemic lupus erythematosus patients. *Clin. Exp. Rheumatol.* **2010**, *28*, 158–168.

69. Zhang, Z.; Song, L.; Maurer, K.; Petri, M.A.; Sullivan, K.E. Global H4 acetylation analysis by CHIP-chip in SLE monocytes. *Genes Immun.* **2010**, *11*, 124–133. [CrossRef]

70. Apostolidis, S.A.; Rauren, T.; Hedrich, C.M.; Tsokos, G.C.; Crispin, J.C. Protein phosphatase 2A enables expression of interleukin 17 (IL-17) through chromatin remodeling. *J. Biol. Chem.* **2013**, *288*, 26775–26784. [CrossRef]

71. Hedrich, C.M.; Rauen, J.; Apostolidis, S.A.; Grammatikos, A.P.; Rodriguez Rodrigues, N.; Ioannidis, C.; Kyttaris, V.C.; Crispin, J.C.; Tsokos, G.C. Stat3 promotes IL-10 expression in lupus T cells through trans-activation and chromatin remodeling. *Proc. Natl. Acad. Sci. USA* **2014**, *111*, 13457–13462. [CrossRef] [PubMed]

72. Sullivan, K.E.; Suriano, A.; Dietzmann, K.; Lin, J.; Goldman, D.; Petri, M.A. The TNF-alpha locus is altered in monocytes from patients with systemic lupus erythematosus. *Clin. Immunol.* **2007**, *123*, 74–81. [CrossRef] [PubMed]

73. Hu, N.; Long, H.; Zhao, M.; Yin, H.; Lu, Q. Aberrant expression pattern of histone acetylation modifier and mitigation of lupus by SIRT1-siRNA in MRL/*lpr* mice. *Scand. J. Rheumatol.* **2009**, *38*, 464–471. [CrossRef] [PubMed]

74. Javierre, B.M.; Fernandez, A.F.; Richter, J.; Al-Shahrour, F.; Martin-Subero, J.I.; Rodriguez-Ubreva, J.; Berdasco, M.; Fraga, M.F.; O'Hanlon, T.P.; Rider, L.G.; et al. Changes in the pattern of DNA methylation associate with twin discordance in systemic lupus erythematosus. *Genome Res.* **2010**, *20*, 170–179. [CrossRef]

75. Ruvkun, G. Molecular biology. Glimpses of a tiny RNA world. *Science* **2001**, *294*, 797–799. [CrossRef]

76. Dai, R.; Ahmed, S.A. MicroRNA, a new paradigm for understanding immunoregulation, inflammation, and autoimmune diseases. *Transl. Res.* **2011**, *157*, 163–179. [CrossRef]

77. Qu, B.; Shen, N. miRNAs in the pathogenesis of systemic lupus erythematosus. *Int. J. Mol. Sci.* **2015**, *16*, 9557–9572. [CrossRef]

78. Chen, J.-Q.; Papp, G.; Szodoray, P.; Zeher, M. The role of microRNAs in the pathogenesis of autoimmune diseases. *Autoimmun. Rev.* **2016**, *15*, 1171–1180. [CrossRef]

79. Le, X.; Yu, X.; Shen, N. Novel insights of microRNAs in the development of systemic lupus erythematosus. *Curr. Opin. Rheumatol.* **2017**, *29*, 450–457. [CrossRef]

80. Long, H.; Wang, X.; Chen, Y.; Wang, L.; Zhao, M.; Lu, Q. Dysregulation of microRNAs in autoimmune diseases: Pathogenesis, biomarkers and potential therapeutic targets. *Cancer Lett.* **2018**, *428*, 90–103. [CrossRef]

81. Lu, M.C.; Lai, N.S.; Chen, H.C.; Yu, H.C.; Huang, K.Y.; Tung, C.H.; Huang, H.B.; Yu, C.L. Decreased microRNA (miR)-145 and increased miR-224 expression in T cells from patients with systemic lupus erythematosus involved in lupus immunopathogenesis. *Clin. Exp. Immunol.* **2013**, *171*, 91–99. [CrossRef] [PubMed]

82. Lu, M.C.; Yu, C.L.; Chen, H.C.; Yu, H.C.; Huang, H.B.; Lai, N.S. Aberrant T cell expression of Ca2+ influx-regulated miRNA in patients with systemic lupus erythematosus promotes lupus pathogenesis. *Rheumatology (Oxford)* **2015**, *54*, 343–348. [CrossRef] [PubMed]

83. Tsai, C.Y.; Hsieh, S.-C.; Lu, M.-C.; Yu, C.-L. Aberrant non-coding RNA expression profiles as biomarker/ biosignature in autoimmune and inflammatory rheumatic diseases. *J. Lab. Preci. Med.* **2018**, *3*, 51. [CrossRef]

84. Su, X.; Ye, L.; Chen, X.; Zhang, H.; Zhou, Y.; Ding, X.; Chen, D.; Lin, Q.; Chen, C. MiR-199-3p promotes ERK-mediated IL-10 production by targeting poly(ADP-ribose)polymerase-1 in patients with systemic lupus eruthematosus. *Chemo-Biol. Interact.* **2019**, *306*, 110–116. [CrossRef]

85. Heegaard, N.H.H.; Carlsen, A.L.; Skovgaard, K.; Heegaard, P.M.H. Circulating extracellular microRNA in systemic autoimmunity. *Exp. Suppl.* **2015**, *106*, 171–195.

86. Turpin, D.; Truchetet, M.E.; Faustin, B.; Augusto, J.F.; Contin-Bordes, C.; Brisson, A.; Blanco, P.; Duffau, P. Role of extracellular vesicles in autoimmune diseases. *Autoimmun. Rev.* **2016**, *15*, 174–183. [CrossRef]

87. Carlsen, A.L.; Schetter, A.J.; Nielsen, C.T.; Lood, C.; Knudsen, S.R.; Voss, A.; Harris, C.C.; Hellmark, T.; Segelmark, M.; Jacobsen, S.; et al. Circulating microRNA expression profiles associated with systemic lupus erythematosus. *Arthritis Rheum.* **2013**, *65*, 1324–1334. [CrossRef]

88. Kim, B.-S.; Jung, J.-Y.; Jeon, J.-Y.; Kim, H.-A.; Suh, C.-H. Circulating hsa-miR-30e-5p, hsa-miR-92a-3p, and hsa-miR-223-3p may be novel biomarkers in systemic lupus erythematosus. *HLA* **2016**, *88*, 187–193. [CrossRef]

89. Ishibe, Y.; Kusaoi, M.; Murayama, G.; Nemoto, T.; Kon, T.; Ogasawara, M.; Kempe, K.; Yamaji, K.; Tamura, N. Changes in the expression of circulating microRNAs in systemic lupus erythematosus patient blood plasma after passing through a plasma absorption membrane. *Ther. Apher. Dial.* **2018**, *22*, 278–289. [CrossRef]

90. Natasha, G.; Gundogan, B.; Tan, A.; Farhatnia, Y.; Wu, W.; Rajadas, J.; Seifalian A.M. Exosomes as immunotheranostic nanoparticles. *Clin. Ther.* **2014**, *36*, 820–829. [CrossRef]

91. Tan, L.; Wu, H.; Liu, Y.; Zhao, M.; Li, D.; Lu, Q. Recent advances of exosomes in immune modulation and autoimmune diseases. *Autoimmunity* **2016**, *49*, 357–365. [CrossRef] [PubMed]

92. Rekker, K.; Saare, M.; Roost, A.M.; Kubo, A.-L.; Zarovni, N.; Chiesi, A.; Salumets, A.; Peters, M. Comparison of serum exosome isolation methods for microRNA profiling. *Clin. Biochem.* **2014**, *47*, 135–138. [CrossRef] [PubMed]

93. Hsieh, S.C.; Tsai, C.Y.; Yu, C.L. Potential serum and urine biomarkers in patients with lupus nephritis and the unsolved problems. *Open Access Rheumatol.* **2016**, *8*, 81–91. [PubMed]

94. Tsai, C.Y.; Lu, M.C.; Yu, C.L. Can urinary exosomal micro-RNA detection become a diagnostic and prognostic gold standard for patients with lupus nephritis and diabetic nephropathy? *J. Lab. Precis. Med.* **2017**, *2*, 91. [CrossRef]

95. Perez-Hernandez, J.; Redon, J.; Cortes, R. Extracellular vesicles as therapeutic agents in systemic lupus erythemaotusus. *Int. J. Mol. Sci.* **2017**, *18*, E717. [CrossRef]

96. Xu, H.; Jia, S.; Xu, H. Potential therapeutic applications of exosomes in different autoimmune diseases. *Clin. Immunol.* **2019**, *205*, 116–124. [CrossRef]

97. Kelemen, E.; Danis, J.; Göblös, A.; Bata-Csörgö, Z.; Szell, M. Exosomal long non-coding RNAs as biomarkers in human diseases. *J. Int. Fed. Clin. Chem. Lab. Med.* **2019**, *30*, 224–236.

98. Wu, G.C.; Hu, Y.; Guan, S.Y.; Ye, D.Q.; Pan, H.F. Differential plasma expression profiles of long non-coding RNAs, reveal potential biomarkers for systemic lupus erythematosus. *Biomolecules* **2019**, *9*, E206. [CrossRef]

99. Gloss, B.S.; Dinger, M.E. The specificity of long noncoding RNA expression. *Biochim. Biophys. Acta* **2016**, *1859*, 16–22. [CrossRef]

100. Derrien, T.; Johnson, R.; Bussotti, G.; Tanzer, A.; Djebali, S.; Tilgner, H.; Guermec, G.; Martin, D.; Merkel, A.; Knowles, D.G.; et al. The GENCODE v7 catalog of human lung noncoding RNAs: Analysis of their gene structure, evolution, and expression. *Genome Res.* **2012**, *22*, 1775–1789. [CrossRef]
101. Cabili, M.N.; Trapnell, C.; Goff, L.; Koziol, M.; Tazon-Vega, B.; Regev, A.; Rinn, J.L. Integrative annotation of human large intergenic noncoding RNAs reveals global properties and specific subclasses. *Genes Dev.* **2011**, *25*, 1915–1927. [CrossRef] [PubMed]
102. Hadjicharalambous, M.R.; Lindsay, M.A. Long non-coding RNAs and the innate immune response. *Noncoding RNA* **2019**, *5*, E34. [CrossRef] [PubMed]
103. Wang, Y.; Chen, S.; Chen, S.; Du, J.; Lin, J.; Qin, H.; Wang, J.; Liang, J.; Xu, J. Long noncoding RNA expression profile and association with SLEDAI score in monocyte-derived dendritic cells from patients with systemic lupus erythematosus. *Arthritis Res. Ther.* **2018**, 20–138. [CrossRef]
104. Mathy, N.W.; Chen, X.-M. Long non-coding RNAs (lncRNAs) and their transcriptional control of inflammatory responses. *J. Biol. Chem.* **2017**, *292*, 12375–12382. [CrossRef] [PubMed]
105. Luo, Q.; Li, X.; Xu, C.; Zeng, L.; Ye, J.; Guo, Y.; Huang, Z.; Li, J. Integrative analysis of long non-coding RNAs and messenger RNA expression profiles in systemic lupus erythematosus. *Mol. Med. Rep.* **2018**, *17*, 3489–3496. [PubMed]
106. Zhao, C.N.; Mao, Y.M.; Liu, L.N.; Li, X.M.; Wang, D.G.; Pan, H.F. Emerging role of lncRNAs in systemic lupus erythematosus. *Biomed. Pharm.* **2018**, *106*, 584–592. [CrossRef]
107. Simchovitz, A.; Hanan, M.; Niederhoffer, N.; Madrer, N.; Yayon, N.; Bennett, E.R.; Greenberg, D.S.; Kadener, S.; Soreq, H. NEAT1 is overexpressed in Parkinson's disease substantia nigra and confers drug-inducible neuroprotection from oxidative stress. *FASEB J.* **2019**, *33*, 11223–11234. [CrossRef]
108. Li, K.J.; Wu, C.H.; Hsieh, S.C.; Lu, M.C.; Tsai, C.Y.; Yu, C.L. Deranged bioenergetics and defective redox capacity in T-lymphocytes and neutrophils are related to cellular dysfunction and increased oxidative stress in patients with active systemic lupus erythematosus. *Clin. Dev. Immunol.* **2012**, *2012*, 548516. [CrossRef]
109. Lee, H.-T.; Lin, C.-S.; Lee, C.-S.; Tsai, C.-Y.; Wei, Y.-H. Increase 8-hydroxy-2′-deoxyguanosine in plasma and decreased mRNA expression of human 8-oxoguanine DNA glycosylase 1, anti-oxidant enzymes, mitochondrial biogenesis-related proteins and glycolytic enzymes in leukocytes in patients with systemic lupus erythematosus. *Clin. Exp. Immunol.* **2014**, *176*, 66–77.
110. Yang, S.K.; Zhang, H.R.; Shi, S.P.; Zhu, Y.Q.; Song, N.; Dai, Q.; Zhang, W.; Gui, M.; Zhang, H. The role of mitochondria in systemic lupus erythematosus: A glimpse of various pathogenetic mechanisms. *Curr. Med. Chem.* **2018**. [CrossRef]
111. Tsai, C.Y.; Shen, C.Y.; Liao, H.T.; Li, K.J.; Lee, H.T.; Lu, C.S.; Wu, C.H.; Kuo, Y.M.; Hsieh, S.C.; Yu, C.L. Molecular and cellular bases of immunosenescence, inflammation, and cardiovascular complications mimicking "inflammaging" in patients with systemic lupus erythematosus. *Int. J. Mol. Sci.* **2019**, *20*, E3878. [CrossRef] [PubMed]
112. Vlassopoulos, A.; Lean, M.E.J.; Combet, E. Oxidative stress, protein glycation and nutrition-interactions relevant to health and disease throughout the lifecycle. *Proc. Nutr. Soc.* **2014**, *73*, 430–438. [CrossRef] [PubMed]
113. McGuire, P.J. Mitochondrial dysfunction and the aging immune system. *Biology (Basel)* **2019**, *8*, E26. [CrossRef] [PubMed]
114. Ye, B.; Hou, N.; Xiao, L.; Xu, Y.; Xu, H.; Li, F. Dynamic monitoring of oxidative DNA double strand break and repair in cardiomyocytes. *Cardiovasc. Pathol.* **2016**, *25*, 93–100. [CrossRef] [PubMed]
115. de Leeuw, K.; Graaff, R.; de Vries, R.; Dullaart, R.P.; Smit, A.J.; Kallenberg, C.G.; Bjil, M. Accumulation of advanced glycation endproducts in patients with systemic lupus erythematosus. *Rheumatology* **2007**, *46*, 1551–1556. [CrossRef] [PubMed]
116. Lin, Y.; Jiang, M.; Chen, W.; Zhao, T.; Wei, Y. Cancer and ER stress: Mutual crosstalk between autophagy, oxidative stress and inflammatory response. *Biomed. Pharmacother.* **2019**, *118*, 109249. [CrossRef] [PubMed]
117. Zhang, Y.; Du, Y.; Le, W.; Wang, K.; Kieffer, N.; Zhang, J. Redox control of the survival of healthy and diseased cells. *Antioxid. Redox Signal.* **2011**, *15*, 2867–2908. [CrossRef]
118. Cao, S.S.; Kaufman, R.J. Endoplasmic reticulum stress and oxidative stress in cell fate decision and human disease. *Antioxid. Redox Signal.* **2014**, *21*, 396–413. [CrossRef]
119. Pervaiz, S. Redox dichotomy in cell fate decision: Evasive mechanism or Achilles heel? *Antioxid. Redox Signal.* **2018**, *29*, 1191–1195. [CrossRef]

120. Prasad, K.N. Oxidative stress and pro-inflammatory cytokines may act as one of the signals for regulating microRNAs expression in Alzheimer's disease. *Mech. Ageing Dev.* **2017**, *162*, 63–71. [CrossRef]

121. Prasad, K.N. Oxidative stress, pro-inflammatory cytokines, and antioxidants regulate expression levels of microRNAs in Parkinson's disease. *Curr. Aging Sci.* **2017**, *10*, 177–184. [CrossRef] [PubMed]

122. Prasad, K.N.; Bondy, S.C. MicroRNAs in hearing disorders: Their regulation by oxidative stress, inflammation and antioxidants. *Front Cell Neurosci.* **2017**, *11*, 276. [CrossRef] [PubMed]

123. Guillaumet-Adkins, A.; Yañez, Y.; Peris-Diaz, M.D.; Calabria, I.; Palanca-Ballester, C.; Sandoval, J. Epigenetics and oxidative stress in aging. *Oxid. Med. Cell Longev.* **2017**, *2017*, 9175806. [CrossRef]

124. Cheleschi, S.; De Palma, A.; Pascarelli, N.A.; Giordano, N.; Galeazzi, M.; Tenti, S.; Fioravanti, A. Could oxidative stress regulate the expression of microRNA-146a and microRNA-34a in human osteoarthritic chondrocyte cultures? *Int. J. Mol. Sci.* **2017**, *18*, E2660. [CrossRef] [PubMed]

125. Zhang, W.; Xu, W.; Feng, Y.; Zhou, X. Non-coding RNA involvement in the pathogenesis of diabetic cardiomyopathy. *J. Cell. Mol. Med.* **2019**, *23*, 5859–5867. [CrossRef]

126. Lan, J.; Huang, Z.; Han, J.; Shao, J.; Huang, C. Redox regulation of microRNAs in cancer. *Cancer Lett.* **2018**, *418*, 250–259. [CrossRef]

127. Esposti, D.D.; Aushev, V.N.; Lee, E.; Cros, M.-P.; Zhu, J.; Herceg, Z.; Chen, J.; Hernandez-Vargas, H. miR-500a-5p regulates oxidative stress response genes in breast cancer and predicts cancer survival. *Sci. Rep.* **2017**, *7*, 15966. [CrossRef]

128. Sangokoya, C.; Telen, M.J.; Chi, J.-T. microRNA miR-144 modulates oxidative stress tolerance and associates with anemia severity in sickle cell disease. *Blood* **2010**, *116*, 4338–4348. [CrossRef]

129. Kim, C.; Kang, D.; Lee, E.K.; Lee, J.S. Long noncoding RNAs and RNA binding proteins in oxidative stress, cellular senescence, and age-related diseases. *Oxid. Med. Cell Longev.* **2017**, *2017*, 2062384. [CrossRef]

130. Tehrani, S.S.; Karimian, A.S.; Parsian, H.; Majidinia, M.; Yousefi, G. Multiple functions of long non-coding RNAs in oxidative stress, DNA damage response and cancer progression. *J. Cell. Biochem.* **2018**, *119*, 223–236. [CrossRef]

131. Mohan, I.K.; Das, U.N. Oxidant stress, anti-oxidants and essential fatty acids in systemic lupus erythematosus. *Prostaglandins Leukot. Essent Fatty Acids* **1997**, *56*, 193–198. [CrossRef]

132. Kudaravalli, J. Improvement in endothelial dysfunction in patients with systemic lupus erythematosus with N-acetylcysteine and atorvastatin. *Ind. J. Pharmacol.* **2011**, *43*, 311–315. [CrossRef] [PubMed]

133. Lai, Z.W.; Hanczko, R.; Bonilla, E.; Caza, T.N.; Clair, B.; Bartos, A.; Miklossy, G.; Jimah, J.; Deherty, E.; Tily, H.; et al. N-aceylcysteine reduces disease activity by blocking mammalian target of rapamycin in T cells from systemic lupus erythematosus patients: A randomized double blind, placebo-controlled trial. *Arthritis Rheum.* **2012**, *64*, 2937–2946. [CrossRef] [PubMed]

134. Tzang, B.S.; Hsu, T.C.; Kuo, C.Y.; Chen, T.Y.; Chiang, S.Y.; Li, S.L.; Kao, S.H. Cytamine attenuates lupus-associated apoptosis of ventricular tissue by suppressing both intrinsic and extrinsic pathways. *J. Cell. Mol. Med.* **2012**, *16*, 2104–2111. [CrossRef] [PubMed]

135. Portal-Nùñez, S.; Esbrit, P.; Alcaraz, M.J.; Largo, R. Oxidative stress, autophagy, epigenetic changes and regulation by miRNAs as potential therapeutic targets in osteoarthritis. *Biochem. Pharmacol.* **2016**, *108*, 1–10. [CrossRef]

136. Dong, D.; Zhang, Y.; Reece, E.A.; Wang, L.; Harman, C.R.; Yang, P. microRNA expression profiling and functional annotation analysis of their targets modulated by oxidative stress during embryonic heart development in diabetic mice. *Reprod. Toxicol.* **2016**, *65*, 365–374. [CrossRef]

Review

MicroRNA and Oxidative Stress Interplay in the Context of Breast Cancer Pathogenesis

Giulia Cosentino [1], Ilaria Plantamura [1], Alessandra Cataldo [1,2,*,†] and Marilena V. Iorio [1,2,*,†]

[1] Molecular Targeting Unit, Research Department, Fondazione IRCCS Istituto Nazionale dei Tumori, 20133 Milan, Italy; giulia.cosentino@istitutotumori.mi.it (G.C.); ilaria.plantamura@istitutotumori.mi.it (I.P.)
[2] IFOM Istituto FIRC di Oncologia Molecolare, 20139 Milan, Italy
*   Correspondence: alessandra.cataldo@istitutotumori.mi.it (A.C.); marilena.iorio@istitutotumori.mi.it (M.V.I.); Tel.: +39-02-2390-5134 (A.C.); +39-02-2390-5134 (M.V.I.)
†   These authors contributed equally to this work.

Received: 27 September 2019; Accepted: 15 October 2019; Published: 17 October 2019

Abstract: Oxidative stress is a pathological condition determined by a disturbance in reactive oxygen species (ROS) homeostasis. Depending on the entity of the perturbation, normal cells can either restore equilibrium or activate pathways of cell death. On the contrary, cancer cells exploit this phenomenon to sustain a proliferative and aggressive phenotype. In fact, ROS overproduction or their reduced disposal influence all hallmarks of cancer, from genome instability to cell metabolism, angiogenesis, invasion and metastasis. A persistent state of oxidative stress can even initiate tumorigenesis. MicroRNAs (miRNAs) are small non coding RNAs with regulatory functions, which expression has been extensively proven to be dysregulated in cancer. Intuitively, miRNA transcription and biogenesis are affected by the oxidative status of the cell and, in some instances, they participate in defining it. Indeed, it is widely reported the role of miRNAs in regulating numerous factors involved in the ROS signaling pathways. Given that miRNA function and modulation relies on cell type or tumor, in order to delineate a clearer and more exhaustive picture, in this review we present a comprehensive overview of the literature concerning how miRNAs and ROS signaling interplay affects breast cancer progression.

Keywords: oxidative stress; miRNAs; breast cancer; ROS

---

1. Oxidative Stress

Reactive oxygen species (ROS) are oxygen-derived small molecules in the form of free radicals (i.e., contains one or more unpaired electrons) or non-radicals [1]. Among the most biologically relevant species there are the superoxide anion radical ($O_2^{-\bullet}$), the hydroxyl radical (OH·) and hydrogen peroxide (H_2O_2). At first, it was thought that these molecules were only metabolic waste, deleterious for nucleic acids, lipids and proteins; scientists, however, discovered that ROS are used by the cell as messages to activate different physiological signaling cascades [2,3]. In fact, in a biological system, the balance between the concentration of ROS and the activation of antioxidant mechanisms is finely tuned [4]. When this equilibrium lacks, the phenomenon of oxidative stress occurs, causing the alteration of intracellular molecules, such as DNA and RNA. A shift towards ROS production, thus, triggers a wide range of cellular responses, even apoptosis or phagocytosis, depending on the amplitude of the shift. Several endogenous and exogenous sources can trigger ROS production. In response to stimuli like cytokines and growth factors, NADPH oxidases (NOXs) and mitochondria produce the larger percentage of ROS. NOXs and metabolic complexes I, II and III present on the mitochondrial inner membrane generate, for example, the radical superoxide starting from a molecule of oxygen. Dangerous levels of ROS can be reached also after prolonged exposure to radiations and carcinogens, along with DNA damaging drugs. The major mutagenic product of DNA oxidation is

8-hydroxyl-2'-deoxyguanosine (8-OHdG). Figure 1 summarizes principal sources producing ROS and main regulators and pathways influenced by ROS production (Figure 1). Cancer cells are usually in a chronic state of oxidative stress, which they are able to exploit to sustain a proliferative and aggressive phenotype. Moreover, due to their detrimental action, ROS can also initiate tumorigenesis [5]. It is thus important not to overlook the impact of such phenomenon on every cellular process and, in particular, on those crucial for the development and progression of a neoplastic disease.

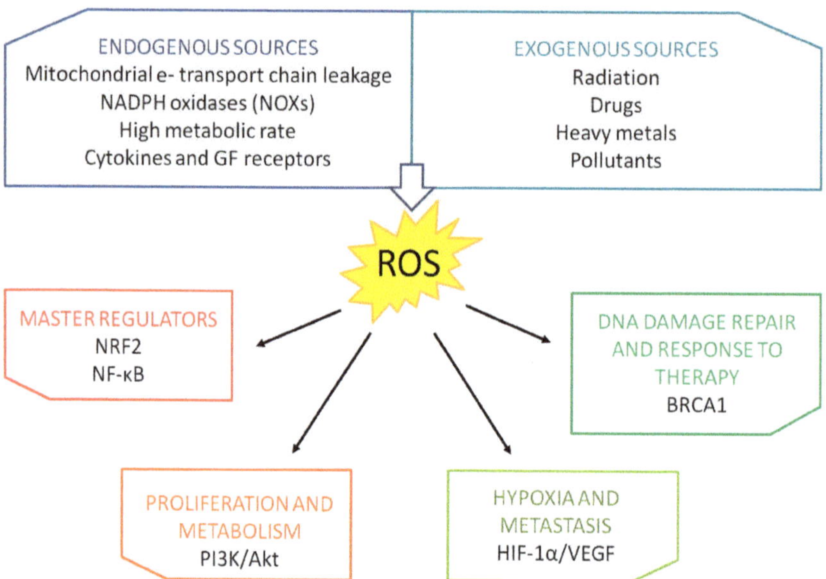

Figure 1. Endogenous and exogenous sources of ROS and pathways influenced by oxidative stress in breast cancer.

2. Breast Cancer

Breast cancer is the second most commonly diagnosed cancer worldwide and the leading cause of cancer death in women [6]. The severity and aggressiveness of breast cancer is evaluated by examining physical and anatomical properties of the disease, in particular by using histological grading and TNM staging, where T (0–4) is used to describe the size and location of the tumor, N (0–3) accounts for the lymph node invasion and M measures the spread of the tumor as distant metastasis [7].

The therapeutic regimen is finally driven by the characterization of the breast cancer subtype according to the immunohistochemical evaluation of three markers: Estrogen receptor (ER), Progesteron Receptor (Pgr) and HER2 (Epidermal Growth Factor Receptor 2) [8]. Tumors lacking the expression of these three markers are called triple negative breast cancers (TNBCs). A major contribution to the increase in survival rate has been provided by the improvement in the therapeutic regimens as well as in early diagnosis. Moreover, it is fundamental to develop always more personalized drugs for different cancers subtypes.

The advent of the genomic era disclosed the complexity of breast cancer. For the first time, in 2000, Perou and colleagues classified the disease in five specific subtypes according to intrinsic gene expression: Luminal-A, Luminal-B, HER2-positive, Basal-like and Normal-like [9]. Further studies later identified a new subtype, the so called claudin-low [10], which accounts for 7–14% of all breast cancers. Moreover, one of the most important applications of the breast cancer molecular classification lies in its ability to identify groups with a different outcome and response to treatments [11–15].

3. Oxidative Stress and Breast Cancer

Breast cancers, in particular estrogen receptor-positive malignancies, are characterized by significant high levels of 8-OHdG, and their detection in blood serum is reported to have prognostic value [16–18].

Estrogen is a major driver of mitochondrial ROS production. It activates redox-sensitive proteins involved in cell proliferation and anti-apoptotic pathways. In order to sustain such signaling without risking cell cycle arrest and apoptosis, estrogen enhances also an antioxidant response by inducing, for example, the transcription factor Nuclear-erythroid-2-related factor 2 (NRF2). This enzyme is the main redox master regulator; under oxidative stress, its inhibitor Kelch-like ECH-associated protein 1 (Keap1) undergoes a conformational change that allows NRF2 dissociation and consequent translocation to the nucleus, where it enhances the transcription of different ROS-counteracting agents [19,20]. Numerous evidence shows that NRF2 is overexpressed in breast cancer, where it promotes cell survival, proliferation, migration and metastasis [21–23].

Additionally, it is important to note the interplay between NRF2 and BRCA1. Gorrini C. et al. demonstrated that BRCA1 enhances and stabilizes NRF2 expression and that estrogen is able to partially mimic this action in BRCA1-null cells [24,25].

Moreover, in 2014, Victorino V. J. et al. analyzed the effect of HER2 overexpression on the oxidative systemic profile in breast cancer patients [26]. The results showed that HER2-overexpressing malignancies are characterized by an enhanced oxidative stress, attenuated by increased SOD and stabilized gluthatione (GSH) levels, which are indicative of an active antioxidant response. In the same year, Kang H. J. et al. reported that also HER2 interacts with NRF2 to promote the transcription of antioxidant and detoxification genes and that this partnership confers drug resistance to human breast cancer cells [27]. Antioxidants can, thus, favor breast neoplastic transformation: by reducing ROS concentrations they can prevent ROS-dependent cell death [28,29]. Therefore, the role of antioxidants in breast cancer is often controversial; for example antioxidant superoxide dismutase 2 (SOD2), which converts the highly toxic radical superoxide into more stable hydrogen peroxide in the mitochondria, can act both as an oncogene and as a tumor suppressor. In fact, it is found downmodulated in early-stage breast cancer while upregulated in advanced tumors [30,31]. Despite these results, SOD mimics have been proposed for therapeutic purposes [32,33]. Catalase, glutathione peroxidases, and peroxiredoxins are among the other antioxidant enzymes which balance ROS production. In 2017, Bao B. et al. demonstrated that the addition of a re-engineered protein form of the catalase enzyme to EGFR-inhibitor erlotinib treatment helps overcoming resistance by specifically targeting the stem-like portion of TNBC cells [34]. Conversely, peroxiredoxin-1 (PRDX1) downmodulation was shown to be beneficial for breast cancer therapy, especially in concomitance with prooxidant agents [35]. Moreover, specific acquaporins allow H_2O_2 to cross cell membranes more rapidly than by sole diffusion [36]. In breast cancer, Aquaporin-3 has been proposed as target for therapy due to its role in CXCL12/CXCR4-dependent cancer cell migration [37].

Finally, damages to RNA molecules have to be considered equally harmful [38]. For example, it is of particular relevance for this review the impact on miRNA biology and the consequent influence on different regulation networks [39].

4. MicroRNAs

MicroRNAs (miRNAs) are small single strand molecules (~18–25 nucleotides), they are non-coding RNAs that are able to control gene expression at post-transcriptional level [40]. MiRNA biogenesis starts when RNA polymerase II/III transcribes for a long primary transcript with a hairpin structure, called pri-miRNAs [41]. The pri-miRNA is the substrate of Drosha and Dicer, two members of the RNase III family enzymes. First, Drosha cleaves the pri-miRNA in a ~70-nucleotide pre-miRNA into the nucleus, which is then exported into the cytoplasm by the Exportin-5 Ran-GTPase, where Dicer catalyzes its conversion to a short miRNA/miRNA* duplex (~20 bp). To complete the miRNA biogenesis, the transactivation-responsive RNA-binding protein (TRBP) leads to the assembling of the miRNA-induced silencing complex (miRISC), mediating the interaction between DICER and Argonaute protein (AGO1, AGO2, AGO3 or AGO4). Finally, the miRISC complex selects one single strand of the duplex (mature miRNA), which recognizes the "seed" region on the target mRNA, usually placed at the 5′ UTR, inducing translational repression or deadenylation and degradation. The small RNA lin-4 was the first no-coding RNA discovered in *Caenorhabditis elegans*, involved in the larval development [42]. Afterwards, several studies have pointed out the importance of these small molecules; currently it is well known that miRNAs are involved in almost every biological process in mammals, including oxidative stress and cancer [43]. Indeed, miRNAs can act as oncosuppressors or oncogenes, which are generally found respectively downregulated and upregulated in tumor cells (e.g., miR-205 and miR-21, respectively). In 2005, Iorio M.V. et al. discovered a panel of dysregulated microRNAs in breast cancer: miR-10b, miR-125b, and miR-145 were down-regulated, and miR-21 and miR-155 were up-regulated, suggesting that they could have a role in breast cancer disease [44].

Recently, we reported that miRNAs have a relevant role in DNA damage response, occurring following an exogenous oxidative stress, such as chemotherapy [45,46]. In fact, miRNAs have the capability to target several genes involved in the DNA repair machinery, regulating therapy responsiveness. Here, we review the literature concerning the role of miRNAs in the regulation of the major actors and principal pathways altered by oxidative stress in breast cancer.

5. MiRNAs Modulate Oxidative Stress Master Regulators: NRF2 and NF-κB

NRF2 is an important transcription factor which induction, or derepression, depends on the redox status of the cell. Normally, NRF2 is found inactive in the cytoplasm bound to its homodimeric repressor Keap1, which anchors the protein Cullin-3 (CUL3) to form an E3 ubiquitin ligase complex; the complex is responsible for NRF2 ubiquitination and consequent proteasomal degradation [47]. When cellular ROS concentrations increase, specific Keap1 cystenyl residues are modified and NRF2 is released and free to translocate into the nucleus, where it recognizes the so called "Antioxidant Responsive Elements (ARE)" sequences on target gene promoters and enhances the transcription process [48]. NRF2 promotes the expression of antioxidants and detoxifying enzymes and, initially, it was thought to act as a defensive agent against tumorigenesis. However, as previously explained for SOD2, an excessive reduction of ROS levels can prove counterproductive. Therefore, it is not unusual to find contradictory literature concerning the prospective of using NRF2 inhibitors for therapeutic purposes [49,50]. NRF2 pathogenic activation and accumulation can be triggered by different events; one of the most frequent alterations concerns Keap1 expression or its ability to stably bind and degrade NRF2 [51]. MiRNAs were found to exert this oncogenic activity, NRF2 induction, in different malignancies [52–54]. In 2011, Eades G. et al. demonstrated for the first time a miRNA-dependent Keap1 regulation in breast cancer: miR-200a targets Keap1 mRNA and induces its degradation [55]. Interestingly, the same group published the same year an additional paper describing NRF2 inhibition by miR-28 in MCF7 breast cancer cell line [56]. Two other miRNAs, miR-93 and miR-153, have been reported to target NRF2 and their overexpression is associated with breast carcinogenesis [57,58]. This evidence validates once more the context-specific value of NRF2 modulation (Figure 2A).

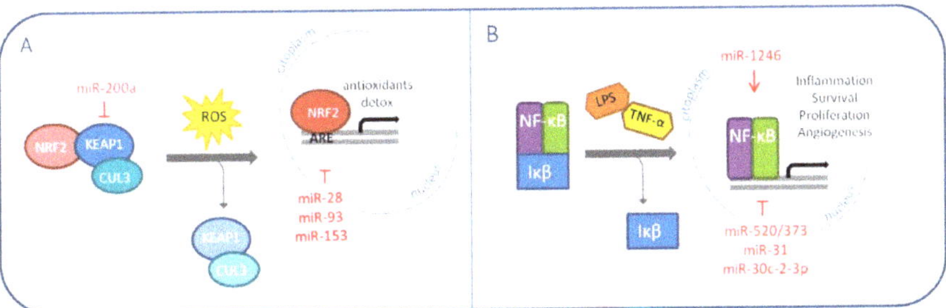

Figure 2. MiRNAs modulating oxidative stress master regulators NRF2 (**A**) and NF-κB (**B**) in breast cancer (The red arrow indicates upmodulation, the red "T" stands for inhibition).

The same concept can be translated to the other redox master regulator, the nuclear factor-kB (NF-κB). NF-κB can be found as both homo- and heterodimer of five distinct proteins, RelA, RelB, c-Rel, p50 and p52. It is inhibited in the cytoplasm by the IκB families, which interfere with the target activity by interacting with its important Rel homology domain (RHD), implicated in the formation of dimers and DNA binding [59].

IκB proteins are generally degraded in response to inflammatory cues like TNFα and lipopolysaccharide (LPS). The consequent NF-κB signaling, modulated by ROS, is cell type and context specific. This is probably due to the transcription factor wide range of action: cell growth, proliferation, migration and apoptosis are among the pathways it influences [60,61]. NF-κB signaling is frequently found dysregulated in human cancers [62]. In breast cancer, the protein is reported as constitutively activated and associated to aggressive and chemoresistant malignances [63–65]. MiRNAs play an important part also in this scenario. First of all, NF-κB favors breast cancer cell invasion by inducing the expression of the oncomiR miR-21 in response to DNA damage [66]. Wiemann S. and his group, instead, published different papers on NF-κB-regulating miRNAs over the years [67–70]. In 2012, they demonstrated the tumor suppressive role of miR-520/373 family in ER-negative breast cancer, through the targeting of NF-κB and TGF-β signaling pathways. In the same context, in 2013, miR-31 was seen to sensitize cancer cells to apoptosis by impairing NF-κB pathway. In 2015, miR-30c-2-3p was shown to reduce proliferation and invasion of MDA-MB-231 cells through the downmodulation of TNFR/NF-κB signaling and cell cycle proteins. Conversely, in 2017, a role as an oncomiR was attributed to miR-1246, which was reported to induce the NF-κB pro-inflammatory signaling in breast cancer cells. The same year, another group discovered that miR-221/222 promote stem-like properties and tumor growth of breast cancer via targeting PTEN and sustained Akt/NF-κB/COX-2 activation (Figure 2B) [71].

Despite having an oscillatory expression in a physiological context, NF-κB thus emerges from the literature presented as a proper oncogene in breast cancer. Moreover, due to its broad spectrum of interactions, numerous are the miRNAs involved in the regulation of the signaling cascade and, consequently, many are the hints for therapeutic interventions.

6. MiRNAs Modulate Pathways Altered by Oxidative Stress

6.1. Metabolism

The main goal of cancer cells is proliferation and survival. Such activities require a great amount of energy in a short period of time. Therefore, cancer cells tend to modify their metabolism in order to respond to this demand. According to the known Warburg effect, cancers prefer a rapid glycolysis to the more efficient mitochondrial oxidative phosphorylation. This switch also allows avoiding an excessive mitochondria-related production of ROS. Interestingly, it has been suggested that the latter could be the primary reason for the metabolic reprogramming [72]. In breast cancers, the metabolic status seems to be linked to the molecular subtype. In fact, the more aggressive TNBCs are characterized by a glycolytic phenotype, while luminal malignancies retain oxidative phosphorylation as the major source of energy [73]. It is important to note that it is not unusual to find heterogeneity also among cells of the same tumor mass, a scenario that can be as deleterious as a predominant Warburg setting. In 2018, our group indeed proposed that, starting from a mixed population of TNBC cells, pushing all the cells towards a glycolytic phenotype could become counterproductive for the tumor. Through the downmodulation of the lactate transporter MCT1, miR-342-3p is able to disrupt the energetic fluxes between neighboring glycolytic and oxidative cells, promoting the shift and ultimately triggering a competition for glucose [74]. It has been demonstrated that glucose deprivation induces oxidative stress in cancer cells [75]. One of the most cited mechanisms of breast cancer cell metabolic reprogramming that involves miRNAs is miR-155 promotion of hexokinase II (HKII) expression, necessary to start glycolysis. This miRNA modulates multiple pathways that control HKII: first, miR-155, through the direct downmodulation of C/EBPβ, reduces miR-143, a HKII inhibitor; second, the miRNA frees STAT3 from its suppressor SOCS1 to enhance HKII; third, miR-155 positively regulates HKII by interfering with the PIK3R1-FOXO3a-cMYC axis [76–78]. Another recent example is the work by Eastlack S. C. et al. that demonstrated miR-27b promotes breast cancer progression by targeting Pyruvate Dehydrogenase Protein X (PDHX), thus altering cell's metabolic configuration [79]. PI3K/Akt pathway, which players are frequently mutated in breast cancers, deeply impacts on metabolism and ROS production by directly regulating mitochondrial bioenergetics and NOX enzymes. Vice versa, oxidative stress activates PI3K and suppresses the activity of PTEN, inhibitor of PI3K/Akt signaling [80–82]. Due to the relevance of the pathway, numerous are the miRNAs found implicated in its regulation in breast cancer. Among the latest reported, there are the tumor suppressor miR-204-5p, which targets PIK3CB, and the PTEN-inhibiting oncomiRs miR-1297 and miR-498 (Figure 3A) [83–85].

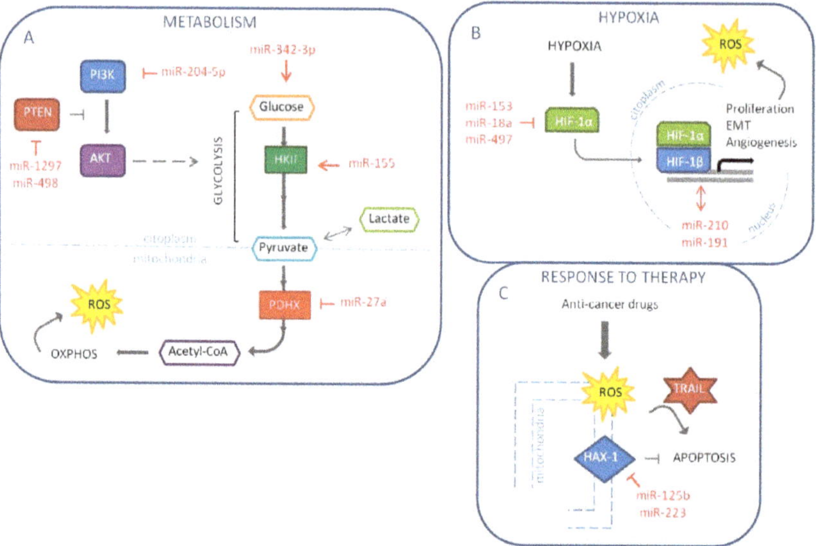

Figure 3. MiRNAs involved in the regulation of hallmarks of cancer influenced by oxidative stress in breast cancer: metabolism (**A**), hypoxia (**B**) and response to therapy (**C**) (The red arrow indicates upmodulation, the red "T" stands for inhibition).

6.2. Hypoxia

Hypoxia refers to a pathological level of oxygen tension, caused by the high proliferative rates of cancer cells and insufficient vasculature. The lack of oxygen supply, thus, induces cancer cells to undergo epithelial-to-mesenchymal transition (EMT), which corresponds to the acquisition of migratory and invasive properties, stem-like features and resistance to apoptosis. The main player in this context is the transcription factor HIF-1α, which stimulates angiogenesis and triggers a positive feedback loop on proliferation pathways. As a consequence, oxidative stress increases upon re-oxygenation and mitochondrial electron leaks. Numerous studies showed a link between the miRNA's role and the hypoxia in the breast cancer initiation and progression. The first miRNA to be pointed out is miR-210. In 2007, this miRNA emerged as part of the miRNA signature of hypoxia and the year after it was elected as an independent prognostic factor in breast cancer [86,87]. Moreover, Liang H. and colleagues investigated miR-153 mechanism of action in breast cancer; showing that this miRNA acts as a tumor suppressor by targeting HIF-1α [88]. In fact, miR-153 inhibits migration, proliferation and tube formation in HUVEC cells and angiogenesis in MDA-MB-231 in vivo model through the inhibition of the HIF-1α/VEGFA axis. In another paper, the high expression of miR-191 in breast cancer cell lines induces a more aggressive tumor under hypoxia [89]. Consequently, the authors suggest that miR-191 inhibition may be exploited as a new therapeutic option for hypoxic breast cancer. In addition, miR-18a targets HIF-1α, which high expression is associated with shorter DMFS (distant metastasis-free survival) in patients with basal-like breast tumors [90]. In metastatic MDA-MB-231 cells, ectopic miR-18a expression reduces both primary tumor and lung metastasis. Another miRNA reported targeting HIF-1α is miR-497, thus, it represses the hypoxic conditions and for this reason it is usually downregulated in breast cancer cells [91]. MiR-497 also targets a pro-angiogenic molecule, VEGF (vascular endothelial growth factor) and its ectopic expression reduces tumor growth and angiogenesis in breast cancer tumor model (Figure 3B). In conclusion, we could support the strategic role of miRNAs in the tumor progression and in particular in hypoxia and metastasis and we could speculate the possibility to use miRNAs as therapeutic tools to reduce tumor aggressiveness and dissemination.

6.3. Response To Therapy

It is well known that miRNAs can influence response to therapy in breast cancer. Moreover, they are under investigation as potential therapeutic tools, alone or in combination with standard therapy to impair cancer progression. Chemotherapy and radiotherapy still represent the standard therapy for breast cancer; miRNAs are able to target different genes reducing drug resistance and promoting therapeutic response. Indeed, in 2016, we reported that miR-302b, by targeting E2F1 and DNA repair, enhances cisplatin response in breast cancer cells [45]. Chemotherapy drugs, such as platinum compounds and anthracyclines, and also ionizing radiation induce oxidative stress generating high levels of ROS [92]. The induction of oxidative stress can lead to the preferential killing of cancer cells. Currently, the main problem of chemotherapy and radiotherapy is the development of resistance mechanisms; recent works report the role of miRNAs in the response to these therapies by targeting oxidative stress molecules. Recently, it was demonstrated that miR-125b is involved in chemotherapy resistance by affecting oxidative stress pathways in breast cancer [93]. MiR-125b, by targeting HAX-1, an anti-apoptotic gene, impacts on doxorubicin resistance. The mechanism behind this phenomenon is a decrease in the levels of MMP following HAX-1 downregulation and the release of ROS from the mitochondria into the cytoplasm. Thus, miR-125b is able to re-sensitize breast cancer cells to doxorubicin treatment using ROS pathway (Figure 3C). Concerning chemoresistance, Roscigno G. et al. have reported that miR-24, up-regulated in breast cancer stem cells, induces resistance to cisplatin by targeting the pro-apoptotic factor BimL [94]. Furthermore, miR-24 targets FIH1 that induces the repression of HIF-1α. Thus, the authors have shown that miR-24 is induced in hypoxic conditions, leading to cancer stem cell growth and consequently inducing chemotherapy resistance. Breast cancer patients often poorly respond to radiotherapy, and the mechanisms of radioresistance have not been elucidated yet. MiR-668 was found increased in breast cancer cells resistant to radiotherapy; this phenomenon occurs because IκBα is a direct target of miR-668, leading to the activation of NF-κB [95]. Generally, drug resistance is an important challenge in the treatment of breast cancer, especially for TNBC, which still don't have target therapy. To date, novel therapeutic strategies have been tested mainly in the treatment of TNBC. MiR-223 is related to resistance to TRAIL-induced apoptosis in cancer stem cells of TNBC [96]. Indeed, reintroduction of miR-223 and treatment with TRAIL in MDA-MB-231 cell line induces a strong generation of ROS, through the targeting of HAX-1 into the mitochondria, and TNBC stem cells are more sensitive to TRAIL treatment. Moreover, the miR-223/HAX-1 axis enhances the sensitivity to doxorubicin and cisplatin in TNBC stem cells (Figure 3C).

7. Conclusive Remarks

In this review, we have illustrated what emerges from the literature about the important role of oxidative stress in the pathogenesis of breast cancer, influencing most of the pathways usually altered in tumors, affecting also response to therapy. Moreover, many of the proteins involved in this process, such as SOD2 and NRF2, can exert opposite roles depending on the context, complicating the scenario. Thus, it is important to explore in more detail the mechanisms behind the regulation of the redox status in relation to a specific scenario in order to better define which pathways can be proposed as therapeutic targets. MiRNAs act as regulative elements in almost every biological process, including oxidative stress and cancer. Here, we have mainly reviewed the literature concerning miRNAs involved in the regulation of oxidative stress players in breast cancer disease. MiRNA role in the regulation of redox status makes them as hypothetical and crucial targets or tools for therapy since they could provide the treatment context specificity. MiRNA general use as therapy option has yet to show relevant results, but an increasing body of evidence has been provided through the years in favor of such a solution, especially in oncology. Additionally, breast cancer is one of the most studied neoplasia and many are the miRNAs which mechanism of action is consolidated in this framework. Hopefully, therefore, it will be soon possible to have major improvements in this research field.

Author Contributions: G.C., I.P. and A.C. wrote the manuscript. M.V.I. and A.C. edited and revised the manuscript.

Funding: Alessandra Cataldo is supported by a fellowship from "Fondazione Umberto Veronesi". Marilena V. Iorio is supported by "Ministry of Health" (GR-2016-02361750) and by "Fondazione Guido Berlucchi" Career Grant.

Conflicts of Interest: The authors declare no conflict of interest.

References

1. Li, R.; Jia, Z.; Trush, M.A. Defining ROS in Biology and Medicine. *React. Oxyg. Species (Apex, NC)* **2016**, *1*, 9–21. [CrossRef] [PubMed]
2. Finkel, T. Signal transduction by reactive oxygen species. *J. Cell Biol.* **2011**, *194*, 7–15. [CrossRef] [PubMed]
3. Zhang, J.; Wang, X.; Vikash, V.; Ye, Q.; Wu, D.; Liu, Y.; Dong, W. ROS and ROS-Mediated Cellular Signaling. *Oxid. Med. Cell. Longev.* **2016**, *2016*, 4350965. [CrossRef] [PubMed]
4. Sies, H.; Berndt, C.; Jones, D.P. Oxidative Stress. *Annu. Rev. Biochem.* **2017**, *86*, 715–748. [CrossRef] [PubMed]
5. Sosa, V.; Moliné, T.; Somoza, R.; Paciucci, R.; Kondoh, H.; LLeonart, M.E. Oxidative stress and cancer: An overview. *Ageing Res. Rev.* **2013**, *12*, 376–390. [CrossRef]
6. Bray, F.; Ferlay, J.; Soerjomataram, I.; Siegel, R.L.; Torre, L.A.; Jemal, A. Global cancer statistics 2018: GLOBOCAN estimates of incidence and mortality worldwide for 36 cancers in 185 countries. *CA Cancer J. Clin.* **2018**, *68*, 394–424. [CrossRef]
7. Giuliano, A.E.; Edge, S.B.; Hortobagyi, G.N. Eighth Edition of the AJCC Cancer Staging Manual: Breast Cancer. *Ann. Surg. Oncol.* **2018**, *25*, 1783–1785. [CrossRef]
8. Waks, A.G.; Winer, E.P. Breast Cancer Treatment: A Review. *JAMA* **2019**, *321*, 288–300. [CrossRef]
9. Perou, C.M.; Sørlie, T.; Eisen, M.B.; van de Rijn, M.; Jeffrey, S.S.; Rees, C.A.; Pollack, J.R.; Ross, D.T.; Johnsen, H.; Akslen, L.A.; et al. Molecular portraits of human breast tumours. *Nature* **2000**, *406*, 747–752. [CrossRef]
10. Herschkowitz, J.I.; Simin, K.; Weigman, V.J.; Mikaelian, I.; Usary, J.; Hu, Z.; Rasmussen, K.E.; Jones, L.P.; Assefnia, S.; Chandrasekharan, S.; et al. Identification of conserved gene expression features between murine mammary carcinoma models and human breast tumors. *Genome Biol.* **2007**, *8*, 76. [CrossRef]
11. Sørlie, T.; Perou, C.M.; Tibshirani, R.; Aas, T.; Geisler, S.; Johnsen, H. Gene expression patterns of breast carcinomas distinguish tumor subclasses with clinical implications. *Proc. Natl. Acad. Sci. USA* **2001**, *98*, 10869–10874. [CrossRef] [PubMed]

12. Rouzier, R.; Perou, C.M.; Symmans, W.F.; Ibrahim, N.; Cristofanilli, M.; Anderson, K.; Hess, K.R.; Stec, J.; Ayers, M.; Wagner, P.; et al. Breast cancer molecular subtypes respond differently to preoperative chemotherapy. *Clin. Cancer Res.* **2005**, *11*, 5678–5685. [CrossRef] [PubMed]

13. Hu, Z.; Fan, C.; Oh, D.S.; Marron, J.S.; He, X.; Qaqish, B.F.; Livasy, C.; Carey, L.A.; Reynolds, E.; Dressler, L.; et al. The molecular portraits of breast tumors are conserved across microarray platforms. *BMC Genom.* **2006**, *7*, 96.

14. Carey, L.A.; Dees, E.C.; Sawyer, L.; Gatti, L.; Moore, D.T.; Collichio, F.; Ollila, D.W.; Sartor, C.I.; Graham, M.L.; Perou, C.M. The triple negative paradox: Primary tumor chemosensitivity of breast cancer subtypes. *Clin. Cancer Res.* **2007**, *13*, 2329–2334. [CrossRef]

15. Shehata, M.; Teschendorff, A.; Sharp, G.; Novcic, N.; Russell, I.A.; Avril, S.; Prater, M.; Eirew, P.; Caldas, C.; Watson, C.J.; et al. Phenotypic and functional characterisation of the luminal cell hierarchy of the mammary gland. *Breast Cancer Res.* **2012**, *14*, R134. [CrossRef]

16. Musarrat, J.; Arezina-Wilson, J.; Wani, A.A. Prognostic and aetiological relevance of 8-hydroxyguanosine in human breast carcinogenesis. *Eur. J. Cancer* **1996**, *32*, 1209–1214. [CrossRef]

17. Jakovcevic, D.; Dedic-Plavetic, N.; Vrbanec, D.; Jakovcevic, A.; Jakic-Razumovic, J. Breast Cancer Molecular Subtypes and Oxidative DNA Damage. *Appl. Immunohistochem. Mol. Morphol.* **2015**, *23*, 696–703. [CrossRef]

18. Nour Eldin, E.E.M.; El-Readi, M.Z.; Nour Eldein, M.M.; Alfalki, A.A.; Althubiti, M.A.; Mohamed Kamel, H.F.; Eid, S.Y.; Al-Amodi, H.S.; Mirza, A.A. 8-Hydroxy-2′-deoxyguanosine as a Discriminatory Biomarker for Early Detection of Breast Cancer. *Clin. Breast Cancer* **2019**, *19*, 385–393. [CrossRef]

19. Tian, H.; Gao, Z.; Wang, G.; Li, H.; Zheng, J. Estrogen potentiates reactive oxygen species (ROS) tolerance to initiate carcinogenesis and promote cancer malignant transformation. *Tumour Biol.* **2016**, *37*, 141–150. [CrossRef]

20. Rojo de la Vega, M.; Chapman, E.; Zhang, D.D. NRF2 and the Hallmarks of Cancer. *Cancer Cell* **2018**, *34*, 21–43. [CrossRef]

21. Zhang, C.; Wang, H.J.; Bao, Q.C.; Wang, L.; Guo, T.K.; Chen, W.L.; Xu, L.L.; Zhou, H.S.; Bian, J.L.; Yang, Y.R.; et al. NRF2 promotes breast cancer cell proliferation and metastasis by increasing RhoA/ROCK pathway signal transduction. *Oncotarget* **2016**, *7*, 73593–73606. [CrossRef] [PubMed]

22. Lu, K.; Alcivar, A.L.; Ma, J.; Foo, T.K.; Zywea, S.; Mahdi, A.; Huo, Y.; Kensler, T.W.; Gatza, M.L.; Xia, B. NRF2 Induction Supporting Breast Cancer Cell Survival Is Enabled by Oxidative Stress-Induced DPP3-KEAP1 Interaction. *Cancer Res.* **2017**, *77*, 2881–2892. [CrossRef] [PubMed]

23. Zhang, H.S.; Zhang, Z.G.; Du, G.Y.; Sun, H.L.; Liu, H.Y.; Zhou, Z.; Gou, X.M.; Wu, X.H.; Yu, X.Y.; Huang, Y.H. Nrf2 promotes breast cancer cell migration via up-regulation of G6PD/HIF-1α/Notch1 axis. *J. Cell. Mol. Med.* **2019**, *23*, 3451–3463. [CrossRef] [PubMed]

24. Gorrini, C.; Baniasadi, P.S.; Harris, I.S.; Silvester, J.; Inoue, S.; Snow, B.; Joshi, P.A.; Wakeham, A.; Molyneux, S.D.; Martin, B.; et al. BRCA1 interacts with Nrf2 to regulate antioxidant signaling and cell survival. *J. Exp. Med.* **2013**, *210*, 1529–1544. [CrossRef]

25. Gorrini, C.; Gang, B.P.; Bassi, C.; Wakeham, A.; Baniasadi, S.P.; Hao, Z.; Li, W.Y.; Cescon, D.W.; Li, Y.T.; Molyneux, S.; et al. Estrogen controls the survival of BRCA1-deficient cells via a PI3K-NRF2-regulated pathway. *Proc. Natl. Acad. Sci. USA* **2014**, *111*, 4472–4477. [CrossRef]

26. Victorino, V.J.; Campos, F.C.; Herrera, A.C.; Colado Simão, A.N.; Cecchini, A.L.; Panis, C.; Cecchini, R. Overexpression of HER-2/neu protein attenuates the oxidative systemic profile in women diagnosed with breast cancer. *Tumour Biol.* **2014**, *35*, 3025–3034. [CrossRef]

27. Kang, H.J.; Yi, Y.W.; Hong, Y.B.; Kim, H.J.; Jang, Y.J.; Seong, Y.S.; Bae, I. HER2 confers drug resistance of human breast cancer cells through activation of NRF2 by direct interaction. *Sci. Rep.* **2014**, *4*, 7201. [CrossRef]

28. Harris, I.S.; Treloar, A.E.; Inoue, S.; Sasaki, M.; Gorrini, C.; Lee, K.C.; Yung, K.Y.; Brenner, D.; Knobbe-Thomsen, C.B.; Cox, M.A.; et al. Glutathione and thioredoxin antioxidant pathways synergize to drive cancer initiation and progression. *Cancer Cell* **2015**, *27*, 211–222. [CrossRef]

29. Monteiro, H.P.; Ogata, F.T.; Stern, A. Thioredoxin promotes survival signaling events under nitrosative/oxidative stress associated with cancer development. *Biomed. J.* **2017**, *40*, 189–199. [CrossRef]

30. Becuwe, P.; Ennen, M.; Klotz, R.; Barbieux, C.; Grandemange, S. Manganese superoxide dismutase in breast cancer: From molecular mechanisms of gene regulation to biological and clinical significance. *Free Radic. Biol. Med.* **2014**, *77*, 139–151. [CrossRef]

31. Wang, Y.; Branicky, R.; Noë, A.; Hekimi, S. Superoxide dismutases: Dual roles in controlling ROS damage and regulating ROS signaling. *J. Cell Biol.* **2018**, *217*, 1915–1928. [CrossRef] [PubMed]

32. Batinić-Haberle, I.; Rebouças, J.S.; Spasojević, I. Superoxide dismutase mimics: Chemistry, pharmacology, and therapeutic potential. *Antioxid. Redox Signal.* **2010**, *13*, 877–918. [CrossRef] [PubMed]

33. Fernandes, A.S.; Saraiva, N.; Oliveira, N.G. Redox Therapeutics in Breast Cancer: Role of SOD Mimics. In *Redox-Active Therapeutics: Oxidative Stress in Applied Basic Research and Clinical Practice*; Batinić-Haberle, I., Rebouças, J., Spasojević, I., Eds.; Springer: Cham, Switzerland, 2016.

34. Bao, B.; Mitrea, C.; Wijesinghe, P.; Marchetti, L.; Girsch, E.; Farr, R.L.; Boerner, J.L.; Mohammad, R.; Dyson, G.; Terlecky, S.R.; et al. Treating triple negative breast cancer cells with erlotinib plus a select antioxidant overcomes drug resistance by targeting cancer cell heterogeneity. *Sci. Rep.* **2017**, *7*, 44125. [CrossRef]

35. Bajor, M.; Zych, A.O.; Graczyk-Jarzynka, A.; Muchowicz, A.; Firczuk, M.; Trzeciak, L.; Gaj, P.; Domagala, A.; Siernicka, M.; Zagozdzon, A.; et al. Targeting peroxiredoxin 1 impairs growth of breast cancer cells and potently sensitises these cells to prooxidant agents. *Br. J. Cancer* **2018**, *119*, 873–884. [CrossRef]

36. Bienert, G.P.; Chaumont, F. Aquaporin-facilitated transmembrane diffusion of hydrogen peroxide. *Biochim. Biophys. Acta* **2014**, *1840*, 1596–1604. [CrossRef] [PubMed]

37. Satooka, H.; Hara-Chikuma, M. Aquaporin-3 Controls Breast Cancer Cell Migration by Regulating Hydrogen Peroxide Transport and Its Downstream Cell Signaling. *Mol. Cell. Biol.* **2016**, *36*, 1206–1218. [CrossRef]

38. Kong, Q.; Lin, C.G. Oxidative damage to RNA: Mechanisms, consequences, and diseases. *Cell. Mol. Life Sci.* **2010**, *67*, 1817–1829. [CrossRef]

39. He, J.; Jiang, B.H. Interplay between Reactive oxygen Species and MicroRNAs in Cancer. *Curr. Pharmacol. Rep.* **2016**, *2*, 82–90. [CrossRef]

40. Lee, Y.; Kim, M.; Han, J.; Yeom, K.H.; Lee, S.; Baek, S.H.; Kim, V.N. MicroRNA genes are transcribed by RNA polymerase II. *EMBO J.* **2004**, *23*, 4051–4060. [CrossRef]

41. Krol, J.; Loedige, I.; Filipowicz, W. The widespread regulation of microRNA biogenesis, function and decay. *Nat. Rev. Genet.* **2010**, *11*, 597–610. [CrossRef]

42. Lee, R.C.; Feinbaum, R.L.; Ambros, V. Elegans heterochronic gene lin-4 encodes small RNAs with antisense complementarity to lin-14. *Cell* **1993**, *75*, 843–854. [CrossRef]

43. Iorio, M.V.; Casalini, P.; Piovan, C.; Braccioli, L.; Tagliabue, E. Breast cancer and microRNAs: Therapeutic impact. *Breast* **2011**, *20*, 63–70. [CrossRef]

44. Iorio, M.V.; Ferracin, M.; Liu, C.G.; Veronese, A.; Spizzo, R.; Sabbioni, S.; Magri, E.; Pedriali, M.; Fabbri, M.; Campiglio, M.; et al. MicroRNA gene expression deregulation in human breast cancer. *Cancer Res.* **2005**, *65*, 7065–7070. [CrossRef] [PubMed]

45. Cataldo, A.; Cheung, D.G.; Balsari, A.; Tagliabue, E.; Coppola, V.; Iorio, M.V.; Palmieri, D.; Croce, C.M. miR-302b enhances breast cancer cell sensitivity to cisplatin by regulating E2F1 and the cellular DNA damage response. *Oncotarget* **2016**, *7*, 786–797. [CrossRef] [PubMed]

46. Plantamura, I.; Cosentino, G.; Cataldo, A. MicroRNAs and DNA-Damaging Drugs in Breast Cancer: Strength in Numbers. *Front. Oncol.* **2018**, *8*, 352. [CrossRef]

47. McMahon, M.; Itoh, K.; Yamamoto, M.; Hayes, J.D. Keap1-dependent proteasomal degradation of transcription factor Nrf2 contributes to the negative regulation of antioxidant response element-driven gene expression. *J. Biol. Chem.* **2003**, *278*, 21592–21600. [CrossRef]

48. Suzuki, T.; Yamamoto, M. Molecular basis of the Keap1-Nrf2 system. *Free Radic. Biol. Med.* **2015**, *88*, 93–100. [CrossRef]

49. Sporn, M.B.; Liby, K.T. NRF2 and cancer: The good, the bad and the importance of context. *Nat. Rev. Cancer* **2012**, *12*, 564–571. [CrossRef]

50. Panieri, E.; Saso, L. Potential Applications of NRF2 Inhibitors in Cancer Therapy. *Oxid. Med. Cell. Longev.* **2019**, *2019*, 8592348. [CrossRef]

51. Taguchi, K.; Yamamoto, M. The KEAP1-NRF2 System in Cancer. *Front. Oncol.* **2017**, *7*, 85. [CrossRef]

52. Kabaria, S.; Choi, D.C.; Chaudhuri, A.D.; Jain, M.R.; Li, H.; Junn, E. MicroRNA-7 activates Nrf2 pathway by targeting Keap1 expression. *Free Radic. Biol. Med.* **2015**, *89*, 548–556. [CrossRef] [PubMed]

53. Shi, L.; Wu, L.; Chen, Z.; Yang, J.; Chen, X.; Yu, F.; Zheng, F.; Lin, X. MiR-141 Activates Nrf2-Dependent Antioxidant Pathway via Down-Regulating the Expression of Keap1 Conferring the Resistance of Hepatocellular Carcinoma Cells to 5-Fluorouracil. *Cell. Physiol. Biochem.* **2015**, *35*, 2333–2348. [CrossRef] [PubMed]

54. Akdemir, B.; Nakajima, Y.; Inazawa, J.; Inoue, J. miR-432 induces NRF2 Stabilization by Directly Targeting KEAP1. *Mol. Cancer Res.* **2017**, *15*, 1570–1578. [CrossRef] [PubMed]

55. Eades, G.; Yang, M.; Yao, Y.; Zhang, Y.; Zhou, Q. miR-200a regulates Nrf2 activation by targeting Keap1 mRNA in breast cancer cells. *J. Biol. Chem.* **2011**, *286*, 40725–40733. [CrossRef] [PubMed]

56. Yang, M.; Yao, Y.; Eades, G.; Zhang, Y.; Zhou, Q. MiR-28 regulates Nrf2 expression through a Keap1-independent mechanism. *Breast Cancer Res. Treat.* **2011**, *129*, 983–991. [CrossRef]

57. Singh, B.; Ronghe, A.M.; Chatterjee, A.; Bhat, N.K.; Bhat, H.K. MicroRNA-93 regulates NRF2 expression and is associated with breast carcinogenesis. *Carcinogenesis* **2013**, *34*, 1165–1172. [CrossRef]

58. Wang, B.; Teng, Y.; Liu, Q. MicroRNA-153 Regulates NRF2 Expression and is Associated with Breast Carcinogenesis. *Clin. Lab.* **2016**, *62*, 39–47. [CrossRef]

59. Hayden, M.S.; Ghosh, S. Shared principles in NF-kappaB signaling. *Cell* **2008**, *132*, 344–362. [CrossRef]

60. Morgan, M.J.; Liu, Z.G. Crosstalk of reactive oxygen species and NF-κB signaling. *Cell Res.* **2011**, *21*, 103–115. [CrossRef]

61. Lingappan, K. NF-κB in Oxidative Stress. *Curr. Opin. Toxicol.* **2018**, *7*, 81–86. [CrossRef]

62. Xia, Y.; Shen, S.; Verma, I.M. NF-κB, an active player in human cancers. *Cancer Immunol. Res.* **2014**, *2*, 823–830. [CrossRef] [PubMed]

63. Smith, S.M.; Lyu, Y.L.; Cai, L. NF-κB affects proliferation and invasiveness of breast cancer cells by regulating CD44 expression. *PLoS ONE* **2014**, *9*, e106966. [CrossRef] [PubMed]

64. Pires, B.R.; Mencalha, A.L.; Ferreira, G.M.; de Souza, W.F.; Morgado-Díaz, J.A.; Maia, A.M.; Corrêa, S.; Abdelhay, E.S. NF-kappaB Is Involved in the Regulation of EMT Genes in Breast Cancer Cells. *PLoS ONE* **2017**, *12*, e0169622. [CrossRef] [PubMed]

65. Kim, J.Y.; Jung, H.H.; Ahn, S.; Bae, S.; Lee, S.K.; Kim, S.W.; Lee, J.E.; Nam, S.J.; Ahn, J.S.; Im, Y.H.; et al. The relationship between nuclear factor (NF)-κB family gene expression and prognosis in triple-negative breast cancer (TNBC) patients receiving adjuvant doxorubicin treatment. *Sci. Rep.* **2016**, *6*, 31804. [CrossRef]

66. Niu, J.; Shi, Y.; Tan, G.; Yang, C.H.; Fan, M.; Pfeffer, L.M.; Wu, Z.H. DNA damage induces NF-κB-dependent microRNA-21 up-regulation and promotes breast cancer cell invasion. *J. Biol. Chem.* **2012**, *287*, 21783–21795. [CrossRef]

67. Keklikoglou, I.; Koerner, C.; Schmidt, C.; Zhang, J.D.; Heckmann, D.; Shavinskaya, A.; Allgayer, H.; Gückel, B.; Fehm, T.; Schneeweiss, A.; et al. MicroRNA-520/373 family functions as a tumor suppressor in estrogen receptor negative breast cancer by targeting NF-κB and TGF-β signaling pathways. *Oncogene* **2012**, *31*, 4150–4163. [CrossRef]

68. Körner, C.; Keklikoglou, I.; Bender, C.; Wörner, A.; Münstermann, E.; Wiemann, S. MicroRNA-31 sensitizes human breast cells to apoptosis by direct targeting of protein kinase C epsilon (PKCepsilon). *J. Biol. Chem.* **2013**, *288*, 8750–8761. [CrossRef]

69. Shukla, K.; Sharma, A.K.; Ward, A.; Will, R.; Hielscher, T.; Balwierz, A.; Breunig, C.; Münstermann, E.; König, R.; Keklikoglou, I.; et al. MicroRNA-30c-2-3p negatively regulates NF-κB signaling and cell cycle progression through downregulation of TRADD and CCNE1 in breast cancer. *Mol. Oncol.* **2015**, *9*, 1106–1119. [CrossRef]

70. Bott, A.; Erdem, N.; Lerrer, S.; Hotz-Wagenblatt, A.; Breunig, C.; Abnaof, K.; Wörner, A.; Wilhelm, H.; Münstermann, E.; Ben-Baruch, A.; et al. miRNA-1246 induces pro-inflammatory responses in mesenchymal stem/stromal cells by regulating PKA and PP2A. *Oncotarget* **2017**, *8*, 43897–43914. [CrossRef]

71. Li, B.; Lu, Y.; Yu, L.; Han, X.; Wang, H.; Mao, J.; Shen, J.; Wang, B.; Tang, J.; Li, C.; et al. miR-221/222 promote cancer stem-like cell properties and tumor growth of breast cancer via targeting PTEN and sustained Akt/NF-κB/COX-2 activation. *Chem. Biol. Interact.* **2017**, *277*, 33–42. [CrossRef]

72. Rodic, S.; Vincent, M.D. Reactive oxygen species (ROS) are a key determinant of cancer's metabolic phenotype. *Int. J. Cancer* **2018**, *142*, 440–448. [CrossRef] [PubMed]

73. Choi, J.; Kim, D.H.; Jung, W.H.; Koo, J.S. Metabolic interaction between cancer cells and stromal cells according to breast cancer molecular subtype. *Breast Cancer Res.* **2013**, *15*, 78. [CrossRef] [PubMed]

74. Romero-Cordoba, S.L.; Rodriguez-Cuevas, S.; Bautista-Pina, V.; Maffuz-Aziz, A.; D'Ippolito, E.; Cosentino, G.; Baroni, S.; Iorio, M.V.; Hidalgo-Miranda, A. Loss of function of miR-342-3p results in MCT1 over-expression and contributes to oncogenic metabolic reprogramming in triple negative breast cancer. *Sci. Rep.* **2018**, *8*, 12252. [CrossRef] [PubMed]

75. Spitz, D.R.; Sim, J.E.; Ridnour, L.A.; Galoforo, S.S.; Lee, Y.J. Glucose deprivation-induced oxidative stress in human tumor cells. A fundamental defect in metabolism? *Ann. N. Y. Acad. Sci.* **2000**, *899*, 349–362. [CrossRef]

76. Jiang, S.; Zhang, L.F.; Zhang, H.W.; Hu, S.; Lu, M.H.; Liang, S.; Li, B.; Li, Y.; Li, D.; Wang, E.D.; et al. A novel miR-155/miR-143 cascade controls glycolysis by regulating hexokinase 2 in breast cancer cells. *EMBO J.* **2012**, *31*, 1985–1998. [CrossRef]

77. Lei, K.; Du, W.; Lin, S.; Yang, L.; Xu, Y.; Gao, Y.; Xu, B.; Tan, S.; Xu, Y.; Qian, X.; et al. 3B, a novel photosensitizer, inhibits glycolysis and inflammation via miR-155-5p and breaks the JAK/STAT3/SOCS1 feedback loop in human breast cancer cells. *Biomed. Pharmacother.* **2016**, *82*, 141–150. [CrossRef]

78. Kim, S.; Lee, E.; Jung, J.; Lee, J.W.; Kim, H.J.; Kim, J.; Yoo, H.J.; Lee, H.J.; Chae, S.Y.; Jeon, S.M.; et al. microRNA-155 positively regulates glucose metabolism via PIK3R1-FOXO3a-cMYC axis in breast cancer. *Oncogene* **2018**, *37*, 2982–2991. [CrossRef]

79. Eastlack, S.C.; Dong, S.; Ivan, C.; Alahari, S.K. Suppression of PDHX by microRNA-27b deregulates cell metabolism and promotes growth in breast cancer. *Mol. Cancer* **2018**, *17*, 100. [CrossRef]

80. Yuan, T.L.; Cantley, L.C. PI3K pathway alterations in cancer: Variations on a theme. *Oncogene* **2008**, *27*, 5497–5510. [CrossRef]

81. Leslie, N.R.; Bennett, D.; Lindsay, Y.E.; Stewart, H.; Gray, A.; Downes, C.P. Redox regulation of PI 3-kinase signalling via inactivation of PTEN. *EMBO J.* **2003**, *22*, 5501–5510. [CrossRef]

82. Koundouros, N.; Poulogiannis, G. Phosphoinositide 3-Kinase/Akt Signaling and Redox Metabolism in Cancer. *Front. Oncol.* **2018**, *8*, 160. [CrossRef] [PubMed]

83. Hong, B.S.; Ryu, H.S.; Kim, N.; Kim, J.; Lee, E.; Moon, H.; Kim, K.H.; Jin, M.S.; Kwon, N.H.; Kim, S.; et al. Tumor Suppressor miRNA-204-5p Regulates Growth, Metastasis, and Immune Microenvironment Remodeling in Breast Cancer. *Cancer Res.* **2019**, *79*, 1520–1534. [PubMed]

84. Liu, C.; Liu, Z.; Li, X.; Tang, X.; He, J.; Lu, S. MicroRNA-1297 contributes to tumor growth of human breast cancer by targeting PTEN/PI3K/AKT signaling. *Oncol. Rep.* **2017**, *38*, 2435–2443. [CrossRef] [PubMed]

85. Chai, C.; Wu, H.; Wang, B.; Eisenstat, D.D.; Leng, R.P. MicroRNA-498 promotes proliferation and migration by targeting the tumor suppressor PTEN in breast cancer cells. *Carcinogenesis* **2018**, *39*, 1185–1196. [CrossRef] [PubMed]

86. Kulshreshtha, R.; Ferracin, M.; Wojcik, S.E.; Garzon, R.; Alder, H.; Agosto-Perez, F.J.; Davuluri, R.; Liu, C.G.; Croce, C.M.; Negrini, M.; et al. A microRNA signature of hypoxia. *Mol. Cell. Biol.* **2007**, *27*, 1859–1867. [CrossRef]

87. Camps, C.; Buffa, F.M.; Colella, S.; Moore, J.; Sotiriou, C.; Sheldon, H.; Harris, A.L.; Gleadle, J.M.; Ragoussis, J. hsa-miR-210 Is induced by hypoxia and is an independent prognostic factor in breast cancer. *Clin. Cancer Res.* **2008**, *14*, 1340–1348. [CrossRef]

88. Liang, H.; Xiao, J.; Zhou, Z.; Wu, J.; Ge, F.; Li, Z.; Zhang, H.; Sun, J.; Li, F.; Liu, R.; et al. Hypoxia induces miR-153 through the IRE1α-XBP1 pathway to fine tune the HIF1α/VEGFA axis in breast cancer angiogenesis. *Oncogene* **2018**, *37*, 1961–1975. [CrossRef]

89. Nagpal, N.; Ahmad, H.M.; Chameettachal, S.; Sundar, D.; Ghosh, S.; Kulshreshtha, R. HIF-inducible miR-191 promotes migration in breast cancer through complex regulation of TGFβ-signaling in hypoxic microenvironment. *Sci. Rep.* **2015**, *5*, 9650. [CrossRef]

90. Krutilina, R.; Sun, W.; Sethuraman, A.; Brown, M.; Seagroves, T.N.; Pfeffer, L.M.; Ignatova, T.; Fan, M. MicroRNA-18a inhibits hypoxia-inducible factor 1α activity and lung metastasis in basal breast cancers. *Breast Cancer Res.* **2014**, *16*, 78. [CrossRef]

91. Wu, Z.; Cai, X.; Huang, C.; Xu, J.; Liu, A. miR-497 suppresses angiogenesis in breast carcinoma by targeting HIF-1α. *Oncol. Rep.* **2016**, *35*, 1696–1702. [CrossRef]

92. Gorrini, C.; Harris, I.S.; Mak, T.W. Modulation of oxidative stress as an anticancer strategy. *Nat. Rev. Drug Discov.* **2013**, *12*, 931–947. [CrossRef] [PubMed]

93. Hu, G.; Zhao, X.; Wang, J.; Lv, L.; Wang, C.; Feng, L.; Shen, L.; Ren, W. miR-125b regulates the drug-resistance of breast cancer cells to doxorubicin by targeting HAX-1. *Oncol. Lett.* **2018**, *15*, 1621–1629. [CrossRef] [PubMed]

94. Roscigno, G.; Puoti, I.; Giordano, I.; Donnarumma, E.; Russo, V.; Affinito, A.; Adamo, A.; Quintavalle, C.; Todaro, M.; Vivanco, M.D.; et al. MiR-24 induces chemotherapy resistance and hypoxic advantage in breast cancer. *Oncotarget* **2017**, *8*, 19507–19521. [CrossRef] [PubMed]

95.   Luo, M.; Ding, L.; Li, Q.; Yao, H. miR-668 enhances the radioresistance of human breast cancer cell by targeting IκBα. *Breast Cancer* **2017**, *24*, 673–682. [CrossRef] [PubMed]

96.   Sun, X.; Li, Y.; Zheng, M.; Zuo, W.; Zheng, W. MicroRNA-223 Increases the Sensitivity of Triple-Negative Breast Cancer Stem Cells to TRAIL-Induced Apoptosis by Targeting HAX-1. *PLoS ONE* **2016**, *11*, e0162754. [CrossRef]

International Journal of
Molecular Sciences

MDPI

Review

MicroRNA Networks Modulate Oxidative Stress in Cancer

Yang-Hsiang Lin

Liver Research Center, Chang Gung Memorial Hospital, Linkou, Taoyuan 333, Taiwan; yhlin0621@cgmh.org.tw;
Tel.: +886-3-3281200 (ext. 7785)

Received: 20 August 2019; Accepted: 9 September 2019; Published: 11 September 2019

Abstract: Imbalanced regulation of reactive oxygen species (ROS) and antioxidant factors in cells is known as "oxidative stress (OS)". OS regulates key cellular physiological responses through signal transduction, transcription factors and noncoding RNAs (ncRNAs). Increasing evidence indicates that continued OS can cause chronic inflammation, which in turn contributes to cardiovascular and neurological diseases and cancer development. MicroRNAs (miRNAs) are small ncRNAs that produce functional 18-25-nucleotide RNA molecules that play critical roles in the regulation of target gene expression by binding to complementary regions of the mRNA and regulating mRNA degradation or inhibiting translation. Furthermore, miRNAs function as either tumor suppressors or oncogenes in cancer. Dysregulated miRNAs reportedly modulate cancer hallmarks such as metastasis, angiogenesis, apoptosis and tumor growth. Notably, miRNAs are involved in ROS production or ROS-mediated function. Accordingly, investigating the interaction between ROS and miRNAs has become an important endeavor that is expected to aid in the development of effective treatment/prevention strategies for cancer. This review provides a summary of the essential properties and functional roles of known miRNAs associated with OS in cancers.

Keywords: oxidative stress; MicroRNA; signal transduction; therapeutic target

1. Introduction

Imbalanced regulation of reactive oxygen species (ROS) and antioxidant factors in cells is known as "oxidative stress (OS)" (Figure 1). OS drives key cellular physiological regulatory responses through signal transduction, transcription factors (TFs) and noncoding RNAs (ncRNAs) [1]. ROS are oxygen-containing products and are formed during cellular oxidative metabolism. ROS, including superoxide anion (O_2^-), hydroxyl radical (OH^-), hydrogen peroxide (H_2O_2), nitric oxide (NO) and singlet oxygen ($^1O_2$), play important roles in cell differentiation, cell death, cell growth, signal transduction, cell apoptosis and chemoresistance [2,3]. Dual roles have been proposed for ROS in biological phenotypes according to their cellular level [4]. High levels of ROS promote cell apoptosis, while low levels of ROS act as a signal transducer to induce cell survival (Figure 1). Recently, excessive ROS production was identified in several cancers where they were significantly correlated with tumorigenesis. However, the underlying mechanism of ROS regulation in cancer development remains unclear.

MicroRNAs (miRNAs) are small ncRNA comprising 18-25-nucleotide functional RNA molecules that play critical roles in the regulation of target gene expression by binding to complementary regions of mRNA and regulating mRNA degradation or inhibiting translation (Figure 2). Previous studies have demonstrated that miRNAs are significantly associated with tumor growth, metastasis and cancer progression [5,6]. Based on these findings, dysregulated miRNA expression is a hallmark of cancer.

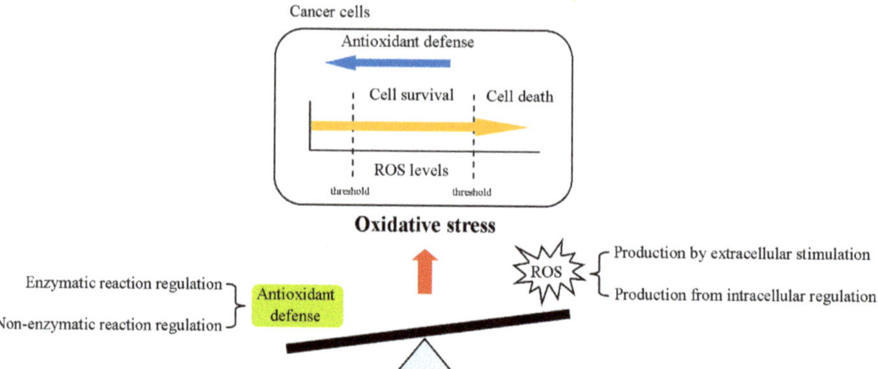

Figure 1. Reactive oxygen species (ROS) production and antioxidant defense in the control of redox homeostasis in cancer cells. Disruption of redox homeostasis by ROS (intra- or extracellular signals) and antioxidant defense (enzymatic or non-enzymatic reactions) induces oxidative stress (OS) and results in various cell functions. The physiological function of ROS is dependent on its concentration. Elevated ROS production and accumulation lead to cell apoptosis. On the other hand, medium levels of ROS promote cell survival and progression.

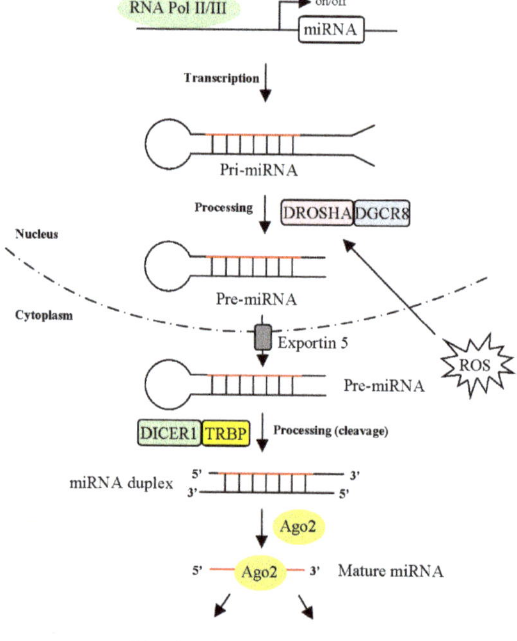

Figure 2. The biogenesis and regulation mechanisms of microRNAs (miRNAs). MiRNAs are transcribed by RNA polymerase II/III and generated the primary miRNA transcript (pri-miRNA). The pri-miRNAs are cleaved into precursor miRNA transcript (pre-miRNA) by the microprocessor complex, a combination of DROSHA and DGCR8. Pre-miRNA is exported to cytoplasm via exportin 5 and further processed by the RNase III enzyme Dicer with the cofactor protein TRBP to generate an approximately 18-25-nt duplex. Either 5p or 3p strand of the mature miRNA (red line) interacts with Argonaute (Ago) protein and forms a miRNA-induced silencing complex (miRISC). There are two models (mRNA degradation and translational repression) of miRNA-mediated gene silencing.

Cross-talk between ROS and miRNAs has been implicated in cancer development, and it is important to identify the nature of this connection. Interestingly, some specific miRNAs, called ROS-miRs or redoximiRs, are regulated by OS and modulate target gene expression in response to ROS [7,8]. Mesenguer et al. [9] demonstrated that the OS/NFκB axis induced miR-9/9* expression and inhibited expression of its target genes, GTPBP3, MTO1 and TRMU, in MELAS cells. On the other hand, a previous study indicated that miR-21 regulated ROS homeostasis and suppressed the antioxidant response in human umbilical vein endothelial cells (HUVECs) [10]. These findings suggest that ROS could be upstream regulators or downstream effectors of miRNAs. In this review, we focus on how ROS affect biological phenotypes through miRNA and how miRNAs regulate ROS-mediated function in cancer.

2. Regulation of ROS Homeostasis in Cells

OS promotes both nuclear and mitochondrial DNA damage and initiates DNA repair pathways [11]. Furthermore, cellular ROS levels can be produced by different mechanisms, such as ionizing radiation, UV radiation, inflammatory cells and chemotherapy. ROS are primarily generated in cells through the byproducts of leaked electrons from the mitochondrial electron transport chain (ETC). Mutations or aberrantly expressed nuclear or mitochondrial genes encoding the ETC components can influence the electron transfer reaction that leads to electron leakage. The electrons are captured by O_2, producing O_2^-, which is usually converted to H_2O_2 by manganese (Mn)-containing mitochondrial superoxide dismutase (MnSOD or SOD2), Cu/Zn-containing cytosolic SOD1 or SOD3 [12]. Subsequently, H_2O_2 can attack chromosomal DNA and subsequently induce DNA damage. On the other hand, O_2^- can be generated through a reaction catalyzed by some enzymes, including the membrane-located NAD(P)H oxidase complex (NOX), which consists of NOX1-4, endoplasmic reticulum-associated xanthine oxidase (XO), cytochrome c oxidase and cyclooxygenase in some cancer cells [13]. In fact, H_2O_2 plays an important role in carcinogenesis because it is capable of diffusing throughout the cell components and producing cellular injury. The injurious effects of ROS in mammalian cells are mediated by the hydroxyl radical (·OH). The generation of OH in vivo is produced in the presence of reduced transition metals, including Co, Cu, Fe, or Ni, mainly through the Fenton reaction [14]. Notably, the ·OH-induced DNA damage includes the generation of 8-hydroxyguanosine (8-OHG), in which the hydrolysis product is 8-hydroxydeoxyguanosine (8-OHdG). 8-OHdG is the most widely used marker of radical attack on DNA. Notably, 8-OHdG is strongly correlated with cancer progression, including that of breast cancer, colorectal cancer, ovarian cancer and hepatocellular carcinoma (HCC) [15–17]. For example, hepatic 8-OHdG levels are useful biomarkers for identifying hepatitis C virus (HCV) infection in patients [18]. Alternatively, cells maintain ROS homeostasis by reducing ROS production and triggering specific antioxidant mechanisms to neutralize ROS or mitigate OS [19]. In fact, antioxidant enzymes include SODs, catalase, peroxiredoxins (PRDXs), thioredoxins, glutathione peroxidase and heme oxygenase. First, SOD converts O_2^- to O_2 or H_2O_2. Then, catalase and glutathione peroxidase subsequently convert H_2O_2 to H_2O and O_2.

3. MiRNAs and Their Roles in Oxidative Stress

Previous studies indicated that ROS can induce or suppress miRNA expression and contribute to downstream biological function through regulation of target genes [20]. Increasing evidence has shown cross-talk between miRNAs and components of redox signaling [21,22]. The transcription, biogenesis, translocation and function of miRNAs are highly correlated with ROS, and miRNAs may regulate the expression of redox sensors and other ROS modulators, such as the key components of cellular antioxidant machinery. Redox sensors have been identified and they include transcription factors (e.g., p53, NFκB, c-Myc and nuclear factor erythroid 2 related factor 2 (NRF2)) and kinases (e.g., Akt and IKK), which trigger cellular redox signaling. Here, we summarize how miRNAs are regulated by ROS at the posttranscriptional and transcriptional levels and how the miRNA/ROS axis controls tumorigenesis.

3.1. MiRNA Processing is Regulated by ROS

Recently, it was reported that miRNAs can be transcribed by RNA polymerase II/III as longer primary transcripts called primary miRNAs (pri-miRNAs). The mature form of miRNA is generated by the two-step processing of pri-miRNA and is subsequently associated with the effector RNA-induced silencing complex (RISC). The biogenesis and function of miRNAs regulated by ROS are described. Two key genes (*Dicer* and *Drosha*) are mediated by the miRNA processing pathway (Figure 2). A report showed that the expression of Dicer was downregulated by aging-related OS in cerebromicrovascular endothelial cells (CMVECs) [23]. Downregulating Dicer dramatically reduced miRNA expression under H_2O_2 treatment compared with the expression of the control. Notably, knocking down Dicer suppressed ROS production in human microvascular endothelial cells (HMECs) [24]. These findings indicated that Dicer expression is part of a feedback loop that modulates ROS production and maintains cellular homeostasis. Upon OS, the expression of pre-miRNA and miRNA in myoblasts is decreased through DGCR8/heme oxygenase-1 (HMOX1) regulation [25]. Heme is required for DGCR8 activity, and the heme-binding domain of DGCR8 plays a crucial role in pri-miRNA recognition for miRNA processing by DROSHA.

3.2. ROS Regulate miRNA Expression through the Modulation of Transcription Factors

Accumulating studies have investigated the miRNAs regulated by ROS/TFs such as c-myc, p53, c-Jun, HIF and NFκB [20,26]. This section summarizes how miRNAs are regulated by ROS/TF at the transcriptional level.

ROS exposure has been shown to be correlated with oncogenic signals such as those transduced by c-Myc and Ras [27,28]. c-Myc, a well-known oncogene, is involved in tumor growth, migration, invasion, metabolism and metastasis through the regulation of gene expression. c-Myc activation induces DNA damage in normal human fibroblasts. This effect has been correlated with ROS generation. Expression levels of miR-15a/16, miR-23a, miR-29 and miR-34 family members were downregulated by c-Myc [29]. Overexpression of miR-15a/16 suppressed cell proliferation, angiogenesis, migration and invasion through inhibition of FGF2 in vitro and in vivo [30]. Furthermore, hypoxia-induced suppression of miR-15/16 expression was directly regulated by c-Myc. By contrast, miR-17-92 and miR-221/222 expression is stimulated by c-Myc [29]. The expression levels of miR-17-92 were remarkably inhibited by triptolide in a c-Myc-dependent manner, which resulted in the induction of target genes, including PTEN, BIM and p21, in HCC cells [31]. Moreover, this suppressive effect contributed to enhanced triptolide-induced cell apoptosis.

P53, a tumor suppressor gene, regulates the cell cycle, apoptosis, growth and metabolism through modulation of target genes. P53 is involved in regulating the drosha-dedicated pri-miRNA processing pathway [32]. In addition, p53 modulates miRNA transcription, such as miR-17-92, miR-34a and miR-200c. Interestingly, stress-regulated miRNAs, namely, miR-34 and miR-200, are upregulated in a p53-dependent manner [33,34]. MiR-34 has been implicated as a tumor suppressor because it suppresses the epithelial-mesenchymal transition (EMT), which promotes cancer cell metastasis. Its expression level is positively associated with p53. Importantly, p53 suppresses Snail expression by interacting with miR-34. A study indicates that miR-200c is upregulated upon H_2O_2 treatment in endothelial cells and that it contributes to cell apoptosis and senescence through inhibition of the target gene ZEB1 [35]. Moreover, knockdown of p53 can reverse H_2O_2-induced miR-200c expression [34].

As mentioned above, exposure to ROS induces chronic inflammation. NFκB acts as a master mediator of the inflammatory response to regulate innate and adaptive immune functions. MiRNA (miR-9, miR-21, miR-30b, miR-146a, miR-155 and miR-17-92 cluster) expression was identified and found to be directly transcriptionally regulated by NFκB [36,37]. Bazzoni et al. [38] indicated that miR-9-1 was induced by lipopolysaccharide (LPS) in a MyD88- and NF-κB-dependent manner. DNA damage activated miR-21 expression through recruitment of NF-κB and signal transducer and activator of transcription 3 (STAT3) to its promoter region and contributed to promoting cell invasion in breast cancer [39]. Another study reported that miR-21 expression and function were mediated by ROS in highly metastatic breast cancer cell lines [40]. In addition, miR-21 induced by ROS via NF-κB activity was involved in arsenic-induced cell transformation [41]. NF-κB bound to the promoter region of the miR-17-92 cluster was identified using chromatin immunoprecipitation (ChIP) assay and was further confirmed by luciferase reporter assay [42]. On the other hand, multiple miRNAs have been identified and have been found to modulate NF-κB activity. MiR-126a was shown to target IκBα, an NFκB inhibitor, and promoted the NFκB signaling pathway [43]. MiR-506 inhibited the expression of the NFκB p65 subunit and led to the production of ROS and p53-dependent apoptosis in lung cancer cells [44]. Notably, miR-506 was regulated by p53. These findings indicated that miR-506 was involved in the p53/NFκB signaling pathway.

NRF2 is a member of the Cap'n'Collar (CNC) family of basic leucine zipper (bZIP) transcription factors [45]. Previously, the actin-binding protein kelch-like ECH-associated protein 1 (KEAP1) was identified as a repressor of NRF2 via proteasomal degradation [46]. NRF2 is involved in antioxidant metabolism, protein degradation, inflammation and radioresistance [47]. Notably, miRNAs can be both indirectly and directly regulated by NRF2 [47,48]. Singh et al. [49] group demonstrated that NRF2 repressed miR-1 and miR-206 expression and led to reprogram glucose metabolism in cancer cells. Furthermore, miR-29 and miR-125b were identified as direct target genes of NRF2 [50,51]. Upregulation of miR-125b by NRF2 resulted in the repression of aryl hydrocarbon receptor repressor and protection of cancer cells from drug-induced toxicity [51]. On the other hand, NRF2 gene was regulated by miRNAs such as miR-28, miR-34a, miR-93 and miR-200a [52–55]. MiR-28 has been shown to interact with NRF2 3′UTR and represses NRF2 expression in breast cancer cells [52]. Overexpression of miR-34a suppressed NRF2 and NRF2 target genes expressions [53]. Functionally, miR-34a was involved in NRF2-dependent antioxidant pathway in liver. These findings suggested that NRF2 and miRNAs formed a regulatory network and regulated cellular functions.

3.3. ROS Regulate miRNA Expression via Epigenetic Regulation

Recently, epigenetic modifications/regulations of the genome have been explored and associated with cancer progression [56]. Changes in the structure or conformation at the nuclear or mitochondrial DNA (nDNA and mtDNA) or RNA level, but not the DNA/RNA sequence, are called epigenetic marks. The main epigenetic alterations in humans are DNA methylation and histone modification, which includes methylation, acetylation and phosphorylation. Aberrant miRNA expression in cancers was discovered and found to be controlled by epigenetic regulation. Promoter regions of miR-125b and miR-199a are hypermethylated through DNMT1 during H_2O_2 treatment, as determined using methylation-specific PCR and bisulfate sequencing [57]. Moreover, these two miRNAs are downregulated by ROS in ovarian cancer cells. The level of histone acetylation has an important role in activating gene expression through chromatin remodeling. In contrast, the gene is silenced by histone deacetylases (HDACs), which promote the deacetylation of lysine residues. MiR-466h-5p acts in a proapoptotic role by directly targeting antiapoptotic genes such as BCL2L2 [58]. ROS induce miR-466h-5p expression through inhibition of HDAC2 and result in increased apoptosis.

4. Interplay between Oxidative Stress, miRNA and Cancer Development

OS has been reported to contribute to neurological disorders, hypertension, diabetes and cancers. This section focuses on the associations of OS and hypoxia, angiogenesis, metastasis, metabolism, cancer stem cell and senescence, which are all involved in cancer progression.

4.1. Association between OS, miRNA and Hypoxia

Hypoxia, known as reduced oxygen availability, mostly occurs in the center of tumors due to the high proliferation ability of cancer cells and abnormal vasculature [59]. Hypoxia-inducible factor 1α (HIF1α) is the master regulator in hypoxia. The activation of HIF1α promotes the expression of several genes, including protein-encoding genes and ncRNAs, and facilitates stem cell renewal, cancer cell survival, metabolism and chemoresistance. The HIF1 transcription factor consists of three hypoxia-induced α subunits (HIF1α/2α/3α) and one β subunit (HIF1β). HIF1α is stabilized and activates a downstream signaling pathway mediated by ROS [60]. Some evidence suggests that telomerase activity is associated with ROS in HCC [61]. Moreover, ROS-mediated telomerase activity is dependent on HIF1α [62]. Expression levels of the human telomerase reverse transcriptase gene (hTERT) are upregulated by HIF1α. Specific binding sites for HIF1α in the hTERT promoter regions were identified by luciferase and ChIP assays. In addition, cancer stem cell (CSC) markers, OCT4 and Notch, are induced by HIF1α and promote stem cell renewal. Expression levels of SOX2 and KLF4 are positively regulated by ROS in glioblastoma cells [63]. HIF1α and ROS activation are responsible for regulating glucose transporter 1 (GLUT1), hexokinase II (HKII) and glutaminase expression and the reprogramming of cancer cell metabolism [64]. Moreover, miR-210 acts in an oncogenic role in cancer development and is induced under hypoxic conditions [65]. HIF1α directly binds to the hypoxia response element (HRE) of the miR-210 promoter. Therefore, miR-210 plays an important role in regulating cellular adaption to hypoxia, suggesting that targeting miR-210 may be a novel approach for the prevention and/or treatment of cancer.

4.2. Association between OS, miRNA and Angiogenesis

Angiogenesis is the process of generating new blood vessels from preexisting vasculature and is required for many functions, such as tissue repair, organ regeneration, cancer development and metastasis [66]. The angiogenesis process is regulated by several cytokines and growth factors, such as vascular endothelial growth factor (VEGF), transforming growth factor β (TGFβ), angiopoietin 1 (Ang-1) and placental growth factor, platelet-derived growth factor (PDGF) β [67–70]. VEGF acts as an effector to control endothelial cell proliferation and new vessel formation. The HIF1α/ROS axis activates tissue-specific angiogenesis through the upregulation of VEGF and its receptors VEGFR1 and VEGFR2. By contrast, VEGF induces ROS production by promoting NADPH oxidase in endothelial cells. ROS can also modulate VEGFR activation, phosphorylation and polymerization. A report indicated that genotoxic stress-induced miR-494 expression suppressed DNA repair and angiogenesis through regulation of MRE11a/RAD50/NBN (MRN) complex in endothelial cells [71]. Moreover, VEGF signaling is regulated by MRN complex in vitro and in vivo. Alternatively, ROS stimulate the MAPK pathway and promote the expression of VEGF. A previous study demonstrated that oxidized phospholipids interact with VEGFR2 and induce angiogenesis through the Src signaling pathway [72]. Other mechanisms of ROS-mediated angiogenesis are the ataxia telangiectasia mutated gene (ATM)/p38α pathway and Sirtuin 1 (SIRT1). Previous studies have indicated that ATM functions in the cell cycle regulation, DNA damage repair and oxidative defense [73]. ATM promotes endothelial cell proliferation and facilitates angiogenesis [74]. Previously, the subtype of histone H2A, called H2AX, can be phosphorylated (γH2AX) and is involved in DNA damage response. Economopoulou et al. [75] group indicated that H2AX is required for endothelial cells to sustain their growth under hypoxia and is important for hypoxia-driven neovascularization. Wilson et al. [76] have shown that miR-103 suppresses developmental and pathological angiogenesis through inhibition of three prime exonucleases 1 in endothelial cells. On the other hand, Yang and co-workers demonstrated that overexpression of miR-328-3p suppressed cell proliferation and promoted radiosensitivity of osteosarcoma cells through suppression of H2AX in vitro and in vivo [77]. Recently, Marampon et al. group demonstrated that NRF2/antioxidant enzymes/H2AX/miRNAs (miR-22, miR-34a, miR-126, miR-146a, miR-210 and miR-375) axis act as potential candidates in radiosensitizing therapeutic strategy for rhabdomyosarcoma clinical treatment [78]. SIRT1, also known as NAD-dependent deacetylase sirtuin-1, has been demonstrated to regulate cellular functions including oxidative stress, apoptosis and aging via deacetylation of a variety of substrates [79]. A report indicates that inhibition of SIRT1 with either an inhibitor or siRNA leads to increased ROS levels, suggesting an association between SIRT1 and ROS. MiR-138, miR-181 and miR-199 have been shown to directly target and inhibit SIRT1 expression in various cell lines [80–82]. MiR-181 is induced by treatment with a high-fat diet and results in repressed SIRT1 expression and insulin sensitivity in the liver [81]. In addition, HIF1α and SIRT1 are upregulated in miR-199a-depleted cells during normoxic conditions [83]. Moreover, SIRT1 is actually a direct target gene of miR-199a and is responsible for suppressing prolyl hydroxylase 2.

4.3. Association between OS, miRNA and Metastasis

Metastasis is a complicated process that includes invasion, intravasation into blood, extravasation to distant organs and growth [84]. Due to these multiple steps, few metastasizing tumor cells can survive and form micrometastases. A typical phenotype that leads to metastasis is EMT, which is a biological event by which epithelial cells undergo alterations that induce the development of a more aggressive mesenchymal phenotype [84]. Increasing evidence suggests that cancer cells during the metastasis process are killed by OS [85,86]. In addition, cancer cells are more sensitive to ROS than normal cells. Reducing ROS levels by treating with antioxidant inhibits tumor promotion of tumor progression in mouse models. ROS-mediated EMT regulation through TGFβ/Smad, E-cadherin, Snail, integrin, β-catenin, matrix metalloproteinases (MMPs) and miRNA has been documented [87–89]. Among these interactions, activation of TGFβ induces ROS production and leads to the promotion of SMAD and ERK1/2 phosphorylation. Moreover, the ROS/TGFβ axis regulates EMT through the interaction of NFκB, HIF1α and cyclooxygenase-2 (COX-2). A previous study indicated that MMP-3, MMP-10 and MMP-13 were directly upregulated by oxidative treatment and promoted cell invasion ability in NMuMG cells [90]. In addition, the activity of MMP-2 and MMP-9 were posttranscriptionally regulated by oxidant treatment [91,92]. These studies suggest that MMP expression or activity is modulated by OS, which is related to chronic inflammation, malignant transformation and the invasive potential of cells. Yoon et al. [93] demonstrated that sustained treatment with H_2O_2 enhances MMP2 activity via the PDGF, VEGF, phosphatidylinositol 3-kinase and NF-κB pathways in HT1080 cell lines. Song and coworkers reported that the expression of miR-509 is significantly more downregulated in breast cancer than it is in normal tissues [94]. Overexpressed miR-509 abrogated cell growth, migration and invasion through inhibition of the target gene SOD2, which is a crucial effector in the production of ROS.

4.4. Association between OS, miRNA and Metabolism

Tumor progression is characterized by the occurrence of metabolic alterations, including those in glycolysis, fatty acid oxidation (FAO) and oxidative phosphorylation [95,96]. The connection and reciprocal regulation between the metabolism and the redox balance of tumor cells have been shown. For this reason, it is important to determine the major metabolic pathways that are the main controllers of the ROS homeostasis of cancer cells. Glucose is converted to glucose-6-phosphate by hexokinase enzyme and triggers a series of downstream enzyme-catalyzed reactions. It is an essential pathway for providing nutrients, metabolites and energy to cells. In 1924, Otto Warburg proposed a theory suggesting that tumor cells tend to exhibit glycolysis regardless of the presence of oxygen [97]. Accumulating evidence has shown that metabolites produced by glucose metabolism are major regulators of the redox homeostasis of tumor cells [98]. Cancer cells demonstrate increased sensitivity to glucose-deprivation-induced cytotoxicity compared with that in normal cells by restricting the burden of ROS. Moreover, inhibition of lactate dehydrogenase-A by a specific inhibitor, FX11, reduced intracellular ATP and promoted OS, which suppressed tumor progression in lymphoma and pancreatic cancer. Sala et al. [10] have shown that miR-21 is upregulated by glucose treatment and inhibits ROS homeostatic genes such as NRF2, SOD2 and KRIT1. Furthermore, other metabolic enzymes, such as TIGAR and ALDH4, decrease ROS production by either inhibiting glycolysis and inducing NAPDH production or enhancing mitochondrial function [99,100]. FAO consists of multiple processes by which fatty acids are broken down by cells to produce ATP and generate biosynthetic pathways. In general, the β-oxidation reaction takes place in mitochondria. FAO causes ROS formation and contributes to the enhanced development of nonalcoholic fatty liver disease (NAFLD) [101]. In hypoxia, HIF-1 suppression of medium-chain acyl-CoA dehydrogenase (MCAD) and light chain acyl-CoA dehydrogenase (LCAD) expression inhibits FAO and ROS production while promoting cell growth of liver cancer cells [102]. The expression levels of LCAD in HCC specimens were analyzed and found to be negatively correlated with survival. These findings indicate the relevance of FAO suppression in the progression of cancer. Previous studies have identified miR-33a/b as an intronic miRNA located with the sterol regulatory element binding factor (SREBP) 1 and 2 genes [103]. These two miRNAs cotranscribe with their host gene and regulate high density lipoprotein (HDL) biosynthesis.

4.5. Association between OS, miRNA and Cancer Stem Cells

Cancer cells are believed to be derived from a small subset of tumor cells that have a high capacity for self-renewal and differentiation—namely, cancer stem cells (CSCs) or tumor-initiating cells [104]. Increasing evidence indicates that miRNAs function as regulators of CSCs and are associated with ROS production during tumor progression and cancer development. Some miRNAs, such as let-7a, miR-21, miR-34a, miR-200 and miR-210, are potentially involved in the modulation of ROS production in CSCs [105–109]. A previous study showed that let-7 acts as a negative regulator of CSC-mediated function by targeting PTEN and LIN28b in prostate and pancreatic cancer. Recently, OS reduced let-7 expression in a p53-dependent manner in various cancer cells. Some experimental studies revealed that the expression of miR-21 is remarkably increased in CSC subpopulations compared to the expression in the hypobromite non-CSC counterparts in vitro and in vivo. Notably, knocking down miR-21 suppressed cell migration, invasion and EMT phenotype in breast cancer CSCs. Moreover, OS induced miR-21 expression and promoted cell migration and self-renewal in prostate and pancreatic CSCs. Another report indicates that miR-21 enhances ROS production via the MAPK pathway and suppresses SOD2, SOD3 and sprouty homolog 2 (SPRY-2) expression [110]. Additionally, a number of studies have revealed that miR-34a suppresses CSC-related genes, such as CD44, and EMT makers and subsequently attenuates cell invasion, metastasis and self-renewal capacity [111]. The interplay between ROS and miR-34 has been documented. The expression of miR-34 is induced by OS in stromal and tumor cells. The first evidence miR-200 was associated with stem cell phenotype, reported in 2009 [112]. Moreover, all five members of the miR-200 family were downregulated in human breast CSCs as well as in normal human and murine mammary stem/progenitor cells [112]. Mechanistically, miR-200 suppresses the expression of B lymphoma Mo-MLV insertion region 1 homolog (Bmil-1), Suz12, and Notch homolog 1 (Notch1), which are known regulators of CSC and EMT phenotypes, and inhibits the CSC self-renewal capacity. MiR-210 expression is enriched in MCF-7 spheroid cells and CD44$^+$/CD24$^-$ MCF7 cells compared with MCF-7 parental cells [113]. Overexpression of miR-210 enhances proliferation, self-renewal capacity, migration and invasion through inhibition of E-cadherin in vitro and in vivo. Thus, these observations indicate that the miRNA/ROS axis plays important roles in multiple events related to CSCs.

4.6. Association between OS, miRNA and Senescence

Cellular senescence is characterized by the expression of senescence-associated β-galactosidase (SA-β-gal), overexpression of the cyclin-dependent kinase (CDK) inhibitor, senescence-associated secretory phenotype (SASP), telomere shortening and persistent DNA damage response (DDR) [114]. ROS cause cell senescence by stimulating the DDR pathway to stabilize p53 and promote CDK inhibitor gene expression. In fact, p53 acts as a master regulator in the cellular response to OS. Mechanically, p53 can decrease ROS levels and repair DNA damage in cells. In contrast, it can also enhance ROS production and promote cell apoptosis or senescence [115]. Several reports indicate that p53 reduces intracellular ROS levels by promoting antioxidant reactions. Several miRNAs, including miR-21, miR-22, miR-29, miR-34a, miR-106b, miR-125b, miR-126, miR-146a, the miR-17-92 cluster, the miR-200 family and miR-210, have been identified to be differentially expressed in senescent cells and to be involved in cellular senescence [116–122]. Notably, miR-34a was found to promote cellular senescence by inhibiting SIRT1 expression in a variety of tissues. Another group indicated that miR-34a and miR-335 promote premature cellular senescence by targeting antioxidative enzymes. Furthermore, miR-217 induces a premature senescence-like phenotype and represses angiogenesis by inhibiting the expression of target gene SIRT1 in endothelial cells [123]. In addition, miR-92a was found to exacerbate endothelial dysfunction under OS exposure by directly targeting SIRT1, Krüppel-like factor 2 (KLF2) and KLF4 genes [124]. Additionally, Liu et al. group demonstrated that knockdown of miR-92a promoted cell growth, decreased caspase 3 activity and ROS through regulation of NRF2-KEAP1/ARE signal pathway [125].

5. ROS-Mediated Therapeutic Strategies in Cancer

OS clearly plays a role in the development of cancer, metastasis and chemotherapeutic resistance. In a strategy to modulate ROS-mediated effects, these biochemical characteristics of tumors are directly impaired. In light of recent studies, the strategy of inhibiting metabolic pathways, targeting NADPH oxidase and ROS scavenging mechanisms represent promising therapeutic options for treatments [126]. The other strategy is to target tumor cells with oxidation-promoting agents that either enhance ROS production or inhibit cellular antioxidants. NADPH oxidase plays an important role in regulating ROS production. Several inhibitors have been demonstrated to reduce NADPH function. In general, diphenylene iodonium (DPI) and apocynin are NADPH inhibitors [127–129]. DPI can inhibit XOD and the proteins of mitochondrial ETC and block flavoprotein. In another strategy, ROS scavenging enzymes are enhanced and used for anticancer therapy [130]. GSH, GST, SOD, GPX, and catalase are able to suppress tumor formation. There are several analogs of GSH drugs, such as *N*-acetylcysteine (NAC), YM737 and Telcyta, used for cancer treatment [131]. NOV-002, an agent containing oxidized GSH, improved the efficacy of cyclophosphamide to treat colon cancer by controlling the ratio of GSH to GSSG and promoting S-glutathionylation [131]. Yang and coworkers indicated that lithocholic acid treatment and bile duct ligation model promoted c-Myc/miR-27/prohibitin 1 axis, with the consequence of repressing NRF2 expression and ARE binding, resulting in decreased suppressed GSH synthesis and antioxidant ability in chronic cholestatic liver injury [132]. Another study reported that the rate-limiting GSH biosynthetic heterodimeric enzyme γ-glutamyl-cysteine ligase (GCL) was regulated by miR-433 [133]. Ectopic of miR-433 in HUVEC inhibited GCL expression in an NRF2-independent manner. Moreover, inhibition of miR-433 prevented TGFβ-mediated GCL downregulation and fibrogenesis in hepatic cells. Recently, Cheng et al. [134] demonstrated that miR-30e expression was suppressed in an atherosclerosis (AS) model. MiR-30e regulates Snai1/TGF-β/Nox4 expression to modulate ROS. These findings provide novel insights on miRNAs in the anti-ROS pathway, in which miRNA-30e may represent a novel target for AS.

6. Conclusions

Overall, many studies have been conducted to elucidate the molecular mechanisms underlying the ROS/miRNA axis and its role in tumorigenesis (Figure 3). Moreover, miRNAs networks that modulate OS in cancer are comprehensively listed in Table 1. Indeed, ROS and miRNAs exhibit overlapping characteristics in tumorigenesis. ROS, as upstream regulators, modulate miRNA expression through transcriptional, posttranscriptional and epigenetic regulation, respectively. On the other hand, miRNAs disrupt ROS production (downstream mediator) and are involved in ROS-mediated functions. MiRNAs and ROS can act either synergistically or antagonistically to regulate cancer progression. However, many details of their interaction remain unclear and need to be further investigated. MiRNAs/ROS-mediated phenotypes depend on the net result of the downstream molecules and multiple signaling pathways in the specific context. There are still many limitations to treatment because ROS play dual roles in cancer progression. As discussed in this review, the functional roles of miRNA in cellular adaptation to ROS are different in cells based on tissue and cell-type specific effects. These observations raise the possibilities to apply specific miRNAs as therapeutic targets in different contexts. Advantages of using miRNA-target therapy include the conservation of miRNA across multiple species with known sequences and the ability to target multiple genes within defined pathways. Notably, several miRNA-based therapies are being developed. For example, the locked nucleic acid (LNA)-modified anti-miR-122 is the first miRNA-targeted therapy to treat HCV in clinical trials. The association between ROS-mediated function and miRNA regulation provides opportunities for developing novel anticancer strategies.

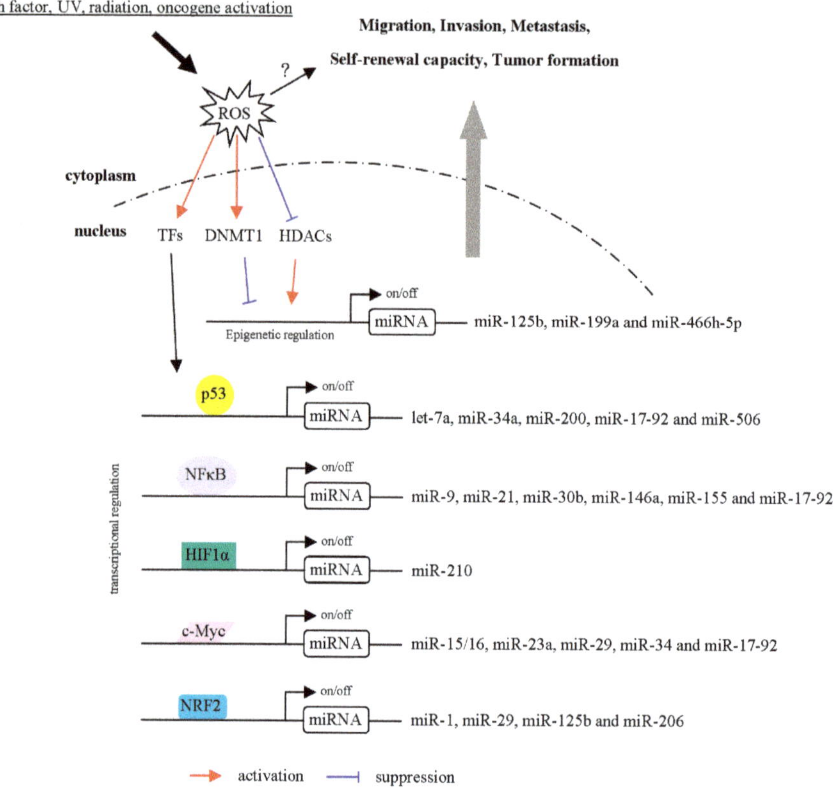

Figure 3. Schematic model showing mechanisms in which ROS regulates the biogenesis and transcription of miRNAs. ROS activate or inhibit epigenetic, transcriptional regulations of miRNA expression. For example, miRNAs are regulated by ROS through modulation of chromatin remodeling factors (DNMT1 and HDACs). In addition, ROS induces or represses transcriptional factor (p53, NFκB, HIF1α, c-Myc and NRF2) to regulate miRNA expressions. Furthermore, ROS/TF/miRNA axis controls cell migration, invasion, metastasis, self-renewal capacity and tumor formation.

Table 1. ROS-related miRNAs and their potential mechanisms in cancers.

miRNA	Regulation Mechanism [a]	ROS Production [b]	Expression in Cancer [c]	Cell/Cancer Types	Molecules, Cellular Processes and Signaling Pathways Involved [d]	References
Let-7a	OS, p53	✓	Down	CSC, prostate cancer, pancreatic cancer	PTEN, LIN28b	[105,135]
miR-1	NRF2, HDAC4	✓	Down	Non-small cell lung cancer	NRF2, KEAP1, glucose metabolism, tumor growth	[49]
miR-15/16	c-Myc	✓	Down	Skin, colon cancer	FGF2, HIF-2α, senescence-like phenotype, angiogenesis, metastasis	[30]
miR-21	Glucose, NFκB, STAT3	✓	Up	CSCs, lung cancer, liver cancer, colorectal cancer	MAPK pathway, cell migration, invasion and EMT phenotype, self-renewal ability	[41,110]
miR-23a	c-Myc	✓	-	Cardiac disease, myeloma	Glutaminase, MnSOD, apoptosis, cell growth	[136–138]
miR-29	c-Myc, H_2O_2, NRF2	✓	Dual role	Ovarian cancer, lung cancer, lymphoma	SIRT1, senescence, proliferation, apoptosis	[50,139–142]
miR-33a/b	-	✓	Down	Liver	HDL biosynthesis, apoptosis, OS resistance	[103]
miR-34	OS, c-Myc, p53	✓	Down	Stromal cells, CSC, bladder cancer, lung cancer	CD44, EMT markers, SIRT1, senescence, metastasis	[33,111,143]
miR-17-92	c-Myc, p53, NFκB	✓	Up	Lung cancer,	Vitamin D, Senescence, apoptosis	[120,144–146]
miR-92a	-	✓	Up	Endothelial cells	SIRT1, KLF2, KLF4	[124,125]
miR-125b	DNMT1, H_2O_2, NRF2	✓	Dual role	Ovarian cancer, liver	Epigenetic regulation	[51,57]
miR-181	-	✓	Up	Macrophagy, HCC	SIRT1, insulin sensitivity, NFκB activity, apoptosis	[80]
miR-199a	DNMT1, H_2O_2	✓	Down, (hypermethylation)	Ovarian cancer	HIF1α, SIRT1, Epigenetic regulation	[57,83]
miR-200	P53, H_2O_2	✓	Down	CSC, breast cancer, liver cancer	Bmil-1, Suz12, Notch-1, self-renewal capacity, EMT markers, senescence	[34,35]
miR-210	Hypoxia	✓	Up	CSCs	E-cadherin, Hypoxia, proliferation, self-renewal capacity, migration and invasion, senescence	[65,109] [113]

Table 1. *Cont.*

miRNA	Regulation Mechanism [a]	ROS Production [b]	Expression in Cancer [c]	Cell/Cancer Types	Molecules, Cellular Processes and Signaling Pathways involved [d]	References
miR-217	-	-	Dual role	Endothelial cells	SIRT1, Angiogenesis, premature senescence-like phenotype	[123]
miR-466h-5p	ROS, HDAC2	-	-	Mouse ovarian epithelial	BCL2L2, apoptosis	[58]
MiR-506	P53	✓	Down	Lung cancer	NFκB signaling pathway	[44]
miR-509	-	✓	Down	Breast cancer	SOD2, Cell growth, migration and invasion	[94]

[a]: MiRNAs are regulated by upstream transcriptional factor, ROS or hypoxia, as indicated. -: Information is unavailable. [b]: ✓: MiRNAs are responsible for producing ROS. -: Information is unavailable. [c]: Expression level of miRNAs in cancer. Up: upregulated in cancer, Down: downregulated in cancer, Dual role: up- or downregulated in cancer. [d]: Downstream molecules, signaling pathways and phenotypes involved in miRNA-mediated functions.

Funding: This work was supported by grants from Chang Gung Memorial Hospital, Taoyuan, Taiwan (CMRPG3H0721, CMRPG3H0722, NZRPG3G0171, NZRPG3G0172, NZRPG3G0173, NMRPG3H0561, NRRPG3J0141 to YHL) and the Ministry of Science and Technology of the Republic of China (MOST 106-2321-B-182A-004-MY3, MOST 107-2320-B-182A-028-, MOST 108-2320-B-182A-004- to YHL).

Conflicts of Interest: The author has no conflict of interest to disclose.

References

1. Dong, Y.; Xu, W.; Liu, C.; Liu, P.; Li, P.; Wang, K. Reactive oxygen species related noncoding RNAs as regulators of cardiovascular diseases. *Int. J. Biol. Sci.* **2019**, *15*, 680–687. [CrossRef] [PubMed]

2. Schieber, M.; Chandel, N.S. ROS function in redox signaling and oxidative stress. *Curr. Biol.* **2014**, *24*, R453–R462. [CrossRef] [PubMed]

3. Engedal, N.; Zerovnik, E.; Rudov, A.; Galli, F.; Olivieri, F.; Procopio, A.D.; Rippo, M.R.; Monsurro, V.; Betti, M.; Albertini, M.C. From oxidative stress damage to pathways, networks, and autophagy via microRNAs. *Oxid. Med. Cell. Longev.* **2018**, *2018*, 4968321. [CrossRef] [PubMed]

4. Castaneda-Arriaga, R.; Perez-Gonzalez, A.; Reina, M.; Alvarez-Idaboy, J.R.; Galano, A. Comprehensive investigation of the antioxidant and pro-oxidant effects of phenolic compounds: A double-edged sword in the context of oxidative stress? *J. Phys. Chem. B* **2018**, *122*, 6198–6214. [CrossRef] [PubMed]

5. O'Brien, J.; Hayder, H.; Zayed, Y.; Peng, C. Overview of microRNA biogenesis, mechanisms of actions, and circulation. *Front. Endocrinol. (Lausanne)* **2018**, *9*, 402. [CrossRef]

6. Tan, W.; Liu, B.; Qu, S.; Liang, G.; Luo, W.; Gong, C. MicroRNAs and cancer: Key paradigms in molecular therapy. *Oncol. Lett.* **2018**, *15*, 2735–2742. [CrossRef]

7. Liu, Y.; Qiang, W.; Xu, X.; Dong, R.; Karst, A.M.; Liu, Z.; Kong, B.; Drapkin, R.I.; Wei, J.J. Role of miR-182 in response to oxidative stress in the cell fate of human fallopian tube epithelial cells. *Oncotarget* **2015**, *6*, 38983–38998. [CrossRef]

8. Fierro-Fernandez, M.; Miguel, V.; Lamas, S. Role of redoximiRs in fibrogenesis. *Redox Biol.* **2016**, *7*, 58–67. [CrossRef]

9. Meseguer, S.; Martinez-Zamora, A.; Garcia-Arumi, E.; Andreu, A.L.; Armengod, M.E. The ROS-sensitive microRNA-9/9* controls the expression of mitochondrial tRNA-modifying enzymes and is involved in the molecular mechanism of MELAS syndrome. *Hum. Mol. Genet.* **2015**, *24*, 167–184. [CrossRef]

10. La Sala, L.; Mrakic-Sposta, S.; Micheloni, S.; Prattichizzo, F.; Ceriello, A. Glucose-sensing microRNA-21 disrupts ROS homeostasis and impairs antioxidant responses in cellular glucose variability. *Cardiovasc. Diabetol.* **2018**, *17*, 105. [CrossRef]

11. Saki, M.; Prakash, A. DNA damage related crosstalk between the nucleus and mitochondria. *Free Radic. Biol. Med.* **2017**, *107*, 216–227. [CrossRef] [PubMed]

12. Case, A.J. On the origin of superoxide dismutase: An evolutionary perspective of superoxide-mediated redox signaling. *Antioxidants (Basel)* **2017**, *6*, 82. [CrossRef] [PubMed]

13. Meitzler, J.L.; Antony, S.; Wu, Y.; Juhasz, A.; Liu, H.; Jiang, G.; Lu, J.; Roy, K.; Doroshow, J.H. Nadph oxidases: A perspective on reactive oxygen species production in tumor biology. *Antioxid. Redox Signal.* **2014**, *20*, 2873–2889. [CrossRef] [PubMed]

14. Collin, F. Chemical basis of reactive oxygen species reactivity and involvement in neurodegenerative diseases. *Int. J. Mol. Sci.* **2019**, *20*, 2407. [CrossRef] [PubMed]

15. Pylvas, M.; Puistola, U.; Laatio, L.; Kauppila, S.; Karihtala, P. Elevated serum 8-OHdG is associated with poor prognosis in epithelial ovarian cancer. *Anticancer Res.* **2011**, *31*, 1411–1415. [PubMed]

16. Plachetka, A.; Adamek, B.; Strzelczyk, J.K.; Krakowczyk, L.; Migula, P.; Nowak, P.; Wiczkowski, A. 8-hydroxy-2'-deoxyguanosine in colorectal adenocarcinoma–is it a result of oxidative stress? *Med. Sci. Monit.* **2013**, *19*, 690–695. [PubMed]

17. Ma-On, C.; Sanpavat, A.; Whongsiri, P.; Suwannasin, S.; Hirankarn, N.; Tangkijvanich, P.; Boonla, C. Oxidative stress indicated by elevated expression of NRF2 and 8-OHdG promotes hepatocellular carcinoma progression. *Med. Oncol.* **2017**, *34*, 57. [CrossRef]

18. Tanaka, H.; Fujita, N.; Sugimoto, R.; Urawa, N.; Horiike, S.; Kobayashi, Y.; Iwasa, M.; Ma, N.; Kawanishi, S.; Watanabe, S.; et al. Hepatic oxidative DNA damage is associated with increased risk for hepatocellular carcinoma in chronic hepatitis c. *Br. J. Cancer* **2008**, *98*, 580–586. [CrossRef]

19. Akhtar, M.J.; Ahamed, M.; Alhadlaq, H.A.; Alshamsan, A. Mechanism of ROS scavenging and antioxidant signalling by redox metallic and fullerene nanomaterials: Potential implications in ROS associated degenerative disorders. *Biochim. Biophys. Acta Gen. Subj.* **2017**, *1861*, 802–813. [CrossRef]

20. He, J.; Jiang, B.H. Interplay between reactive oxygen species and microRNAs in cancer. *Curr. Pharmacol. Rep.* **2016**, *2*, 82–90. [CrossRef]

21. Gong, Y.Y.; Luo, J.Y.; Wang, L.; Huang, Y. MicroRNAs regulating reactive oxygen species in cardiovascular diseases. *Antioxid. Redox Signal.* **2018**, *29*, 1092–1107. [CrossRef] [PubMed]

22. Lan, J.; Huang, Z.; Han, J.; Shao, J.; Huang, C. Redox regulation of microRNAs in cancer. *Cancer Lett.* **2018**, *418*, 250–259. [CrossRef] [PubMed]

23. Ungvari, Z.; Tucsek, Z.; Sosnowska, D.; Toth, P.; Gautam, T.; Podlutsky, A.; Csiszar, A.; Losonczy, G.; Valcarcel-Ares, M.N.; Sonntag, W.E.; et al. Aging-induced dysregulation of dicer1-dependent microRNA expression impairs angiogenic capacity of rat cerebromicrovascular endothelial cells. *J. Gerontol. A Biol. Sci. Med. Sci.* **2013**, *68*, 877–891. [CrossRef] [PubMed]

24. Shilo, S.; Roy, S.; Khanna, S.; Sen, C.K. Evidence for the involvement of miRNA in redox regulated angiogenic response of human microvascular endothelial cells. *Arterioscler. Thromb. Vasc. Biol.* **2008**, *28*, 471–477. [CrossRef] [PubMed]

25. Kozakowska, M.; Ciesla, M.; Stefanska, A.; Skrzypek, K.; Was, H.; Jazwa, A.; Grochot-Przeczek, A.; Kotlinowski, J.; Szymula, A.; Bartelik, A.; et al. Heme oxygenase-1 inhibits myoblast differentiation by targeting myomirs. *Antioxid. Redox Signal.* **2012**, *16*, 113–127. [CrossRef] [PubMed]

26. Markopoulos, G.S.; Roupakia, E.; Tokamani, M.; Alabasi, G.; Sandaltzopoulos, R.; Marcu, K.B.; Kolettas, E. Roles of NF-kB signaling in the regulation of miRNAs impacting on inflammation in cancer. *Biomedicines* **2018**, *6*, 40. [CrossRef]

27. Graves, J.A.; Metukuri, M.; Scott, D.; Rothermund, K.; Prochownik, E.V. Regulation of reactive oxygen species homeostasis by peroxiredoxins and c-myc. *J. Biol. Chem.* **2009**, *284*, 6520–6529. [CrossRef]

28. Ferro, E.; Goitre, L.; Retta, S.F.; Trabalzini, L. The interplay between ROS and Ras GTPases: Physiological and pathological implications. *J. Signal. Transduct.* **2012**, *2012*, 365769. [CrossRef]

29. Jackstadt, R.; Hermeking, H. MicroRNAs as regulators and mediators of c-Myc function. *Biochim. Biophys. Acta* **2015**, *1849*, 544–553. [CrossRef]

30. Xue, G.; Yan, H.L.; Zhang, Y.; Hao, L.Q.; Zhu, X.T.; Mei, Q.; Sun, S.H. c-MYC-mediated repression of miR-15-16 in hypoxia is induced by increased HIF-2alpha and promotes tumor angiogenesis and metastasis by upregulating FGF2. *Oncogene* **2015**, *34*, 1393–1406. [CrossRef]

31. Li, S.G.; Shi, Q.W.; Yuan, L.Y.; Qin, L.P.; Wang, Y.; Miao, Y.Q.; Chen, Z.; Ling, C.Q.; Qin, W.X. c-Myc-dependent repression of two oncogenic miRNA clusters contributes to triptolide-induced cell death in hepatocellular carcinoma cells. *J. Exp. Clin. Cancer Res.* **2018**, *37*, 51. [CrossRef] [PubMed]

32. Gurtner, A.; Falcone, E.; Garibaldi, F.; Piaggio, G. Dysregulation of microRNA biogenesis in cancer: The impact of mutant p53 on drosha complex activity. *J. Exp. Clin. Cancer Res.* **2016**, *35*, 45. [CrossRef] [PubMed]

33. Baker, J.R.; Vuppusetty, C.; Colley, T.; Papaioannou, A.I.; Fenwick, P.; Donnelly, L.; Ito, K.; Barnes, P.J. Oxidative stress dependent microRNA-34a activation via pi3kalpha reduces the expression of sirtuin-1 and sirtuin-6 in epithelial cells. *Sci. Rep.* **2016**, *6*, 35871. [CrossRef] [PubMed]

34. Xiao, Y.; Yan, W.; Lu, L.; Wang, Y.; Lu, W.; Cao, Y.; Cai, W. P38/p53/miR-200a-3p feedback loop promotes oxidative stress-mediated liver cell death. *Cell Cycle* **2015**, *14*, 1548–1558. [CrossRef] [PubMed]

35. Magenta, A.; Cencioni, C.; Fasanaro, P.; Zaccagnini, G.; Greco, S.; Sarra-Ferraris, G.; Antonini, A.; Martelli, F.; Capogrossi, M.C. MiR-200c is upregulated by oxidative stress and induces endothelial cell apoptosis and senescence via zeb1 inhibition. *Cell Death Differ.* **2011**, *18*, 1628–1639. [CrossRef] [PubMed]

36. Wu, J.; Ding, J.; Yang, J.; Guo, X.; Zheng, Y. MicroRNA roles in the nuclear factor kappa b signaling pathway in cancer. *Front. Immunol.* **2018**, *9*, 546. [CrossRef] [PubMed]

37. Ma, X.; Becker Buscaglia, L.E.; Barker, J.R.; Li, Y. MicroRNAs in NF-kappaB signaling. *J. Mol. Cell Biol.* **2011**, *3*, 159–166. [CrossRef] [PubMed]

38. Bazzoni, F.; Rossato, M.; Fabbri, M.; Gaudiosi, D.; Mirolo, M.; Mori, L.; Tamassia, N.; Mantovani, A.; Cassatella, M.A.; Locati, M. Induction and regulatory function of miR-9 in human monocytes and neutrophils exposed to proinflammatory signals. *Proc. Natl. Acad. Sci. USA* **2009**, *106*, 5282–5287. [CrossRef] [PubMed]

39. Niu, J.; Shi, Y.; Tan, G.; Yang, C.H.; Fan, M.; Pfeffer, L.M.; Wu, Z.H. DNA damage induces NF-kappaB-dependent microRNA-21 up-regulation and promotes breast cancer cell invasion. *J. Biol. Chem.* **2012**, *287*, 21783–21795. [CrossRef] [PubMed]

40. Chao, J.; Guo, Y.; Li, P.; Chao, L. Role of kallistatin treatment in aging and cancer by modulating miR-34a and miR-21 expression. *Oxid. Med. Cell. Longev.* **2017**, *2017*, 5025610. [CrossRef]

41. Pratheeshkumar, P.; Son, Y.O.; Divya, S.P.; Wang, L.; Zhang, Z.; Shi, X. Oncogenic transformation of human lung bronchial epithelial cells induced by arsenic involves ROS-dependent activation of STAT3-miR-21-PDCD4 mechanism. *Sci. Rep.* **2016**, *6*, 37227. [CrossRef] [PubMed]

42. Zhou, R.; Hu, G.; Gong, A.Y.; Chen, X.M. Binding of NF-kappaB p65 subunit to the promoter elements is involved in LPS-induced transactivation of miRNA genes in human biliary epithelial cells. *Nucleic Acids Res.* **2010**, *38*, 3222–3232. [CrossRef] [PubMed]

43. Feng, X.; Tan, W.; Cheng, S.; Wang, H.; Ye, S.; Yu, C.; He, Y.; Zeng, J.; Cen, J.; Hu, J.; et al. Upregulation of microRNA-126 in hepatic stellate cells may affect pathogenesis of liver fibrosis through the NF-kappaB pathway. *DNA Cell Biol.* **2015**, *34*, 470–480. [CrossRef] [PubMed]

44. Yin, M.; Ren, X.; Zhang, X.; Luo, Y.; Wang, G.; Huang, K.; Feng, S.; Bao, X.; Huang, K.; He, X.; et al. Selective killing of lung cancer cells by miRNA-506 molecule through inhibiting NF-kappaB p65 to evoke reactive oxygen species generation and p53 activation. *Oncogene* **2015**, *34*, 691–703. [CrossRef] [PubMed]

45. Sykiotis, G.P.; Bohmann, D. Stress-activated cap'n'collar transcription factors in aging and human disease. *Sci. Signal.* **2010**, *3*, re3. [CrossRef] [PubMed]

46. Bellezza, I.; Giambanco, I.; Minelli, A.; Donato, R. Nrf2-Keap1 signaling in oxidative and reductive stress. *Biochim. Biophys. Acta Mol. Cell Res.* **2018**, *1865*, 721–733. [CrossRef]

47. Zhang, C.; Shu, L.; Kong, A.N. MicroRNAs: New players in cancer prevention targeting Nrf2, oxidative stress and inflammatory pathways. *Curr. Pharmacol. Rep.* **2015**, *1*, 21–30. [CrossRef] [PubMed]

48. Cheng, X.H.; Ku, C.H.; Siow, R.C.M. Regulation of the Nrf2 antioxidant pathway by microRNAs: New players in micromanaging redox homeostasis. *Free Radic. Biol. Med.* **2013**, *64*, 4–11. [CrossRef]

49. Singh, A.; Happel, C.; Manna, S.K.; Acquaah-Mensah, G.; Carrerero, J.; Kumar, S.; Nasipuri, P.; Krausz, K.W.; Wakabayashi, N.; Dewi, R.; et al. Transcription factor Nrf2 regulates miR-1 and miR-206 to drive tumorigenesis. *J. Clin. Investig.* **2013**, *123*, 2921–2934. [CrossRef]

50. Kurinna, S.; Schafer, M.; Ostano, P.; Karouzakis, E.; Chiorino, G.; Bloch, W.; Bachmann, A.; Gay, S.; Garrod, D.; Lefort, K.; et al. A novel Nrf2-miR-29-desmocollin-2 axis regulates desmosome function in keratinocytes. *Nat. Commun.* **2014**, *5*, 5099. [CrossRef]

51. Joo, M.S.; Lee, C.G.; Koo, J.H.; Kim, S.G. Mir-125b transcriptionally increased by nrf2 inhibits AHR repressor, which protects kidney from cisplatin-induced injury. *Cell Death Dis.* **2013**, *4*, e899. [CrossRef] [PubMed]

52. Yang, M.; Yao, Y.; Eades, G.; Zhang, Y.; Zhou, Q. Mir-28 regulates Nrf2 expression through a Keap1-independent mechanism. *Breast Cancer Res. Treat.* **2011**, *129*, 983–991. [CrossRef] [PubMed]

53. Huang, X.; Gao, Y.; Qin, J.; Lu, S. The role of mir-34a in the hepatoprotective effect of hydrogen sulfide on ischemia/reperfusion injury in young and old rats. *PLoS ONE* **2014**, *9*, e113305. [CrossRef] [PubMed]

54. Singh, B.; Ronghe, A.M.; Chatterjee, A.; Bhat, N.K.; Bhat, H.K. MicroRNA-93 regulates Nrf2 expression and is associated with breast carcinogenesis. *Carcinogenesis* **2013**, *34*, 1165–1172. [CrossRef] [PubMed]

55. Eades, G.; Yang, M.; Yao, Y.; Zhang, Y.; Zhou, Q. Mir-200a regulates Nrf2 activation by targeting Keap1 mRNA in breast cancer cells. *J. Biol. Chem.* **2011**, *286*, 40725–40733. [CrossRef] [PubMed]

56. Nebbioso, A.; Tambaro, F.P.; Dell'Aversana, C.; Altucci, L. Cancer epigenetics: Moving forward. *PLoS Genet.* **2018**, *14*, e1007362. [CrossRef] [PubMed]

57. He, J.; Xu, Q.; Jing, Y.; Agani, F.; Qian, X.; Carpenter, R.; Li, Q.; Wang, X.R.; Peiper, S.S.; Lu, Z.; et al. Reactive oxygen species regulate ERBB2 and ERBB3 expression via miR-199a/125b and DNA methylation. *EMBO Rep.* **2012**, *13*, 1116–1122. [CrossRef] [PubMed]

58. Druz, A.; Betenbaugh, M.; Shiloach, J. Glucose depletion activates mmu-miR-466h-5p expression through oxidative stress and inhibition of histone deacetylation. *Nucleic Acids Res.* **2012**, *40*, 7291–7302. [CrossRef]

59. Muz, B.; de la Puente, P.; Azab, F.; Azab, A.K. The role of hypoxia in cancer progression, angiogenesis, metastasis, and resistance to therapy. *Hypoxia* **2015**, *3*, 83–92. [CrossRef]

60. Azimi, I.; Petersen, R.M.; Thompson, E.W.; Roberts-Thomson, S.J.; Monteith, G.R. Hypoxia-induced reactive oxygen species mediate n-cadherin and serpine1 expression, EGFR signalling and motility in mda-mb-468 breast cancer cells. *Sci. Rep.* **2017**, *7*, 15140. [CrossRef]

61. Cardin, R.; Piciocchi, M.; Sinigaglia, A.; Lavezzo, E.; Bortolami, M.; Kotsafti, A.; Cillo, U.; Zanus, G.; Mescoli, C.; Rugge, M.; et al. Oxidative DNA damage correlates with cell immortalization and miR-92 expression in hepatocellular carcinoma. *BMC Cancer* **2012**, *12*, 177. [CrossRef]

62. Yatabe, N.; Kyo, S.; Maida, Y.; Nishi, H.; Nakamura, M.; Kanaya, T.; Tanaka, M.; Isaka, K.; Ogawa, S.; Inoue, M. HIF-1-mediated activation of telomerase in cervical cancer cells. *Oncogene* **2004**, *23*, 3708–3715. [CrossRef] [PubMed]

63. Jung, N.; Kwon, H.J.; Jung, H.J. Downregulation of mitochondrial UQCRB inhibits cancer stem cell-like properties in glioblastoma. *Int. J. Oncol.* **2018**, *52*, 241–251. [CrossRef] [PubMed]

64. Corcoran, S.E.; O'Neill, L.A. HIF1alpha and metabolic reprogramming in inflammation. *J. Clin. Investig.* **2016**, *126*, 3699–3707. [CrossRef] [PubMed]

65. Bavelloni, A.; Ramazzotti, G.; Poli, A.; Piazzi, M.; Focaccia, E.; Blalock, W.; Faenza, I. Mirna-210: A current overview. *Anticancer Res.* **2017**, *37*, 6511–6521.

66. Fallah, A.; Sadeghinia, A.; Kahroba, H.; Samadi, A.; Heidari, H.R.; Bradaran, B.; Zeinali, S.; Molavi, O. Therapeutic targeting of angiogenesis molecular pathways in angiogenesis-dependent diseases. *Biomed. Pharmacother.* **2019**, *110*, 775–785. [CrossRef]

67. Hoeben, A.; Landuyt, B.; Highley, M.S.; Wildiers, H.; Van Oosterom, A.T.; De Bruijn, E.A. Vascular endothelial growth factor and angiogenesis. *Pharmacol. Rev.* **2004**, *56*, 549–580. [CrossRef]

68. Ferrari, G.; Cook, B.D.; Terushkin, V.; Pintucci, G.; Mignatti, P. Transforming growth factor-beta 1 (TGF-beta1) induces angiogenesis through vascular endothelial growth factor (VEGF)-mediated apoptosis. *J. Cell. Physiol.* **2009**, *219*, 449–458. [CrossRef]

69. Fagiani, E.; Christofori, G. Angiopoietins in angiogenesis. *Cancer Lett.* **2013**, *328*, 18–26. [CrossRef]

70. Raica, M.; Cimpean, A.M. Platelet-derived growth factor (PDGF)/PDGFreceptors (PDGFR) axis as target for antitumor and antiangiogenic therapy. *Pharmaceuticals* **2010**, *3*, 572–599. [CrossRef]

71. Espinosa-Diez, C.; Wilson, R.; Chatterjee, N.; Hudson, C.; Ruhl, R.; Hipfinger, C.; Helms, E.; Khan, O.F.; Anderson, D.G.; Anand, S. MicroRNA regulation of the MRN complex impacts DNA damage, cellular senescence, and angiogenic signaling. *Cell Death Dis.* **2018**, *9*, 632. [CrossRef] [PubMed]

72. Chu, L.Y.; Ramakrishnan, D.P.; Silverstein, R.L. Thrombospondin-1 modulates VEGF signaling via CD36 by recruiting SHP-1 to VEGFR2 complex in microvascular endothelial cells. *Blood* **2013**, *122*, 1822–1832. [CrossRef] [PubMed]

73. Guo, Z.; Kozlov, S.; Lavin, M.F.; Person, M.D.; Paull, T.T. Atm activation by oxidative stress. *Science* **2010**, *330*, 517–521. [CrossRef] [PubMed]

74. Okuno, Y.; Nakamura-Ishizu, A.; Otsu, K.; Suda, T.; Kubota, Y. Pathological neoangiogenesis depends on oxidative stress regulation by atm. *Nat. Med.* **2012**, *18*, 1208–1216. [CrossRef] [PubMed]

75. Economopoulou, M.; Langer, H.F.; Celeste, A.; Orlova, V.V.; Choi, E.Y.; Ma, M.; Vassilopoulos, A.; Callen, E.; Deng, C.; Bassing, C.H.; et al. Histone h2ax is integral to hypoxia-driven neovascularization. *Nat. Med.* **2009**, *15*, 553–558. [CrossRef] [PubMed]

76. Wilson, R.; Espinosa-Diez, C.; Kanner, N.; Chatterjee, N.; Ruhl, R.; Hipfinger, C.; Advani, S.J.; Li, J.; Khan, O.F.; Franovic, A.; et al. MicroRNA regulation of endothelial TREX1 reprograms the tumour microenvironment. *Nat. Commun.* **2016**, *7*, 13597. [CrossRef]

77. Yang, Z.; Wa, Q.D.; Lu, C.; Pan, W.; Lu, Z.; Ao, J. Mir3283p enhances the radiosensitivity of osteosarcoma and regulates apoptosis and cell viability via H2AX. *Oncol. Rep.* **2018**, *39*, 545–553. [PubMed]

78. Marampon, F.; Codenotti, S.; Megiorni, F.; Del Fattore, A.; Camero, S.; Gravina, G.L.; Festuccia, C.; Musio, D.; De Felice, F.; Nardone, V.; et al. NRF2 orchestrates the redox regulation induced by radiation therapy, sustaining embryonal and alveolar rhabdomyosarcoma cells radioresistance. *J. Cancer Res. Clin. Oncol.* **2019**, *145*, 881–893. [CrossRef] [PubMed]

79. Salminen, A.; Kaarniranta, K.; Kauppinen, A. Crosstalk between oxidative stress and SIRT1: Impact on the aging process. *Int. J. Mol. Sci.* **2013**, *14*, 3834–3859. [CrossRef]

80. Luo, J.; Chen, P.; Xie, W.; Wu, F. MicroRNA-138 inhibits cell proliferation in hepatocellular carcinoma by targeting SIRT1. *Oncol. Rep.* **2017**, *38*, 1067–1074. [CrossRef]

81. Zhou, B.; Li, C.; Qi, W.; Zhang, Y.; Zhang, F.; Wu, J.X.; Hu, Y.N.; Wu, D.M.; Liu, Y.; Yan, T.T.; et al. Downregulation of miR-181a upregulates sirtuin-1 (SIRT1) and improves hepatic insulin sensitivity. *Diabetologia* **2012**, *55*, 2032–2043. [CrossRef] [PubMed]

82. Yamakuchi, M. MicroRNA regulation of SIRT1. *Front. Physiol.* **2012**, *3*, 68. [CrossRef] [PubMed]

83. Rane, S.; He, M.; Sayed, D.; Vashistha, H.; Malhotra, A.; Sadoshima, J.; Vatner, D.E.; Vatner, S.F.; Abdellatif, M. Downregulation of miR-199a derepresses hypoxia-inducible factor-1alpha and Sirtuin 1 and recapitulates hypoxia preconditioning in cardiac myocytes. *Circ. Res.* **2009**, *104*, 879–886. [CrossRef] [PubMed]

84. Lambert, A.W.; Pattabiraman, D.R.; Weinberg, R.A. Emerging biological principles of metastasis. *Cell* **2017**, *168*, 670–691. [CrossRef] [PubMed]

85. Peiris-Pages, M.; Martinez-Outschoorn, U.E.; Sotgia, F.; Lisanti, M.P. Metastasis and oxidative stress: Are antioxidants a metabolic driver of progression? *Cell Metab.* **2015**, *22*, 956–958. [CrossRef] [PubMed]

86. Gill, J.G.; Piskounova, E.; Morrison, S.J. Cancer, oxidative stress, and metastasis. *Cold Spring Harb. Symp. Quant. Biol.* **2016**, *81*, 163–175. [CrossRef] [PubMed]

87. Hsieh, C.L.; Liu, C.M.; Chen, H.A.; Yang, S.T.; Shigemura, K.; Kitagawa, K.; Yamamichi, F.; Fujisawa, M.; Liu, Y.R.; Lee, W.H.; et al. Reactive oxygen species-mediated switching expression of MMP-3 in stromal fibroblasts and cancer cells during prostate cancer progression. *Sci. Rep.* **2017**, *7*, 9065. [CrossRef] [PubMed]

88. Cichon, M.A.; Radisky, D.C. ROS-induced epithelial-mesenchymal transition in mammary epithelial cells is mediated by NF-kB-dependent activation of snail. *Oncotarget* **2014**, *5*, 2827–2838. [CrossRef]

89. Zeller, K.S.; Riaz, A.; Sarve, H.; Li, J.; Tengholm, A.; Johansson, S. The role of mechanical force and ROS in integrin-dependent signals. *PLoS ONE* **2013**, *8*, e64897. [CrossRef]

90. Mori, K.; Shibanuma, M.; Nose, K. Invasive potential induced under long-term oxidative stress in mammary epithelial cells. *Cancer Res.* **2004**, *64*, 7464–7472. [CrossRef]

91. Sawicki, G. Intracellular regulation of matrix metalloproteinase-2 activity: New strategies in treatment and protection of heart subjected to oxidative stress. *Scientifica* **2013**, *2013*, 130451. [CrossRef] [PubMed]

92. Dezerega, A.; Madrid, S.; Mundi, V.; Valenzuela, M.A.; Garrido, M.; Paredes, R.; Garcia-Sesnich, J.; Ortega, A.V.; Gamonal, J.; Hernandez, M. Pro-oxidant status and matrix metalloproteinases in apical lesions and gingival crevicular fluid as potential biomarkers for asymptomatic apical periodontitis and endodontic treatment response. *J. Inflamm.* **2012**, *9*, 8. [CrossRef]

93. Yoon, S.O.; Park, S.J.; Yoon, S.Y.; Yun, C.H.; Chung, A.S. Sustained production of H(2)O(2) activates pro-matrix metalloproteinase-2 through receptor tyrosine kinases/phosphatidylinositol 3-kinase/NF-kappa B pathway. *J. Biol. Chem.* **2002**, *277*, 30271–30282. [CrossRef] [PubMed]

94. Song, Y.H.; Wang, J.; Nie, G.; Chen, Y.J.; Li, X.; Jiang, X.; Cao, W.H. MicroRNA-509-5p functions as an anti-oncogene in breast cancer via targeting sod2. *Eur. Rev. Med. Pharmacol. Sci.* **2017**, *21*, 3617–3625. [PubMed]

95. Yu, L.; Chen, X.; Sun, X.; Wang, L.; Chen, S. The glycolytic switch in tumors: How many players are involved? *J. Cancer* **2017**, *8*, 3430–3440. [CrossRef] [PubMed]

96. Pacella, I.; Procaccini, C.; Focaccetti, C.; Miacci, S.; Timperi, E.; Faicchia, D.; Severa, M.; Rizzo, F.; Coccia, E.M.; Bonacina, F.; et al. Fatty acid metabolism complements glycolysis in the selective regulatory t cell expansion during tumor growth. *Proc. Natl. Acad. Sci. USA* **2018**, *115*, E6546–E6555. [CrossRef]

97. Schwartz, L.; Supuran, C.T.; Alfarouk, K.O. The warburg effect and the hallmarks of cancer. *Anticancer Agents Med. Chem.* **2017**, *17*, 164–170. [CrossRef] [PubMed]

98. Kim, J.; Kim, J.; Bae, J.S. ROS homeostasis and metabolism: A critical liaison for cancer therapy. *Exp. Mol. Med.* **2016**, *48*, e269. [CrossRef]

99. Hu, W.; Zhang, C.; Wu, R.; Sun, Y.; Levine, A.; Feng, Z. Glutaminase 2, a novel p53 target gene regulating energy metabolism and antioxidant function. *Proc. Natl. Acad. Sci. USA* **2010**, *107*, 7455–7460. [CrossRef]

100. Budanov, A.V. The role of tumor suppressor p53 in the antioxidant defense and metabolism. *Subcell. Biochem.* **2014**, *85*, 337–358.

101. Ipsen, D.H.; Lykkesfeldt, J.; Tveden-Nyborg, P. Molecular mechanisms of hepatic lipid accumulation in non-alcoholic fatty liver disease. *Cell. Mol. Life Sci.* **2018**, *75*, 3313–3327. [CrossRef] [PubMed]

102. Mello, T.; Simeone, I.; Galli, A. Mito-nuclear communication in hepatocellular carcinoma metabolic rewiring. *Cells* **2019**, *8*, 417. [CrossRef] [PubMed]

103. Horie, T.; Nishino, T.; Baba, O.; Kuwabara, Y.; Nakao, T.; Nishiga, M.; Usami, S.; Izuhara, M.; Sowa, N.; Yahagi, N.; et al. MicroRNA-33 regulates sterol regulatory element-binding protein 1 expression in mice. *Nat. Commun.* **2013**, *4*, 2883. [CrossRef] [PubMed]

104. Ayob, A.Z.; Ramasamy, T.S. Cancer stem cells as key drivers of tumour progression. *J. Biomed. Sci.* **2018**, *25*, 20. [CrossRef] [PubMed]

105. Saleh, A.D.; Savage, J.E.; Cao, L.; Soule, B.P.; Ly, D.; DeGraff, W.; Harris, C.C.; Mitchell, J.B.; Simone, N.L. Cellular stress induced alterations in microRNA let-7a and let-7b expression are dependent on p53. *PLoS ONE* **2011**, *6*, e24429. [CrossRef] [PubMed]

106. Zhang, X.; Ng, W.L.; Wang, P.; Tian, L.; Werner, E.; Wang, H.; Doetsch, P.; Wang, Y. MicroRNA-21 modulates the levels of reactive oxygen species by targeting sod3 and tnfalpha. *Cancer Res.* **2012**, *72*, 4707–4713. [CrossRef] [PubMed]

107. Li, X.J.; Ren, Z.J.; Tang, J.H. MicroRNA-34a: A potential therapeutic target in human cancer. *Cell Death Dis.* **2014**, *5*, e1327. [CrossRef]

108. Balzano, F.; Cruciani, S.; Basoli, V.; Santaniello, S.; Facchin, F.; Ventura, C.; Maioli, M. Mir200 and miR302: Two big families influencing stem cell behavior. *Molecules* **2018**, *23*, 282. [CrossRef]

109. Kim, J.H.; Park, S.G.; Song, S.Y.; Kim, J.K.; Sung, J.H. Reactive oxygen species-responsive miR-210 regulates proliferation and migration of adipose-derived stem cells via ptpn2. *Cell Death Dis.* **2013**, *4*, e588. [CrossRef]

110. Mei, Y.; Bian, C.; Li, J.; Du, Z.; Zhou, H.; Yang, Z.; Zhao, R.C. Mir-21 modulates the erk-mapk signaling pathway by regulating spry2 expression during human mesenchymal stem cell differentiation. *J. Cell. Biochem.* **2013**, *114*, 1374–1384. [CrossRef]

111. Yu, G.; Yao, W.; Xiao, W.; Li, H.; Xu, H.; Lang, B. MicroRNA-34a functions as an anti-metastatic microRNA and suppresses angiogenesis in bladder cancer by directly targeting cd44. *J. Exp. Clin. Cancer Res.* **2014**, *33*, 779. [CrossRef] [PubMed]

112. Peter, M.E. Let-7 and miR-200 microRNAs: Guardians against pluripotency and cancer progression. *Cell Cycle* **2009**, *8*, 843–852. [CrossRef] [PubMed]

113. Tang, T.; Yang, Z.; Zhu, Q.; Wu, Y.; Sun, K.; Alahdal, M.; Zhang, Y.; Xing, Y.; Shen, Y.; Xia, T.; et al. Up-regulation of miR-210 induced by a hypoxic microenvironment promotes breast cancer stem cells metastasis, proliferation, and self-renewal by targeting e-cadherin. *FASEB J.* **2018**, *32*, fj201801013R. [CrossRef] [PubMed]

114. Maciel-Baron, L.A.; Moreno-Blas, D.; Morales-Rosales, S.L.; Gonzalez-Puertos, V.Y.; Lopez-Diazguerrero, N.E.; Torres, C.; Castro-Obregon, S.; Konigsberg, M. Cellular senescence, neurological function, and redox state. *Antioxid. Redox Signal.* **2018**, *28*, 1704–1723. [CrossRef] [PubMed]

115. Liu, D.; Xu, Y. P53, oxidative stress, and aging. *Antioxid. Redox Signal.* **2011**, *15*, 1669–1678. [CrossRef] [PubMed]

116. Dellago, H.; Preschitz-Kammerhofer, B.; Terlecki-Zaniewicz, L.; Schreiner, C.; Fortschegger, K.; Chang, M.W.; Hackl, M.; Monteforte, R.; Kuhnel, H.; Schosserer, M.; et al. High levels of oncomiR-21 contribute to the senescence-induced growth arrest in normal human cells and its knock-down increases the replicative lifespan. *Aging Cell* **2013**, *12*, 446–458. [CrossRef] [PubMed]

117. Xu, D.; Takeshita, F.; Hino, Y.; Fukunaga, S.; Kudo, Y.; Tamaki, A.; Matsunaga, J.; Takahashi, R.U.; Takata, T.; Shimamoto, A.; et al. Mir-22 represses cancer progression by inducing cellular senescence. *J. Cell Biol.* **2011**, *193*, 409–424. [CrossRef]

118. Hu, Z.; Klein, J.D.; Mitch, W.E.; Zhang, L.; Martinez, I.; Wang, X.H. MicroRNA-29 induces cellular senescence in aging muscle through multiple signaling pathways. *Aging* **2014**, *6*, 160–175. [CrossRef]

119. He, X.; Yang, A.; McDonald, D.G.; Riemer, E.C.; Vanek, K.N.; Schulte, B.A.; Wang, G.Y. Mir-34a modulates ionizing radiation-induced senescence in lung cancer cells. *Oncotarget* **2017**, *8*, 69797–69807. [CrossRef]

120. Hong, L.; Lai, M.; Chen, M.; Xie, C.; Liao, R.; Kang, Y.J.; Xiao, C.; Hu, W.Y.; Han, J.; Sun, P. The miR-17-92 cluster of microRNAs confers tumorigenicity by inhibiting oncogene-induced senescence. *Cancer Res.* **2010**, *70*, 8547–8557. [CrossRef]

121. Nyholm, A.M.; Lerche, C.M.; Manfe, V.; Biskup, E.; Johansen, P.; Morling, N.; Thomsen, B.M.; Glud, M.; Gniadecki, R. Mir-125b induces cellular senescence in malignant melanoma. *BMC Dermatol.* **2014**, *14*, 8. [CrossRef] [PubMed]

122. Olivieri, F.; Lazzarini, R.; Recchioni, R.; Marcheselli, F.; Rippo, M.R.; Di Nuzzo, S.; Albertini, M.C.; Graciotti, L.; Babini, L.; Mariotti, S.; et al. Mir-146a as marker of senescence-associated pro-inflammatory status in cells involved in vascular remodelling. *Age* **2013**, *35*, 1157–1172. [CrossRef] [PubMed]

123. Menghini, R.; Casagrande, V.; Cardellini, M.; Martelli, E.; Terrinoni, A.; Amati, F.; Vasa-Nicotera, M.; Ippoliti, A.; Novelli, G.; Melino, G.; et al. MicroRNA 217 modulates endothelial cell senescence via silent information regulator 1. *Circulation* **2009**, *120*, 1524–1532. [CrossRef] [PubMed]

124. Shang, F.; Wang, S.C.; Hsu, C.Y.; Miao, Y.; Martin, M.; Yin, Y.; Wu, C.C.; Wang, Y.T.; Wu, G.; Chien, S.; et al. MicroRNA-92a mediates endothelial dysfunction in ckd. *J. Am. Soc. Nephrol.* **2017**, *28*, 3251–3261. [CrossRef] [PubMed]

125. Liu, H.; Wu, H.Y.; Wang, W.Y.; Zhao, Z.L.; Liu, X.Y.; Wang, L.Y. Regulation of miR-92a on vascular endothelial aging via mediating nrf2-keap1-are signal pathway. *Eur. Rev. Med. Pharmacol. Sci.* **2017**, *21*, 2734–2742. [PubMed]

126. Sanchez-Sanchez, B.; Gutierrez-Herrero, S.; Lopez-Ruano, G.; Prieto-Bermejo, R.; Romo-Gonzalez, M.; Llanillo, M.; Pandiella, A.; Guerrero, C.; Miguel, J.F.; Sanchez-Guijo, F.; et al. Nadph oxidases as therapeutic targets in chronic myelogenous leukemia. *Clin. Cancer Res.* **2014**, *20*, 4014–4025. [CrossRef]

127. Altenhofer, S.; Radermacher, K.A.; Kleikers, P.W.; Wingler, K.; Schmidt, H.H. Evolution of nadph oxidase inhibitors: Selectivity and mechanisms for target engagement. *Antioxid. Redox Signal.* **2015**, *23*, 406–427. [CrossRef]

128. Suzuki, S.; Pitchakarn, P.; Sato, S.; Shirai, T.; Takahashi, S. Apocynin, an nadph oxidase inhibitor, suppresses progression of prostate cancer via rac1 dephosphorylation. *Exp. Toxicol. Pathol.* **2013**, *65*, 1035–1041. [CrossRef]

129. Fuji, S.; Suzuki, S.; Naiki-Ito, A.; Kato, H.; Hayakawa, M.; Yamashita, Y.; Kuno, T.; Takahashi, S. The nadph oxidase inhibitor apocynin suppresses preneoplastic liver foci of rats. *Toxicol. Pathol.* **2017**, *45*, 544–550. [CrossRef]

130. Kumari, S.; Badana, A.K.; G, M.M.; G, S.; Malla, R. Reactive oxygen species: A key constituent in cancer survival. *Biomark. Insights* **2018**, *13*, 1177271918755391. [CrossRef]

131. Traverso, N.; Ricciarelli, R.; Nitti, M.; Marengo, B.; Furfaro, A.L.; Pronzato, M.A.; Marinari, U.M.; Domenicotti, C. Role of glutathione in cancer progression and chemoresistance. *Oxid. Med. Cell. Longev.* **2013**, *2013*, 972913. [CrossRef] [PubMed]

132. Yang, H.; Li, T.W.; Zhou, Y.; Peng, H.; Liu, T.; Zandi, E.; Martinez-Chantar, M.L.; Mato, J.M.; Lu, S.C. Activation of a novel c-Myc-miR27-prohibitin 1 circuitry in cholestatic liver injury inhibits glutathione synthesis in mice. *Antioxid. Redox Signal.* **2015**, *22*, 259–274. [CrossRef] [PubMed]

133. Espinosa-Diez, C.; Fierro-Fernandez, M.; Sanchez-Gomez, F.; Rodriguez-Pascual, F.; Alique, M.; Ruiz-Ortega, M.; Beraza, N.; Martinez-Chantar, M.L.; Fernandez-Hernando, C.; Lamas, S. Targeting of gamma-glutamyl-cysteine ligase by miR-433 reduces glutathione biosynthesis and promotes tgf-beta-dependent fibrogenesis. *Antioxid. Redox Signal.* **2015**, *23*, 1092–1105. [CrossRef] [PubMed]

134. Cheng, Y.; Zhou, M.; Zhou, W. MicroRNA-30e regulates tgf-beta-mediated nadph oxidase 4-dependent oxidative stress by snail1 in atherosclerosis. *Int. J. Mol. Med.* **2019**, *43*, 1806–1816. [PubMed]

135. Cho, K.J.; Song, J.; Oh, Y.; Lee, J.E. MicroRNA-let-7a regulates the function of microglia in inflammation. *Mol. Cell. Neurosci.* **2015**, *68*, 167–176. [CrossRef] [PubMed]

136. Fulciniti, M.; Amodio, N.; Bandi, R.L.; Cagnetta, A.; Samur, M.K.; Acharya, C.; Prabhala, R.; D'Aquila, P.; Bellizzi, D.; Passarino, G.; et al. Mir-23b/sp1/c-myc forms a feed-forward loop supporting multiple myeloma cell growth. *Blood Cancer J.* **2016**, *6*, e380. [CrossRef] [PubMed]

137. Chen, Q.; Zhang, F.; Wang, Y.; Liu, Z.; Sun, A.; Zen, K.; Zhang, C.Y.; Zhang, Q. The transcription factor c-myc suppresses miR-23b and miR-27b transcription during fetal distress and increases the sensitivity of neurons to hypoxia-induced apoptosis. *PLoS ONE* **2015**, *10*, e0120217. [CrossRef]

138. Gao, P.; Tchernyshyov, I.; Chang, T.C.; Lee, Y.S.; Kita, K.; Ochi, T.; Zeller, K.I.; De Marzo, A.M.; Van Eyk, J.E.; Mendell, J.T.; et al. C-myc suppression of miR-23a/b enhances mitochondrial glutaminase expression and glutamine metabolism. *Nature* **2009**, *458*, 762–765. [CrossRef]

139. Mott, J.L.; Kurita, S.; Cazanave, S.C.; Bronk, S.F.; Werneburg, N.W.; Fernandez-Zapico, M.E. Transcriptional suppression of miR-29b-1/miR-29a promoter by c-myc, hedgehog, and nf-kappab. *J. Cell. Biochem.* **2010**, *110*, 1155–1164. [CrossRef]

140. Hou, M.; Zuo, X.; Li, C.; Zhang, Y.; Teng, Y. Mir-29b regulates oxidative stress by targeting sirt1 in ovarian cancer cells. *Cell. Physiol. Biochem.* **2017**, *43*, 1767–1776. [CrossRef]

141. Mazzoccoli, L.; Robaina, M.C.; Apa, A.G.; Bonamino, M.; Pinto, L.W.; Queiroga, E.; Bacchi, C.E.; Klumb, C.E. Mir-29 silencing modulates the expression of target genes related to proliferation, apoptosis and methylation in burkitt lymphoma cells. *J. Cancer Res. Clin. Oncol.* **2018**, *144*, 483–497. [CrossRef] [PubMed]

142. Wu, D.W.; Hsu, N.Y.; Wang, Y.C.; Lee, M.C.; Cheng, Y.W.; Chen, C.Y.; Lee, H. C-myc suppresses microRNA-29b to promote tumor aggressiveness and poor outcomes in non-small cell lung cancer by targeting fhit. *Oncogene* **2015**, *34*, 2072–2082. [CrossRef] [PubMed]

143. Okada, N.; Lin, C.P.; Ribeiro, M.C.; Biton, A.; Lai, G.; He, X.; Bu, P.; Vogel, H.; Jablons, D.M.; Keller, A.C.; et al. A positive feedback between p53 and miR-34 miRNAs mediates tumor suppression. *Genes Dev.* **2014**, *28*, 438–450. [CrossRef] [PubMed]

144. Mihailovich, M.; Bremang, M.; Spadotto, V.; Musiani, D.; Vitale, E.; Varano, G.; Zambelli, F.; Mancuso, F.M.; Cairns, D.A.; Pavesi, G.; et al. Mir-17-92 fine-tunes myc expression and function to ensure optimal b cell lymphoma growth. *Nat. Commun.* **2015**, *6*, 8725. [CrossRef] [PubMed]

145. Yan, H.L.; Xue, G.; Mei, Q.; Wang, Y.Z.; Ding, F.X.; Liu, M.F.; Lu, M.H.; Tang, Y.; Yu, H.Y.; Sun, S.H. Repression of the miR-17-92 cluster by p53 has an important function in hypoxia-induced apoptosis. *EMBO J.* **2009**, *28*, 2719–2732. [CrossRef] [PubMed]

146. Borkowski, R.; Du, L.; Zhao, Z.; McMillan, E.; Kosti, A.; Yang, C.R.; Suraokar, M.; Wistuba, I.I.; Gazdar, A.F.; Minna, J.D.; et al. Genetic mutation of p53 and suppression of the miR-17 approximately 92 cluster are synthetic lethal in non-small cell lung cancer due to upregulation of vitamin d signaling. *Cancer Res.* **2015**, *75*, 666–675. [CrossRef] [PubMed]

International Journal of
Molecular Sciences

Review

Development and Clinical Trials of Nucleic Acid Medicines for Pancreatic Cancer Treatment

Keiko Yamakawa, Yuko Nakano-Narusawa, Nozomi Hashimoto, Masanao Yokohira and Yoko Matsuda *

Oncology Pathology, Department of Pathology and Host-Defense, Faculty of Medicine, Kagawa University, 1750-1 Ikenobe, Miki-cho, Kita-gun, Kagawa 761-0793, Japan
* Correspondence: youkoh@med.kagawa-u.ac.jp; Tel.: +81-87-891-2109; Fax: +81-87-891-2112

Received: 26 July 2019; Accepted: 27 August 2019; Published: 29 August 2019

Abstract: Approximately 30% of pancreatic cancer patients harbor targetable mutations. However, there has been no therapy targeting these molecules clinically. Nucleic acid medicines show high specificity and can target RNAs. Nucleic acid medicine is expected to be the next-generation treatment next to small molecules and antibodies. There are several kinds of nucleic acid drugs, including antisense oligonucleotides, small interfering RNAs, microRNAs, aptamers, decoys, and CpG oligodeoxynucleotides. In this review, we provide an update on current research of nucleic acid-based therapies. Despite the challenging obstacles, we hope that nucleic acid drugs will have a significant impact on the treatment of pancreatic cancer. The combination of genetic diagnosis using next generation sequencing and targeted therapy may provide effective precision medicine for pancreatic cancer patients.

Keywords: nucleic acid medicine; pancreatic cancer; clinical trial; siRNA; antisense oligonucleotide

1. Introduction

Despite advances in diagnostics and therapeutics, the prognosis of pancreatic cancer remains poor with an overall five-year survival rate of 6%, due in part to difficulties in treating carcinoma at an advanced stage. Mutations of *KRAS, CDKN2a, TP53,* and *SMAD4* are driver mutations in pancreatic cancer; however, a targeted approach for those molecules has not been successful yet. Precision medicine for individual patient has been greatly expected to improve pancreatic cancer patients' outcomes. Recent advances of comprehensive gene analysis using next-generation sequencers can provide a wealth of information of genetic abnormalities of cancers [1,2]. There have been several candidates for treatment targets in pancreatic cancer. Approximately 30% of pancreatic cancer patients harbor druggable mutations; for example, *KRAS, BRCA1* and *2, PALB2, ATM, HER2, MET, MLH1, MSH2, MSH6, PMS2, PI3CA, PTEN, CDKN2A, BRAF,* and *FGFR1* [2]. However, there has been no clinical therapy targeting these molecules, because it is difficult to inhibit target RNA in humans.

RNA interference (RNAi) is a biological process in which RNA molecules inhibit gene expression or translation by neutralizing targeted mRNA molecules. Nucleic acid medicine consists of natural or chemically modified nucleotides that can act directly without changes in gene expression [3]. These drugs show high specificity and can target mRNA and noncoding RNAs. Nucleic acid medicine is considered the next-generation treatment next to small molecules and antibodies. There are several aspects of nucleic acid therapy that are potentially advantageous over traditional drugs. These include the ability to generate specific inhibitors of targets that were previously inaccessible, with the only limit being the genetic information available. Inhibition of mRNA expression has the potential to produce faster and longer-lasting responses than protein inhibition by conventional targeted therapy. Moreover, the side-effects of nucleic acid medicine might be less than those of conventional therapy [4]. Lastly, oligonucleotides can be chemically synthesized and thus their development duration is relatively short compared to antibodies.

There are several kinds of nucleic acid drugs, including antisense oligonucleotides (ASOs), small interfering RNAs (siRNAs), microRNAs (miRNAs), aptamers, decoys, and CpG oligodeoxynucleotides (CpG oligos) (Table 1). They can be classified as either extracellular or intracellular according to their site of function; ASOs, siRNAs, miRNAs, and decoys act in the nucleus or cytoplasm, while aptamers bind to extracellular proteins and CpG oligos act on Toll-like receptor 9 (TLR9) in the endosome. The drugs also have different targets; ASOs, miRNAs, and siRNAs target RNA, whilst aptamers, decoys, and CpG oligos target proteins. Nucleic acid drugs are suited for coextinction or therapeutic synergy, which may represent an important step to overcome compensatory effects typically observed in cancer cells following knockdown of a single target. In this review, we provide an update on the current research of nucleic acid-based therapies, focusing on ASO and siRNA for pancreatic cancer, and summarize the outcomes from published data.

Table 1. Nucleic acid medicines.

	Antisense Oligonucleotides	siRNAs	Antisense miRNAs	miRNA Mimics	Decoys	Aptamers	CpG Oligodeoxynucleotides
Structure	Single strand DNA/RNA	Double strand RNA	Single strand DNA/RNA	Double strand RNA	Double strand DNA	Single strand DNA/RNA	Single strand DNA
Length (base pairs)	12–21 20–30	20–25	12–16	20–25	20	26–45	20
Site	Intracellular (nucleus, cytoplasm)	Intracellular (cytoplasm)	Intracellular (cytoplasm)	Intracellular (cytoplasm)	Intracellular (nucleus)	Extracellular	Extracellular (endosome)
Target	mRNA pre-mRNA miRNA	mRNA	miRNA	mRNA	Protein (transcription factor)	Protein	Protein (TLR9)
Function	mRNA degradation Translational inhibition miRNA inhibition Splicing inhibition	mRNA degradation	miRNA degradation	mRNA degradation Translational inhibition	Transcriptional inhibition	Inhibition of protein function	Activation of natural immunity via TLR9
Drug delivery system	Modified or unnecessary	Necessary	Necessary	Necessary	Necessary	PEGylation	Antigen

TLR9, toll like receptor 9.

2. Functions

2.1. Antisense Oligonucleotides

ASOs are single strands of DNA or RNA that are complementary to a chosen sequence. In the case of antisense RNA, they prevent protein translation of certain messenger RNA strands by binding to them [5].

Antisense DNA can be used to target a specific, complementary (coding or noncoding RNA). If binding takes place, this DNA/RNA hybrid can be degraded by the enzyme RNase H. After crossing the cell membrane, ASOs target mRNA directly in the nucleus or cytosol, thus blocking and neutralizing the targeted miRNA, with the help of the enzyme RNase H1. Furthermore, ASOs have various functions, including the inhibition of translation, miRNA, and splicing. ASOs have been investigated for more than 20 years and their use is now a standard technique in developmental biology and they are used to study altered gene expression and gene function. Recently, several ASOs have been modified for an unnecessary drug delivery system (DDS).

2.2. siRNAs

siRNAs are double-stranded RNAs with a length of 20–25 base pairs. siRNAs can suppress the gene expression via sequence specific inhibition of RNA expression (RNA interference, RNAi). The cellular process of RNAi occurs in almost all eukaryotic organisms [6]. After being processed by the ribonuclease III-like DICER enzyme, siRNA interacts with RNA-induced silencing complex to block and neutralize the target mRNA [7]. siRNA libraries have been created to dissect the function of independent genes since they show high sequence specificity. The application of siRNAs allows researchers to discover novel targets and pathway mediators.

2.3. Aptamers

Nucleic acid aptamers are short single-stranded DNA or RNA oligonucleotides that fold into unique three-dimensional structures and bind to a wide range of targets, including proteins, small molecules, metal ions, viruses, bacteria, and whole cells [8]. Aptamers have high specificity and binding affinities (in the low nanomolar to picomolar range) similar to those of antibodies and are frequently referred to as 'chemical antibodies'. Proteins constitute by far the largest class of aptamer targets. The high stability of aptamer–protein complexes, frequently characterized by a Kd in the low nanomolar range, combined with an excellent specificity of interaction make aptamers valuable tools for various applications, such as affinity purification, bio-sensing, imaging, and enzyme inhibition [9].

2.4. Decoys

Decoys are double-stranded molecules that mimic the consensus DNA binding site of a specific transcription factor in the promoter region of its target genes [10]. The regulation of transcription of disease-related genes in vivo has important therapeutic potential. Gene expression controlled by the transcription factor is effectively prevented, thereby effectively silencing gene expression and preventing protein production. Therefore, being less specific in comparison with the siRNA or ASO method, the decoy technique can be considered a gene silencing approach.

2.5. CpG Oligos

CpG oligodeoxynucleotides (CpG oligos) are short single-stranded synthetic DNA molecules that contain cytosine triphosphate deoxynucleotide followed by a guanine triphosphate deoxynucleotide [11]. Synthetic phosphorothioate oligodeoxynucleotides bearing unmethylated CpG motifs can mimic the immune-stimulatory effects of bacterial DNA and are recognized by Toll-like receptor 9 (TLR9), which is constitutively expressed only in B cells and plasmacytoid dendritic cells. Nucleotide modifications at positions at or near the CpG dinucleotides can severely affect immune modulation. CpG oligos induce type I interferon, cytokines, B cell proliferation, dendritic cell maturation, and natural killer cell activation. CpG oligos have been applied for antiallergenic or anticancer treatment.

3. Modifications of Nucleic Acid Drugs

Although, the function of nucleic acid drugs is promising, several challenges have been identified, including lack of stability against extracellular and intracellular degradation by nucleases, poor uptake and low potency at target sites of nucleic acid drugs, and off-target effects [12]. Off-target effects are nonspecific suppressive effects of nucleic acid drugs. Although it has been considered that nucleic acid drugs possess high specificity, several nucleic acid drugs can affect gene expression of multiple genes. Furthermore, nucleic acid drugs are quickly degraded by RNase in vivo. In humans, naked nucleic acid drugs preferentially accumulate in the liver and kidneys, which causes the nucleic acid drugs to be rapidly cleared from circulation with poor tissue distribution [13]. The pursuit of clinically viable antisense drugs has led to the development of various types of strategies, such as carriers or chemical modifications. Apart from structural modification of oligonucleotides, different cell-penetrating peptides and ligands conjugated to oligonucleotide-based DDS are normally adopted following the conjugation.

3.1. Structural Modifications of Nucleic Acid Drugs

Important modifications have been implemented to improve the therapeutic potential of nucleic acid medicines. However, the properties of the modifications have also led to some decreased affinity for the target sequence, with associated nonhybridization toxicities such as complement activation, increased coagulation times, or immune activation (Table 2). Another concern relates to the hybridization-dependent toxicity, caused by exaggerated action of the drug or off-target hybridization.

Table 2. Modifications of nucleic acid drugs.

Structural Modifications	Contents	Stability	Cellular Uptake	Gene Silencing Effect	Cytotoxicity	Binding Affinity
Diester modification	Phosphorothioate	superior	superior	inferior	superior	
Ribose modification	2′-O-Me, 2′-O-A, 2′-F	superior		inferior		
Base modification	Adenine methylation and deamination, cytosine methylation, hydroxy methylation and carboxy substitution, Guanine oxidation			superior		
Oligonucleotide analogues replacement	Peptide nucleic acid, locked nucleic acid, morpholino phosphamide	superior		superior		inferior
Conjugation to cell-penetrating peptides	Cysteine, transactivator of transcription peptide, gelatin		superior	superior	inferior	
Aptamer	20–100 nucleotides		superior	superior		

The first of the modifications included phosphorothioate backbone modification, which defined the first-generation nucleic acid drugs [5]. One of the nonbridge oxygen atoms in the diester bond is replaced by sulfur. Chemical modification can help enhance cellular uptake and increase the bioavailability of the modified nucleic acid drugs. Resistance to circumscribed nucleases is also effectively increased. However, although the modified siRNA is found to be significantly stable in the body, it increases the cytotoxicity and decreases the gene silencing effect. Modification of phosphorylated phosphate ester in the phosphorylation location damages RISC activity [14].

The second-generation nucleic acid drugs included the nucleoside analogues containing a modified sugar moiety, such as 2′-O-methyl-modified or 2′-O-methoxyethyl. The 2′ modifications inhibit the ability of RNase H to cleave the bound sense RNA strand within the heteroduplex formed between the nucleic acid drugs and the target RNA [15]. The widespread use of thiophosphate modifications results in a certain cytotoxicity, but the 2′-O-methylation improves the siRNA activity and is nontoxic to normal cells [16]. The activity of siRNA depends on the position of the modified parts.

Base modification plays an important role in the function of nucleic acid drugs; for example, it can improve the function of siRNA and increase the ability of the siRNA interaction with the target mRNA. The modification increases the ability of RISC to recognize and cleave the mRNA. The modifications on the base include adenine methylation and deamination, cystosine methylation, hydroxymethylation and carboxyl substitution, and guanine oxidation, etc. [17]. The modified bases are related to the changes of functional groups, which is the basis of triggering the functional changes through the modification of structure of nucleic acid drugs.

Oligonucleotide analogs' replacement includes peptide substitution, and the resulting materials typically include peptide nucleic acid, locked nucleic acid, and morpholino phosphamide. They can reduce the degradation of oligonucleotides by nucleases, and have low toxicity and a slight decrease in affinity compared with unmodified sequences [18]. These nucleotide analogs do not support the cleavage of RNase H-mediated target mRNA in ASPs; thereby, they primarily exhibit their reflective activity by steric hindrance to prevent gene expression during transcription or translation. This method further enhances the binding affinity, nuclease resistance, and targeted effect compared with several other chemical modifications.

3.2. Conjugation of Ligand or Cell-Penetrating Peptides

Cell-penetrating peptides are a class of short peptides that are rich in cations and can efficiently enter cells through penetrating biofilms. Based on these properties, cell-penetrating peptides are used to modified DNA, RNA, and oligonucleotides and are loaded on nanocarriers for therapy. The conjugation of oligonucleotides and cell-penetrating peptides can overcome the deficiencies of cytotoxicity and enhance the efficiency in eukaryotic cells. Complexes formed by cationic cell-penetrating peptides and anionic oligonucleotides which are formed through electrostatic interaction can promote oligonucleotides' entry into cells and initiate RNA interfering, leading to silencing of endogenous genes [19]. Cell-penetrating peptides include cysteine, transactivator of transcription peptide [20], and gelatin [21].

4. Aptamers

Aptamers are synthetic single-stranded oligonucleotides of short length (20–100 nucleotides) whose three-dimensional disposition confers high avidity for their target DNA or RNA. They shows high stability, lack of immunogenicity, flexible structure, and small size, which increases their penetration strength [22]. Aptamer-based targeted delivery of siRNAs using aptamer–siRNA chimeras are becoming a very useful tool for targeting gene-knockdown in cancer therapy [23]. Aptamer–siRNA chimeras bind the aptamer's receptor and upon engagement, the chimera–receptor complex is embedded into an endocytosis vesicle. The chimera reaches the cytoplasm and the duplex siRNA is recognized by Dicer and loaded into Dicer and RNA-induced silencing complex (RISC). Several aptamers have been reported for treatment of prostate, breast, and colon cancer, melanoma, lymphoma, and glioblastoma, for example *PSMA, 4-1BBm EpCAP, CTLA4, PDGFRβ, HER2,* and *HER3* [23].

5. Drug Delivery Systems of Nucleic Acid Drugs

DDS has been necessary to regulate the drug distribution in the body in terms of quantity and spatiotemporal aspects. Several kinds of DDSs have been developed based on the diameter of medicine, specific antibody for tumor, sustained release, and percutaneous absorption. They are expected to improve the specificity, effects, usability, and economy of drug as well as to suppress the side-effects.

Various carriers of siRNAs have become increasingly available because RNAi can integrate short hairpin RNA into the cell genome, leading to stable siRNA expression and long-term knockdown of a target gene. Nonviral carriers have been increasingly preferred owing to lower toxicity compared with other carrier methods. These carriers typically involve a positively charged vector (cationic cell-penetrating peptides, cationic polymers, and lipids), small molecules (cholesterol, bile acids, lipids, and PEGylated lipids), polymers, antibodies, aptamers, and lipid and polymer-based nanocarriers encapsulating the siRNA [24]. Specific delivery of siRNAs to hepatocytes has been accomplished by conjugation to *N*-acetylgalactosamine in order to target an asialoglycoprotein receptor present in the liver [25].

Different nanocarrier strategies are still needed in practical applications to make them more effective in diagnosing and treating diseases. A combination of chemical modification and a nanoparticle-based DDS is likely to be more effective for oligonucleotide delivery. For example, the siRNA can be modified with the free thiol group of the amino acid cysteine on cell-penetrating peptides, then they are encapsulated into ultrasound-sensitive nanomicrobubbles. When nanomicrobubbles reach the target site, they disintegrate under external ultrasonic irradiation, releasing siRNA to achieve cytoplasmic delivery [26].

Liposomes are widely used as oligonucleotide delivery systems (Table 3). Cationic liposomes include monovalent lipids such as DODMA and DOTAP [27]. Oligonucleotides are negatively charged and easy to encapsulate into cationic liposomes. Neutral liposomes are primarily constructed by neutral lipids, which include PC, PE, cholesterol, and DOPE [28]. Neutral liposomes have good biocompatibility and excellent pharmacokinetic characteristics, but they cannot interact with oligonucleotides to adsorb them and encapsulate them into the liposomes efficiently. Neutral liposomes are adopted to modified cationic liposomes to enhance particle stability. Ionizable liposomes are important for siRNA delivery. They can protonated and deprotonated according to the acidity of the environment [28]. Under hypoxic conditions, tumor tissues are more acidic and pH-responsive liposomes have more positive charges. Cationic liposomes are the most widely used form of liposomes.

Table 3. Drug delivery systems.

	Materials
Liposomes	
Cationic liposome	DOTAP, DODMA, DOGS, DC-Chol
Neutral liposome	PC, Chol, DOPE
Ionizable liposome	DODMA, DODAP
Micelles	
Polymeric micelles	Amphiphilic copolymer, PEG, polyamino acid, polylactic or glycolic acid, polycaprolactone, and short phospholipid chains
Cationic polymer micelles	PEG-PLL-PLLeu, PEI-CG-PEI, PgP
Nanoparticles	
Albumin-based	thiol, arginine-glycine-aspartic acid peptide
Metal-based	gold, silver, magnetic

Polymeric micelles have promising applications in drug delivery including extending the drug cycle time, changing the drug release curve, and easily connecting targeted ligands [29]. Cationic polymer micelles can ensure good oligonucleotide loading capacity through electrostatic adsorption. They show long circulation times, tumor passive targeting by the enhanced permeability and retention effect, and efficient oligonucleotide endosome release by the proton sponge effect [30]. Furthermore, the suitable carrier should can deliver oligonucleotides and chemotherapy drugs together to the tumor tissue and release the two drugs simultaneously, for example polymeric micelles with doxorubicin and siRNA targeting P-glycoprotein [31].

Nanoparticles using albumin, metals, and polymers have been used for drug delivery. Tumor cells can take up human serum albumin through endocytosis; therefore, albumin-based nanoparticles can show high stability without cytotoxicity [32]. Metallic nanometer-sized particles, such as silver, gold, and magnetic metals show the property of the enhanced surface to volume ratio; therefore, they have good applications in oligonucleotide delivery [33].

Another challenge to overcome in the DDS for pancreatic cancer is intratumoral injection [34] or implantation [35,36] of siRNAs in the pancreas (Figure 1). Implantation of Local Drug EluterR

(LODER), can release siRNAs targeting KRAS over months in pancreatic cancer in vivo [36]. LODER is a biodegradable polymeric matrix that shields drugs against enzymatic degradation. EUS have enabled researchers to obtain pancreatic tissue samples and inject medicines into the pancreas repeatedly; therefore, DDS using EUS may improve the effectiveness of siRNA treatment for pancreatic cancer. In an animal model, we have reported that administration of siRNA by intratumoral injection with atelocollagen [37] and intravenous injection [38]. Both settings were effective to reduce targeted mRNA expression in vivo without severe side effects in the short term. Clinical trials are necessary to determine the long-term effects and safety of nucleic acid medicines.

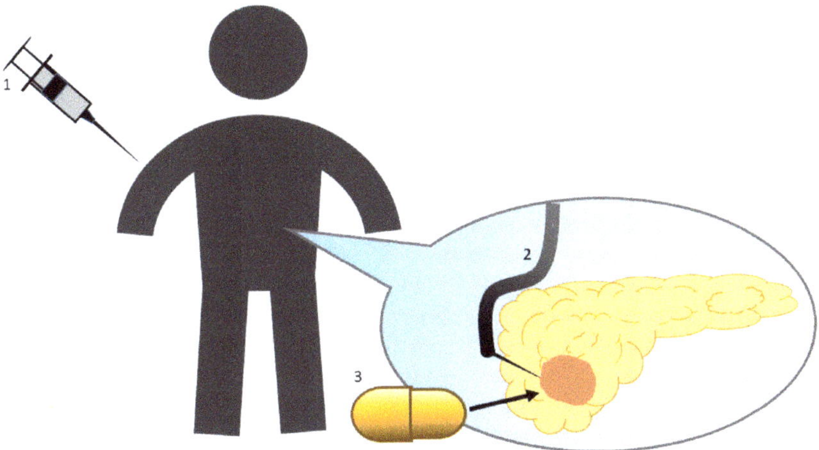

Figure 1. Delivery of nucleic acid medicines. (1) Intravenous injection, (2) intratumoral injection under EUS, and (3) intratumoral implantation.

6. Clinical Trials

6.1. Antisense Oligonucleotide

Eight nucleic acid medicines have been approved by the FDA (Table 4), five of which are ASOs used to treat nervous muscular diseases and familial metabolic diseases.

There have been a lot of reports about ASOs for pancreatic cancer treatment in preclinical studies. *KRAS* is the most common target because approximately 90% of pancreatic cancer harbor *KRAS* mutation. AZD-4785, a high-affinity constrained ethyl-containing therapeutic ASO targeting *KRAS* mRNA, potently depleted *KRAS* mRNA in *KRAS*-mutant colon, pancreatic, and lung cancer cell lines, with no feedback activation of MAPK signaling. Significant antitumor activity was obtained in mice bearing *KRAS*-mutant lung cancer xenografts [39].

ASOs have been tested in more than 1000 clinical trials. Various ASOs have reached clinical trials for the treatment of pancreatic cancer. The targets of these molecules were related to cell proliferation (X-linked inhibitor of apoptosis protein, *XIAP* [40]; Protein Kinase A, *PKA* [41]), cell signaling (*HRAS*) [42], resistance to chemotherapy (heat shock protein 27, *Hsp27*) [43], or cancer stroma (*TGFβ2*) [44]. However, few ASOs have shown antitumor effects in clinical trials.

Table 4. Food and Drug Administration (FDA)-approved nucleic acid medicines.

Drug	Nucleic Acid	Disease	Modification	Administration	Company
Vitravene [45]	ASO	Cytomegalovirus retinitis	Phosphorothioated	Intravitreous	Isis Pharmaceuticals, Carlsbad, CA
Macugen [46]	Aptamer	Age-related macular degeneration	PEGylation 2′-F 2′-OMe	Intravitreous	Valeant Pharmaceuticals, Laval, Canada
Kynamro [47]	ASO	Homozygous familial hypercholesterolemia	Phosphorothioated 2′-MOE	Subcutaneous	Kastle Therapeutics, Chicago, IL
Exondys 51 [48]	ASO	Duchenne muscular dystrophy	Morpholino nucleic acid	Intravenous	Sarepta Therapeutics, Cambridge, MA
Spinraza [49]	ASO	Myelopathic muscular atrophy	Phosphorothioated 2′-MOE	Intraspinal	Biogen, Cambridge, MA
Heplisav-B [50]	CpG oligo	Hepatitis B	Phosphorothioated	Intramuscular	Dynavax Technologies, Berkeley, CA
Tegsedi [51]	ASO	Hereditary transthyretin-mediated amyloidosis	Phosphorothioated 2′-MOE	Subcutaneous	Akcea Therapeutics, Boston, MA
Onpattro [52]	siRNA	Hereditary transthyretin-mediated amyloidosis	2′-MOE	Intravenous	Alnylam Pharmaceuticals, Cambridge, MA

FDA, Food and Drug Administration; ASO, antisense oligonucleotide; CpG oligo, CpG oligodeoxynucleotide; 2′-MOE, 2′-O-methoxyethyl; 2′-OMe, 2′-O-Methyl; 2′-F, 2′-Fluoro.

ISIS 2503 (ASO targeting *XIAP*) showed evidence of growth inhibition when combined with gemcitabine in locally advanced or metastatic pancreatic cancer in first-line treatment [40]. In that study, 58% of patients who received the combination survived 6 months or longer. Addition of apatorsen, the *Hsp27*-targeting antisense oligonucleotide, to chemotherapy did not improve outcomes in unselected patients with metastatic pancreatic cancer in the first-line setting, although a trend toward prolonged overall survival in patients with high baseline serum Hsp27 suggests that this therapy may warrant further evaluation in this subgroup.

6.2. Clinical Trials for siRNAs

Fourteen years after the first clinical trial using RNAi was entered (2004), the FDA approved the first therapeutic RNAi, ONPATTRO (patisiran), a lipid complex injection for treatment of peripheral nerve disease caused by hereditary transthyretin-mediated amyloidosis in adults [52] (Table 4). However, there is no clinically available therapeutic RNAi for pancreatic cancer.

Some siRNAs have already entered clinical trials for the treatment of locally advanced pancreatic cancer. siRNA targeting mutated *KRAS* is the most common [35,36]. The vast majority of *KRAS* mutations in pancreatic cancer are gain-of-function mutations, most of which occur in codon 12 with substitution of the Glycine for Aspartate (G12D). Golan et al. implanted siRNA targeting *KRAS* (G12D) in the pancreatic tumor using LODER in combination with Gemcitabine treatment [35]. The majority of patients (83%) demonstrated stable disease and 17% of patients showed partial response. Decrease in CA19-9 was observed in 70% of patients. The most frequent adverse events observed were grade 1 or 2 severity (89%); transient abdominal pain, diarrhea, and nausea. They concluded that the combination of mutated *KRAS*-targeting siRNAs and chemotherapy is well tolerated, safe, and demonstrated potential efficacy in pancreatic cancer patients [53].

Nishimura et al. have shown that EUS-guided fine-needle injection (EUS-FNI) of a synthetic double-stranded RNA oligonucleotide directed against *CHST15* (STNM01), an extracellular matrix component, was safe and feasible [34]. There were no adverse effects. STNM01 is also directly injected by endoscopy to treat ulcerative colitis.

Atu027 is a liposomally formulated siRNA with antimetastatic activity, which silences protein kinase N3 (PKN3) expression in the vascular endothelium [54]. PKN3 acts as a Rho effector downstream of PI3K. Combination of Atu027 and gemcitabine for the treatment of advanced pancreatic cancer was safe and well tolerated.

TKM-080301 is a lipid nanoparticle formulation of an siRNA against Polo-like kinase 1 (PLK1), which regulates critical aspects of tumor progression [55]. Preliminary antitumor efficacy for advanced pancreatic cancer has been observed. A potential molecular therapeutic context of increased PLK1 expression with inactivation of p53 or NF1 was observed in a remarkable responder.

However, these data must be interpreted with caution because they are early-phase trials and some are still recruiting patients. The best responses observed so far have been tumor stabilization, with very few complete or partial responses documented. siRNAs were well tolerated but one death and a few grade 3–4 toxic effects due to elevation of liver enzymes were observed [56]. Several trials with different combinations including siRNAs are ongoing, and the combination of several nucleic acid medicines may be explored in the coming years.

7. Conclusions

Despite the challenging obstacles, we hope that nucleic acid drugs will have a significant impact on the treatment of pancreatic cancer. The combination of genetic diagnosis using next-generation sequencing and targeted therapy may provide effective precision medicine for pancreatic cancer patients.

Acknowledgments: We thank Sanae Kushida for preparing the manuscript.

Conflicts of Interest: The authors declare no conflict of interest.

References

1. Torres, C.; Grippo, P.J. Pancreatic cancer subtypes: A roadmap for precision medicine. *Ann. Med.* **2018**, *50*, 277–287. [CrossRef] [PubMed]
2. Gleeson, F.C.; Kerr, S.E.; Kipp, B.R.; Voss, J.S.; Minot, D.M.; Tu, Z.J.; Henry, M.R.; Graham, R.P.; Vasmatzis, G.; Cheville, J.C.; et al. Targeted next generation sequencing of endoscopic ultrasound acquired cytology from ampullary and pancreatic adenocarcinoma has the potential to aid patient stratification for optimal therapy selection. *Oncotarget* **2016**, *7*, 54526–54536. [CrossRef] [PubMed]
3. Barata, P.; Sood, A.K.; Hong, D.S. RNA-targeted therapeutics in cancer clinical trials: Current status and future directions. *Cancer Treat. Rev.* **2016**, *50*, 35–47. [CrossRef] [PubMed]
4. Jansen, B.; Zangemeister-Wittke, U. Antisense therapy for cancer—The time of truth. *Lancet Oncol.* **2002**, *3*, 672–683. [CrossRef]
5. Moreno, P.M.; Pego, A.P. Therapeutic antisense oligonucleotides against cancer: Hurdling to the clinic. *Front. Chem.* **2014**, *2*, 87. [CrossRef]
6. Lee, R.C.; Feinbaum, R.L.; Ambros, V. The C. elegans heterochronic gene lin-4 encodes small RNAs with antisense complementarity to lin-14. *Cell* **1993**, *75*, 843–854. [CrossRef]
7. Elbashir, S.M.; Harborth, J.; Lendeckel, W.; Yalcin, A.; Weber, K.; Tuschl, T. Duplexes of 21-nucleotide RNAs mediate RNA interference in cultured mammalian cells. *Nature* **2001**, *411*, 494–498. [CrossRef]
8. Hori, S.I.; Herrera, A.; Rossi, J.J.; Zhou, J. Current Advances in Aptamers for Cancer Diagnosis and Therapy. *Cancers (Basel)* **2018**, *10*, 9. [CrossRef]
9. Dausse, E.; Da Rocha Gomes, S.; Toulme, J.J. Aptamers: A new class of oligonucleotides in the drug discovery pipeline? *Curr. Opin. Pharmacol.* **2009**, *9*, 602–607. [CrossRef]
10. Hecker, M.; Wagner, A.H. Transcription factor decoy technology: A therapeutic update. *Biochem. Pharmacol.* **2017**, *144*, 29–34. [CrossRef]
11. Vollmer, J.; Weeratna, R.D.; Jurk, M.; Davis, H.L.; Schetter, C.; Wullner, M.; Wader, T.; Liu, M.; Kritzler, A.; Krieg, A.M. Impact of modifications of heterocyclic bases in CpG dinucleotides on their immune-modulatory activity. *J. Leukoc. Biol.* **2004**, *76*, 585–593. [CrossRef] [PubMed]

12. Wu, S.Y.; Yang, X.; Gharpure, K.M.; Hatakeyama, H.; Egli, M.; McGuire, M.H.; Nagaraja, A.S.; Miyake, T.M.; Rupaimoole, R.; Pecot, C.V.; et al. 2′-OMe-phosphorodithioate-modified siRNAs show increased loading into the RISC complex and enhanced anti-tumour activity. *Nat. Commun.* **2014**, *5*, 3459. [CrossRef] [PubMed]

13. Geary, R.S.; Norris, D.; Yu, R.; Bennett, C.F. Pharmacokinetics, biodistribution and cell uptake of antisense oligonucleotides. *Adv. Drug Deliv. Rev.* **2015**, *87*, 46–51. [CrossRef] [PubMed]

14. Soutschek, J.; Akinc, A.; Bramlage, B.; Charisse, K.; Constien, R.; Donoghue, M.; Elbashir, S.; Geick, A.; Hadwiger, P.; Harborth, J.; et al. Therapeutic silencing of an endogenous gene by systemic administration of modified siRNAs. *Nature* **2004**, *432*, 173–178. [CrossRef] [PubMed]

15. Yu, R.Z.; Grundy, J.S.; Geary, R.S. Clinical pharmacokinetics of second generation antisense oligonucleotides. *Expert Opin. Drug Metab. Toxicol.* **2013**, *9*, 169–182. [CrossRef] [PubMed]

16. Amarzguioui, M.; Holen, T.; Babaie, E.; Prydz, H. Tolerance for mutations and chemical modifications in a siRNA. *Nucleic Acids Res.* **2003**, *31*, 589–595. [CrossRef] [PubMed]

17. Yi, C.; Pan, T. Cellular dynamics of RNA modification. *Acc. Chem. Res.* **2011**, *44*, 1380–1388. [CrossRef] [PubMed]

18. Fattal, E.; Barratt, G. Nanotechnologies and controlled release systems for the delivery of antisense oligonucleotides and small interfering RNA. *Br. J. Pharmacol.* **2009**, *157*, 179–194. [CrossRef]

19. Simeoni, F.; Morris, M.C.; Heitz, F.; Divita, G. Insight into the mechanism of the peptide-based gene delivery system MPG: Implications for delivery of siRNA into mammalian cells. *Nucleic Acids Res.* **2003**, *31*, 2717–2724. [CrossRef]

20. Xie, X.; Yang, Y.; Lin, W.; Liu, H.; Liu, H.; Yang, Y.; Chen, Y.; Fu, X.; Deng, J. Cell-penetrating peptide-siRNA conjugate loaded YSA-modified nanobubbles for ultrasound triggered siRNA delivery. *Colloids Surf. B Biointerfaces.* **2015**, *136*, 641–650. [CrossRef]

21. Arami, S.; Mahdavi, M.; Rashidi, M.R.; Yekta, R.; Rahnamay, M.; Molavi, L.; Hejazi, M.S.; Samadi, N. Apoptosis induction activity and molecular docking studies of survivin siRNA carried by Fe3O4-PEG-LAC-chitosan-PEI nanoparticles in MCF-7 human breast cancer cells. *J. Pharm. Biomed. Anal.* **2017**, *142*, 145–154. [CrossRef] [PubMed]

22. Keefe, A.D.; Pai, S.; Ellington, A. Aptamers as therapeutics. *Nat. Rev. Drug Discov.* **2010**, *9*, 537–550. [CrossRef] [PubMed]

23. Soldevilla, M.M.; Meraviglia-Crivelli de Caso, D.; Menon, A.P.; Pastor, F. Aptamer-iRNAs as Therapeutics for Cancer Treatment. *Pharmaceuticals (Basel)* **2018**, *11*, 108. [CrossRef] [PubMed]

24. Ramot, Y.; Rotkopf, S.; Gabai, R.M.; Zorde Khvalevsky, E.; Muravnik, S.; Marzoli, G.A.; Domb, A.J.; Shemi, A.; Nyska, A. Preclinical Safety Evaluation in Rats of a Polymeric Matrix Containing an siRNA Drug Used as a Local and Prolonged Delivery System for Pancreatic Cancer Therapy. *Toxicol. Pathol.* **2016**, *44*, 856–865. [CrossRef]

25. Nair, J.K.; Willoughby, J.L.; Chan, A.; Charisse, K.; Alam, M.R.; Wang, Q.; Hoekstra, M.; Kandasamy, P.; Kel'in, A.V.; Milstein, S.; et al. Multivalent N-acetylgalactosamine-conjugated siRNA localizes in hepatocytes and elicits robust RNAi-mediated gene silencing. *J. Am. Chem. Soc.* **2014**, *136*, 16958–16961. [CrossRef]

26. Jing, H.; Cheng, W.; Li, S.; Wu, B.; Leng, X.; Xu, S.; Tian, J. Novel cell-penetrating peptide-loaded nanobubbles synergized with ultrasound irradiation enhance EGFR siRNA delivery for triple negative Breast cancer therapy. *Colloids Surf. B Biointerfaces* **2016**, *146*, 387–395. [CrossRef]

27. Rabbani, P.S.; Zhou, A.; Borab, Z.M.; Frezzo, J.A.; Srivastava, N.; More, H.T.; Rifkin, W.J.; David, J.A.; Berens, S.J.; Chen, R.; et al. Novel lipoproteoplex delivers Keap1 siRNA based gene therapy to accelerate diabetic wound healing. *Biomaterials* **2017**, *132*, 1–15. [CrossRef]

28. Wang, Y.; Miao, L.; Satterlee, A.; Huang, L. Delivery of oligonucleotides with lipid nanoparticles. *Adv. Drug Deliv. Rev.* **2015**, *87*, 68–80. [CrossRef]

29. Amjadi, M.; Mostaghaci, B.; Sitti, M. Recent Advances in Skin Penetration Enhancers for Transdermal Gene and Drug Delivery. *Curr. Gene Ther.* **2017**, *17*, 139–146. [CrossRef]

30. Yin, T.; Wang, L.; Yin, L.; Zhou, J.; Huo, M. Co-delivery of hydrophobic paclitaxel and hydrophilic AURKA specific siRNA by redox-sensitive micelles for effective treatment of breast cancer. *Biomaterials* **2015**, *61*, 10–25. [CrossRef]

31. Shen, J.; Wang, Q.; Hu, Q.; Li, Y.; Tang, G.; Chu, P.K. Restoration of chemosensitivity by multifunctional micelles mediated by P-gp siRNA to reverse MDR. *Biomaterials* **2014**, *35*, 8621–8634. [CrossRef] [PubMed]

32. Ji, S.; Xu, J.; Zhang, B.; Yao, W.; Xu, W.; Wu, W.; Xu, Y.; Wang, H.; Ni, Q.; Hou, H.; et al. RGD-conjugated albumin nanoparticles as a novel delivery vehicle in pancreatic cancer therapy. *Cancer Biol. Ther.* **2012**, *13*, 206–215. [CrossRef] [PubMed]

33. Kahalekar, V.; Gupta, D.T.; Bhatt, P.; Shukla, A.; Bhatia, S. Fully covered self-expanding metallic stent placement for benign refractory esophageal strictures. *Indian J. Gastroenterol.* **2017**, *36*, 197–201. [CrossRef] [PubMed]

34. Nishimura, M.; Matsukawa, M.; Fujii, Y.; Matsuda, Y.; Arai, T.; Ochiai, Y.; Itoi, T.; Yahagi, N. Effects of EUS-guided intratumoral injection of oligonucleotide STNM01 on tumor growth, histology, and overall survival in patients with unresectable pancreatic cancer. *Gastrointest Endosc.* **2018**, *87*, 1126–1131. [CrossRef] [PubMed]

35. Golan, T.; Khvalevsky, E.Z.; Hubert, A.; Gabai, R.M.; Hen, N.; Segal, A.; Domb, A.; Harari, G.; David, E.B.; Raskin, S.; et al. RNAi therapy targeting KRAS in combination with chemotherapy for locally advanced pancreatic cancer patients. *Oncotarget* **2015**, *6*, 24560–24570. [CrossRef] [PubMed]

36. Shemi, A.; Khvalevsky, E.Z.; Gabai, R.M.; Domb, A.; Barenholz, Y. Multistep, effective drug distribution within solid tumors. *Oncotarget* **2015**, *6*, 39564–39577. [CrossRef]

37. Yamahatsu, K.; Matsuda, Y.; Ishiwata, T.; Uchida, E.; Naito, Z. Nestin as a novel therapeutic target for pancreatic cancer via tumor angiogenesis. *Int. J. Oncol.* **2012**, *40*, 1345–1357.

38. Matsuda, Y.; Ishiwata, T.; Yoshimura, H.; Yamashita, S.; Ushijima, T.; Arai, T. Systemic Administration of Small Interfering RNA Targeting Human Nestin Inhibits Pancreatic Cancer Cell Proliferation and Metastasis. *Pancreas* **2016**, *45*, 93–100. [CrossRef]

39. Ross, S.J.; Revenko, A.S.; Hanson, L.L.; Ellston, R.; Staniszewska, A.; Whalley, N.; Pandey, S.K.; Revill, M.; Rooney, C.; Buckett, L.K.; et al. Targeting KRAS-dependent tumors with AZD4785, a high-affinity therapeutic antisense oligonucleotide inhibitor of KRAS. *Sci. Transl. Med.* **2017**, *9*. [CrossRef]

40. Mahadevan, D.; Chalasani, P.; Rensvold, D.; Kurtin, S.; Pretzinger, C.; Jolivet, J.; Ramanathan, R.K.; Von Hoff, D.D.; Weiss, G.J. Phase I trial of AEG35156 an antisense oligonucleotide to XIAP plus gemcitabine in patients with metastatic pancreatic ductal adenocarcinoma. *Am. J. Clin. Oncol.* **2013**, *36*, 239–243. [CrossRef]

41. Goel, S.; Desai, K.; Macapinlac, M.; Wadler, S.; Goldberg, G.; Fields, A.; Einstein, M.; Volterra, F.; Wong, B.; Martin, R.; et al. A phase I safety and dose escalation trial of docetaxel combined with GEM231, a second generation antisense oligonucleotide targeting protein kinase A R1alpha in patients with advanced solid cancers. *Investig. New Drugs* **2006**, *24*, 125–134. [CrossRef] [PubMed]

42. Alberts, S.R.; Schroeder, M.; Erlichman, C.; Steen, P.D.; Foster, N.R.; Moore, D.F., Jr.; Rowland, K.M., Jr.; Nair, S.; Tschetter, L.K.; Fitch, T.R. Gemcitabine and ISIS-2503 for patients with locally advanced or metastatic pancreatic adenocarcinoma: A North Central Cancer Treatment Group phase II trial. *J. Clin. Oncol.* **2004**, *22*, 4944–4950. [CrossRef] [PubMed]

43. Ko, A.H.; Murphy, P.B.; Peyton, J.D.; Shipley, D.L.; Al-Hazzouri, A.; Rodriguez, F.A.; Womack, M.S.; Xiong, H.Q.; Waterhouse, D.M.; Tempero, M.A.; et al. A Randomized, Double-Blinded, Phase II Trial of Gemcitabine and Nab-Paclitaxel Plus Apatorsen or Placebo in Patients with Metastatic Pancreatic Cancer: The RAINIER Trial. *Oncologist* **2017**, *22*, 1427–e129. [CrossRef] [PubMed]

44. Jaschinski, F.; Rothhammer, T.; Jachimczak, P.; Seitz, C.; Schneider, A.; Schlingensiepen, K.H. The antisense oligonucleotide trabedersen (AP 12009) for the targeted inhibition of TGF-beta2. *Curr. Pharm. Biotechnol.* **2011**, *12*, 2203–2213. [CrossRef] [PubMed]

45. Vitravene Study Group. A randomized controlled clinical trial of intravitreous fomivirsen for treatment of newly diagnosed peripheral cytomegalovirus retinitis in patients with AIDS. *Am. J. Ophthalmol.* **2002**, *133*, 467–474.

46. Ng, E.W.; Shima, D.T.; Calias, P.; Cunningham, E.T., Jr.; Guyer, D.R.; Adamis, A.P. Pegaptanib, a targeted anti-VEGF aptamer for ocular vascular disease. *Nat. Rev. Drug Discov.* **2006**, *5*, 123–132. [CrossRef] [PubMed]

47. Raal, F.J.; Santos, R.D.; Blom, D.J.; Marais, A.D.; Charng, M.J.; Cromwell, W.C.; Lachmann, R.H.; Gaudet, D.; Tan, J.L.; Chasan-Taber, S.; et al. Mipomersen, an apolipoprotein B synthesis inhibitor, for lowering of LDL cholesterol concentrations in patients with homozygous familial hypercholesterolaemia: A randomised, double-blind, placebo-controlled trial. *Lancet* **2010**, *375*, 998–1006. [CrossRef]

48. Stein, C.A. Eteplirsen Approved for Duchenne Muscular Dystrophy: The FDA Faces a Difficult Choice. *Mol. Ther.* **2016**, *24*, 1884–1885. [CrossRef]

49. Hua, Y.; Sahashi, K.; Hung, G.; Rigo, F.; Passini, M.A.; Bennett, C.F.; Krainer, A.R. Antisense correction of SMN2 splicing in the CNS rescues necrosis in a type III SMA mouse model. *Genes Dev.* **2010**, *24*, 1634–1644. [CrossRef]

50. Splawn, L.M.; Bailey, C.A.; Medina, J.P.; Cho, J.C. Heplisav-B vaccination for the prevention of hepatitis B virus infection in adults in the United States. *Drugs Today (Barc.)* **2018**, *54*, 399–405. [CrossRef]

51. Gales, L. Tegsedi (Inotersen): An Antisense Oligonucleotide Approved for the Treatment of Adult Patients with Hereditary Transthyretin Amyloidosis. *Pharmaceuticals (Basel)* **2019**, *12*, 78. [CrossRef] [PubMed]

52. Adams, D.; Gonzalez-Duarte, A.; O'Riordan, W.D.; Yang, C.C.; Ueda, M.; Kristen, A.V.; Tournev, I.; Schmidt, H.H.; Coelho, T.; Berk, J.L.; et al. Patisiran, an RNAi Therapeutic, for Hereditary Transthyretin Amyloidosis. *N. Engl. J. Med.* **2019**, *379*, 11–21. [CrossRef] [PubMed]

53. Weng, Y.; Xiao, H.; Zhang, J.; Liang, X.J.; Huang, Y. RNAi therapeutic and its innovative biotechnological evolution. *Biotechnol. Adv.* **2019**, *37*, 801–825. [CrossRef] [PubMed]

54. Suzuki, K.; Yokoyama, J.; Kawauchi, Y.; Honda, Y.; Sato, H.; Aoyagi, Y.; Terai, S.; Okazaki, K.; Suzuki, Y.; Sameshima, Y.; et al. Phase 1 Clinical Study of siRNA Targeting Carbohydrate Sulphotransferase 15 in Crohn's Disease Patients with Active Mucosal Lesions. *J. Crohns Colitis* **2017**, *11*, 221–228. [CrossRef] [PubMed]

55. Schultheis, B.; Strumberg, D.; Kuhlmann, J.; Wolf, M.; Link, K.; Seufferlein, T.; Kaufmann, J.; Gebhardt, F.; Bruyniks, N.; Pelzer, U. A phase Ib/IIa study of combination therapy with gemcitabine and Atu027 in patients with locally advanced or metastatic pancreatic adenocarcinoma. *J. Clin. Oncol.* **2016**, *34*, 385. [CrossRef]

56. Demeure, M.J.; Armaghany, T.; Ejadi, S.; Ramanathan, R.K.; Elfiky, A.; Strosberg, J.R.; Smith, D.C.; Whitsett, T.; Liang, W.S.; Sekar, S.; et al. A phase I/II study of TKM-080301, a PLK1-targeted RNAi in patients with adrenocortical cancer (ACC). *J. Clin. Oncol.* **2016**, *34*, 2547. [CrossRef]

MDPI

St. Alban-Anlage 66

4052 Basel

Switzerland

Tel. +41 61 683 77 34

Fax +41 61 302 89 18

www.mdpi.com

International Journal of Molecular Sciences Editorial Office

E-mail: ijms@mdpi.com

www.mdpi.com/journal/ijms

www.ingramcontent.com/pod-product-compliance
Lightning Source LLC
LaVergne TN
LVHW071533170726
843515LV00008B/2136